Spectral Music Design

Spectral Music Design

A COMPUTATIONAL APPROACH

Victor Lazzarini

OXFORD
UNIVERSITY PRESS

OXFORD
UNIVERSITY PRESS

Oxford University Press is a department of the University of Oxford. It furthers the University's objective of excellence in research, scholarship, and education by publishing worldwide. Oxford is a registered trade mark of Oxford University Press in the UK and in certain other countries.

Published in the United States of America by Oxford University Press
198 Madison Avenue, New York, NY 10016, United States of America.

Library of Congress Control Number: 2021937269

ISBN 978-0-19-752401-5 (hbk.)
ISBN 978-0-19-752402-2 (pbk.)

DOI: 10.1093/oso/9780197524015.001.0001

FOREWORD

Victor Lazzarini's *Spectral Music Design: A Computational Approach* is a superb contribution to computer music literature. It is a timely book that focuses on spectral audio processing and synthesis, a burgeoning field that is well served by this comprehensive and clear offering. Indeed, spectral audio processing is increasingly prevalent in audio software and provides the foundation for many new developments in both commercial and experimental digital audio. Despite the seemingly ubiquitous incorporation of spectral techniques in our technological devices, there remains a dearth of books that adequately address this realm of audio processing. This welcome book addresses this need by presenting the topic in a thorough manner that will provide a wealth of information for computer scientists and computer musicians alike.

As a computer scientist, computer music researcher, composer, and professor, Lazzarini offers great expertise in this book. *Spectral Music Design* is informed by his work as a lead developer of Csound and the author of the AuLib C++ object-oriented audio programming library. The reader also benefits from his ongoing creative research in developing new audio synthesis techniques such as modified FM, Adaptive FM, Feedback AM, and Vector Phase Shaping, all of which are lucidly covered in this text. This book also synthesises and coalesces many of the ideas presented in his previous publications on spectral audio found in his numerous papers and contributions in *The Csound Book* and *The Audio Programming Book*. Finally, this book can also be seen as companion work that complements Lazzarini's recently authored *Computer Music Instruments I* and *Computer Music Instruments II*.

The first part of Spectral Music Design provides a systematic and succinct introduction to the fundamentals of the spectrum, a brief history of the mathematical foundations related to spectrum, and a discussion of the principles of audio signals. The next four chapters focus on audio signal processing techniques that are based on Fourier's mathematical foundations. The book provides clear explanations of the math and computational developments as well as the many variants of spectral signal processing made available via Fourier transforms. This second section covers topics ranging from classic waveforms and convolution to filter design and analysis/synthesis paradigms that permeate current computer audio strategies.

The following chapters focus on aspects of spectral design that include non-linear synthesis techniques as well as strategies for working with noise. In addition to more traditional non-linear techniques of FM, AM, and waveshaping, Lazzarini provides an extensive look at feedback modulation, complex phase modulation, exponential FM, vector phase shaping, modified FM, adaptive FM, and other non-linear techniques that can be used to create dynamic and expressive spectra. This section also deals with strategies for handling *noise* as an integral component of spectral design. Noise topics covered include random numbers in audio computing, granular synthesis, linear prediction, and the role that residuals play in various computer audio processing techniques.

The book concludes by focusing on design and composition strategies. These final chapters include discussions of the emergence of spectral color as a structural determinant and the development of these ideas in both popular and experimental music. He includes classic examples from seminal figures such as Jean Claude Risset and John Chowning. His chapter on computer sound design includes an abundance of creative ideas for implementing spectral strategies within a computer music context, while his final chapter discusses his own composition, *Mouvements*, thus providing a prime example demonstrating the implementation of spectral ideas in a musical composition.

I am struck by how clearly and systematically Lazzarini presents these materials. The book could easily be used as a course textbook, while it also provides numerous ideas that create an incredibly rich resource for computer musicians. Indeed, the entire book is interspersed with exquisite and elegant code examples that many will find not only helpful in understanding spectral audio concepts, but also inspiring for their own creative work. While I have been composing computer music for more than three decades, this book reminds me how many more exciting techniques and strategies remain to be discovered and explored.

Denton, Texas, August 2020

Jon Christopher Nelson

PREFACE

This book is the product of thirty years of experience of spectral music practice. My journey started as a undergraduate battling with a synthi AKS to coerce musically interesting sounds out of it. Oscillators, filters, envelopes, editing tape, it was all a fascinating new world of possibilities to explore. In time, I would have my first brief encounters with a computer, and a glimpse of software sound synthesis. However, it was only as a postgraduate that I fully immersed myself in computer music, thanks to the Composer's Desktop Project software and Csound, first running on an Atari ST, then under the new UNIX graphical workstations that were starting to appear at the time. This was in the midst of the microprocessor revolution of the early 1990s, and suddenly computer music was available to more people than the elected few who were lucky enough to work at a very selective group of research centers. It was also the dawn of the internet age, and with it, the emergence of online communities of like-minded individuals, who would collaboratively build a formidable collection of ideas, concepts, and software for music making. The work in this book owes much to this movement and to the sharing spirit that went along with it.

So what's this book about? From the title, there are a number of concurrent topics, which are threaded into it. We can read it as *Spectral* music design, *Spectral Music* design, spectral *Music* design, or even as spectral *Music Design*. That is, we can look at it from a technical perspective of the study of spectral signal processing, with an emphasis on the methods and their justification. We can also read it as a thorough description of a way of making music, that is of a spectral nature and based on computational methods. Above all, this is a book about music, and the emphasis on it should never be lost, even when we find ourselves deeply involved in describing a technique or an aspect of the spectrum. It is also a book about designing things, both music itself and the means to make it, which we can simply call *instruments*. To help our goal, we make use of many devices that are not traditionally employed in music theory, such as mathematical formulae and concepts, graphs, computer programs, and so on. These are elements that have become essential for the new breed of musicians who have emerged at the start of this new century. Formally or informally, we have grown used to having to deal with questions that once were thought to be purely of a technical and engineering interest, but now form part of our métier.

From a mathematical perspective, all of the concepts and expressions used are accessible from the starting point of high-school mathematics (e.g., algebra, trigonometry, complex numbers, and basic calculus). For those who need reminders, basic explanations are given in the text itself. As noted, this is not a book about mathematics, but music. The language of mathematics is but one of the many perspectives used to illustrate a point. It is an important one, because it can be very precise and succinct. However, it is always accompanied by one or more alternative explanations, either in textual or graphical forms, or via a computer program. Those who are not so mathematically inclined may skim the more complex formulae and look at the accompanying text or code. In any case, most of the discussion is built from first principles, and I have attempted to avoid any sleight of hand by fully justifying and deriving the key concepts introduced throughout the book. Sometimes, if an idea is dependent on others that have not been introduced yet, the full explanation is not given immediately, but it is returned to at a later stage in the text.

As the title implies, computer programming is an essential element of the book. The perspective I take in relation to programming, both in the practice and in the teaching of it, is best described by Donald Knuth, in the preface to his magnum opus *The Art of Computer Programming:*

> The process of preparing programs for a digital computer is specially attractive, not only because it can be economically and scientifically rewarding, but also because it can be an aesthetic experience much like composing poetry or music. (Knuth, 1997, v)

It is, however, beyond the scope of the discussion to explore programming techniques. Therefore, it may be useful for the reader to have had some acquaintance with some form of coding. This could be just through a simple music patching language of some description, or through scripting. The two basic languages used in the book, Python and Csound, are fairly accessible, and tutorials for these are widely available. These two languages were chosen for their complementarity: Python can cover the fundamental signal processing algorithms, program logic, and provide input/output in the form of graphs and soundfiles; and Csound provides the efficient sound synthesis and processing environment. They are also complementary in the sense that Csound can be embedded in Python, and Python can also be embedded in Csound. Together they form a complete set that allow us to approach spectral music design from the highest to the lowest level (well, the lowest *sane* level, at least).

This book follows a straightforward plan. We begin by an extremely gentle and very general look at what is meant by the word *spectrum*, in relation to sound and music, and our perception of it. This sets the tone for the following, ever more detailed, definitions of the spectrum. In Chapter 2, we tell an abridged history of the study of the spectrum, from the first steps in classical Greece to the 20th century. This is followed by an exploration of the fundamental elements that support the study of spectral audio processing, from a sound and music computing perspective. This completes Part I, which is the foundation for the rest of the book.

The following four chapters are dedicated to the wide-ranging ideas, concepts, and techniques that stem from Fourier's theorem. We first look at it in its original form, and explore its basic attributes, within the continuum of time and frequency. Then we move to the world of discrete mathematics and start using forms of the theorem that allow us to process audio in the computer. The question of time and dynamic spectra soon becomes a concern, and we then develop ways to manipulate sounds within a Fourier-based framework that involves both domains of time and frequency. The highly significant topic of filters is discussed next, by which stage we are the outskirts of the Fourier theory, and ready to move beyond it.

We then explore, in the remaining chapters of Part II, key aspects of sound synthesis and processing that are nominally not within the bounds of the Fourier approach. We look at non-linear methods, which include various ways of creating and modifying signals using the means of amplitude distortion. Within these, we find the classic techniques of frequency modulation synthesis and waveshaping, besides a number of other more recent ideas. Finally, we cannot ignore the importance of *noise*, in its various incarnations, which are fundamental for the theory and practice of spectral music. A complete chapter is dedicated to exploring random processes, the modeling, synthesis, and manipulation of non-deterministic spectra, which completes the discussion of spectral audio processing.

Part III starts by tracking back to the historical narrative in order to discuss aesthetically relevant issues of spectral music. We re-examine the role of early 20th century composers in a movement to conquer the spectrum, which has been a characteristic of musical practice since then. We argue that the invention of sound recording and reproduction has been the single most important element in the development of an essentially spectral way of making and listening to music. We conclude the book by applying the ideas developed earlier in the book to sound design and composition.

It is my hope that this book can be used as an essential guide to working with spectral music, from a practical to a research perspective. This text was written to serve readers who are interested in composing, researching, designing, and producing music, as well as anyone who wants to understand what the adjective *spectral* means. I would like to repeat here the final words of the book, which are a full and simple statement of my intentions: *the spectral playground is open to all, anyone is welcome to share its delights and exercise their creativity within it.*

Maynooth, September 2020 *Victor Lazzarini*

ACKNOWLEDGMENTS

The ideas and concepts in this book have been developed through many years of research and practice of spectral music composition. Much of it has also been prompted by discoveries made as a result of over two decades of postgraduate level teaching and supervision at Maynooth University. So I am deeply indebted to all of my students and colleagues who have played an important part in my academic life.

I would also like to acknowledge the help and encouragement of the computer music community, as well as the various contributions to software development, ideas, and concepts that have arisen from them. I would like also to thank my colleagues in the Ubiquitous Music Group, in particular Damián Keller and Nuno Otero, who have been partners in various projects throughout the years.

Special thanks should go to colleagues in the Csound development team John ffitch, and Steven Yi, and in the wider Csound community, Rory Walsh, Øyvind Brandtsegg, Richard Boulanger, Tarmo Johannes, Joachim Heintz, Stephen Kyne, François Pinot, Alex Hoffmann, Bernt Isak Waerstad, and Alessandro Petrolatti, among many others, for their input into the open-source project and also for the enlightening discussions on all matters to do with audio programming and beyond. In particular, I would like to note the passing of our colleague Eugenio Giordani at the time of the writing of this book, who was a key figure in the computer music community of Italy and Europe.

I am very grateful for the endorsement given by Jon Nelson, who very kindly wrote the foreword for this book. I was delighted to have the support of such a wonderful composer. I would also like to thank Maynooth University, whose sabbatical scheme provided support for the completion of this project, and the Arts and Humanities Institute, and its director, Prof. Thomas O'Connor, for providing a home during this period. I am also very grateful for the support of my editors at OUP, Norman Hirschy and Lauralee Yeary, who have been extremely helpful during the preparation of this book.

As ever, the work for this book has been thoroughly supported by the patience and help I get from my wife Alice, and our children Danny, Ellie, and Chris. They are an integral part of any achievements I might be in a position to claim.

CONTENTS

PART I

Background

1

What Is the Spectrum?

As indicated by its title, this book is about spectral music design; that is, it explores the spectrum of music as an object of manipulation. It is therefore important that we start the discussion with a general description of our object of study, which will be refined incrementally as we explore it from different perspectives. Firstly, we should start by assuming an understanding of what constitutes sound, as mechanical disturbances traveling in a medium (which, unless stated otherwise, is the air). These take place in the form of *waves*, which are physical deformations of the medium, observable as its constituent particles are put into motion. As far as this text is concerned, sound becomes synonymous with audio if it can be captured and decoded by our hearing mechanism. With that very fundamental understanding, we can proceed to construct all concepts used in this book one by one, in successive approximations. Before we are in a position to attempt a first definition of spectrum as the subject matter of this text, we need to introduce and discuss some basic notions that underpin it. This is our starting point.

1.1 Functions and Signals

In its simplest form, a *function* is a means of mapping a *domain* into an *image*; that is, given an input, an output is found as a result of function application. This is a fundamental mathematical tool that has many practical applications. More important, for us here, they can be used to define a *signal*, which may be described as a carrier of messages from a sender to a receiver. For example, traffic lights provide a signal composed of three discrete states, red, yellow, and green, which is sent to drivers to control a street crossing. We can describe this signal as a function with three discrete values (red, yellow, green) depending on an input variable, *time*. We can plot it: at time t_0, the function outputs red; yellow is output at time t_1; and red at time t_2 (Fig. 1.1). If this is a simple fixed set of lights, then it will follow this pattern repeatedly, in regular periods of time, making this a *periodic* function.

From this simple example, we can see that a signal such as the one produced by traffic lights can be described by a function of time. Signals can take many different forms: the light emitted by the traffic lamp is itself a function of time, whose output are different values of electromagnetic fields. Our eyes are equipped to detect the signal denoted by these oscillations and translate it into a given color. Likewise, we can talk about acoustic signals, which take the form of mechanical waves, traveling from source to receiver. In this case, our ears translate these into a form that is intelligible, and they become audio signals. Functions can also be used to define these waves, often using time as a

Spectral Music Design. Victor Lazzarini, Oxford University Press. © Oxford University Press 2021.
DOI: 10.1093/oso/9780197524015.003.0001

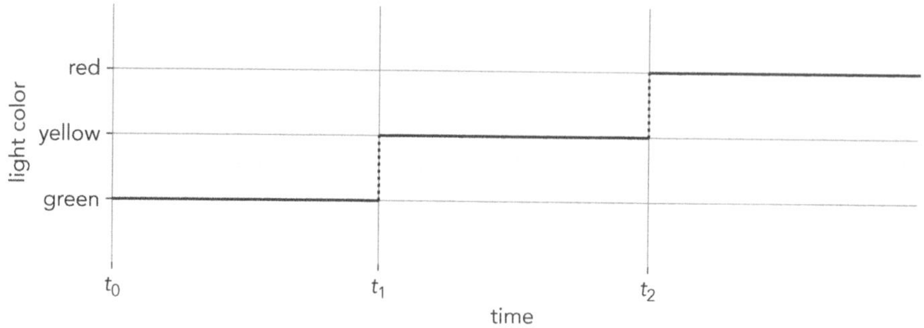

Figure 1.1
A graph of a function representing the signal from a set of traffic lights.

variable, and measuring its displacement amplitude, pressure, or velocity, depending on what quantity we want to represent.

Signals and functions are closely associated, and sometimes we might talk about them interchangeably in some contexts. This will happen in particular when we want to describe audio signals in general terms, where the distinction between the signal and the function representing it is not particularly necessary. However, it is important to remember that they refer to two different concepts: a signal is a message carrier, and a function is a means to describe the nature of a signal in a precise way.

1.1.1 Functions of Time

As indicated by the examples shown earlier, and in particular within the scope of this book, signals are commonly represented by functions of time. This means that the underlying function that describes a signal has time as one (or the only one) of its variables. So, when we speak of an audio wave as our signal, the function representing it, also called a waveform, is a unidimensional function of time, whose domain is the set of real numbers[1] and its image is generally a subset thereof. In some places, particularly when considering sounds traveling in a medium, a waveform might be represented by a function of space or distance. However, by measuring the displacement, velocity, or pressure at a single point it is possible to describe the same signal as a function of time. Also, if the speed of the traveling wave is known, it is possible to go from a function of space to one of time. For most of the discussion in this book, we can think of the two forms as equivalent, and treat waveforms solely as functions of time.

Thus the waveform as studied here can be conceptually thought as the measurement of pressure at a point in space, above and below the atmospheric pressure. For instance, we could think of a sensor placed at the eardrum, which would measure the push and pull caused by the movement in the medium on outside of it. We could equally think of the pressure placed on an omnidirectional microphone by the air around it, or conversely, by the displacement of a loudspeaker cone, forwards and backwards off its rest position. All of these measurements of a waveform are analogous (which is not to say they are the same, as we have to take into account the interference of distortion and noise that may be present). At this point, we can safely abstract the waveform from its physical existence to make a number of general points about it. By doing that, we will assume that this function of time outputs varying values of amplitude, which in the real world may refer to pressure, displacement, or velocity of air particles.

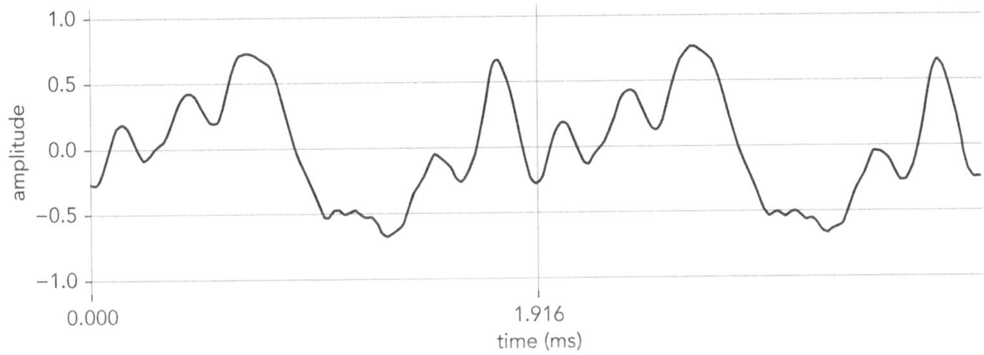

Figure 1.2
A graph of a segment of a viola waveform playing a 523 Hz note.

The first thing that we can say is that waveforms will be restricted to a given range (of amplitudes), which is largely dependent on the system involved (be it mechanical, electrical, digital, etc.). Often it is the case that for a given system, we can speak of a nominal working range, above which the behavior of the waveform might be unspecified. Again, we can capture this in another abstraction, that the range of a waveform falls in the interval $[-1.0, 1.0]$. In some places, this is also known as the *normal* range of a signal. Within this range, it is possible that waveform does not actually reach its maximum values; in any case, the maximum absolute value (that is, disregarding its sign) of a signal is known as the *peak amplitude*. It is an important parameter, which we will often need to pay attention to, in order to keep the waveform within the normal range.

A characteristic of audio signals, and their waveforms, is that they are oscillating, fluctuating between various values within their range. Therefore any instantaneous value of their output will not be particularly indicative of the overall amplitude variation that takes place over a period of time. An important quantity is the overall amount of intensity in a signal, which although related to the peak amplitude, is more closely associated with the average amount of amplitude variation of a waveform. A simple average of values is not helpful, since audio signals are bipolar (their range includes positive and negative numbers), therefore the most common method is to use squared measurements at regular intervals, averaging them and calculating the square root of the result. This is called the *root-mean-square* (RMS) amplitude, and it is proportional to the energy of the signal over the total period of measurement. The relationship between RMS and peak amplitude is dependent on the function that defines the waveform.

At this point, it is worth looking at very specific types of waveforms that may represent some audio signals. This on one hand restricts us to consider only functions under certain conditions, but on the other allows us to build the foundations to understand waveforms more generally. The first limitation we will impose is that the function is periodic; that is, as we noted before in the case of traffic light signals, it has a pattern that repeats in time. In this case, we are able to segment the function into separate cycles. This is shown in Fig. 1.2, where a segment of a viola waveform is plotted. Note that while the wave cycles of a real instrument signal may not be exactly similar, once they are clearly defined, as it is the case here, they characterize a periodic waveform in general terms. We will see later on what is the meaning of these slight (or sometimes not so slight) variations in wave shape.

The concept of periodic signals allows us to introduce a couple of parameters that are related to each other. Since the waveform repeats after a certain time is elapsed, then

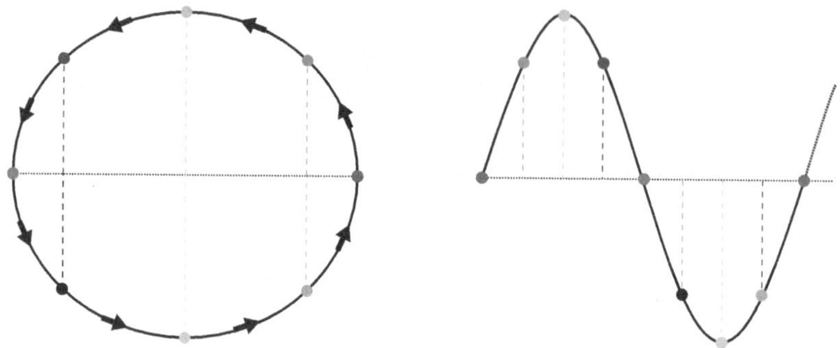

Figure 1.3 A sinusoidal wave and its relationship to circular motion.

we can measure this, and we call it the *period* of the function. Equally, if we had been measuring the waveform in space, each cycle would have a certain length. As we noted earlier, wavelength and period, as observed in functions of space and time, respectively, are therefore related to each other through the speed of sound in a medium.

Conversely, in any given length of time, we might have a number of complete periods (or, more commonly, *cycles*) of the waveform, and this is called its *frequency*. Conventionally, we measure periods in seconds and frequencies in cycles per second (cps). More generally, we will use the SI unit Hertz (Hz) to denote *per second* in the context of periodic functions, whether we are talking about cycles, clock pulses, samples, or any other periodic patterns. In the case of Fig. 1.2, the signal frequency is 523 Hz, and the waveform period is 1/523 sec.

While the frequency (and period) parameters are helpful to characterize any periodic waveform, it is worth imposing one further limitation on the signals so that we can arrive at further definitions. Periodic waveforms may have any shape (as plotted in a graph, as we can see in the example of Fig. 1.2), as long as it repeats in time. The next limitation we impose is that only a certain shape is allowed, one which can be derived from circular or pendular movements, called the *sinusoid* shape. Figure 1.3 shows how a sinusoidal function can be derived from plotting the height of a spoke on a rotating wheel.

Sinusoid signals allow us to introduce another key parameter, which is used to determine timing relationships between two or more such waveforms. This is called the *phase*. It measures the time offset or delay relationships between two signals, and we can define it in terms of fractions of a cycle (e.g., 1/4 cycle, 1/2 cycle, etc.). We can also measure phase by means of an angle offset: as one full cycle of sinusoid is equivalent to one whole revolution of a circle (equivalent to a 360° angle), fractions of it may be characterized by angles between 0° and 360°. An example of phase offset is shown in Fig. 1.4, where two signals are separated by a 90° offset. More precisely, we will use linear angle measurements in *radians*, but we will leave these to be discussed at a later stage.

1.1.2 Functions of Frequency

Frequency, amplitude, and phase define a unique sinusoidal waveform. These waves are fundamental to audio signals, as they can represent elementary functions, with which we can construct other waveforms. A common way to express this is to say that sinusoids are building blocks of audio signals. They also allow us to look at a different way to describe these, by considering how they can be broken down, or *analysed*, into a set of constituent

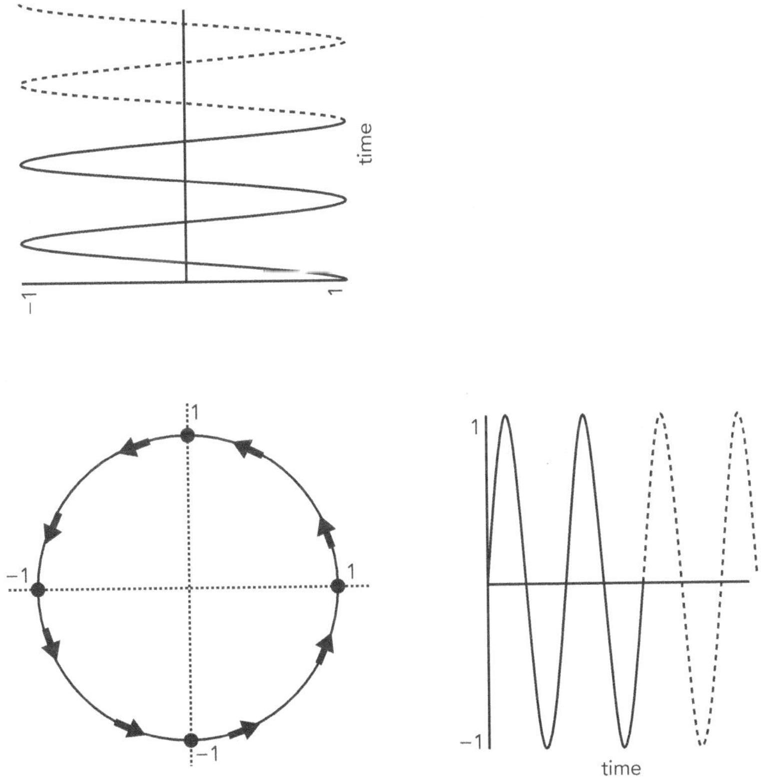

Figure 1.4
Two sinusoid waveforms with the same frequency, separated by a phase offset of 90°.

sinusoidal waves having different frequencies. At this point, we are able to move from a function of time to a function of frequency, and determine the mixtures of sinusoids that make up the waveform, each individual one at a specific frequency.

Functions of frequency allow us, among other things, to determine how energy in a waveform is distributed at different frequencies. By analyzing a signal into its sinusoidal amplitudes and then providing these as a function of frequency, we are able to see the components that make up a given signal. Figure 1.5 shows a graph of a function of frequency, which was produced from the analysis of the waveform of a harpsichord note. In this plot, we can see that the signal is composed of many sinusoidal components, which are called *partials*. Some of these are very prominent, others are faint. The tone quality of a waveform, which is what we call *timbre*, will be closely linked to the presence or absence of partials, and their relative amplitudes.[2]

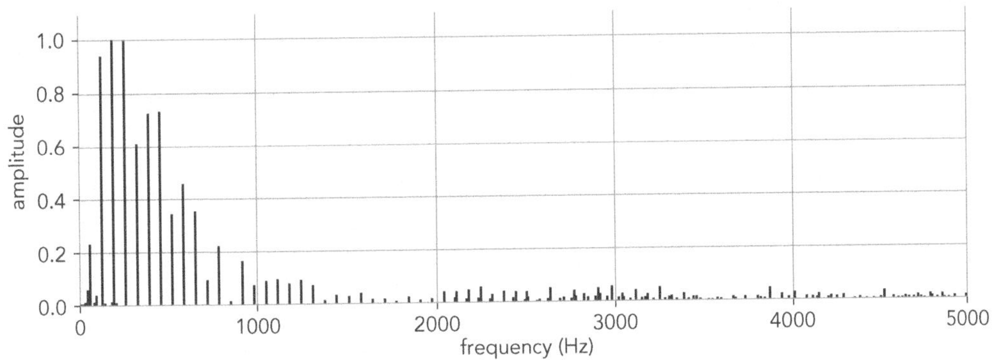

Figure 1.5
A function of frequency obtained from the analysis of a harpsichord waveform.

Note that in order to obtain such a function of frequency out of a waveform (a function of time), we will need to pick a fixed time position as a starting point, and a time slice of a given duration, in order to analyze it. In the case of Fig. 1.5, we made a 7.5 sec recording of a harpsichord note and used that as the analysis time slice. Any fluctuations of the tone in terms of its partial amplitudes are averaged over this duration. We may think of a function of frequency such as this one as a snapshot of a waveform, freezing its partials in time. If these are more or less stable, as in the case of the harpsichord, the resulting picture is well defined, and the signal components appear clearly in the graph. In other cases, varying degrees of smearing or spread may be present in the analysis.

Other functions of frequency can be obtained from similar analyses. For instance, instead of looking for sinusoidal amplitudes, we may be interested in the way the phases of components are distributed in frequency. We can also represent both amplitude and phase bundled together as the levels of two sinusoids 90° apart (see Fig. 1.4) at each frequency. Any such representations are also described as belonging to the *frequency domain*, as opposed to the *time domain* in the case of functions of time.

1.2 Fundamental Concepts of Spectrum

At this point, we are ready to attempt to define the term *spectrum* (pl. *spectra*). In general usage, it may refer to several methods of classification of physical properties. We will investigate the historical use of the term in more detail on Chapter 2, but for now we can say that it was brought into the physical sciences originally to describe optical phenomena, such as the variation in color due to length or frequency of light waves. In relation to audio signals, the term will be related to the various frequency-domain representations we can obtain and manipulate. These representations may be applied directly to waveforms, as we have seen for instance in Fig. 1.5, and also to processes. For instance, we may speak of the spectrum of a *filter*, which is a signal processing operation that can be used to modify audio signals. In that case we are accessing a representation of the process that is applied to an audio signal, rather than one that denotes the signal itself.

Rather than forcing a specific definition of spectrum, we should loosely connect it with the idea of measuring or describing the *frequency content* of signals, or also the way in which this can be modified by a process. Of course, the concept of frequency content itself is tied, again in a loose way, to the sinusoid partials that make up a waveform. We do not want to be too forceful at this stage because there are significant questions about how such representation may or may not be meaningful in every situation. We ought to remember that we arrived at it by looking at particular subsets of the totality of possible audio signals, and things may become muddier once we move away from these simple waveforms. In this section, we will introduce some of these questions, partially answering them, with a view that a more complete discussion is of course the subject of the remaining chapters of this book.

1.2.1 Periodicity and Pitch

One of the important characteristics of some audio signals, from a musical point of view, is that they evoke the sensation of pitch, that is, we are able to assign a specific level (in a scale from low to high) to a sound. Pitch is not easily definable, and often we use its associations to music in order to describe it. The Acoustical Society of America, for example, defined it as a characteristic of sound perception that allows the construction of musical scales. Experimentally, the pitch of an audio signal may be determined by comparing it to a pure sinusoid, and ascribed to the frequency of that waveform. A more general definition proposes that pitch is

> that attribute of auditory sensation in terms of which sounds may be ordered on a scale extending from low to high. Pitch depends primarily on the frequency content of the sound stimulus, but it also depends on the sound pressure and the waveform of the stimulus. (ANSI, 1994, 34)

In any case, pitch is a perceptual concept that arises from a number of given conditions. The simplest of these is that of periodic waveforms, whose *fundamental* frequency will generally be proportional to pitch. In a signal with many partials at various frequencies, the fundamental frequency is given by the inverse of the period of one complete cycle. Under this condition, we observe an important characteristic in the spectrum: partial frequencies are aligned as integer multiples of the fundamental frequency (and their period or wavelength are whole-number fractions of the fundamental period or wavelength). Each one of these is called a *harmonic* partial (or just harmonic), and the set of such components is often called the *harmonic series*. Harmonics may have different amplitudes, some may be missing altogether, but the presence of only a few will be enough to denote a given fundamental frequency. This is often the first and lowest partial in the spectrum, but may also be missing even though we would ascribe a pitch to its frequency. An example of a simple periodic waveform, its harmonics and spectrum is shown in Fig. 1.6.

Beyond these very simple cases, there are a number of other situations where the sensation of pitch may arise. It is often useful to look at the spectrum in order to understand these. Firstly, the presence of a few low-amplitude or short-lived partials that are not harmonic (also called *inharmonic partials*) will possibly not destroy the overall sensation of pitch in a mostly harmonic spectrum. It is possible to note this in the harpsichord note example (Fig. 1.5), where some very low amplitude components are present, against the harmonics (the evenly spaced partials most prominent in the graph). Likewise, if partial frequencies deviate slightly from the whole-number harmonic series, the sensation of pitch might still be preserved. The piano is a well-known case where the measured partial frequencies are slightly stretched in comparison to the harmonic series. In some other situations, where the inharmonic mixture of partials feature one or two strong components against a few less prominent ones, a sense of pitch might be assigned to these stronger components (Fig. 1.7). Finally, pitch might be a fleeting sensation, while some sounds may not normally be perceived as pitched (as the various forms of non-periodic random signals, often called *noise*), depending on the situation, pitch may be ascribed to them, at various levels of definition. The range of definitely pitched to unpitched audio signals is a continuum, and the transition stages may be very fluid.

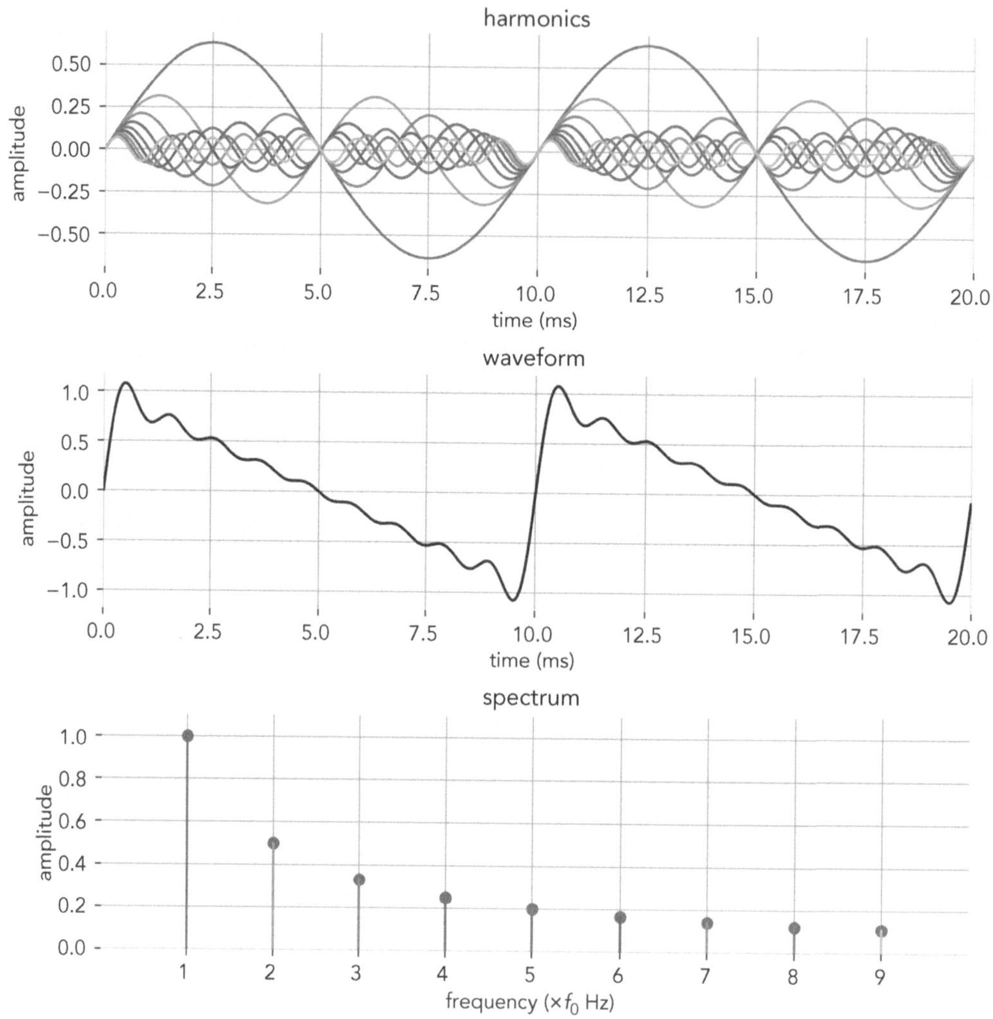

Figure 1.6
A periodic waveform with fundamental frequency $f_0 = 100$ Hz (middle panel), its harmonic partials (top panel), and spectrum (bottom panel).

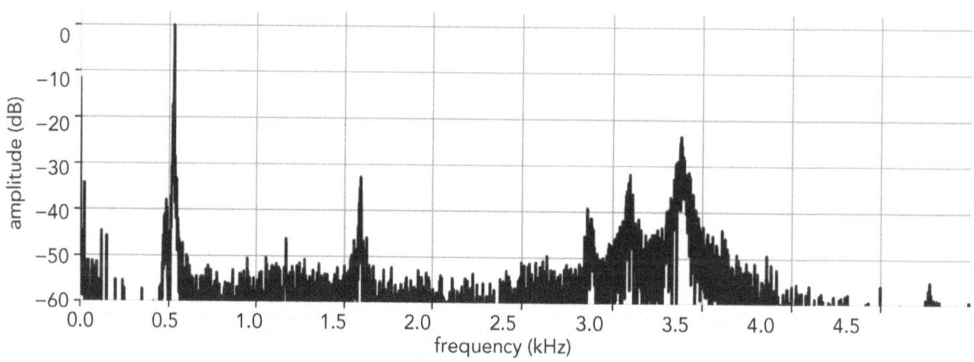

Figure 1.7
A graph of the spectrum of a xylophone note (C5, 523 Hz), showing a strong component at the perceived f_0 and some low-level peaks at other inharmonic frequencies. The plot uses amplitudes in the decibel (dB) scale (see Chapter 3), which may provide a more perceptually meaningful picture of the spectrum.

1.2.2 Distributed Spectra

Some audio signals exhibit clearly defined individually defined partials. Most so-called *pitched* musical instrument sounds feature these (at least in their stable states, see Sec. 1.2.3), while the vast majority of other sounds do not. In general, we observe a variety of spectra with different combinations of inharmonic partials, and in some cases a multitude of components concentrated on specific bands of frequencies, where it is not possible to separate each component in a clear way. These types of spectra are known as *distributed*. With them, the kind of analysis into partials that we have introduced in Sec. 1.1.2 may fail to provide meaningful results. Among these cases, we find the various types of *colored* noise sounds, rumbles, scrapes, etc. Within the realm of traditional musical instruments, we find distributed spectra in cymbals, some drum sounds, and percussion in general. As we noted earlier, some of these might evoke pitch even though their spectra do not feature clearly separated partials, let alone a harmonic series. These sensations might have to do with the concentration of partials within a very narrow band, which coupled with other conditions might allow us to assign pitch to them with various degrees of confidence.

Distributed spectra make the task of frequency-domain representation somewhat more complex. The principle under which we may say that all waveforms are composed of sinusoids of different frequencies, each with a given amplitude and phase, is still valid, but may be of little practical use. For example, an audio signal called *white* noise is defined as the sum of an infinite number of components all with the same amplitude, within the auditory range of frequencies. An analysis of this and other similar signals in terms of its constituent sinusoids is not practical, therefore we need to develop other notions that may involve statistical representations, which consider parts of the spectrum as not resolvable into partials, but as separate indivisible objects with given characteristics and behavior.

1.2.3 Dynamic Spectra

Complicating matters further, it is very seldom that we find that spectra are static over time. Figure 1.2, where we see slight variations in the shape of each wave cycle, gave us a hint in this direction. The fluctuations in shape are due to slight changes in amplitude or frequency of each partial (very minimal in this case). In fact, completely static waveforms, as for instance produced by electronic means, are not very appealing. The dynamic (that is, time-varying) characteristics of spectra are essential for the majority of musical sounds, regardless of their origin or source. Therefore, being able to capture, understand, and manipulate the dynamic aspect of spectra is essential.

In the simplest forms of dynamic spectra, partial frequencies are more or less stable, but the amplitudes vary with time. This is the case with many instrumental tones, where their onset (i.e., the start portion) generally contains a rich set of components whose amplitude decreases over time. Some sounds exhibit a set of very short-lived components at onset time known as *transients*. These may be inharmonic, or even distributed, and are very important for our identification of the nature or source of the sound waveform. For this reason, transients are often treated separately to the stable portions of the spectra of audio signals.

In any case, the changes in spectra of mechanically produced sounds are dependent on the physical nature of their sources, and the way these sounds are generated. The mode of production, striking, blowing, bowing, etc., as well as the vibrating objects, and

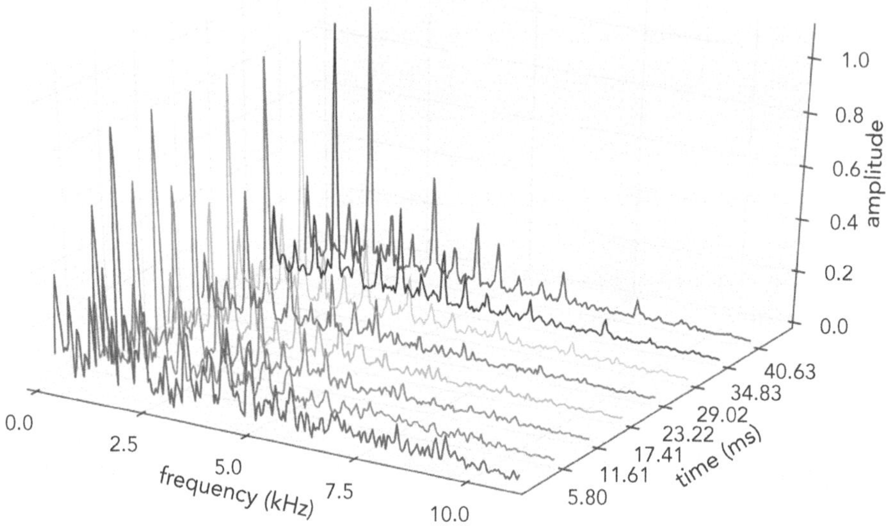

Figure 1.8
A three-dimensional graph showing the amplitudes of the onset of a viola note at different frequencies, in successive time slices.

any resonance apparatus, are significant in shaping the spectra of sounds. In order to capture such changes, we will need to reintroduce time in our spectral representation. This means that we will require functions that depend on two variables, time and frequency. In this case, instead of a single snapshot of the whole sound, we will have several slices, at different times, providing a picture of the spectral evolution.

An example of such representation is called the *spectrogram*, where we have amplitudes at different frequencies as they change over time. The difficulty is that such a graph needs two dimensions, one for time and another for frequency. Therefore in order to present it, we need to decide how to display the plot. Two alternatives are common: a two-dimensional graph of frequency versus time, where the amplitudes are shown in varying degrees of color; or a three-dimensional *waterfall* plot, where the successive slices are shown with frequency, time, and amplitude in separate axes.

Figure 1.8 shows an example of a waterfall plot, where time is placed on the y (front-to-back) axis, frequency on the x (left-to-right) axis, and the amplitudes are displayed in the vertical (z) axis. In contrast, a spectrogram of the same signal is shown on Fig. 1.9, where a graph of frequency versus time is displayed, with amplitudes presented as varying colors whose magnitudes (in decibels in this case) are to be interpreted via a color bar legend.

1.2.4 The Uncertainty Principle

When considering the dynamic aspects of spectrum, we are faced with the duality of the time and frequency representations of an audio signal, whose relationship requires close attention. From a measurement stance, in order to capture more precisely each component present in a signal (say, for instance, to try to get a better picture of noisy sounds), we may attempt to increase the frequency resolution of the analysis so that it becomes fine-grained enough to resolve them. As we do this, we will notice that the time

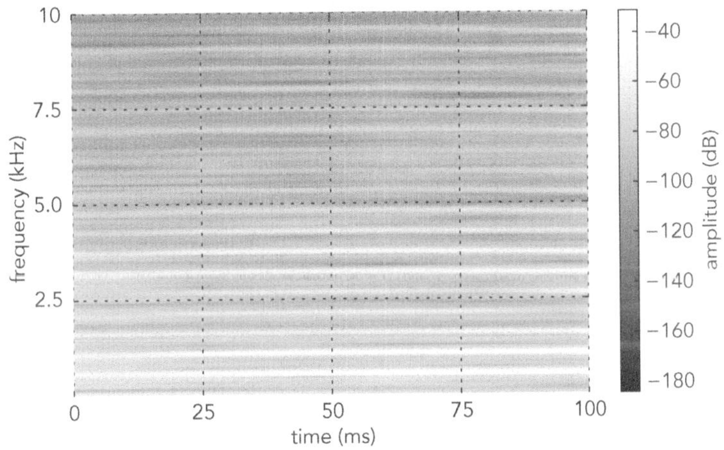

Figure 1.9
A spectrogram plot of the same signal in Fig. 1.8, where the amplitudes (in decibels) are shown in varying colors as per the color legend.

resolution, that is, how the spectral parameters change in time, is progressively degraded. Conversely, if we try to improve how we capture the dynamic aspects of a spectrum, we will not be able to represent its frequency content as well. This is related to what has been described in particle physics as the *uncertainty principle*.

Rather than an artifact of the methods we employ to analyze the spectrum, this is actually a fundamental characteristic the time-frequency duality of an audio signal, in a similar way to how the position and momentum of a sub-atomic particle are tied to each other. In the quantum-mechanical wave function, there is a trade-off between precisely defined positions and wavelengths (which are related to the momentum),[3] just as in the case of the sound waveform, which can be represented in terms of time and frequency. This leads to measurements that may be very precise for one or the other, but not for both at the same time. A well-resolved frequency analysis will be fairly smeared over time, and likewise, with very well-defined (i.e., short) time slices, the degree of frequency uncertainty will be increased.

This principle will manifest itself in many ways, and we will be coming back to it at different times in this book, but we can look at some simple examples to provide an idea of how it applies to the ideas developed so far. In order for us to be able to define the frequency content of a signal, we need to observe it for a certain duration so that the different components that it may contain can be fully resolved. As we do this, inevitably, the measurement process will average out any changes to the signal over that period. From another perspective, if we try to focus in on a very short period, in order for us to be able to increase the time resolution of the measurement, we may simply not be able to be certain of the precise frequency content of the signal. The fact that the analyses of data plotted in Figs. 1.8 and 1.9 are somewhat smeared in both frequency and time is a good demonstration of this principle.

Theoretically, a most well-defined, frequency-wise, signal is a sinusoidal wave with infinite duration, whose start and end times cannot be determined. Conversely, a pulse, described by a *Dirac delta* (δ) function[4] that is precisely defined in time, has a spectrum that is spread (or smeared) throughout all frequencies, which is the same to say that it cannot be precisely determined by any specific frequency. Figure 1.10 demonstrates this using the time and frequency plots of a pulse and an infinitely long DC signal (which is

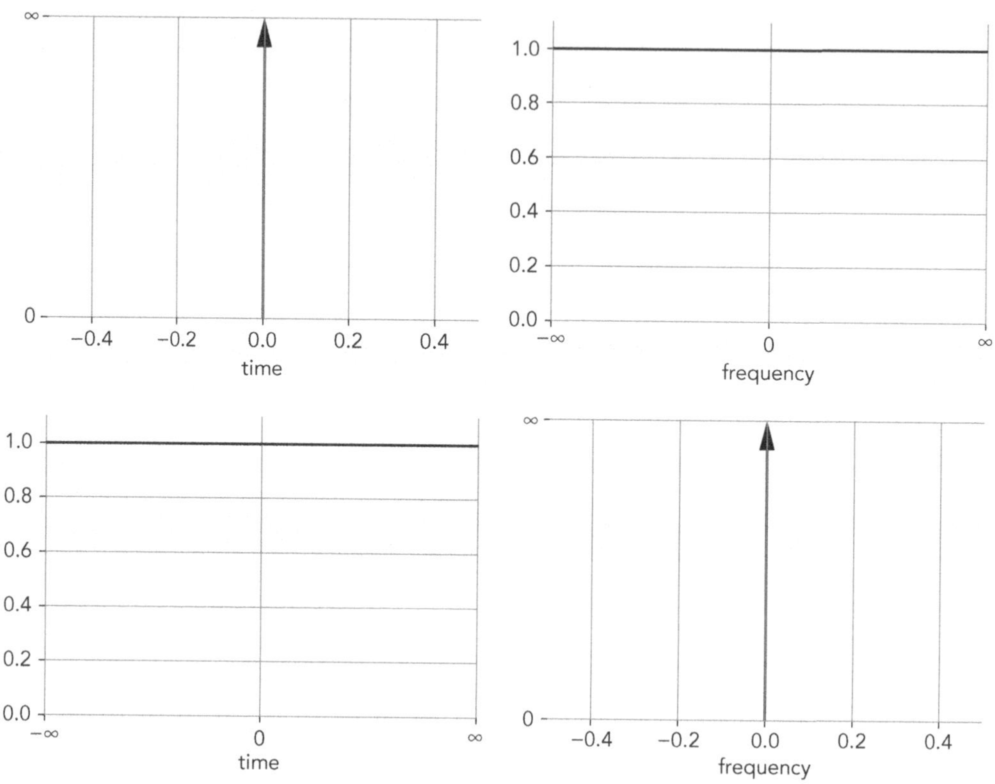

Figure 1.10 The duality of time and frequency and the uncertainty principle. On the two top panels, a pulse in its time and frequency representations. The bottom panels show the corresponding time and frequency plots of a 0 Hz time-domain signal.

equivalent to a cosine wave at 0 Hz). While the former is precisely defined in time and not in frequency, the latter is, conversely, exactly placed in frequency, but not in time.

We can experience this particular characteristic of the uncertainty principle with a simple experiment, which may be carried out with a synthesiser and sequencer. To do this, we program a sequence of notes to be played using any periodic waveform (a sinusoid would do), and then we continuously adjust the duration of each tone, shortening them by increasing the length of rests between the notes. At the start, we should be able to discern the constituent pitches fairly well; as we increase the rests, at one point the sensation of pitch will be less defined, becoming at some point closer to a timbral variation, then just nondescript clicks, when the tones approach short pulses (their amplitude may also need to be adjusted for them to be audible). Tones made up of only a few wave cycles do not have a well-defined frequency content. In fact, the very act of switching an audio waveform on and off abruptly will cause it to have a smeared frequency spectrum at the transition points. This simple experiment bridges the theoretical extremes portrayed by Fig. 1.10, providing an insight into nature of the uncertainty principle.

1.3 Psychoacoustic Aspects

The relevance of spectral representations of audio signals is underpinned by the way the hearing mechanism translates mechanical waves into neural data for processing by the brain. Part of this mechanism works as a spectral analyser, responding to energy at different frequency bands. This is combined with a phase locking process that is sensitive

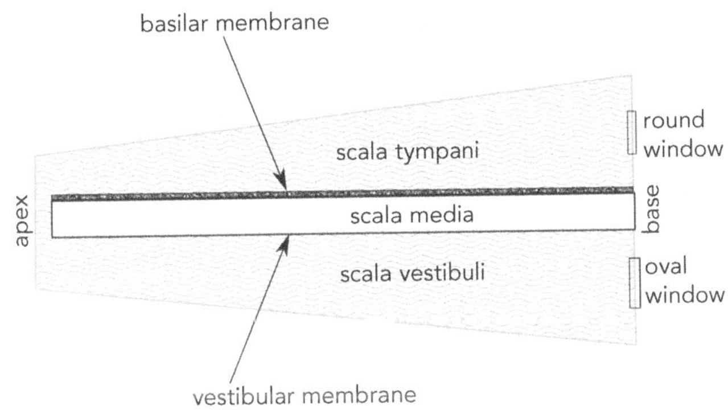

Figure 1.11
A schematic view of the major structures of the cochlea.

to time aspects of sound waveforms to provide very precise neural signals for the sound perception subsystem of the brain. The anatomy of the hearing mechanism is generally subdivided in three sections: the outer ear, made up of the external sound-focusing organ, the pinna, plus the auditory canal and the eardrum; the middle ear, composed of three ossicles that transmit the mechanical vibrations from the outer ear, providing protection and impedance conversion;[5] and the middle ear, whose main organ is the cochlea. For now, we will concentrate on this part, as it has the most importance for the subject of this book, and return more fully to the subject acoustics and psychoacoustics in Chapter 3.

1.3.1 The Cochlear Mechanism

The cochlea is a spiral-shaped organ, with a tube-like cavity that is connected to the middle ear ossicles via a thin membrane that covers an aperture called the oval window. This cavity is composed of two connected sections, called the *scala vestibuli* and *scala tympani*, which envelop a middle structure known as the *scala media*. The incompressible fluid that fills the scala vestibuli and scala tympani, called *perilymph*, can flow from one to the other via a small opening at one of the ends of the tube, called the apex. The other end, the base of the cochlea, is adjacent to the middle ear, and contains both the oval window, connected the scala vestibuli section, and its pressure-compensating counterpart, the round window, that is at the base end of the scala tympani (Rask-Andersen et al., 2012). Figure 1.11 shows a schematic of these different structures in an idealized, uncoiled cochlea.

The scala media is separated from the scala tympani by the *basilar membrane*, and from the scala vestibular by the *vestibular membrane*. Of these two, the former is the most important for this discussion, as it is responsible for carrying out a frequency detection of an input signal represented by a wave traveling in the fluid-filled cavity of the cochlea. The membrane has a key characteristic: it is thin and narrow at one end, thickening and widening gradually as it extends toward the other end. This allows it to respond to higher frequencies at one extreme and to low frequencies at the other. The change in thickness and width is continuous along its length; therefore, at different points it will detect different sinusoidal components of a complex waveform. The thin end of the basilar membrane sits at the base, whereas the widest and thickest part is at the end of the coil, the apex.

As a waveform travels in the perilymph, it may displace different parts of the basilar membrane; for a sinusoidal signal, the maximum point of displacement varies with the

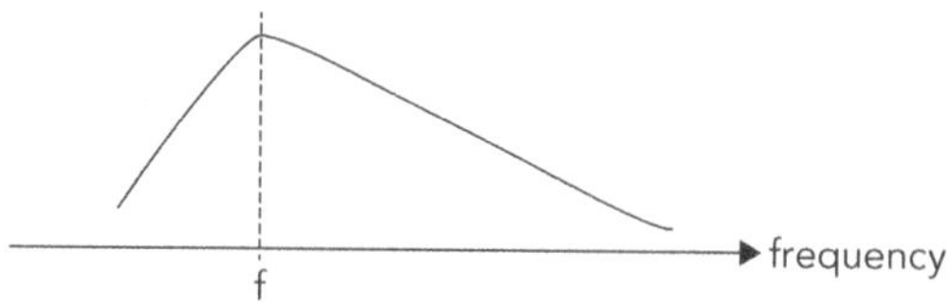

Figure 1.12 An idealized basilar membrane displacement envelope for an input frequency f.

input frequency. The position of this point, as measured from the apex end of the basilar membrane (which responds to low frequencies) is directly proportional to the logarithm of the sinusoidal wave frequency.[6] This provides a psychoacoustic explanation for the logarithmic characteristic of our frequency (and pitch) perception. It is also important to note that the envelope of displacement in the basilar membrane is not pulse-like, but has a certain asymmetric spread over a given region, having a more abrupt decay toward the lower frequencies, and a more gradual tail-off toward the higher frequencies. This is illustrated in Fig.1.12, where an idealized. basilar membrane displacement envelope is shown, for an arbitrary sinusoidal frequency.[7] When the basilar membrane is displaced, it causes the hair cells sitting underneath it to produce neural signals that are sent out of the ear via the auditory nerve. This is the function of the organ of Corti, which occupies the scala media, providing the interface between the cochlea and the rest of the auditory system.

Another important aspect of the frequency and pitch detection process has to do with the timing of the neural firing that is caused when a point in the basilar membrane is excited by an input waveform.[8] It has been shown that such neural impulses occur at maximum once per cycle, and above 300 Hz they occur at longer intervals as the nerves cannot fire at higher rates (Pickles, 2013). Additionally, each firing will occur always at the same cycle phase for a given position at the basilar membrane, that is, there is a phase-locking process that ties in the neural pulses with the input signal phase (Johnson, 1980). For a multi-component signal, this is one of the mechanisms that helps the very sharp pitch discrimination observed in psychoacoustics experiments. Therefore, both basilar membrane location and timing are tightly involved in pitch and frequency perception.

1.3.2 Critical Bandwidth

As indicated by Fig. 1.12, the displacement at the basilar membrane is not punctual, or pulse-like, and is spread over a certain range of frequencies. This indicates that two sinusoidal components that are very close in frequency will be captured by the same region, and will possibly not be finely discriminated. In that case, we can speak of a combined signal that is exciting the basilar membrane, rather than a single sinusoid. In order to define this region, we can introduce the concept of a *critical bandwidth*, which is a psychoacoustic phenomenon and needs to be defined through qualitative perceptual measurements. Once these are in place, we can determine experimentally the critical bandwidths at various frequencies.

The common definition of a critical bandwidth is that of the "bandwidth at which subjective responses abruptly change" (Scharf, 1970). From a physical point of view, this is the bandwidth at which two displacements in the basilar membrane do not interfere with each other, and are captured by two distinct and separate displacements. From

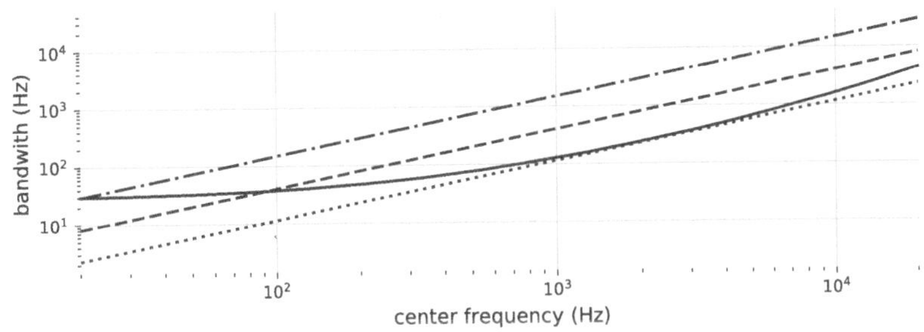

Figure 1.13 A plot of the equivalent rectangular bandwidth of a filter that models critical bandwidths at various center frequencies (solid line), and three interval bandwidths: a major second (2 semitones, dots), a perfect fifth (7 semitones, dashes), and two octaves (24 semitones, dot-dashes).

a psychoacoustics perspective, there are possibly two regions within one critical bandwidth. If we consider, for instance, two input sinusoidal tones separated by a certain frequency difference, one of these regions is characterized by a beating percept, caused by an amplitude modulation effect. The other region, reached as the frequency difference is increased, is where beating is absent, but a rough sensation persists. At the point when this gives away to a smooth feeling, the two input sounds are spaced by one critical bandwidth.

One way to model the displacements in the basilar membrane that are due to multiple sinusoidal components is to think of it as a bank of overlapping filters, each tuned to a given center frequency and being able to remove all components but the ones within its bandwidth. Each auditory filter will only respond to these frequencies, therefore two sinusoidal tones mixed together that are widely separated will be captured by two distinct filters. Conversely, tones that are very close together will not be resolved by separate filters and therefore cannot be distinctively separated. From this, it is possible to appreciate that this psychoacoustic phenomenon is very relevant to our discussion of spectral music design.

It is possible to experimentally measure and model these auditory filters via the principle of an equivalent rectangular bandwidth filter, which would be defined to pass as much energy as an auditory filter centered at given frequency. From this, critical bandwidths for a range of frequencies can be derived (Fig. 1.13).[9] It has been observed that the bandwidth of these filters increases as we move toward the lower end of the frequency spectrum, and that below 50 Hz, different frequencies do not lead to differentiated displacements in the basilar membrane. We can observe from the plots in Fig. 1.13 that the equivalent rectangular bandwidth varies from about two octaves at 20 Hz to be as narrow as two semitones above 2 kHz and then widening slightly at the higher end of the spectrum.

1.3.3 Loudness Perception

Another important aspect of the auditory mechanism is that the perceived loudness as a function of frequency is not flat or linear. Due to the action of the various parts of the outer and inner ear, there is considerable difference in how we are able to perceive individual sinusoidal components in the spectrum. Again, this has been measured experimentally and is another important psychoacoustic phenomenon for this discussion. From these measurements, a number of equal-loudness curves have been derived as functions of frequency for sinusoidal tones of the same amplitude at different levels.[10]

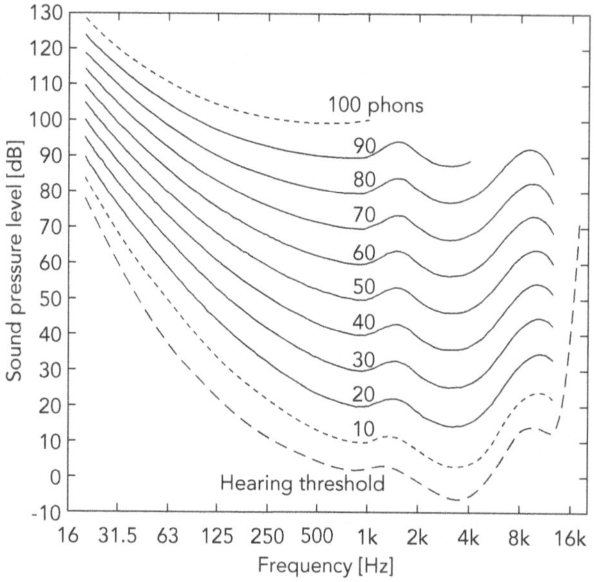

Figure 1.14 Equal-loudness contours (ISO 226:2003). Reproduced under permission by the International Organization for Standardization through the National Standards Association of Ireland (NSAI). A copy of the standard can be purchased from ISO at https://www.iso.org/standard/34222.html

Figure 1.14 shows the estimated equal loudness contours, estimated from a dataset of 12 studies, which defines the ISO226:2003 standard.

Each equal-loudness contours defines a different perceptual loudness level defined in the *phon* scale, which is equivalent to the sound pressure level measurement at 1kHz. There are two significant points about these: our ability to perceive sounds at the lower end and at the higher end of the spectrum is diminished; and different sound intensity levels have specific loudness functions. The first point indicates that the middle to middle-high range of frequencies would appear louder to us than the other ranges, for sinusoidal tones of the same measured amplitude. The second implies that the difference in perceived loudness will vary with intensity levels, and for low amplitude tones, this difference is greater. As we increase the amplitude of the input, there is a leveling of the equal-loudness perception contours, as observed in Fig. 1.14.

1.4 A Provisional Definition

In this chapter, we have provided a gentle introduction to a number of key concepts that will be employed in this book. In particular, we have sought to define in general lines what we mean by the term *spectrum* and the ways we can describe and study it. Now we are ready to attempt a provisional definition based on the ideas developed so far:

> Spectrum refers to the acoustic and/or psychoacoustic aspects of sound that take place concomitantly at different regions of frequency, which may have infinitesimally small bandwidths. These aspects may evolve dynamically in time and be decomposed analytically in separate components. Similarly, they may be composed to make up a sound waveform.

This basic definition will be refined as we progress through the book. It is important at this stage to develop intuitive ideas about the subject, and introduce further formalisms to support these well-understood principles in a progressive manner. In the next chapter, we will enhance our notion of the spectrum by tracing back a history of ideas about the nature of sound, music, and acoustics.

Notes

1 This means in this case that time, theoretically, extends from $-\infty$ to ∞, although we will in many circumstances limit this to, at least, a conventional start time of 0, often providing a finite duration as well. However, this will have consequences, which are discussed in Sec. 1.2.4.

2 In fact, it has been shown that timbre is a multi-dimensional property, therefore the presence or absence of partials can be seen as one of the elements defining it, but not the sole one. Pioneer experimental work in timbre perception was carried out by Grey (1978), and much work on the topic followed, including for instance Jensen's model (Jensen, 2002). Manipulation of timbre is of course one of the central concerns of this book.

3 Griffiths introduces this concept in a very intuitive way: "[...] you're holding the end of a very long rope, and you generate a wave by shaking it up and down rhythmically [...]. If someone asked you, 'Precisely where *is* the wave?' you'd think he was a little nutty: The wave isn't precisely *anywhere*—it's spread out [...]. On the other hand, if he asked you what is its wavelength, you could give him a precise answer [...]. By contrast, if you gave the rope a sudden jerk [...], you get a relatively narrow bump traveling down the line. This time the first question is a sensible one, and the second seems nutty—it isn't vaguely periodic, how can you assign a wavelength to it?" (Griffiths, 1995). Wishart also has some similar intuitive insights in relation to sound waveforms, pointing out that before a waveform completes a few cycles, it is uncertain what frequency characteristics it might have (Wishart, 1996).

4 We will study the Dirac δ more carefully in Part II of this book.

5 As the mechanical resistance to motion in the inner ear fluid is higher than that of the air moving the eardrum, it is necessary to convert the pressure exerted onto the eardrum into a level that would overcome this higher resistance. This makes the middle ear mechanism an impedance transformer. Together with this, there is a protection that come into effect at high levels of sound intensity, making the transmission process less efficient therefore reducing harm to the delicate structures of the inner year (Pickles, 2013).

6 The experimental studies leading to the description of the frequency-detection mechanism of the cochlea were carried out reasonably recently and reported in von Békesy (1961). These results were further improved in Johnstone and Boyle (1967).

7 Based on displacement envelopes as reported by von Békesy (1961).

8 A temporal theory of pitch perception was first put forward by Schouten (1940) against the prevailing idea that the detection of pitch was solely due to the ear acting as a frequency analyzer.

9 A formula for the equivalent rectangular bandwidth as a function $B()$ of center frequency f_c has been proposed as $B(f_c) = 0.00000623f_c^2 + 0.009339f_c + 28.52$, which we apply to get the plot in Fig. 1.13, in Moore and Glassberg (1983).

10 This was originally described by Fletcher and Munson, who provided a set of equal-loudness curves that have been used in various applications, including the derivation of weighted scales for sound intensity measurement (Fletcher and Munson, 1933). Since the original measurements, more precise experiments led to a better set of equal-loudness contours that have been adopted by the International Organization for Standardization as ISO 226:2003 2003.

2

A History of the Spectrum

The idea of spectrum, which we attempted to introduce partially in Chapter 1, is a complex one, as it emerged only very gradually once we gained more understanding about the phenomena of sound and our own perception of it. If we look at the history of music and science in the Western world,[1] we will see that the modern concepts around spectral audio were likewise slowly developed along centuries of study. This was mostly due to two intertwined lines of study, one in music theory and another in physical sciences. The former was largely informed by the practical abilities of musicians, who also demanded a better explanation of the properties of what they classed as musical sounds. The latter was born out of the desire to understand physical phenomena more generally. Throughout the years, these two areas have found many meeting points, and in a way, this book is also inspired by their overlap.

In this brief historical account of spectral audio, we will proceed by looking first at how early studies in acoustics related to aspects of the spectrum (of instrumental tones, mostly) and how it informed music theory. The story will move swiftly from musical concerns toward the investigation of the physical world, as we enter the era of classical physics. Many important discoveries will be made, culminating on the work of Fourier, which is of fundamental importance to spectral audio. At the end of this period, musical acoustics would become an important field of study, with significant influence in music theory. From the late 19th century and into the 20th century, developments in electrical engineering provide first an improved theory, then a practical means to manipulate the spectra of sounds. In parallel with this, musical aesthetics will enter a phase of expansion, where composers move away from the simple harmonic spectra of instruments and the singing voice to conquer all sounds. The narrative then culminates with the emergence of computer music, which provides the environment, the tools, and the motivation for this book.[2]

2.1 Principles of Pitch and Scale

The great classical tradition of inquiry promoted by the Ancient Greeks gave us the first steps into the study of spectrum, even though these were not directly aimed at other aspects which interest us today, such as timbre or spectral color. Much of their focus was put on the establishment of relationships between musical tones, which for them meant spectra that were primarily pitched. Little or no attention was given to sound qualities that went beyond this, and a lot of emphasis was given to what could be sung and how tone relationships determined the elements of music theory. Instruments, however,

Spectral Music Design. Victor Lazzarini, Oxford University Press. © Oxford University Press 2021.
DOI: 10.1093/oso/9780197524015.003.0002

played an important part in terms of measurements that could be objectively established. While the relevance of defining pitch relationships is limited to the understanding of the spectrum in general, the findings that originated at this stage have some bearing on important concepts such as the harmonic series and complex periodic vibrations.

The first aspects of the study of acoustics and music theory that have some relevance to us are attributed to Pythagoras (c. 570–495 BCE), or more generally, to the Pythagorean School.[3] They established that simple ratios determined the fundamental pure (i.e., exactly in tune) intervals, 2:1, 3:2, and 4:3. These, we now know, stem directly from the frequency ratios between the first four successive tones of the harmonic series, although the concept of frequency as a parameter of musical tones was a much later discovery. Much of the music theory in the ancient and medieval times was derived from these basic principles, which were established in an experimental way. A favored instrument used for this type of measurement was the single-string *kanōn*, also known as the monochord. With this, it was possible to be precise about ratios of string lengths, which were shown to be equivalent to given tone relationships, or in modern terminology, *intervals*.

2.1.1 The Pythagorean Scale

Employing these three simple ratios, a seven-step set of tones, or a *scale*, can be constructed from a given starting point.[4] The method we employ is based on a fairly simple recursive algorithm, defined by the following steps:

1. From a base tone, find another that is tuned to the ratio of 3:2 (a tone that is the same pitch as of a string tuned to the base tone and stopped 2/3 of the way along its length). If this is the fifth tone generated, then go to 4.
2. Replace the base tone with the newly generated one and find another that is tuned to the ratio of 2:3 (see the inversion, this is a tone below the starting one, but a similar approach can be used).
3. Replace the base tone with the newly generated one and repeat from 1.
4. From the starting tone, find another that is tuned to the ratio of 4:3.
5. Stop.

We can then reorder the tones in ascending pitch, and notice that the scale is composed of two step sizes, based on the ratios 9:8 (e.g., between the first two tones, and the second and third) and 256:243 (e.g., between the third and the fourth), which were given the name *tone* and *semitone*, respectively. The tone-tone-semitone sequence was significant in early music theory as it determined a division of the scale called the *tetrachord*. The scale (and tuning system) generated by this method is named after Pythagoras, the *Pythagorean scale*. The modern naming of intervals is derived from the steps of the seven-note scale: second, third, fourth, fifth, sixth, seventh, and octave. By employing purely tuned fifths (3:2) and fourths (4:3), the Pythagorean scale allows such intervals to agree with the lower tones of the harmonic series, making them completely consonant (Fig. 2.1). However, the difficulty is that it renders other intervals (thirds and sixths), based on the next harmonic ratios, relatively out of tune.

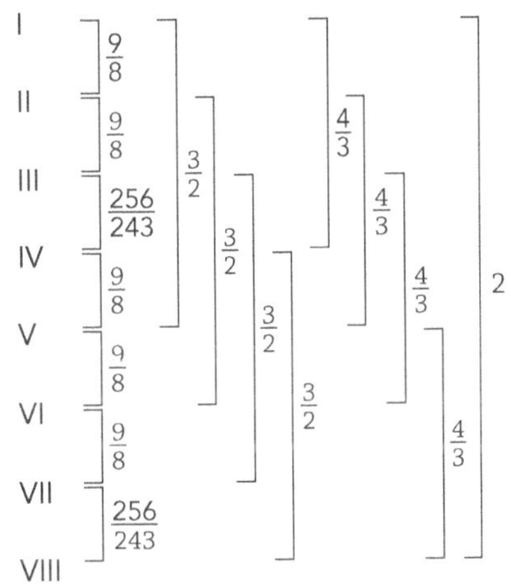

Figure 2.1 Frequency ratios between steps in the Pythagorean scale.

Thus, if we progress to the next interval found in the harmonic series, determined by the ratio 5:4, we will not find this between any two tones in the Pythagorean scale. The closest we will get to is between the first and the third step of the tetrachord (a modern major third), by applying the tone interval twice, $(9{:}8)^2 = 81{:}64$, which is wider than 5:4 by the ratio 81:80. That ratio is called the *syntonic comma* and is a significant quantity in the various methods to create tunings that would attempt to be true to all low intervals in the harmonic series.

2.1.2 Just Intonation

Solving this issue to provide a way to improve the tuning of certain intervals was the motivation behind the just intonation scale, which originated from Ptolemy (c. 100–170 CE) (Barker, 1989, Hunt, 1978). It tuned the tetrachord thirds to the ratio of 5:4, defining a ratio 4:5:6 for the first, third, and fifth tones of the scale, and 3:4:5 for first, fourth, and sixth, and the second, fifth, and seventh. While this tunes some steps of the scale to intervals of the harmonic series up to the sixth harmonic, it also creates two sizes for the tone interval (9:8 and 10:9) as a side effect (Fig. 2.2). The result is that some intervals within the scale deviate significantly from their closest counterparts in the lower harmonic series. In particular, the fifth between the second and sixth step of the scale is very narrow (ratio 1.481 as opposed to 1.5), and the minor third between the second and fourth steps is also short of what it should be (1.185 instead of 1.2).

In general, the study of tuning and scales is a preoccupation that prevailed for many centuries until in fact the modern times, and the attempt to resolve the difficulties outlined above generated many solutions. Also, as scales were expanded beyond seven steps and music practice incorporated notions such as tonal modulation, difficulties emerged in the design of fixed-tuning instruments. Both Pythagorean and just intonation approaches start to break down if frequencies that are further away from the base reference are added to the scale. In the case of the former, it is quite easy to extend the scale-generation algorithm by modifying the endpoint check in step one, so that we progress further using the same cycle of intervals. However, as we do that the system becomes a spiral that never returns back to the original reference frequency. The closest step, the 12th tone generated, is higher in frequency by small but significant ratio

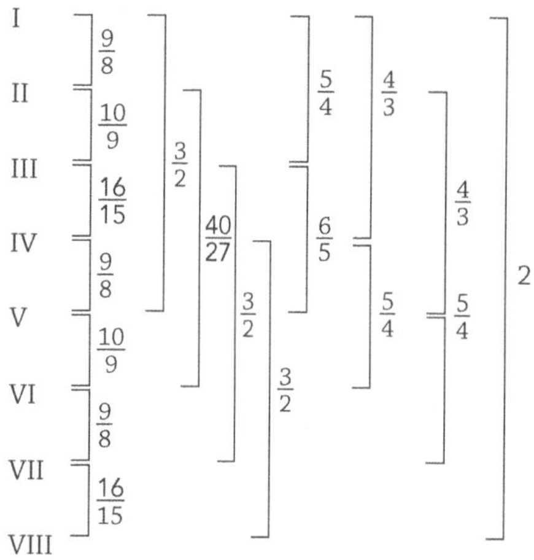

Figure 2.2 Frequency ratios between steps in the just intonation scale.

(531441:524288 $\approx$ 1.036).[5] The just intonation method does not give us a clear way to find tones beyond the seven steps, leading to artificial solutions such subdividing the tones to provide semitones to complete a 12-step scale.

The questions raised by studies of pitch and scales have implications for musical instrument design and, and also, indirectly to considerations of timbre and spectra. In particular, very early on, they affected the decisions taken in the construction of fixed-tuning instruments such as organs, harpsichords, and pianos. Moreover, there is an interdependence between instrumental timbre and scales, which has been the object of study in recent times.[6] The perception of what we class as consonance and dissonance is influenced both by the tuning employed and the spectrum of the instrument used. This can be clearly seen as a main preoccupation of these early studies, where tuning alignment with the lower partials of the harmonic series was a fundamental requirement.

2.1.3 Musical Instruments

These principles of tuning developed by Greek and, later, medieval theorists provided a foundation for the development of musical instruments based on strings and pipes. The organ is an important example of this, whose basic principles of construction greatly evolved during the medieval era. It was very early on established that, like strings, the length of organ pipes was inversely proportional to their pitch. While open flue pipes were first used, the stopped variety was introduced later on, producing pitches in the lower octave (if compared to an open pipe) with a somewhat hollow timbre, an effect that was only completely elucidated in the 19th century. Variation of tone color becomes a sought-after quality in organs, and by the late 15th century, the concept of *registers* dividing the organ into sets of pipes becomes established (Williams, 1980, 64). Combinations of pipes in octaves ($8'$ and $4'$ footage),[7] twelfths ($2\frac{2}{3}'$), and other intervals relating to smaller sizes became a typical technique to produce a variety of timbres; likewise, the addition of pipes with other tone qualities, which, also in different registrations, could add a significant variety to the sonic palette of the instrument. In time, the organ will become a formidable spectral synthesizer.[8]

Other instruments were also subject to significant development that was mostly empirical, although later their construction was also informed by the scientific discoveries of the day. In relation to their spectra, and tonal quality, from the medieval times onward, a particular purity of timbre seemed to pursued, which was related to clearly defined harmonics in the lower range with a more subdued (i.e., less buzzy) higher register. This was influenced by the musical aesthetics of the Western tradition, which favored the tonal relationships emphasized by the spectra of these tones. At times, practical needs, such as for instance a requirement for better projection and loudness, would have lasting influence in the tonal character of the instrument. For example, in the move away from harpsichords toward pianofortes, which was motivated partly by a need for finer dynamic control and sound body, the specific sound of keyboard music changed forever. Besides the different form of excitation (striking versus plucking), the high tension and the materials used in piano strings produced a completely different spectrum, if compared to the light harpsichord string. Similarly, the power requirements involved in projecting the voice to a large theater forged the development of the *bel canto* technique, which modified the natural timbre of singing in very striking ways (such as to render difficult the recognition of vowels in the sung text in some situations).

2.2 Classical Physics

The understanding of key concepts relating to the spectrum took a significant leap at the start of the era of classical physics. From the late 16th century to the end of the 19th century, many philosophers, mathematicians, and scientists made a number of discoveries that are at the foundation of modern acoustics. Some of these were directly related to the concepts introduced in Chapter 1, and it is worth tracing these particular developments, which will lead to a clearer understanding of the basic principles of spectral audio.

2.2.1 Frequency and Pitch

At this stage, a number of scientists started to make the connection between frequency of vibrations and pitch. Benedetti (1530–1590) appears to be the first to put forward ideas in this respect, proposing that ratios of tone pitches are equivalent to ratios of frequencies in the vibrations produced.[9] Such ideas were reinforced by Beeckman (1588–1637), who established that a string will have twice the frequency if its length is halved, and that a low-pitched string can cause a higher one to resonate if they are tuned perfectly to an octave, or to an octave and a fifth (Truesdell, 1960, 25–27). Mersenne (1588–1649) followed these discoveries, establishing that the frequency ν of vibration of a stretched string is inversely proportional to its length l, and proportional to the square root of the ratio of tension T to cross-section area A (thickness), providing the expression[10]

$$\nu \quad \alpha \quad \frac{1}{l}\sqrt{\frac{T}{A}}. \tag{2.1}$$

Mersenne also postulated that a bell would sound three tones simultaneously, its original, plus an octave and a fifth, and that in general other instruments have similar properties (multiple tones produced at the same time). These ideas were significant as one of the first characterizations of spectra arising among scientists and philosophers of

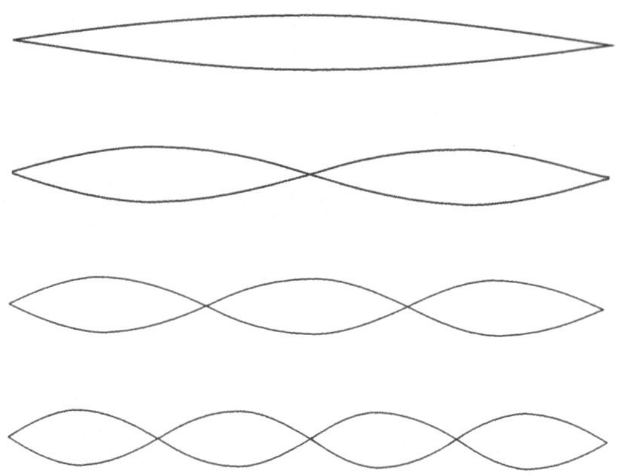

Figure 2.3 The first four vibration modes of a stretched string.

the era. Finally, Galileo (1564–1642) discussed the matter at some length in his *Dialogues*. He was able to determine quite precisely that the frequency vibrations of a body were indeed connected to the perceived pitch and that simple ratios related to consonances, whereas more complex ones were less consonant.[11]

2.2.2 Harmonics

Toward the end of the 17th century and in the following century, many discoveries in the field of acoustics were made by a variety of scientists and mathematicians. The very name of the discipline was first introduced at around the turn of the century.[12] An understanding of the nature of the spectrum of musical tones started to become better formed. Confirming the earlier findings, Wallis (1616–1703) observed that strings vibrate in different simultaneous modes, with some points showing no movement at all, while others have wide displacements.[13] This is illustrated in Fig. 2.3, where the first four modes are shown as divisions of the string in 1, 2, 3, and 4 parts, as described by Wallis in his letter to the Royal Society.

Sauveur (1653–1716) made similar discoveries, calling the points of no movement *nodes* and determining that each mode constituted an integral multiple of the base frequency, for which he introduced the word *fundamental*. He was also responsible for using the word *harmonics* for each one of these partials, and noted that several of these were produced by a single string at the same time. Daniel Bernoulli (1700–1782) also described how strings can vibrate in several harmonic oscillations at the same time, each of which is a component of the total motion. A complete mathematical solution for this was only forthcoming following the publishing of Fourier's (1768–1830) work in 1822 (Fourier, 1822).

2.2.3 Strings and the Wave Equation

Also during this era, following the fundamental definitions in Newtonian mechanics, studies of the acoustical properties of strings and pipes were made by several authors, in particular attempting to find the differential equations that modeled string movement and fluid displacement in pipes. Taylor (1685–1731) provided the first dynamical solution for the vibrating string function[14] and was able to provide a formula for the fundamental frequency, ν that agreed with the earlier observations by Mersenne,

$$\nu = \frac{1}{2l}\sqrt{\frac{T}{\sigma}}, \tag{2.2}$$

where σ is the mass, T the tension and l the length of the string. From this, we can deduce that the quantity $\sqrt{\frac{T}{\sigma}}$ also denotes the speed[15] of sound on the string, as the frequency is related to how fast the wave moves from one end of the string to the other.

Following this, to d'Alembert (1717–1783) it is attributed the first differential equation of the wave propagation on a string (d'Alembert, 1971),

$$\frac{\partial^2 y}{\partial t^2} = c^2 \frac{\partial^2 y}{\partial x^2}, \tag{2.3}$$

where c is the speed of sound, x is the position along the string, and t is time. This equation became known as the *wave equation*, in this case in a single dimension, as it pertains to the movement along the string. A function of two variables, $y(x,t)$, the vertical displacement, is the solution to this equation.

The wave equation models the movement of the string in time and space. The second-order partial derivatives $\frac{\partial^2 y}{\partial t^2}$ and $\frac{\partial^2 y}{\partial x^2}$ can be thought of as describing the temporal and spatial components of this movement, respectively. One interpretation we can give to a derivative is that it gives us the rate of change of a function with respect to a variable or parameter. A second-order derivative is then the rate of change applied to the rate of change of that function. So, for example, $\frac{dy}{dt}$ is the rate of change of vertical displacement with respect to time, or the velocity. The second-order derivative $\frac{d^2y}{dt^2}$ is then the rate of change of velocity, or acceleration. Similarly, we can interpret the derivative $\frac{d^2y}{dx^2}$ as the second-order rate of change of displacement with respect to the position x along the string. As the string movement is happening both in time and in space, derivatives with respect to each one of these parameters are needed. In this case partial derivatives are employed. Conceptually, it says we are applying a derivative to one variable, while holding the other fixed (and vice-versa).

The significance of the wave equation to our understanding of the spectrum can be demonstrated further, and for this we will deviate slightly from the historical narrative. It can be shown that sinusoidal functions are present in the solutions to this equation. We start by isolating the spatial from the temporal part in the wave equation, which will allow us to see how the modes of vibration (Fig. 2.3) can come about.[16] The former involves the x variable and the second, t. By setting $y(x,t) = S(x)T(t)$ and substituting this into Eq. 2.3, we have

$$T(t)\frac{\partial^2 S(x)}{\partial x^2} = c^2 S(x)\frac{\partial^2 T(t)}{\partial t^2} \tag{2.4}$$

At this stage, we can make an assumption, backed by what we know so far, that a solution to the temporal part is a sinusoid[17] with a fundamental frequency ν, which is defined in Eq. 2.2 with respect to the physical properties of the string. The phase of the sinusoid is not particularly crucial, so we can define $T(t) = \cos(2\pi\nu t + \phi)$, and by substituting it in Eq. 2.4, we have an an ordinary differential equation,[18]

$$c^2 \cos(2\pi\nu t + \phi)\frac{d^2 S(x)}{dx^2} = (2\pi\nu)^2 \cos(2\pi\nu t + \phi)S(x), \tag{2.5}$$

which, by setting $c^2 = \frac{T}{\sigma}$, simplifies to

$$\frac{d^2S(x)}{dx^2} = -kS(x), \tag{2.6}$$

with $k = (\frac{\pi}{l})^2$.

This equation tells us that $S(x)$ is a function whose second derivative is proportional to it, with an inverted sign. There are two known functions that qualify for this, $\sin(x)$ and $\cos(x)$.[19] Since, in the string case, both ends are clamped, we have $S(0) = 0$ and $S(l) = 0$, where l is the string length, the most appropriate sinusoid is $\sin(x)$, and we have a solution in the form of

$$S(x) = \sin\left(\frac{\pi x}{l}\right). \tag{2.7}$$

Applying this allows us to generate the first mode discovered by Wallis (as shown in Fig. 2.3). However, all integer multiples of $k/2x$ are equally good solutions to this equation, therefore the other modes are also provided by a similar approach. We can now restore the temporal part of the wave equation to get a full solution to it,

$$y(x,t) = \cos(2\pi\nu t + \phi)\sin\left(\frac{\pi x}{l}\right). \tag{2.8}$$

We can interpret $y(x,t)$ as describing a vibrating sinusoidal wave bound by another sinusoid due to the string being clamped at its two ends. As we will soon see, this actually is only one element of an infinite set of solutions to the wave equation in the string case. Alongside d'Alembert's contributions, other notable discoveries in this area were made by Euler (1707–1783) and Lagrange (1736–1813).

2.2.4 Chladni Figures

Another important contribution to the understanding of vibrating bodies was made by Ernst Chladni (1756–1827), in his studies of plates (Chladni, 1787, 1971). His innovative experimental idea was to sprinkle sand onto a vibrating plate, then excite it with a bow, making a continuous vibration. After a little while, the sand would be deposited along the points of no movement, the nodal lines of the plate. The resulting figures, some examples of which are shown in Fig. 2.4, were later named after him.

While the modes of vibrations in plates had already been mathematically determined by Euler for a number of cases, Chladni's experimental approach allowed a more immediate impression of their geometry. In addition, he was particularly interested in the frequencies of the partials generated in the different cases. He clearly determined that these were generally not in a harmonic relationship, and were generated by the various concomitant vibration modes marked by different nodal lines. While most of the studies until then had concentrated on tones with a harmonic spectrum, Chladni's investigations demonstrated how inharmonic spectra can come about.

2.2.5 Fourier's Theorem

In Bernoulli's work,[20] we already find the suggestion that each harmonic is a sinusoid and that the solution for the complete function describing a tone is a trigonometric series of

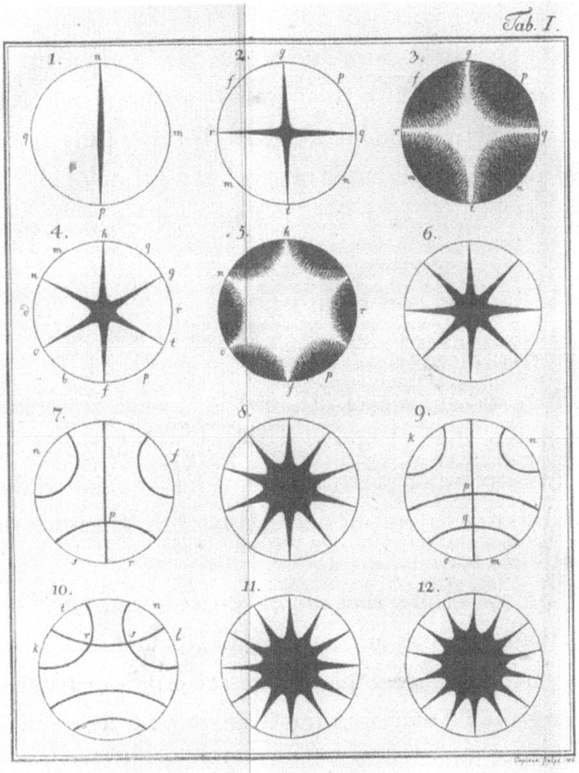

Figure 2.4 Chladni figures showing nodal lines, from Chladni's *Discoveries in the Theory of Sound* (1787).

the form

$$f(x) = \frac{1}{2}a_0 + \sum_{n=1}^{\infty} a_n \cos(nx) + b_n \sin(nx) \qquad -\pi \leq x \leq \pi, \tag{2.9}$$

and that the coefficients a_n and b_n contribute to the tonal characteristics of the sound (i.e., its timbre). We should of course be able to identify this as the basis for the waveform described in Chapter 1, Fig. 1.6, where each partial is scaled by the coefficient $b_n = \frac{1}{n}$ (and $a_n = 0$).

Although Bernoulli made an important step in describing the string movement as a sum of harmonics, he did not have the mathematical tools necessary to determine their values. As we noted in Sec. 2.2.2, this problem was solved later by Fourier in his work on heat propagation. On Chapter III of his book, Fourier notes:

> If we apply these principles to the problem of the motion of vibrating strings, we can solve difficulties which first appeared in the researches of Daniel Bernoulli. The solution given by this geometrician assumes that any function whatever may always be developed in a series of sines and cosines of multiple arcs. (Fourier, 2005, 551)

The principles he refers to here provide the method to find the solution for the value of the coefficients in Eq. 2.9. These are determined by the following integrals[21]

$$a_n = \frac{1}{\pi}\int_{-\pi}^{\pi} f(x)\cos(nx)dx, \tag{2.10}$$

and

$$b_n = \frac{1}{\pi}\int_{-\pi}^{\pi} f(x)\sin(nx)dx. \tag{2.11}$$

Thus once we know the form of $f(x)$, we can apply this method to decompose it as an infinite sum of cosines and sines. The coefficients a_n and b_n describe the spectrum of that function.[22] With this, Fourier accomplished something quite extraordinary[23] that will provide the fundamental tools for the study of the spectrum until today. The Fourier transform, which is based on a generalization of his theorem, will form the basis for many tools used in spectrum analysis and synthesis.

2.2.6 Partials and Hearing

Following the previous work of Savart (1791–1841), who studied the limits of human hearing, and of Seebeck (1805–1849), who explored the relationship between pitch and periodicity, Ohm (1787–1854) determined, based on Fourier's theorem, that the ear is sensitive to pure tones at different frequencies, and that the quality of tone of different sounds is due to the our perception of the presence or absence of these components of a sound waveform (Ohm, 1971). He also determined that pure sinusoidal contain no upper partials (something that can of course be surmised from Fourier's theorem).

His theory was also responsible for defining how the pitch of two or more tones of different frequencies may be perceived. Somewhat controversially, his work indicated that we should attribute the fundamental frequency of periodic tone to its lowest partial, something that was later demonstrated not to be true in all cases. In relation to this, what he proposed was a much simplified version of what we now understand about auditory perception, as it tended to ignore other effects, in particular the contribution of the timing of neural impulses (which is related to the phase of signals). In general, the relationship between pitch and frequency is not a simple one, as noted in Chapter 1. However, Ohm's work is significant for our understanding of the spectrum as it ties together the frequency content of a signal, its constituent partials, and our perception of the whole.

2.3 Helmholtzian Theory

In the second half of the 19th century, we see the first complete treatment of modern musical acoustics appear, in the form of Helmholtz's (1821–1894) treatise *On the Sensations of Tone as a Psychological Basis for the Theory of Music* (Helmholtz, 1954). This was a towering achievement from one of the leading German physicists of the era, not only summarizing the current knowledge on acoustics, but also introducing some new experimental approaches, including a discussion on the anatomy and physiology of the ear, and a theory of the auditory system. Another important aspect of the work is that it appears to be written both for scientists and for musicians, as well as the general audience. The main body of the text contains no equations, and the bare minimum of numerical data, tables, and graphs. Contrasting with another magnum opus of the time, Rayleigh's (1842–1919) *Theory of Sound* (Rayleigh, 1945), this could be approached from different starting points. For the technically minded, the key mathematical points were made in a series of appendices, which provided support for the main assertions of Helmholtz's theory. Its very detailed descriptions, on the other hand, were accessible to a more general audience not interested in proofs for each point made.

Helmholtz used as a point of departure to his investigations the fundamental basis given by Fourier's theorem, which arrived to him via Ohm's studies. He adopts and attempts to prove the points made by Ohm, which describe the ear as a spectral analyzer in the sense of Fourier's decomposition of complex tones into pure sinusoids (called by Helmholtz *pendular vibrations*).[24] From this, and from his experimental observations, a theory of musical timbre is constructed, with significant portions of it linked to music theory and practice. This work was highly influential to our modern understanding of the spectrum. However, it also provided a skewed view of what sounds are worth our attention, from a musical acoustics perspective, which left a number of important aspects of spectral audio unresolved.

2.3.1 Musical Tones and Noise

At the onset of his book, Helmholtz makes a clear distinction between what he defines as his object of study, *musical tones*, in opposition to all other sounds that we can perceive, or *noise*. He notes that these two categories can for a certain extent be mixed up and we may move from one to another transparently. However, at the extremes they are completely distinct. As simple examples of musical tones, he gives the sounds of pitched musical instruments, and for noise, he cites the sound of the wind, water splashing, and similar types of audio signals. While we may criticize the very limited view taken by Helmholtz in establishing these categories, it is important to consider that his aim was to explain aspects of the Western musical practice of his time. Music theory was fundamentally turned to determining the rules that governed composition in terms of these *musical* tones, even though the actual practice included all sorts of sounds that could well fall within the noise category.[25]

From an acoustics standpoint, the definition of a musical tone is linked with uniform-looking vibrations, of which periodic waves are the best example. This is contrasted with disorganized, complex, non-periodic fluctuations that are present in noise waveforms. Helmholtz puts this succinctly as follows:

> The sensation of a musical tone is due to a rapid period motion of the sonorous body; the sensation of a noise to non-periodic motions. (Helmholtz, 1954, 8)

Given that the main objective of his study is musical tones, we are soon directed at an explanation of their nature, based on the principles of Fourier's theorem. Some difficulty is encountered when the subject of sounds with inharmonic partials is discussed, as some of these may be classed as noise or musical tones. Following Chladni, Helmholtz demonstrates how the different modes of vibration come about and describes them in terms of pitches over a prime tone. He notes that in some cases, like that of the tuning fork, these inharmonic partials are very high in frequency and only fleeting, thus not disturbing the proper pitch of the instrument. He also discusses bar instruments, where the sounding elements are mounted in such a way to dampen the more problematic inharmonic partials and emphasize their prime tone. This is not so much the case in bells and drums, where partials of similar intensity can be much closer in frequency. Helmholtz surmises:

> The common character of the instruments hitherto described is, that, when struck they produce inharmonic partials. If these are of nearly the same pitch

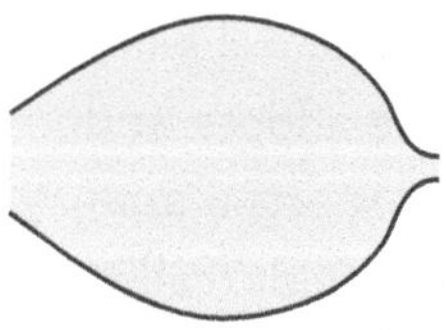

Figure 2.5 The cross-section of a Helmholtz glass resonator.

> as the prime tone, their quality of sound is in the highest degree unmusical, bad, and tinkettly. If the secondary tones are of very different pitch from the prime, and weak in force, the quality of sound is more musical (. . .); and such tones are applicable for marches and other boisterous music, principally intended to mark time. But for really artistic music, such instruments as these have always been rejected, as they ought to be, for the inharmonic secondary tones, although they rapidly die away, always disturb the harmony most unpleasantly, renewed as they are at every fresh blow. (Helmholtz, 1954, 73)

Therefore a host of perfectly interesting sounds are only worth four pages of his treatise, marked as they are as only musical in certain contexts, all of which are not seen as important.

2.3.2 Resonators and Other Analytical Instruments

Helmholtz's study of the spectrum, while based on Fourier's theorem, is grounded on experimental observation. To enable this, he will look for a variety of methods, the most significant of which is the use of sympathetic resonances, that is, how vibrations can be induced through an external sound excitation. He dedicates a complete chapter on these, examining how they can take place in a number of sonorous bodies, such as strings, boxes, and different vessels. These studies lead him to define a type of resonator to which later his name was attached, the *Helmholtz resonator*. It can take various forms, but the essence of it is an enclosed body of air, which is connected to the exterior via a small aperture. In one of his appendices, details and dimensions are given to allow these to be constructed and tuned to a variety of frequencies. The principal characteristic of these resonators is the capacity of emphasizing a single pure tone, while at the same time mostly removing others. In modern terms, it is what we call a narrow *bandpass filter*.

Equipped with a selection of these, quite an effective means of spectral analysis is possible. Helmholtz utilizes this for much of his studies of the upper harmonic tones. The particular version employed in this process is a round glass vessel with one end open to the outside and another containing a very small spout that is applied directly to the entrance of the ear canal, for listening (see Fig. 2.5). With these, he was able to observe experimentally the relative strengths of upper partials in the different studies undertaken. Spectral analyses of this type were quite unique and will also inspire the building of other apparatus for the investigation of sound.

His other use for resonators is applied to the amplification of faint vibrations for sound generation. For some of his experiments, he designed an electromechanical instrument based on tuning forks for sound generation and resonators for amplification. The forks were made to vibrate by inducing a current around them. In front of each one of

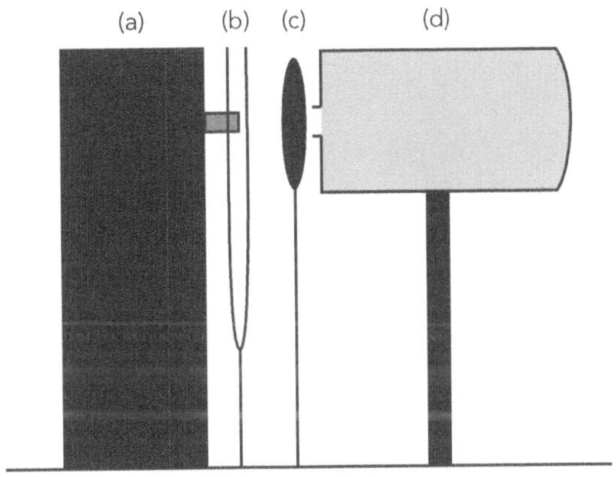

Figure 2.6
A schematic view of one of the tone generators in Helmholtz's additive synthesis instrument: (a) electric induction circuit; (b) tuning fork; (c) tongue used to close or open resonator aperture; and (d) resonator.

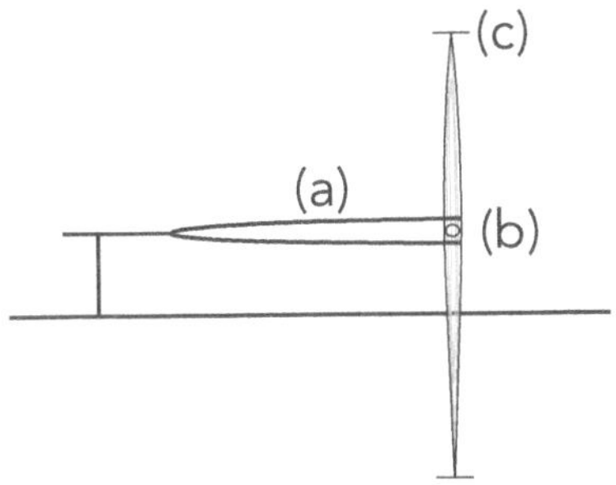

Figure 2.7
A schematic view of one of Helmholtz's Lissajous *sound microscope*: (a) tuning fork; (b) lens; (c) vibrating string. The electromagnetic element that makes the tuning fork vibrate is omitted.

these stood a resonator, tuned to the same frequency, whose opening was directly facing the main vibration body. Each opening had a small tongue, which, like a window, could open or close the aperture. When this was open, a loud sinusoidal tone would be produced, as the tuning fork vibrations were amplified; if the hole was closed, then the sound of the forks was very weak and barely audible. Figure 2.6 shows schematically one such tone generator. An instrument containing eight of these was created, with each sound generating element perfectly tuned to a harmonic above a given fundamental. This was possibly the first purposefully designed additive synthesizer ever created.

Another interesting experimental instrument designed and employed by Helmholtz was a device that used light to draw up waveforms from vibrating strings. This employed a principle developed by Lissajous,[26] where the combination of movement in two perpendicular directions resulted in particular patterns if the two motions were synchronized. To implement the instrument, he employed a moving lens, to create the impression of movement in a stationary object, such as point source of light. This was set in motion by an electrically induced tuning fork, and using a small reflective point attached to a moving string vibrating at a right angle to the tuning fork motion, he was able to observe the resulting pattern. This instrument was referred to as a *sound microscope*, and is shown schematically in Fig. 2.7.

When the two movements were related by a simple integer ratio, a fixed figure would emerge. From these, it was possible to deduce the shape of the waveform generated by the string, which otherwise would have been too fast to visualize. Figure 2.8 shows, on the left side, three waveforms studied by Helmholtz and, on the right, their corresponding Lissajous patterns. This is an electro-mechanical form of the modern electronic oscilloscope, which is used to display the waveform of an input signal.

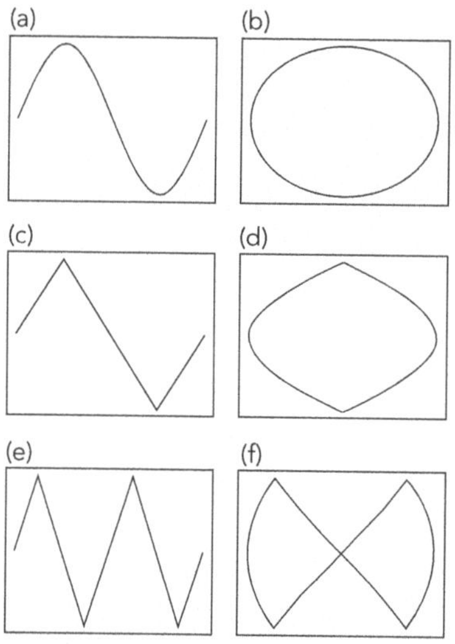

Figure 2.8 Lissajous figures (on the right side, b, d, f) generated by their corresponding different waveforms (on the left side, a, c, d).

2.3.3 Theory of Spectral Hearing

A fundamental contribution of Helmholtz, for better or for worse, is to characterize the ear as a spectral analyzer, sensitive to the pure sinusoids that compose a musical tone.[27] An additional point was that the phase of a waveform, or of its constituent partials, played no part in the perceptual of the tonal quality of musical tones. That was a standpoint introduced by Ohm, which was defended, with an attempted proof, by Helmholtz. The main issue with this was that the cases that were studied and proven never included more complex situations. The very narrow definition of a musical tone, for one, did not allow a wider study of the problem. So, for the cases employed, steady, unchanging, tones with harmonic partials, changes in phase have little effect on the perceived tone quality, but that is not a general proof of the proposition.

Also trained in medicine, Helmholtz had the knowledge and capacity to study the anatomy and physiology of the hearing system, to the limits of the established science of his era. He was able to establish the function of the cochlea as a transducer of mechanical to neural signals in the hearing system, with an idea that the role of the basilar membrane and the organ of Corti were the basic structures involved in this:

> On reviewing the whole arrangement [of the anatomical structures in the inner ear], there can be no doubt that Corti's organ is an apparatus adapted for receiving the vibrations of the membrana basilaris, and for vibrating of itself, but our present knowledge is not sufficient to determine with accuracy the manner in which these vibrations take place. (Helmholtz, 1954, 141)

From his understanding of resonances, Helmholtz deduces that sympathetic vibrations are induced in the inner ear and that through these sine tones are detected. However, he is unaware of any other effect that may arise due to the timing of signals (i.e., phases). We should probably conclude that the limitations of his theory of hearing were both due to the status of the medical sciences of his time and the methodological problems he had in his own physical investigations. These were induced partly due to the choices made at the very beginning, between what were worthy objects of study. The more difficult and interesting cases would have always been beyond the scope of his research.

2.3.4 Musical Timbre

From an overall perspective, Helmholtz's concept of musical timbre is fairly straightforward: musical tones are those generated by periodic waveforms, and their tonal quality, or timbre, is due solely to the relative strengths of individual harmonics, their presence or absence. Other tones, from non-periodic waveforms (or even periodic but with exceedingly long wavelengths), are not musical, therefore lie outside his theory of timbre. These conclusions, in themselves, are already a significant achievement, as for the first time tones are explored in terms of their spectra, rather than for other more immediate physical qualities, such as frequency and intensity. A downside is that his theory was so influential that, as incomplete as it is, continued to be the basis for the understanding of the spectrum over one hundred years later, at times making a clearer understanding of the subject more difficult.

We should not discount, however, that much of the detail of Helmholtz studies is, although limited, correct. For example, in his studies of instrumental tones, he looks closely at how vibrations are set into motion, and concludes how this can influence the resulting tone. In the case of struck, plucked, or bowed strings, he is able to tell that the position of the strike (etc.) determines the presence or absence of some of the vibration modes, as it creates an antinode at a given point, dampening all modes that have nodes at the position. He also shows how the shape of the waveform in the wave is influenced by this (and therefore the overall quality of tone). In studying pipes, he likewise notes the mode in which air is set into motion, and how this, together with the physical construction and materials employed, determines the spectra of tones. Phenomena such as overblowing, which excites higher modes in certain conditions, and dampening modes by lightly touching on string nodes, are also discussed and explained. Other innovative investigations of his included a close look at the singing voice and the way different vowels are linked to different spectra, which were also experimentally recreated using his additive synthesis instrument.

One of the key points missed in Helmholtz's theory of musical timbres is temporal evolution. Most of his attention is directed toward what we can call the *steady-state* portion of a tone, even though in some of the cases he studies there is no such thing.[28] Moreover, it does not appear that he was unaware of the fact that the spectra of sounds change over time. When discussing inharmonic tones, he clearly notes that some of the problematic (from a musical point of view) upper partials die off quite quickly in the more usable non-periodic instrument tones (Helmholtz, 1954, 70–73). At another point, when discussing his Lissajous-figure observations of bowed-string waveforms he actually shows how their shape can change with bowing position (Helmholtz, 1954, 83), but there is no further discussion of the dynamic aspects of this process.

This indicates that his unwillingness to explore these dynamic aspects of the spectrum are related more to aesthetic choices rather than to scientific method. Given that in proper musical tones what matters is the steady-state, because out of that we are able to make harmonic and melodic relationships, and no art music can arise of any other type of audio structuring, there is no point on wasting time with the uncertain and complex motions happening at the onset of sounds. Likewise, we should not bother with how upper tones change in amplitude as a sound evolves, for example, at the decay of plucked or struck strings, because as long as the upper partials are harmonic, this change in timbre is of no musical consequence.

2.3.5 Rayleigh's Theory of Sound

Following Helmholtz's seminal work, Rayleigh's *The Theory of Sound* (Rayleigh, 1945), first published in 1877, is an equally important treatise. In this book, the focus is on the physical properties of sound, from an objective point of view; aesthetical and musical considerations are left outside the discussion. Likewise, Rayleigh does not attempt to discuss aspects of sound perception or auditory physiology, such as covered by Helmholtz.[29] Instead, he keeps to two basic topics, forming the two volumes of the work, sound generation and propagation. He acknowledges the principle introduced by Ohm and developed by Helmholtz that pure tones are detected separately by the ear, and that the if a partial is absent it cannot be heard in the resulting sound.[30] He equally determines that sinusoidal tones are the fundamental components of a sound wave, as these waves cannot be further decomposed,[31] and from this Fourier's theorem follows to describe a complex vibration.

Rayleigh's approach to sound production, the most relevant portion of his work to the topic of spectral audio, is systematic. He first examines simple harmonic motions in a general way, then explores systems with one degree of freedom, which would for instance describe tuning forks, one-dimensional waves in strings, etc. At the outset, he introduces an equation to describe these motions, accounting for frictional losses, which takes the form[32]

$$f(t) = ae^{-kt}\cos(\omega t + \phi). \tag{2.12}$$

The interesting aspect of this equation is that it includes not only something related to the steady state of the system (the sinusoid that solves the differential equation as seen in Sec. 2.2.3), but also a *decaying envelope*, represented by the exponential function that multiplies it. The question of envelopes is scantly treated in Helmholtz, but here in a purely physical systems investigation, it starts to be considered. Dynamic aspects of sound production are looked at, even if they are fleeting. A few pages further, while discussing forced vibrations (e.g., the ones in bodies subject to continued forces), Rayleigh mentions beating phenomena relating to the onset of these motions, which decay fairly quickly. This is something of course connected to transient behavior, which is also very important in the study of spectra. From the examination of simple behavior, he moves to general vibrating systems, and then to specific cases: strings (transversally and longitudinally), bars (longitudinally, torsionally, and laterally), membranes, and plates (straight and curved). He also takes a look at "electrical vibrations," which are of course a different type of phenomena. Finally, in volume two, aerial vibrations in pipes and resonators are considered. His treatise provides the basic foundations for acoustics as a branch of physics in the 20th century.

2.4 Twentieth Century

Spectral audio effectively becomes a relevant consideration in the 20th century, motivated by a revolution in the aesthetics of music and the growing interest of composers in aspects of tone quality, and in sounds in general, not specifically the ones dictated to be musical in character. Of course, it has always been the case that the timbre of instruments, and variations of it, have always been important in the performance of music. However, this traditionally has been only a secondary aspect in terms of the notation composers developed to transmit their music. At the dawn of the new century, it started to become a consideration, which was more systematically built into composition. This interest will not only spark new ways of making music, but lead to further investigation into the spectral qualities of sound, especially as electronic music and sound synthesis became a practical proposition.

Alongside these musical developments, which will be further discussed in Chapter 10, the study of acoustics, and the nascent area of psychoacoustics, also underwent significant development, especially as experimental fields of research. New instrumentation, both mechanic and electronic, played a fundamental part in allowing new discoveries. The Acoustical Society of America, founded in 1929, and, later, the Audio Engineering Society (1948), provided significant support for the dissemination of new ideas in the area of acoustics and audio. Both societies sponsored publications in which key research relating to the spectrum, and our perception of spectral phenomena, were first reported. The work of Fletcher and Munson on the frequency dependency of amplitude levels (published in 1933),[33] is an example of the kind of fundamental ideas that had long-lasting impact on our understanding of auditory perception.

2.4.1 Electronic Instruments and Signal Processing

The advent of electricity as part of everyday life allowed the construction of instruments that would employ it for sound generation. One of the first of these, the *Telharmonium* (1896), conformed very well to Helmholtz's model of timbre, composed of sustained pure tones with little variation over time, whose onset and decay was more or less immediate and unchanging. Each individual harmonic tone was generated by electrical induction, and given that a practical means of amplification was yet to be invented, a significant amount of power was needed at the sound generation stage so that it could be transmitted to loudspeakers (the telephone network was employed as a means of distribution). This made the whole instrument equivalent to a small power plant.

The Hammond organ (1935) employed a similar principle, but with amplification already present, it could be made into a smaller, self-contained unit of equivalent size to existing organ consoles. It was designed to generate harmonic partials of a note, via specially designed tone wheels which rotated in synchrony as they were connected to a common shaft powered by an electric motors. Each wheel would induce a current in a magnetic pickup, producing a (very nearly) sinusoidal tone (a practical realization of Fig. 1.3 in Chapter 1). A typical organ would have 96 such wheels, covering 12 notes and their harmonics, with up to 9 harmonics per note. Drawbars could be used to adjust their intensity, and the sound envelope was just of the start-stop variety, although percussive onsets could be added to it in some models.

Other electronic instruments of the early 20th century were more adventurous in their design. Of these, the *Theremin* (1920) was particularly interesting, as it used hand gestures to control the amplitude and pitch of a sinusoidal tone. This allowed significant flexibility in creating inflections of pitch and also different types of amplitude envelope and modulation. Similarly, the *ondes Martenot* (1928) employed a ribbon controller for pitch and a button for amplitude, allowing a similar approach, with a little more limitation with regard to amplitude control, but easier to learn as a pitch reference could be provided for players. In time, the ondes Martenot became a part of some orchestras in France and a significant repertory was built for it.

At this stage, electronic instruments for acoustic measurement also become an important tool for the study of spectra. The technology microphones, also enabled by the introduction of amplification, could be used to capture sound for further processing or recording. The invention of electronic filters allowed the isolation of portions of the spectrum, as we saw in the case of acoustical resonators. Lowpass and highpass filters, for example, could be used to eliminate high and low partials of a sound. Bandpass filters could isolate particular partials of interest. A good example of their application were the *voder* and *vocoder* developed at Bell Labs,[34] which employed a bank of parallel bandpass filters to reproduce the spectra of different vowels, using envelopes to articulate them. This was a first attempt at an electronic voice synthesizer.

Work in telecommunications, which had important commercial applications, helped the development of various innovations in signal processing (filters, for instance, are examples of this). In fact, it is probably fair to say that, in relation to our understanding of the spectrum and of ways to manipulate it, the new field of signal processing (as an area of electrical engineering) was the most important contributor of fundamental ideas. Among its many innovations, two are of particular significance to the subject matter of this book: the definition of the bandwidth of a communications channel, first by Nyquist, then by Shannon; and the principle of localizing the Fourier transform in time, introduced by Gabor.

In 1924, Nyquist introduced the notion that the amount of information that could be transmitted over a communications channel was related to the speed of transmission.[35] Or, from another perspective, that the number of signals that could be transmitted were related to the bandwidth of the channel. This was a significant discovery, as it determines the width of the spectrum of digital audio signals, which we will study in detail in Chapter 3. From Nyquist's, and later Shannon's,[36] principles we have that the maximum frequency that can be encoded in a discrete-time signal is half the frequency of transmission, or in other words, the communication channel has to be twice the width of the highest frequency employed.[37]

One particular aspect of Fourier's theorem that often caused discomfort for researchers in the early days of telecommunications theory is that it excluded the concept of time. Even though the Fourier series defines a time-domain function, the theorem actually applies to sinusoids that are infinite in time, in other words, which have existed forever and will continue indefinitely. The action of starting and stopping a sinusoid will actually have an effect on its Fourier spectrum. To demonstrate this, we plot the spectra

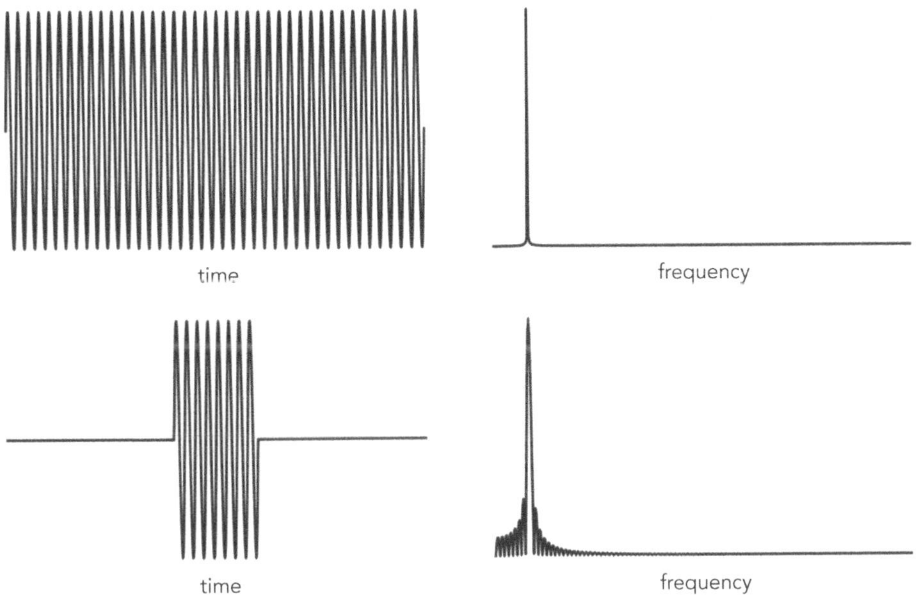

Figure 2.9
Comparing the spectra of two signals: a forever-lasting sinusoid (top) and another that starts and stops at given points in time (bottom).

of two waveforms in Fig. 2.9. The first one of these, at the top, shows a sinusoid that matches the theoretical limits of the Fourier series, lasting forever. Its analysis shows a single component at a a given frequency, just as we expected. Below it, we see a sinusoid that is abruptly started and stopped; its spectrum does not show a single line at a specific frequency, but it is smeared around a central point. The reason for this is that we are not analyzing a pure sinusoid anymore, but one that has a shape imposed on it. The effect of adding time to our considerations has changed the signal we are looking at. A sinusoid that lasts forever is represented by a partial found at a single *infinitesimally small* frequency band,[38] which can actually be represented as such as we are removing its time localisation: exactly defined in frequency, not at all defined in time.

Fourier's theorem therefore referred to a static, never changing, representation of the frequency spectrum. This of course is a source of the difficulties in its direct use by Helmholtz to define the timbre of instruments. Changes in the spectrum, for instance in the amplitude of components, or their frequency simply cannot be captured by it. In particular, this poses difficulties for us to develop the idea of an *instantaneous frequency*, if we are trying to examine for instance, frequency modulation.[39] While the Fourier transform was considered mathematically sound, and was routinely employed in the study of signals, these difficulties put some limitation to its usefulness.

In order to overcome these, Gabor (1940) provided a way forward by proposing to localize the Fourier transform in time, introducing an analysis that is dependent both on time and frequency. This later became known as the *Gabor transform*. In his paper,[40] he looked to results in quantum physics (in particular applying Heisenberg's uncertainty principle, as well as Dirac's delta function, already introduced in Chapter 1) to provide a theoretical basis on which to re-introduce time as a variable, alongside frequency, in the study of the spectrum. This is highly significant, as such an approach would form the basis for a variety of practical implementations of spectral analysis that are commonly used today.[41]

Gabor appears also to be one of the first authors to employ the word *spectrum* in relation to sound (although his paper was about signals in general, he does mention sound and music as examples, and in part II he discusses hearing); after presenting the formulae for the forward and inverse Fourier transforms of signals, he notes:

> [w]e will refer to $S(f)$ as the "spectrum" of $s(t)$. (Gabor, 1940, 431)

It is hard to assert exactly when the term becomes more current, but it seems to have migrated first from its original use (in the 18th century) meaning the spectrum of visible light as split by a prism, to the whole electromagnetic range of frequencies (of which light is a portion) around the end of the 19th century. From there, electrical engineers started to use it more generically for any signals, and the term then became commonly used for audio processing.

2.4.2 Electronic Music

Electronic music as a concerted effort, in both aesthetic and practical terms, only started in the postwar years with Pierre Schaeffer's work in Paris at the Radio France studio, then with Herbert Eimert in Cologne at the West-German radio, followed by other studios elsewhere in Europe. To Schaeffer, in particular, we owe much of the development of experimental interest in composing the spectrum of sound that characterized electronic music practice. He approached the problem of creating a new music based on electronic manipulations of recorded sound, called *Musique Concrète*, from an experimental standpoint, using ear as a primary guide. Based on his and his colleagues' studio and musical activities, from the late 1940s to the 1960s, he developed a solid theoretical work that covered aspects of acoustics and psychoacoustics interspersed with considerations on musical theory and aesthetics.

In particular, two areas of his theoretical work were significant to the history of spectral audio: his studies of the dynamic evolution of sounds in time (and envelopes), and his development of a typology of musical objects. In Book 3 of his large treatise Schaeffer (2017) discusses his experimental results in splicing the recording of various instrumental sounds, exploring the importance of transients to the recognition of some sounds with inharmonic components (such as the vibraphone) and the possibility of creating artificial onsets for other sounds with mostly harmonic spectra (such as the piano). Likewise, he discusses how modifications of attack shape may modify the characteristics of a sound. In exploring instrumental timbre, he also applies modifications to these, such as variable-speed playback transposition and filtering, to demonstrate how modified spectra can be recognized in some situations and not in others. From these studies, he builds a number of important theoretical notions, which will allow him to develop a classification of musical objects based on their morphology, in Book 5 of his treatise. Much of this work is informed by a spectral approach, using the electroacoustic studio as the laboratory. His classification of musical objects will be very influential in music theory, leading to the development of *spectromorphology*[42] as an area of research, to which we will return in Chapter 10.

From an opposing perspective, both methodologically and musically, we find the work of the studio in Cologne, led by Eimert and Stockhausen. They adopted an approach that may be seen as the culmination of Helmholtz's timbral model.[43] Using electronically-generated sinusoidal tones as their basis, they would construct the musical material for their work, *tone mixtures*, in an additive synthesis fashion. These were originally conceived as unchanging in time (apart from crude enveloping applied to them), and were composed based on various serial permutation operations. Stockhausen's *Studie I* (1953) is probably the work that epitomizes this approach. Moving away from this musical dead end, however, composers in the Cologne studio started to look to other, more dynamic ways, to synthesize and manipulate sounds with more interesting spectra. Stockhausen again led the way, with very intricate synthesis procedures in pieces such as *Kontakte* (1960). From the beginning of the 1960s onward, many composers in Europe and elsewhere will develop various spectrally informed methods to compose both electronic and instrumental music, as we shall see in detail in Chapter 10.

2.4.3 Computer Music

As noted at the start, the story of spectral audio culminates with the emergence of sound and music computing. This credited to the work of Max Mathews at Bell Labs in 1957,[44] with his MUSIC I and, in quick succession, MUSIC II programs. Generating simple waveforms with limited means of controls, these programs were a proof of concept that sound could be computed by a general-purpose computer such as the IBM704 for which they were written. A major step, however, only came with Mathews's MUSIC III in the early 1960s (Mathews, 1961). In this program, two highly influential ideas were introduced: the concept of a compiler for sound synthesis programs (an *acoustic compiler*, as Mathews described it), and a fundamental component of all music programming languages to come, the *table lookup oscillator*.

From his experience of writing two dedicated synthesis programs, Mathews realized that such software, which employed a hard-wired means of creating audio waveforms, were not general enough for musical and research applications. In other to facilitate a more flexible use, his idea that providing fundamental building blocks that could be connected together to make different types of sound was the best way forward. He then provided an incipient language (using what was available to him, assembler macros)[45] that could be used to build sound synthesis instruments through the patching of *unit generators*, his basic computing modules. Once these instruments were defined, they could be compiled into efficient sound synthesis programs, which could be fed control parameters as they were executed to create anything from simple acoustic stimuli (for use in psychological experiments)[46] to whole compositions. Such a groundbreaking idea was the basis for many music programming systems that followed.

The table-lookup oscillator, his second fundamental contribution, provided the means of electronically producing the spectra of any arbitrary harmonic tone that could be described by a Fourier series. In addition, it was general enough to support the generation of other time functions, which could be used as means of parametric control. Thus it could be used to create envelopes as well as audio signals. Provided as a unit generator in his MUSIC III system, it required the definition of a waveform (or another function) in a memory location, called the function table. This was then read (or looked up) by the oscillator routine. To make a pitched tone, the lookup procedure would wrap around the end of the table periodically, creating a steady tone.

Oscillators had parameters for amplitude and frequency, as well as initial phase. They could receive control inputs from other unit generators (e.g., other oscillators), patched up to amplitude or frequency. A single cycle of a waveform of an arbitrary shape could be created by specifying its harmonics and stored in a table. Any oscillator could then access this to produce a signal. This produced a fixed waveform spectrum, but with the use of envelopes and other oscillators in parallel, time-varying spectra could be realized. As we noted earlier, oscillators could themselves also be used to generate envelopes, by specifying the desired functions of time as tables, making them read these very slowly in one single pass. The compiler allowed a number of these (depending on the computing resources) to be connected together in various configurations. Unit generators for summing and multiplying signals were also provided, along with a random number generator (for the generation of noise).

James Tenney one of the first composers to work with Mathew's program, provided an interesting insight into the use of the computer for electronic music composition (Tenney 1963).[47] He describes the processes involved in good detail, including some of the more technical aspects, using a language that was accessible to musicians. He discusses the use of graphic flowcharts to help in the design of instruments and how these can be translated into code, with some interesting examples. These already include some new unit generators that would be central to manipulating spectra: bandpass filters and band-limited noise generators. Among other things, Tenney explains how modulation (the periodic or aperiodic variation of a parameter) can be used to generate dynamic spectra. His early work with sound synthesis was seminal in showing that the computer provided a practical means of realizing a wide range of sounds for electronic music composition.

MUSIC III was eventually developed into MUSIC IV by Mathews and Joan Miller (1964). This new system incorporated the first full music programming language for designing sound synthesis.[48] It was a much more expanded program, including various means of manipulating the input data that controlled the synthesis process, and in that way introducing further support for an algorithmic approach to composition. The sound synthesis capabilities were expanded: in addition to oscillators, the new bandpass filters and noise generators were made available, along with oscillators whose function table could vary over time, a formant oscillator, dedicated envelope generators, and new routines to compute function tables of different forms. With these, a huge variety of dynamic spectra could be created, limited only by the composer skills and imagination.

Computer music also provided the environment for the advance of sound synthesis and processing techniques. An early example of this is provided by John Chowning's research into frequency modulation as a means of generating dynamic spectra of various forms, based on theory developed for telecommunications in the 1920s. The motivation was to produce interesting and malleable sounds by an efficient means, avoiding the costly additive synthesis methods, with an intentional application in electronic music composition. Many other techniques that we commonly use today will be likewise developed, arising out of musical necessity.

From Mathews's and Miller's MUSIC IV, a number of other systems were developed,[49] including MUSIC V, the final system to emerge from Bell Labs.[50] Barry Vercoe implemented his own version, MUSIC 360 (for the IBM360 computer, 1968), and then expanded it in MUSIC 11 (Vercoe, 1983) (for the PDP11, 1977), and Csound (Lazzarini et al., 2016) (written in the C language, 1986). The latter is now very comprehensive sound and music computing system, which will be employed in many of the examples used in this book. Even though Csound already provided a great variety of possibilities for spectral sound synthesis in its early versions, it was only with the addition of the phase vocoder in 1990 that it became one of the first widely available systems with a practical means of spectral analysis-synthesis. The phase vocoder itself was a technique developed in the 1960s (Flanagan and Golden, 1966), for which an efficient implementation became possible due to the development of another important tool: the fast Fourier transform (FFT) algorithm.

The computation of the Fourier transform had always posed difficulties in situations where the number of primitive operations (additions, multiplications) was large. With the advent of computers, the most common of the transforms was its discrete-time discrete-frequency form, called the discrete Fourier transform (DFT).[51] The DFT requires N^2 operations, where N is the size of the input data, for each transform to be calculated, which becomes a significant burden as N gets large. In 1965, Cooley and Tukey (1965) proposed a method[52] that, for certain sizes of N, could cut down the number of operations to $N \log_2 N$, making it very efficient for large sizes of N. This advance opened up a significant range of practical applications for the DFT in many fields of research. For sound and music computing, it provided the basis for the various methods of spectral analysis, which we will see in Part II.

One of these is the aforementioned phase vocoder, which has been used widely for many applications. Its principle is to employ time-localized Fourier transforms (as in Gabor's case, but using different forms of these), then from these, extract time-varying functions of frequency for amplitude and frequency of partials. Its link to the original vocoder is that it can be described as a bank of parallel bandpass filters, but this is only one interpretation of the phase vocoder operation. The details of this will of course be discussed at length in later chapters. The presence of this and other methods of analysis-synthesis, together with the wide availability of sound and music computing systems such as Csound, opened up the scope of spectral audio for music and sound design. This was concomitant with the microcomputer revolution and the emergence of the personal computer as practical music instrument.

Research into digital signal processing methods continued to evolve and new ideas, which went beyond the traditional Fourier-based approach, started to appear. For a long time, and perhaps at times not in a very informed way, Fourier methods had been subject to criticism.[53] Much of it was directed to the original theorem in its time-eternal form, which is beside the point, since most of its applications include both frequency and time dimensions (bound, of course, by the uncertainty principle). Other criticisms focused on frustration with technology not being able to deliver the promise of a "method to manipulate all possible and imaginable sounds," which is also again coming from a misinformed perspective. However, despite this, it is important to assert that the methods that followed from Fourier's theorem have practical limitations, and other possibilities to manipulate spectra are needed. Approaches such as stochastic techniques, pioneered by

Xenakis (1992), alongside granular methods, and other more recent ideas can be added to the very large existing toolset for spectral audio and music design.

Notes

1 While a number principles and ideas about spectrum have also appeared in other world cultures, these are not directly influential on the subject matter of this book, therefore we will keep the discussion to the concepts and terminology developed in the West, from the ancient Greeks onward.
2 A moment of personal epiphany occurred to me as I started to make music with computers. A sudden realization, that the numbers I was computing became these different sounds as they were played back, delivered a strong feeling of working directly with musical matter from the ground up, in ways that I could have never managed before.
3 While we do not have direct sources relating to this, mostly due to the secrecy of the Pythagoreans (Hunt, 1978), there are enough secondary accounts that allow us to make this attribution. In particular, from Ptolemy (Barker, 2004) and Boethius (Boethius, 1971), we have a discussion of the theory put forward by Pythagoras and his followers.
4 The first full description of this scale was found in a fragment attributed to Philolaus of Croton (c. 470–385 BCE), a philosopher of the Pythagorean tradition, contemporary of Socrates (Hunt, 1978, 15–16).
5 This can be easily verified by applying the ratio 3:2 12 times ($(\frac{3}{2})^{12}$), comparing it to applying the octave (2) 7 times (2^7), and verifying that they are close but not the same. The ratio between these tones is called the *Pythagorean comma*.
6 This has been a central tenet of the work of Sethares (2005).
7 The figure of eight feet is used because it is historically associated with the length of a pipe used for the lowest note (C) of the organ keyboard. This is an approximation of the exact length of open pipes used in modern tuning.
8 Howard and Angus discuss the organ as a synthesizer at some length in Howard and Angus (2009, 253–257).
9 Lindsay (1966) notes that the topic was covered on a treatise on musical intervals published in 1588 in Turin.
10 Truesdell (1960) cites the source for this as p. 616 of Mersenne's *Verité des Sciences* (1625).
11 Galileo used a visualizazion of pendular movement to describe the ratios that are consonant and the ones that are discordant, reasoning that simple ratios gave more regular patterns that are pleasing to the eye and more complex ratios of frequencies gave rise to less coherent patterns (Lindsay, 1966).
12 This was proposed by Sauveur (1701).
13 This was described in a letter to the Royal Society from 1676 (Wallis, 1677).
14 However, he felt short of describing it fully as a partial differential equation (Lindsay, 1966).

15 Both words *speed* and *velocity* are used here to denote how fast something moves, the first as an absolute value (or magnitude), and the second has the notion of the direction (e.g., positive or negative, forwards or backwards) built into it.

16 See also Helmholtz (1954, 375). A similar approach is also taken in Steiglitz (1996)

17 Why a sinusoid? We can intuitively assume that the movement is simple, therefore a sinusoid might be appropriate, but we can back this up with some physics. Consider a spring-mass system, when set in motion, it oscillates back and forth since the original pull moves it one direction with a force $F = ma$ (m is the mass of the body and a the acceleration), and the stiffness of the spring resists this, pulling it back with a force $F = -kx$, where x is a measure of the displacement and k is a constant. Since $a = \frac{d^2x}{dt^2}$, the second-order derivative of x with respect to time, we have the equation

$$\frac{d^2x}{dt^2} = -\frac{k}{m}x$$

describing the movement. A function $x(t)$ (the displacement measured as time passes) that satisfies this equation is a sinusoid. We come back to the original question, why a sinusoid? Because sinusoids have the property of being proportional to their second derivative,

$$\frac{d^2\cos(ax + \phi)}{dx^2} = -a^2\cos(x + \phi).$$

Therefore the displacement function $x(t)$ can be described by a sinusoid. This is called simple *harmonic* motion, and it is involved in the vibrations of tuning forks, in the movement of a pendulum, etc. A movable point x along the string will also behave in similar fashion.

18 That is, involving a derivative with respect only one variable; see the previous note for an explanation of how the second derivative of $T(t)$ is solved so that we can accomplish this.

19 See the earlier note on harmonic motion, which explains this.

20 In Truesdell (1960, 258) we read that one of the reasons why Daniel Bernoulli did not get further with these ideas was the fact that partial differential equations were not part of his repertory. Therefore, neither d'Alembert's wave equation nor Fourier's theorem could have been arrived at, as both depend on these.

21 In Hawking (2005, 498), we find an interesting summary of the process that led Fourier to the solution. This involved, for the a_n coefficients, multiplication by *cos*() followed by integration over $[-\pi, \pi]$, as well as the assumption that the function $f(x)$ was integrable, and that integral of the infinite sums was the same as the infinite sum of the integrals.

22 The problem that Fourier was investigating has a strong connection to the discussion of the wave equation in Sec. 2.2.3. He was looking at heat propagation on an infinite rectangular solid, in which two opposite sides were kept a temperature 0, and to the side perpendicular to it a constant source of heat kept it at temperature 1 (the other side, opposite to this, was open). He studied this as a problem where two spatial dimensions, x (the width) and y (the length), applied with the following differential equation:

$$\frac{\partial^2 v}{\partial x^2} + \frac{\partial^2 v}{\partial y^2} = 0.$$

So from the temperature conditions given above, $v = 0$ for $x = 0$ or $x = r$ (similar to a clamped string). The function $v = \phi(x,y) = F(y)f(x)$ is shown by Fourier to be satisfied by $ae^{-my}\sin(mx)$, where $m = i\frac{\pi}{r}$ ($i \in \mathbb{Z}$) and r is the width of the heated end of the solid (the domain of x). This means that the general solution is

$$v = \sum_{i=1}^{\infty} a_i e^{-\pi\frac{iy}{r}} \sin\left(\frac{i\pi}{r}x\right).$$

Fourier then assumes $y = 0$, eliminating the y-related scaling factors,

$$f(x) = \sum_{i=1}^{\infty} a_i \sin\left(\frac{i\pi}{r}x\right),$$

and using Eq. 2.11 can find the values of a_i and the solution to v. Note that the scaling by $\frac{2}{r}$ comes from the fact that we are integrating over the interval $[0,r]$ instead of $[-\pi,\pi]$,

$$v = \frac{2}{r}\sum_{i=1}^{\infty} e^{-\pi\frac{yi}{r}} \sin\left(\frac{i\pi}{r}x\right)\left[\int_0^r f(x)\sin\left(\frac{i\pi}{r}\right)\right]$$

The way this was developed is analogous to the approach we used to solve the wave equation for strings, although there we have refrained from giving the full treatment demonstrated by Fourier. Other similar cases can also be worked through the use of this method.

23 In relation to this, Lord Kelvin stated: "Fourier's Theorem is not one of the most beautiful results of modern analysis, but it is said to furnish an indispensable instrument in the treatment of nearly every recondite question in modern physics . . . . Fourier is a mathematical poem" (Hawking, 2005, 499).

24 Interestingly, given that his theory provides the foundation for all the work here, Fourier is mentioned by name only in a couple of sections of the main text of the book. The first of these, at p. 35, appears to doubt whether there is any reality in the theorem, before nodding in confirmation of it: "We have rather to inquire, do these partial constituents of a musical tone, such as the mathematical theory distinguishes and the ear perceives, really exist in the mass of air external to the ear? Is this means of analysing forms of vibration fiction, permissible for facilitating calculation, but not necessarily having any corresponding actual meaning in the things themselves? What makes us hit upon pendular vibrations, and none other, as the simplest element of all motions possible? We can conceive a whole to be split into parts in very different and arbitrary ways (...). Just as little does the mathematical possibility, proved by Fourier, of compounding all periodic vibrations out of simple vibrations, justify us in concluding that this is the only permissible form of analysis, if we cannot establish that this analysis has also an essential meaning in nature. That is indeed the case, that this analysis has a meaning in nature, independently of theory (. . .)." Helmholtz was right of course in doubting and then confirming through his experimental investigations, although in reality this confirmation turns out to be also limited, as his theory of auditory perception was imperfect.

In the other passage, however, Helmholtz is more assertive: "in one sense this theorem [Fourier's] may be considered as the prime source of the theory of harmony" (Helmholtz, 1954, 229). In any case, the mathematical form of Fourier's theorem itself is never quoted anywhere, except indirectly through, for instance, the solutions of the wave equation, and in graphs showing the addition of tones.

25 We should also note that this traditional separation between "noise" and "music" still exists even within certain practices of electronic music, which would appear to us incongruous given the wide range of spectral possibilities afforded. Audiences, and musicians themselves, seem to maintain in their heads a separation between pitched and unpitched material (sometimes even between simple melodic/harmonic elements and more tonally challenging or atonal material, the latter being equated to noise). These are very pervasive impressions, which are difficult to remove completely. So we should not look down at the limited notions we observe in Helmholtz, since they are still around us.

26 Lissajous (1857) originally employed mirrors attached to tuning forks, whereas in Helmholtz's apparatus a lens was used.

27 Since it was of no interest, the question of how noise is perceived was not investigated; presumably it would be covered by similar principles, but in a more complicated way.

28 The steady-state is associated with a sustained tone, thus in plucked and struck strings, which produce a sound with a decaying amplitude, there is no such state. Even in sustained sounds, in many cases it is hard to argue for the existence of a steady portion, as their spectra may changing dynamically at all times, especially if we consider effects such as amplitude and frequency modulation (tremolo and vibrato) that are usually employed by performers.

29 Helmholtz reviewed the book positively for *Nature* and suggested the author to include a study of physiological acoustics in future editions, which was not followed up (Lindsay, 1945, xxvii).

30 This of course was defended very strongly by Helmholtz, even if experimentally he knew about combination tones, such as difference tones, appearing with certain combinations of sounds. He attempted to resolve the difficulties of combinatorial tones via a theory that these vibrations are induced in the ear, therefore actually appearing as partials, even though a Fourier analysis of the waveform would not include these. Rayleigh seems to accept this: "[i]t is found by experiment that, whenever the simple vibration is absent then the tone cannot be heard" (Rayleigh, 1945, 18).

31 An idea also originally introduced by Ohm, and asserted in Rayleigh (Rayleigh, 1945, 25).

32 This is shown in a slightly more expanded fashion as $u = Ae^{-\frac{kt}{2}} \cos(\sqrt{n^2 - k^2/4}t - \alpha)$ in Rayleigh (1945, 45).

33 See Chapter 1, Sec. 1.3.3.

34 Developed by Dudley (1940).

35 Nyquist was looking at the question of signals in telegraphy, and the number of symbols that could be transmitted at a given unit of time (Nyquist, 1924). This work was latter expanded in Nyquist (1928)

36 Shannon (1949) elaborated Nyquist's in a form that is more closely related to discrete-time signals.

37 The so-called Nyquist-Shannon sampling theorem is the fundamental basis for all sound computation employed in this book, so we will take our time to explore it in detail.

38 See our definition of spectrum at the end of Chapter 1. We will continue to explore this idea later on in this book.

39 Carson discussed these difficulties in the development of his theory of frequency modulation (Carson, 1922).

40 This article, in three parts, covers a number of areas and includes an application of his time-frequency principle to hearing and data compression (Gabor, 1940).

41 There are many connections between the theoretical approaches developed for quantum physics and the spectral theories we will explore in this book. At times, results in both signal processing and physics were shared. This is the case, for instance, of the Wigner-Ville distribution (Cohen, 1989), which is an alternative approach to dealing with time and frequency. The methods and techniques in this book will mostly use the theory that followed Gabor's approach to the problem.

42 First proposed by Smalley (1986).

43 Perhaps perversely so, as the serial music proposed by Eimert was in opposition to the ideals of tonal harmony that are at the basis of Helmholtz's theory.

44 Although there had been previous efforts to make computers play sounds with various direct means (e.g., connecting speakers to registers etc.), none of these was systematic in employing the principles of digital signal processing, which provided the practical basis for manipulating audio waveforms in the computer. This was done for the first time by Mathews at Bell Labs, which was was at the time the only place with the right kind of hardware to convert a digital waveform into an analogue signal Roads and Mathews (1980).

A general description of his achievements was published in Mathews (1963), which was a seminal paper introducing sound and music computing to a wide audience. This had the consequence to capture the attention of the composers John Chowning and Jean-Claude Risset, who then decided to spend time at Bell Labs with Mathews Roads and Mathews (1980). As a result, some significant developments in computer music ensued.

45 Macros are a form of computing short hand. Code that is repeatedly used can be predefined by a given token text, and then all instances of this can be replaced automatically with the long form of the code.

46 This was the justification Mathews gave to dedicate his time to create a program to compute sound; whether this was an actual subterfuge to be allowed to work with music we can only speculate.

47 This 1963 paper in the *Journal of Music Theory* was cited by Moog in his first description of a modular *synthesizer* (Moog, 1967). Here we can see how the design of MUSIC III not only preceded analogue modular systems, but also influenced their inception.

48 This assertion is based on the fact that MUSIC III was only a collection of assembler macros, and this did not constitute a language as such.

49 For a more complete perspective, see Lazzarini (2013).

50 Partly written in FORTRAN (Mathews et al., 1969), this system was shared and used widely.

51 Computers have finite memory; therefore, functions are programmed in their discrete, rather than continuous, form, requiring this variation of the original Fourier transform expression.

52 Other authors had arrived at a similar approach, although in incomplete and less general ways, most notably Danielsson and Lanczos (1942).

53 We find an example of this in Xenakis's writings, some of which have been included in Xenakis (1992). In a very bombastic form, he asserts that those employing Fourier-based methods are completely misguided, and that he has a better approach, based on his stochastic techniques. Despite this, he still praises Mathews's MUSIC V program for allowing composers to manipulate the spectrum in a flexible way (via a Fourier approach, naturally).

3

Fundamental Aspects of Audio and Music Signals

In this chapter, we would like to develop many of the concepts introduced earlier, with a view to provide a foundation for the next part of this book. We will proceed as before, putting our ideas through in increasingly more formal terms, but not overwhelmingly so. We begin by examining the nature of audio signals, what makes them different from other types of signals. This allows us to focus on the specific aspects that are relevant to spectral manipulation and ignore some of the more general signal processing theory that does not apply to the case in hand. This discussion, as always, is informed by how audio signals are handled in sound and music computing, but it will be directed at general characteristics of what we call *analogue* signals at first. Once the key concepts regarding these are introduced, we will turn our attention to the form of audio most commonly manipulated in computer applications: digital signals. In order to get there, we will also have to examine the fundamental distinction between discrete and continuous representations, which are at the basis of digital and analogue audio, respectively.

3.1 The Nature of Audio Signals

The fundamental form of an audio signal has to be the one that can be captured by our senses. We have already noted this at the outset, in Chapter 1, when we spoke about vibrations traveling in a medium, the air, that surrounds us. These can be characterized as a pressure wave, which is the form of the audio signal coming from an arbitrary source that is then picked up by our hearing mechanism, which allows us to perceive it. As part of this, the signal undergoes a *transduction* process, that is, it changes from one form to another. First, as it enters our ears the air molecule vibrations are carried through the middle ear via the movement of three small bones, or ossicles. Then, this movement excites the liquid that fills the cochlea, and the audio signal, which comes in as a mechanical wave is then transduced into neural impulses.

Transduction may also take place at the source of audio signals, when for instance a loudspeaker cone is put into motion through the application of an electric current to a magnet that is attached to it. We can also capture mechanical vibrations via transducers such as microphones or piezo-electric pickups. The electrical form of an audio signal allows it to be stored and transmitted over long distances. The process of transduction does not alter the nature of the signal; therefore, any of the modes of transmission or

Spectral Music Design. Victor Lazzarini, Oxford University Press. © Oxford University Press 2021.
DOI: 10.1093/oso/9780197524015.003.0003

storage cited earlier (as electrical currents or as mechanical vibrations) are equally valid to represent an audio signal at its fundamental form.

As we have also seen in Chapter 1, signals can be represented by functions, and we can exchange one for the other, depending on the context. Functions representing audio signals of the fundamental form we have described some primary characteristics, such as being *continuous* and *unidimensional*. Some secondary aspects also include the fact that, while audio signals might not have a limited spectrum, our hearing mechanism can only deal with part of it; and, likewise, the range of amplitudes that are within our perceptual scope is also limited. We can start our exploration of audio signals by looking in some detail at these two primary aspects of a sound waveform, to gain further insight into its nature.

3.1.1 Real Signals

We should first try to understand the unidimensional nature of an audio signal, and the type of functions that represent it. By now, we should be used to looking at a sound waveform as a function of time, although as we have seen before, we could describe it as a function of space as well. Why time? It is simpler for us to abstract the audio waveform as something that is happening in one point in space (say, right in the middle of our eardrum), instead of all the points it occupies in space at the same time. This latter representation is useful, for instance, if we want to model vibrations in a solid, but less so if we want to characterize the spectrum of an audio signal.

Therefore, a function of time will be sufficient to represent a sound in any of the equivalent forms described earlier: the waveform emitted by a string vibrating, the movement of a point at a microphone diaphragm, the electric signal passing through an amplifier, etc. We have already characterized these as a signal in the time domain. Moreover, we are measuring a single quantity, which can be the displacement of a point along the string (see Sec. 2.2.3), the pressure at the eardrum, the voltage at a point in a circuit, and so on. Although what we are measuring in each of these cases might be different, it is still a single quantity, so we have a unidimensional functional.[1]

There is another concept that is associated with a time-domain audio waveform in any of the modes of measurement or transmission mentioned earlier. We may call these *real signals*.[2] That is, we are representing them by real-valued functions of a real variable (time). That is, whatever quantity we are measuring can be described mathematically by a *real number*.

While the continuum of real numbers had been used in algebra for a long time, it was only in the 19th century that we had a complete description of what these numbers are. Integers were more readily understood as counting numbers extended to include negative values, and rational numbers as any that could be defined from an integer ratio a/b, but there were numbers that were known to be beyond these definitions, such as the square root of two and π. Dedekind (2005, 921) provided a fundamental definition for any real number a as a *cut* in the continuum, dividing two sets of numbers, A_1 and A_2, such that $A_1 < a$ and $A_2 > a$. Note that such definition equally works for integers, rationals, and irrational numbers, and that arithmetic operations are also preserved, so that we can continue to apply these to the continuum. Why is being real-valued a significant characteristic? There are several reasons, but two are particularly important here: a real number can represent a continuous quantity (whereas, for instance, an integer cannot), and an audio waveform can be described by a real function in a single dimension, as opposed to being multidimensional. While complex-valued functions and signals, which are two-dimensional, will play an important part in spectral audio processing, it is important to emphasize that an audio signal is described by a real-valued function. In particular, we will also see that this characteristic allows us to treat it in specific ways when we start exploring the frequency domain.

3.1.2 Instantaneous Frequency and Phase

In the introduction to periodic waveforms given in Chapter 1, we defined frequency and phase in very simple terms, which nonetheless were useful for the points we were making then. At this point, it is worth exploring these two parameters of audio signals a little further. While frequency can still be generally defined as the number of wave cycles per time period, and this definition is easily conferred when dealing with fixed-frequency waveforms, it is worth considering the cases where it varies over time, as for example in a chirp or in a gliding tone. More generally, we would like to know what is the frequency of the signal at a given point in time, which may be somewhat contradictory. However, let's see where we can get to following this line of thought.

What we would like to know, therefore, is the *instantaneous* frequency of a signal. In order to determine this, we will have to revisit our definition of phase. Simply stated, the phase was put in terms of the starting point of a waveform, measured in relation to some reference point. Again, there is nothing wrong with this definition, but we should expand it a little further. We can now say that the phase of a signal is a position in the cycle of the waveform, and we can measure it in terms of fractions of cycles or in *radian* angles with reference to a full circle $[0 - 2\pi)$, and for a sinusoid, such as a cosine, it is given by an inverse function such as the $\arccos()$, also notated as $\cos^{-1}()$. So, more generally, phase is a function of time $\phi(t)$, which in the case of a periodic signal is applied to produce the various values of an arbitrary waveform. For instance, we can typically apply it to a fixed waveshape, such as $\cos()$, to get an output,

$$s(t) = \cos(\phi(t)). \tag{3.1}$$

This definition of a sinusoid allows us to make the frequency vary in time, whereas if we had made the argument a simple product of frequency and time (e.g., $\cos(2\pi ft)$), we would require it to have a fixed value.

We may wonder why that is the case, and the answer is in the relationship between frequency and phase. The fact is that we cannot make frequency vary without first defining what an instant value of it is, otherwise we will probably come up with the wrong conclusions. This is related to the fact that what the cosine function takes as an argument is a phase angle, not a frequency. To get the correct phase angle for a given time t, we can use an expression that employs a value for the frequency of the signal, such as $2\pi ft$, but that f cannot be simply turned into a function of time without any further consideration. Actually, we need to define what a frequency function $\omega(t)$ is first, from which we can get a phase function $\phi(t)$.

So what is $\omega(t)$? We may define it by looking at how the phase is changing over time, since we know that the frequency measures the number of cycles over a period of time. A signal whose phase is changing 2π radians over 1 second has therefore a frequency of 1 Hz. If we want to know its instantaneous frequency then we need to make the phase change infinitesimally, and from that we can obtain its rate of change, which is its derivative. Simply put, we can say that[3]

$$\omega(t) = \frac{d\phi(t)}{dt}. \tag{3.2}$$

That is a very useful definition, which means that now we can reinstate frequency in the sinusoidal signal expression with confidence. From Eq. 3.2, we should infer that

$$\phi(t) = \int \omega(t)dt, \tag{3.3}$$

and so substituting back in Eq. 3.1, we have

$$s(t) = \cos\left(\int \omega(t)dt\right). \tag{3.4}$$

These ideas will have some important applications at various points in the study of spectral audio. For example, a common method of sound synthesis is *frequency modulation* (FM), whose name implies that we are somehow varying the frequency over time. This is often implemented via the modulation of phase, so understanding the relationships between the frequency and phase parameters of a signal is fundamental for exploring FM. Another application is found when we are able to measure the phase of signals at different points in time, from which we can immediately estimate their frequency through differentiation. Conversely, if we have a time-varying frequency signal, we can apply integration to update the phases for driving a sinusoid.

An example of a time-varying frequency sinusoid, Fig. 3.1, plots the frequency, phase, and resulting waveform of a cosine function whose frequency glides from 1 to 100 Hz. The plots show a linearly rising function of time describing the frequency, which is integrated to provide the signal phase. This signal is then finally used to look up values of a cosine, shown in the lower graph. Note that the phase function $\phi(t)$ is multivalued as it wraps around the full circle (2π), so we plot it within its range of *principal* values, $[-\pi, \pi)$.

3.2 Manipulating Analogue Audio Signals

The type of continuous signals we have been exploring so far is best described by the term *analogue*, although this tends to be specifically employed when an electrical form of the signal is found somewhere between source and subject. The basic meaning of this term is related to the fact that an electric signal is an analogue of a mechanical waveform, and the two are linked together by the process of transduction, which we have already spoken about. It is worth, however, exploring a little further the details of this analogical relationship.

Electronic audio signals are characterized by a varying voltage, which is commonly called alternating current (AC) electricity. As the name indicates, the electricity flow, or

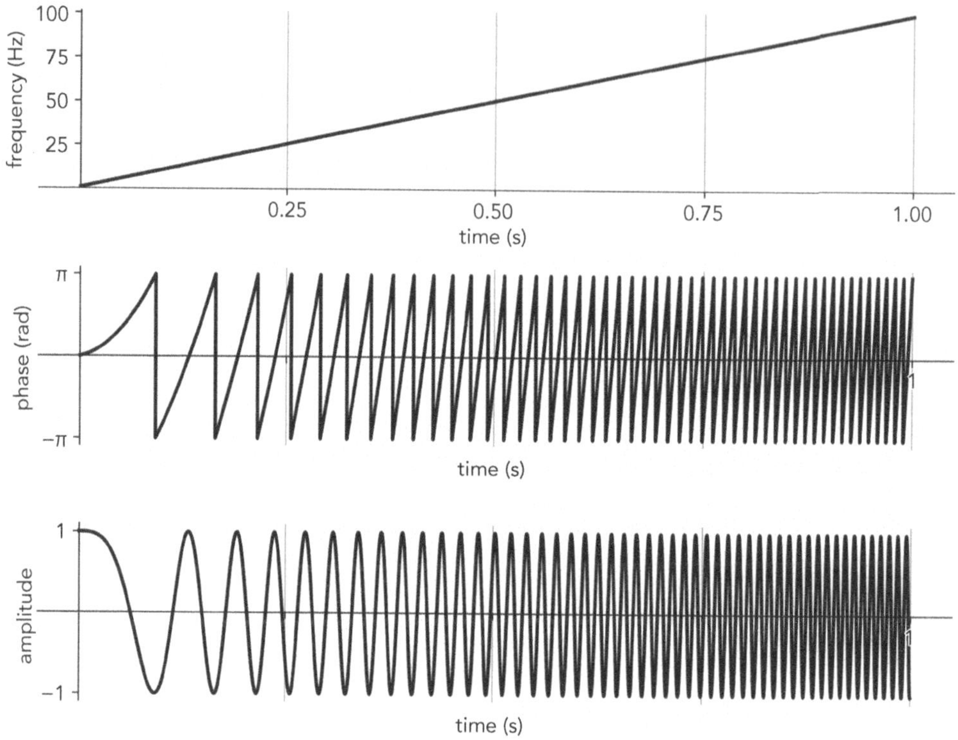

Figure 3.1 Plots of $f(t) = 99t + 1$, $\phi(t) = \int \omega(t)dt$, and $\cos(\phi(t))$ (top, middle, and bottom graphs, respectively), with $\omega(t) = 2\pi f(t)$, demonstrating a sinusoid whose frequency glides from 1 to 100 Hz.

current, periodically changes from a positive (e.g., forward) to a negative (e.g., backward, reverse) movement. This causes the voltage[4] to alternate accordingly. In contrast, the other common form of electricity is called direct current (DC), where there is no reversal of flow. Audio signals are generally treated as bipolar, with the AC voltage varying above and below a mean value of 0.

To produce a sound wave, an electric audio signal is applied to a loudspeaker, where the rising voltage will push a membrane forward and a falling voltage will pull it backward. This is the basic principle of transduction that turns an electric signal into a mechanical wave for listening. The inverse transducer to the loudspeaker is a microphone, which takes the air vibrations and induces an electrical current. Microphones and loudspeakers can be designed in different ways depending on the particular application requirements, but their transduction roles as outlined here do not change.

In general, microphones tend to produce very small amounts of electrical power,[5] which would be unsuitable for further processing, or for driving a loudspeaker. For the signal to be used in any practical application it needs to be amplified. For this, an active circuit is employed to boost the amount of power contained in it. An amplifier applies a scalar constant called the *gain* as a multiplier to allow a change in the amplitude of the electrical waveform. This defined simply by

$$y(t) = gx(t), \tag{3.5}$$

where $y(t)$ and $x(t)$ are the input and output of the circuit and g is the gain. When $0 < g \leq 1$, the signal will be attenuated, and when $g > 1$, it will be boosted (or amplified). Note that if the amplification is strictly characterized by Eq. 3.5, then we have a linear system.

Gain scaling is one of the basic operations on signals. The other fundamental operation is mixing, where two or more inputs are summed together to produce a composite output. This is often combined with amplification so that the amounts or levels of each input can be adjusted accordingly, as in

$$y(t) = g_1 x_1(t) + g_2 x_2(t) + \ldots + g_n x_n(t), \tag{3.6}$$

where several inputs $x_n(t)$, scaled by the gain factors g_n, are mixed together linearly. In practical situations, it is hard to guarantee that the system will be fully linear in all conditions, although it is possible to approximate a linear behavior very closely through good electronic design.

3.2.1 Non-Linear Distortion

At this point, it is important to characterize what is meant by a non-linear system, as this is part of the fundamental terminology we will employ in the book. In the most general terms, we can say that a non-linear response can be characterized by

$$y(t) = f(x(t)), \tag{3.7}$$

where $f()$ is an arbitrary non-linear function, commonly called a *transfer function*. In the trivial examples of Eq. 3.5 and 3.6, this is not the case, as the transfer functions employed are linear (i.e., $f(x) = gx$). Non-linear distortion may occur anywhere in the signal chain (e.g., transduction, amplification, recording, broadcasting, etc.), and in many systems considerable effort is put into minimizing it, although we can also find useful sound processing applications for it.

Consider for instance a situation where an amplifier can only work with inputs within a given range, for example, not scaling any voltages beyond the $(-1, +1)$ interval. The *transfer function* could be characterized as

$$f(x) = \begin{cases} x, & -1 < x < 1 \\ 1, & x \geq 1 \\ -1, & x \leq -1 \end{cases}, \tag{3.8}$$

from which we can see that anything that if the amplification exceeds the stated interval, the output will be *clipped* at $|1|$. The transfer function graph (Fig. 3.2) shows that, indeed, this is not a linear function, as it has a couple of kinks that break its curve. Its effect on a sinusoidal input that exceeds the maximum peak amplitude is shown in Fig. 3.3, where we can see that the smooth function has been somewhat changed around its peaks. This is a typical example of *hard-clip* distortion, which will alter the signal by adding to it several high-order harmonics, as the waveform is not sinusoidal anymore. Note that the distortion only occurs in this case if the signal exceeds a certain amplitude, which is often the case for system implementations, where the designers try to guarantee a linear response within a certain range of inputs. Therefore, we generally take distortion into consideration when determining the maximum signal level a system can handle.

Different types of clipping exist, all of which will add components to a signal. Some of these are employed in effects such as fuzz and overdrive, which impart typical timbral modifications to an input signal. Non-linear distortion provides an important class

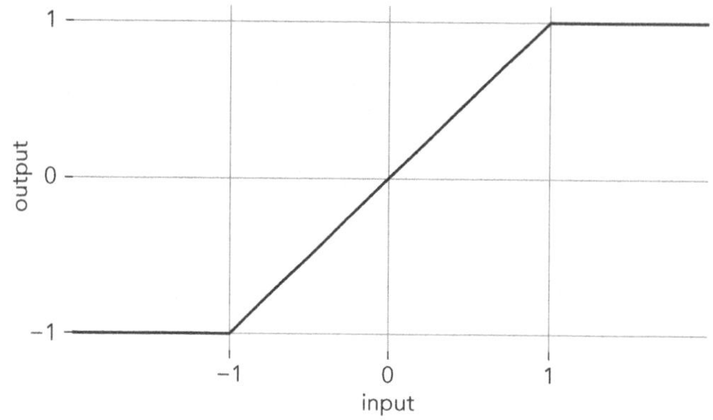

Figure 3.2 A hard-clipping transfer function (Eq. 3.8). The horizontal axis refers to the input signal, which is mapped to the output (vertical axis) according the function curve.

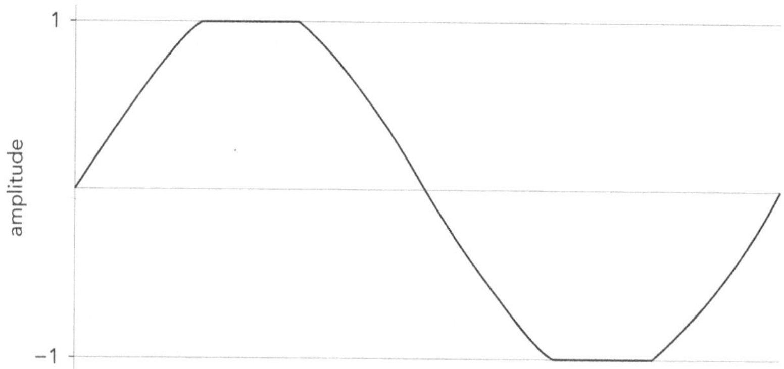

Figure 3.3 The output of Eq. 3.8 (with $g = 1$) when fed with a sinusoid with peak amplitude of 1.2. Note that the peaks of the waveform are clipped, distorting the sinusoidal shape.

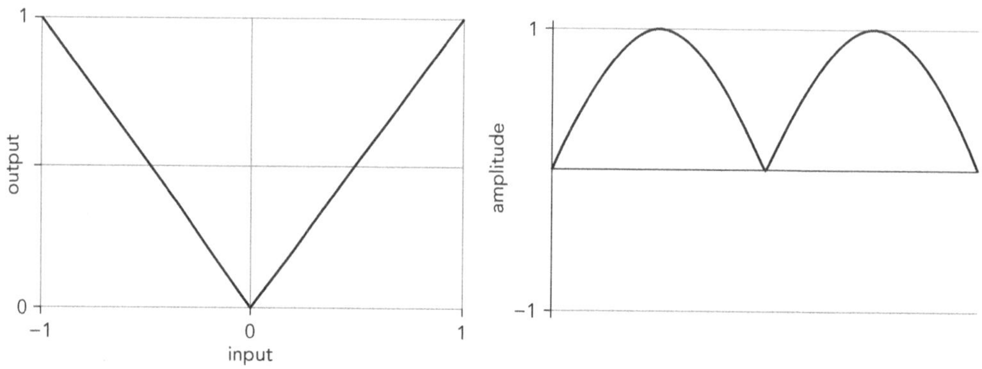

Figure 3.4 A rectifier transfer function (*right*) and its effect on a bipolar sinusoid (*left*).

of operations for spectral design, which we will explore later in the book. In synthesis applications, the transfer function is also called a *waveshaper*.

Another example of distortion is given by

$$y(t) = |x(t)|, \tag{3.9}$$

where a bipolar input signal is *rectified*. Applying this transfer function to any waveform has the effect of making it non-negative (Fig. 3.4). This type of operation has a number of practical applications in signal processing.

All the examples discussed so far can be classed under the name of *amplitude distortion*. This is because we are looking at how the transfer function is applied directly to the waveform of the signal. It is, in essence, a time-domain process. From a complementary angle, we can look at how signals can be distorted in terms of frequency. From that perspective, the *frequency response* of a system is what we need to examine. We can divide this in terms of magnitude (the response to the amplitude of signals at different frequencies) and phase (the response to the timing of signals in the spectrum). Since it involves a number of concepts of spectral audio, which are yet to be introduced, we will delay considering this in more detail until later.

3.2.2 Noise and Signal Level

Alongside distortion, another important element affecting the quality of audio signals is *noise*. While the term may have a number of different meanings, in this context we are considering any unwanted signal that is *added* to the system. Unlike distortion, noise does not depend on the input signal level, it is always present at a given level. It is generally broadband in nature, that is, it contains components spread throughout a wide range of frequencies, but we could also consider narrowband signals such as ground hum[6] as an example of an unwanted component that is independent of the input. The main consideration regarding noise is its level, particularly relative to the amount of signal. The ratio between these levels is a quantity called *signal-to-noise ratio*.

Since our perception of amplitude levels is logarithmic,[7] a linear scale has been constructed to allow us to measure these in a more practical way. This is called the *decibel* (dB) scale, which can be defined as

$$\mathrm{dB}(a,b) = 20\log_{10}\frac{a}{b} \tag{3.10}$$

where a and b are the measured amplitudes of two signals, and $\mathrm{dB}(a,b)$ is their level difference in the dB scale. If we examine this expression, we will note that

- A positive dB change implies that a is greater than b, and a negative change is likewise the inverse.
- Since $20\log_{10} 2 \approx 6$, each 6 dB change implies a doubling of amplitude levels. Equally, each -6 dB change is related to a halving of levels.

Note that Eq. 3.10 determines a relative difference of levels. For an absolute scale, we need to use a reference value. There are two common forms of these that are employed for distinct applications: the sound pressure level (SPL) and the 0 dB full-scale (0dBFS) scales. The first one uses a reference value that is close to the threshold of hearing and so all levels above 0 dB up to 120 dB are generally within the audible range (higher levels approaching and then exceeding the threshold of pain). This is normally used for environmental acoustics measurements and systems specifications. The other scale uses a maximum system amplitude (set for instance by taking distortion into consideration) and calculates all levels as negative values below this, with 0 dB set as the maximum amplitude.

In some situations, we may want to find the signal *power* difference rather than amplitude. In this case, we need to use a different expression for the decibel calculation,

$$\mathrm{dB}_{power}(a,b) = 10\log_{10}\frac{a}{b}. \tag{3.11}$$

The reason behind this is that power W and amplitude A have a quadratic relationship,

$$A = \sqrt{W}. \tag{3.12}$$

From this we can also infer that a 3 dB change is equal to a doubling or halving of signal power. An equivalent sound power level (SWL) scale can be constructed using a low power threshold as a reference value.

In order to get the correct signal levels, we need to determine how the measurement should be made. We have already seen in Chapter 1 that we can measure the maximum absolute value of a signal, the peak amplitude of a signal, as a means of determining its level. While this is useful to make sure we do not exceed the stated system range, it is not always the best to determine a longer-term signal amplitude amount. For this, as we also noted in Chapter 1, we need to find a means of averaging the signal, which is given by the RMS method. Since we have acquired the necessary tools, we can define this more precisely now as

$$\mathrm{RMS}(s(t)) = \sqrt{\frac{\int_{t_0}^{t_1} s(t)^2 dt}{t_1 - t_0}} \tag{3.13}$$

where t_0 and t_1 determine the time interval of the RMS measurement of a signal $s(t)$. An integral is used, as we have already seen, because the signal is continuous and so the sum needs to be over all the infinitesimal time steps. We can interpret this as the square root of the square mean of the signal, which gives the method its name: root mean square.

Alternatively, if we take several closely spaced periodic measurements of the signal amplitude, we could redefine the RMS expression using a (discrete) sum,

$$\sqrt{\frac{\sum_1^N a_n^2}{N}}, \tag{3.14}$$

where a_n are the values of the signal at regular time intervals.

Using these tools, it is possible to determine the maximum signal-to-noise ratio of a system, which is important as it defines its dynamic range. For this we need to know what is the maximum amplitude before any distortion sets in, which defines the highest level that can be achieved, as well as the measurement of the inherent noise in the system, called the *noise floor*. The difference in dB between these levels gives us the dynamic range available, since any sound whose amplitude level is below the noise floor will always be masked by it, and any sound that exceeds the maximum level will not be reproduced faithfully.

3.2.3 Modulation

An important class of manipulations of audio signals is provided by the various means of *modulation*. This term can be best defined as the continuous modification of a signal parameter by another signal. This can be periodic or aperiodic, depending on the

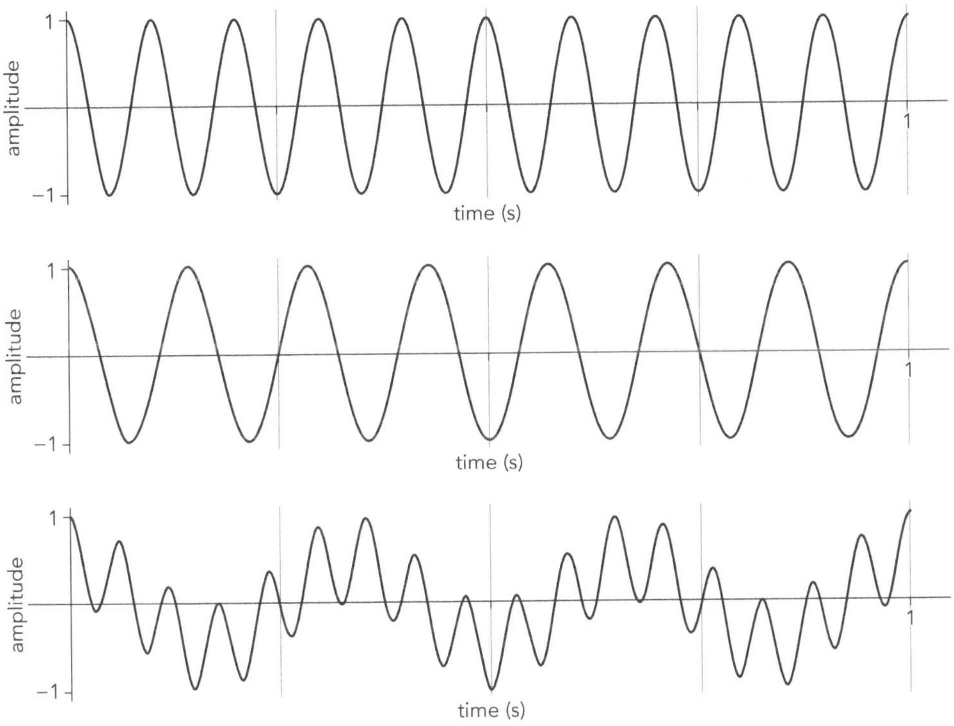

Figure 3.5 Plots of two sinusoids with frequencies of 10 and 7 Hz (top and middle, respectively) and the resulting ring modulation of the two signals (bottom).

modulation source. An example of modulation involves replacing the gain constant of an amplifier, discussed in Sec. 3.2.1, by a signal,

$$y(t) = m(t)x(t), \tag{3.15}$$

which is called *ring modulation*. This is a special type of the more general set of *amplitude modulation* operations. The resulting effect of ring modulation depends on the type of signals employed. A simple case employing sinusoids can be worked out using a trigonometric identity,

$$\cos(\omega_1)\cos(\omega_2) = \frac{\cos(\omega_1 - \omega_2) + \cos(\omega_1 + \omega_2)}{2}, \tag{3.16}$$

which shows that two sinusoids in ring modulation will produce an output that consists of the mix of two sinusoids whose frequencies are the sum and the difference of the input frequencies (Fig. 3.5). This is also known in some spectral processing contexts as *heterodyning*. If one of the signals contains several components, then the output will contain partials with the sums and differences of the frequencies of the modulator and each component. In general, amplitude modulation by a sub-audio frequency wave results in a fluctuation of signal level called *tremolo*.

While sinusoidal ring modulation produces sums and differences of two frequencies, it is also possible to modulate the amplitude of a signal in such a way that only one or the other is produced. This is called *single sideband* (SSB) modulation. For this we need to employ complex-valued signals by combining a cosine and a sine wave of the same frequency. Such signals are then said to be in *quadrature* as their phases are 1/4 of a cycle or $\pi/2$ apart. The resulting complex-valued function built from these represents an *analytic* signal. Therefore, the complex sinusoid

$$e^{j\omega t} = \cos(\omega t) + j\sin(\omega t) \tag{3.17}$$

constitutes an analytic signal. SSB modulation is now based on the product of two such signals. We can confirm it by showing that

$$e^{j\omega}e^{j\theta} = e^{j\omega+\theta} = \cos(\omega+\theta) + j\sin(\omega+\theta), \tag{3.18}$$

where, as we can see, only the sums of the frequencies of the signals are present at the output. The real part of such an expression can then be used as an audio signal. This forms the basis for the frequency shifter effect, where two signals in quadrature are multiplied together.

Two other forms besides amplitude modulation are commonly used in spectral sound synthesis. These are frequency and phase modulation, where the resulting effect may be a change in pitch (if the modulation is slow, for example in *vibrato*) or a sound with multiple partials (with fast modulation rates). In the former case, as we modulate the frequency, we have an integration step,[8]

$$s_{fm}(t) = \cos\left(\omega t + \int m(t)dt\right), \tag{3.19}$$

and in the latter, we modulate the phase directly (Fig. 3.6),

$$s_{pm}(t) = \cos(\omega t + m(t)). \tag{3.20}$$

As we can see, the two forms of modulation are closely related, but not exactly equivalent. The theory behind these methods and the identities involved is more complex than amplitude modulation, but in general we should note that they are capable of producing interesting multi-component spectra from just two sinusoidal signals. However, they are more difficult to control in a purely analogue synthesis context, as the signals need to be precisely calculated, and tightly synchronized in phase, for the results to be well defined. Therefore, it is simpler to realize these in digital contexts as a powerful means of sound synthesis, which we will explore later in Chapter 8.

3.2.4 DC Offset

We have so far considered the multiplication of a signal by a constant (amplification), the multiplication of two signals (ring modulation), and the addition of two or more signals (mixing). To complete the basic operations on signals, we should consider the addition of a constant to a signal,

$$y(t) = x(t) + c, \tag{3.21}$$

called a *DC offset*. The effect of this is twofold:

- In the time domain, a bipolar signal that was centred at 0 is now offset by c.
- In the frequency domain, we now have added a certain amount of amplitude at 0 Hz.

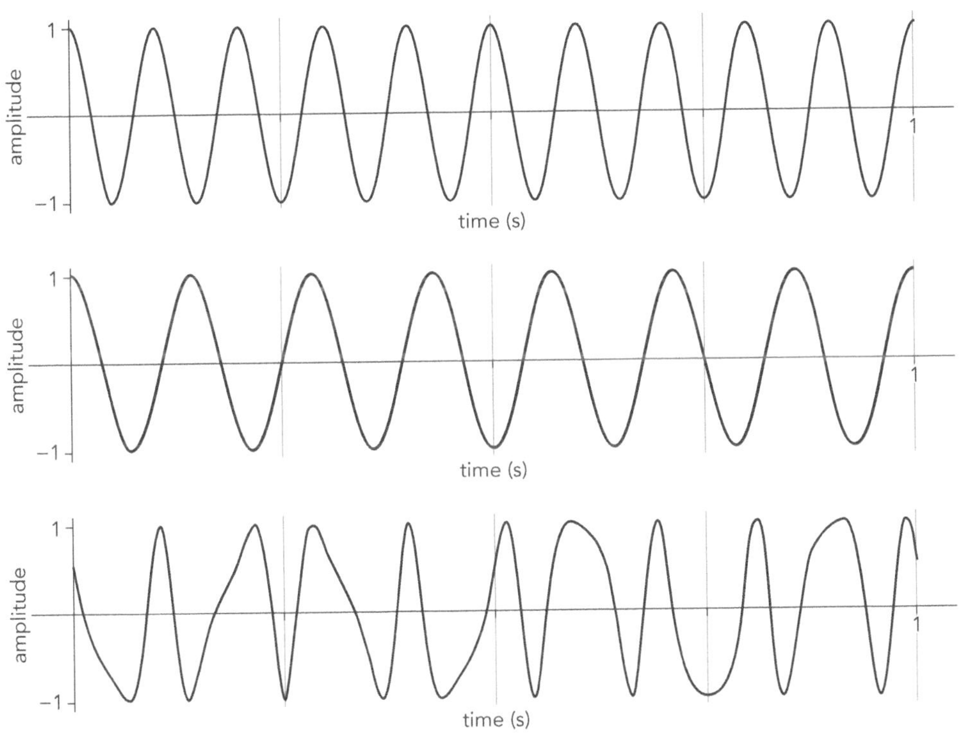

Figure 3.6
Plots of two sinusoids with frequencies of 10 and 7 Hz (top and middle, respectively) and the resulting phase modulation of the two signals (bottom).

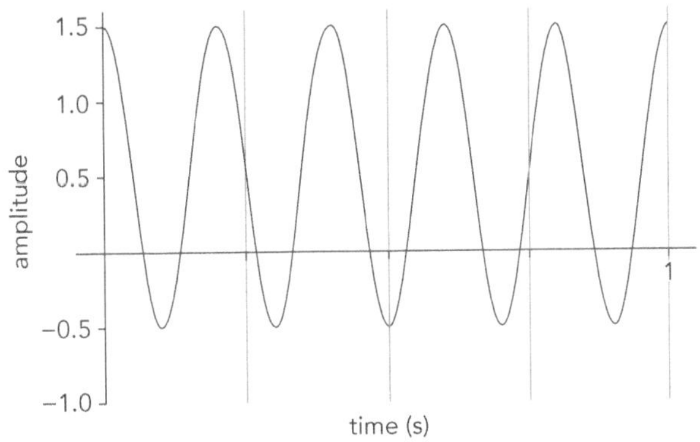

Figure 3.7
A sinusoid with a DC offset of 0.5.

Since DC does not refer to varying electrical signal, it is equivalent to a frequency of 0 Hz. Any such offsets in a waveform will imply low frequency components. For example, the rectified signal in Fig. 3.4 will exhibit a certain amount of DC. An example of a sinusoid with a positive DC offset of 0.5 is shown in Fig. 3.7.

Such offsets may have implications to the signal chain, as it will lower the available headroom (how much the signal amplitude can grow before it exceeds the maximum amplitude). It can also cause problems in transduction, as a device will be biased in one direction, and full movement might be restricted. DC offset can be removed with the use of special lowpass filters.

3.3 Discrete Signals

While an understanding of the fundamentals of analogue audio signals is significant for the topic of this book, most of the work will be done within the context of a (digital) computer. Therefore, it is important that we turn our attention to the type of signals we will be manipulating both in the time and frequency domain. The nature of these is fundamentally different: they are not continuous, but discrete. This means that we are going to consider functions that are defined for specific points in their domain, but not in between these steps. We need these functions because digital computers cannot deal directly with the continuum, given that their memory and numerical representations are finite. Therefore the first question we may ask at this point is, how do we go from the continuous to the discrete?

Digital signals, which we will be using in our exploration of spectral audio, expect both the domain and the image of functions to be discrete. However, we can start by looking at how we could discretize functions without playing attention at their range. In fact, that simplification allows us to continue using real numbers to represent signals, which is more meaningful than having to translate everything into integer mathematics. We can introduce the problem of discretization by looking at how a continuous function that is given by a differential equation can be computed for discrete steps of time. The original method for this was provided by Euler (1768), where a differential equation with a given initial value $y_0 = y(t_0)$ is redefined as

$$\frac{dy(t)}{dt} = f(t, y(t)) \tag{3.22}$$

and its value for each discrete step is calculated as

$$y(t_{n+1}) = y(t_n) + hf(t_n, y(t_n)) \tag{3.23}$$

with h determining the step size.

This method can be applied, for instance, in a computational model of harmonic motion, which was introduced in Sec. 2.2.3 in relation to a string model. Let's consider an even simpler system, that of a spring attached to a mass, which is set into motion (Fig. 3.8). The force acting on this system can be defined by

$$F = -kx, \tag{3.24}$$

where x is the displacement, and k is a constant related to the spring acting on a mass. We can use the fact that force is also the product of the mass and the acceleration, $F = ma$,

$$a = \frac{-kx}{m}, \tag{3.25}$$

to get an expression for the acceleration.

Now that we have the system well described, we need to calculate its values as it is put into motion, where we apply Euler's method. This will gives the next value of a differential equation from the current one plus the product of the derivative and the step. So we can calculate first the velocity, whose derivative is the acceleration given by Eq. 3.25,

$$v(n+1) = v(n) + ah. \tag{3.26}$$

With this we can complete the model[9] by calculating the position x,

$$x(n+1) = x(n) + vh. \tag{3.27}$$

Given an initial velocity and position, this model will be sufficient to re-create a spring-mass system, which can produce an audio signal, a sinusoid, by tracking the position[10] x.

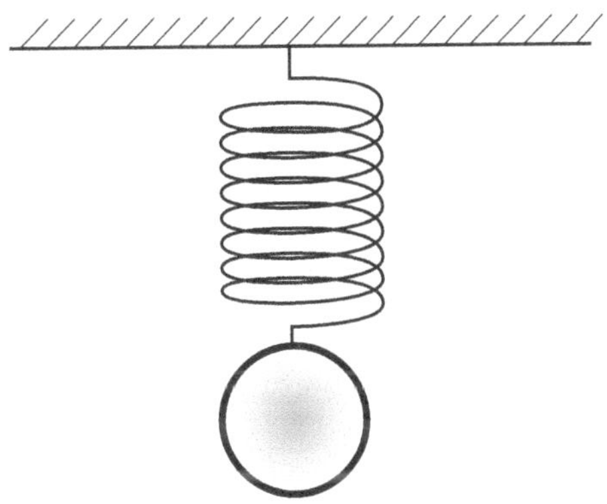

Figure 3.8
A spring-mass model.

How precise is it? That will depend on the step h, and how finely graded it is. Since this is a time increment (after all, the spring-mass system is moving as time passes), the step size is related to how small the period is between each time point. This brings us to the central question in discretization, how often should we produce a new output value? We will attempt to answer this as thoroughly as possible in the next section, as this touches a number of fundamental points of spectral audio processing.

3.3.1 Sampling

Sampling a signal, taking regular measurements of a continuous signal, or, alternatively, producing a sequence of values, each relative to a given time step, is the basis of the process we employ to generate a time-discrete function for further manipulation in a computer system. In order to obtain a sequence of values s_n from a continuous function $f(t)$, which we call *samples*, we take values at evenly spaced periods in time, T_s. How can we justify this in mathematical terms? The most precise way is to use the mechanism of the Dirac pulse $\delta(t)$, which can be conventionally defined by

$$\delta(t) = \begin{cases} \infty, & t = 0, \\ 0, & t \neq 0 \end{cases} \tag{3.28}$$

and

$$\int_{-\infty}^{\infty} \delta(t)dt = 1. \tag{3.29}$$

Among other applications,[11] the Dirac delta conveniently allows us to pick the value of a function at $t = 0$. From this we can define another expression, called a *Dirac comb*, denoted by III_{T_s}, which can be thought of as defining a stream of pulses,

$$\mathrm{III}_{T_s}(t) = \sum_{-\infty}^{\infty} \delta(t - kT_s), \tag{3.30}$$

spaced by the period T_s. We can then apply this to the input signal, $\mathrm{III}_{T_s}(t)f(t)$, to select values from each sampling point, which are separated by T_s, the *sampling period* and produce the signal samples $s_n = \{f(0), f(T_s), f(2T_s), \ldots, f(nT_s)\}$. The rate at which we generate a new output value is called the *sampling rate* (or frequency) f_s, the reciprocal of T_s. The process of sampling is shown in Fig. 3.9.

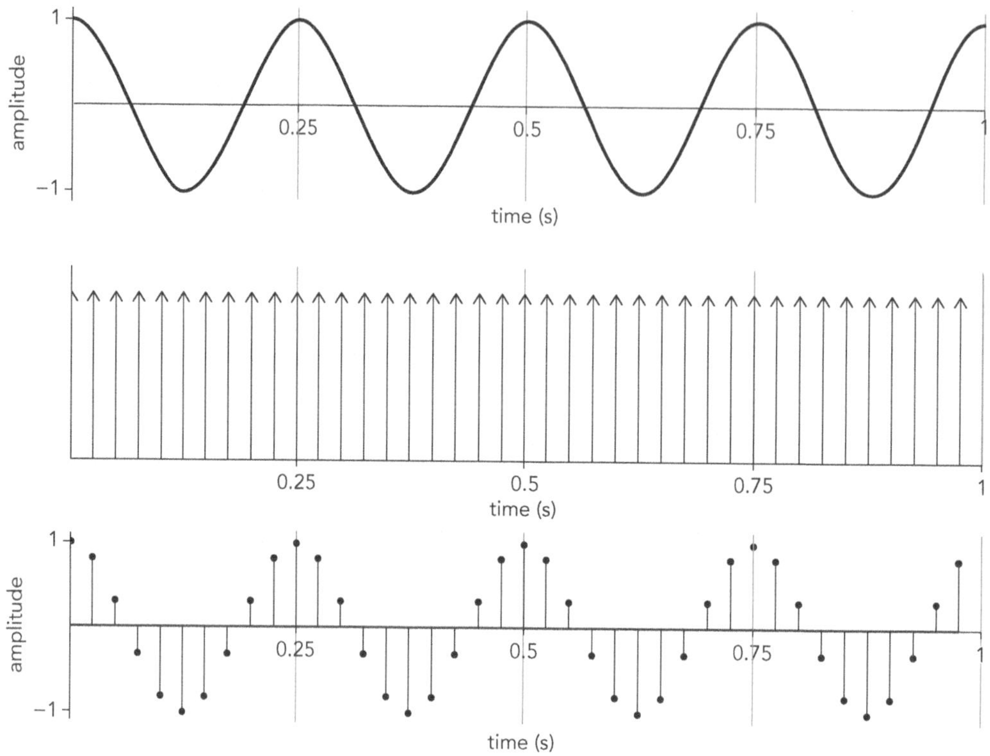

Figure 3.9
A continuous-time sinusoid (*top*), the Dirac comb (*middle*), and a discrete-time sinusoid (*bottom*), produced as a result of sampling with $T_s = 0.025s$ (sampling rate of 40 Hz).

This provides a general method to go from a continuous-time function $f(t)$ to a discrete-time sequence s_n, which is what we need to get an audio signal ready for computing. What about the reverse operation? Starting from a discrete-time sequence, for instance the output of the discretized spring-mass model, how can we obtain the continuous-time analogue signal? For this, we must first introduce another important tool for signal processing, the sinc() function, defined as[12]

$$\mathrm{sinc}(t) = \begin{cases} 1, & t = 0, \\ \frac{\sin(\pi t)}{\pi t}, & t \neq 0 \end{cases}. \tag{3.31}$$

A plot of this function is shown in Fig. 3.10. As we can see, an interesting characteristic of sinc() is that it is always zero at integer values of t, and it has most a positive main lobe between −1 and 1. Such properties are very useful for its applications.

With this definition in hand, the continuous-time reconstruction of $f(t)$ from s_n can be achieved by[13]

$$f(t) = \sum s_n \mathrm{sinc}\left(\frac{t - nT_s}{T_s}\right), \tag{3.32}$$

where T_s is the sampling period. We should note two things that stand out from this equation:

- It places a sinc() function centered at each sampling point, scaled by respective the sample value.
- The sinc() argument is scaled by T_s, which means that the function is zero at all other sample time points. The main lobe is bounded by the two neighboring sampling points.

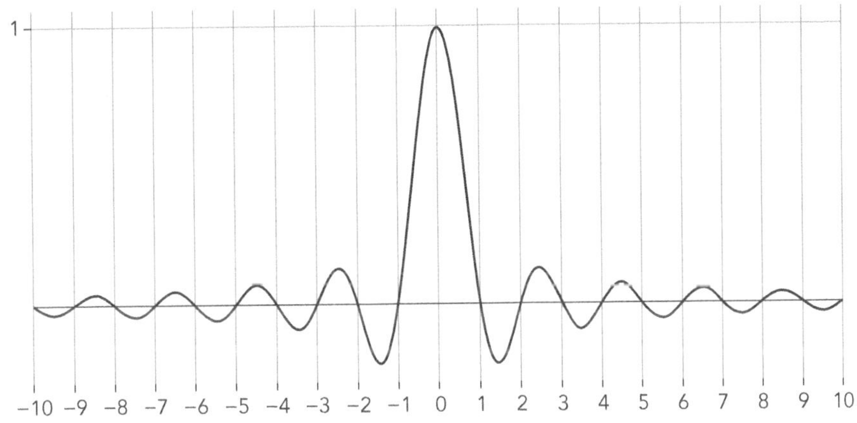

Figure 3.10
A plot of the sinc() function in the interval $[-10, 10]$

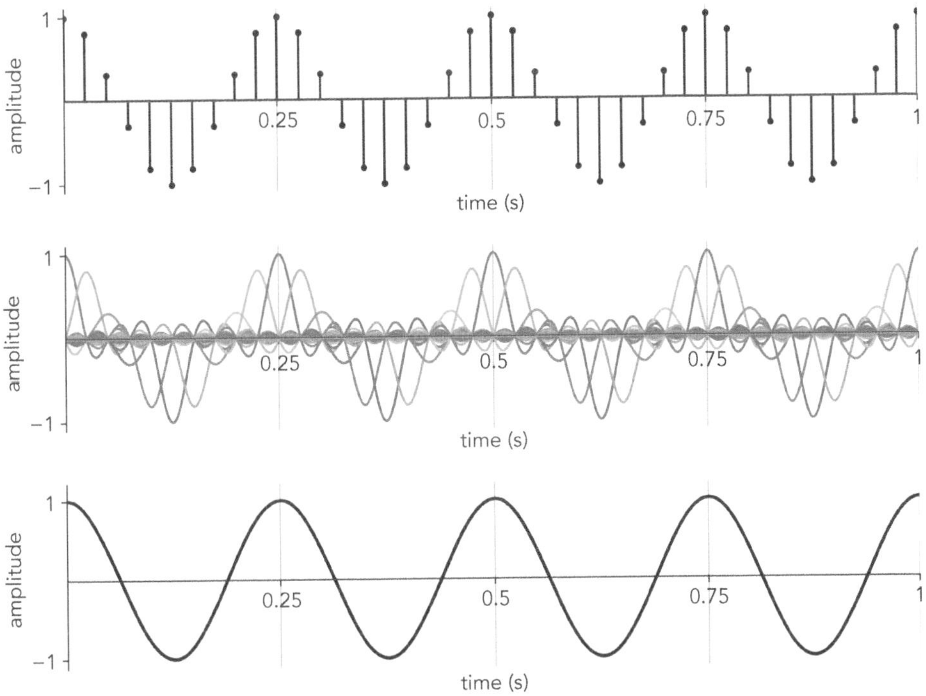

Figure 3.11
A discrete-time sinusoid (*top*), to which a series of sinc() functions are applied (*middle*), and the resulting sum of these, which recover the original signal.

We can demonstrate that the reconstruction indeed works by applying it to a sample sinusoid, reversing what we did in Fig. 3.9, to get the continuous-time sinusoid. This is shown in Fig. 3.11, where we start with the samples, apply the sinc() expression to each one of them, and by effecting their sum we recover the original signal. With this, we can now go from continuous to discrete time and backward, painlessly. It actually seems to indicate that we can continue to treat signals in mostly the same way as before, as we can always use these tools to convert from one form to the other. This is in fact true, but within limits: the reconstruction can only be perfect if we are careful enough with the types of signals we are discretizing or generating from scratch. These limits are of course related to the original question of how often we should sample a signal.

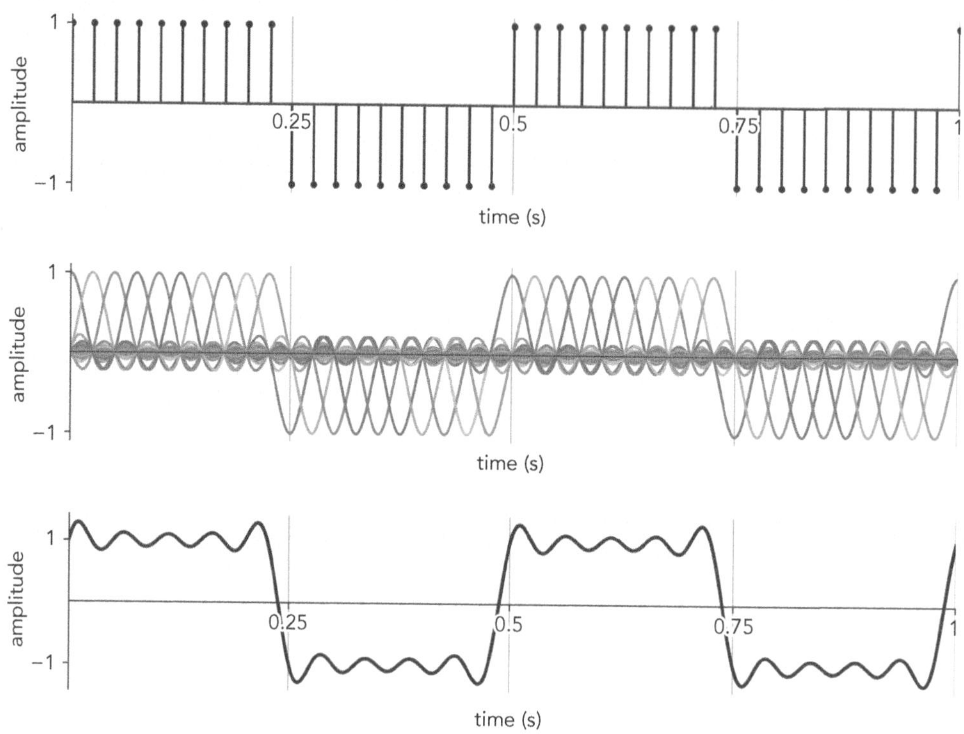

Figure 3.12 A 2 Hz ideal square wave sampled with a sampling rate of 40 Hz (*top*), its reconstruction with individual sinc() functions (*middle*) and the resulting continuous-time signal (*bottom*).

3.3.2 The Discrete-Time Baseband

In order to determine the limits of our discretization process, let's consider a different continuous-time signal. Unlike the sinusoid we have used so far, we should attempt to sample a waveform with several harmonics, such as for instance the ideal bipolar square wave, which alternates between two values (positive and negative). To simplify things, we will tune the sampling rate to an exact multiple of its (fundamental) frequency, so that there are as many positive samples as negative ones. From this sampled sequence, we can now try to recover the continuous-time signal, using the method given by Eq. 3.32. This process, from sampled signal to analogue waveform, is shown in Fig. 3.12.

As we can see from this example, we started with a non-bandlimited square (all harmonics up to infinity), converted this to a discrete representation, then as we recovered it, we ended up with a different waveform. This approximates the ideal square wave, but it is clearly *not* the original signal. What happened? The result of this process resembles a bandlimited waveform. So we can conclude that the high-order harmonics, the ones above a certain frequency, have been removed.

Before we look into exactly what frequencies were removed in this case, let's look another example. Here we start again with an ideal square wave, but this time we do not take care to tune in the sampling rate to a multiple of the fundamental frequency. This is illustrated in Fig. 3.13. As we can see, the result is anything but a square wave and it is not even close to a bandlimited approximation of it. The root cause of this poor reconstruction is the same as in the previous case where we had selected a particular sampling rate to suit our needs.

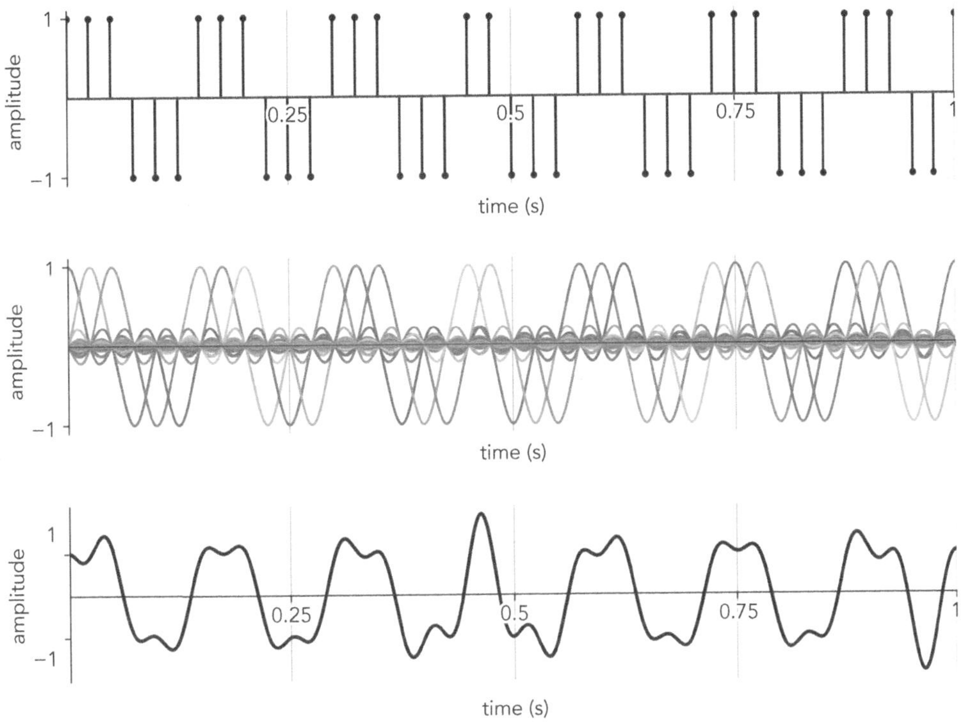

Figure 3.13
A 7 Hz ideal square wave sampled with a sampling rate of 40 Hz (*top*), its reconstruction with individual sinc() functions (*middle*), and the resulting continuous-time signal (*bottom*).

These two examples demonstrate that a discrete-time representation of a non-bandlimited signal is not accurate: in the first case, we lost high frequencies; in the second case, we also ended up with an unrecognizable signal. Therefore we will have to determine what is the frequency limit for signals to be accurately discretized. One way to do this is to look more closely at the sinc() function, which is what we used to render each sample back to continuous-time. Consider a one-sample pulse signal, called a *unit impulse*, which is a represented by $u_n = \{1, 0, 0, ...\}$. When converted using Eq. 3.32 we have a signal that is defined by $\text{sinc}(t/T_s)$. If we look at its spectrum, we might be able to determine the extent of the discrete-time frequency band. Each sinc() impulse has a bandlimited spectrum that is bound to half the sampling frequency ($f_s/2$) (Bracewell, 2000, 138), which is known as the *Nyquist* frequency.[14] From this, we can infer that the signal reproduced by mixing several of these impulses together, as per Eq. 3.32, therefore cannot exceed that frequency. This determines the extent of the discrete-time *baseband* spectrum. This explains the first result: the signal is bandlimited to frequency components extending only to the Nyquist.

We have not explained fully, however, how the second example is such a mess. We know that the discrete representation can only be reconstructed with frequencies up to 1/2 sampling rate, so a 7 Hz square wave has its second component (the third harmonic) at 21 Hz, already outside the 20 Hz baseband (since the sampling rate is 40 Hz). If all harmonics except for the fundamental are located outside the available range, where are they in the discrete-time signal? The answer is that they are misrepresented by partials present inside the baseband, which are *aliases* of the original signal components. Since sampling occurs at T_s time intervals, any variations in the continuous-time signal that occur in a shorter space of time will never be detected. Instead, these will appear as components with frequencies that are less or equal to $|1/2T_s|$. One intuitive way of

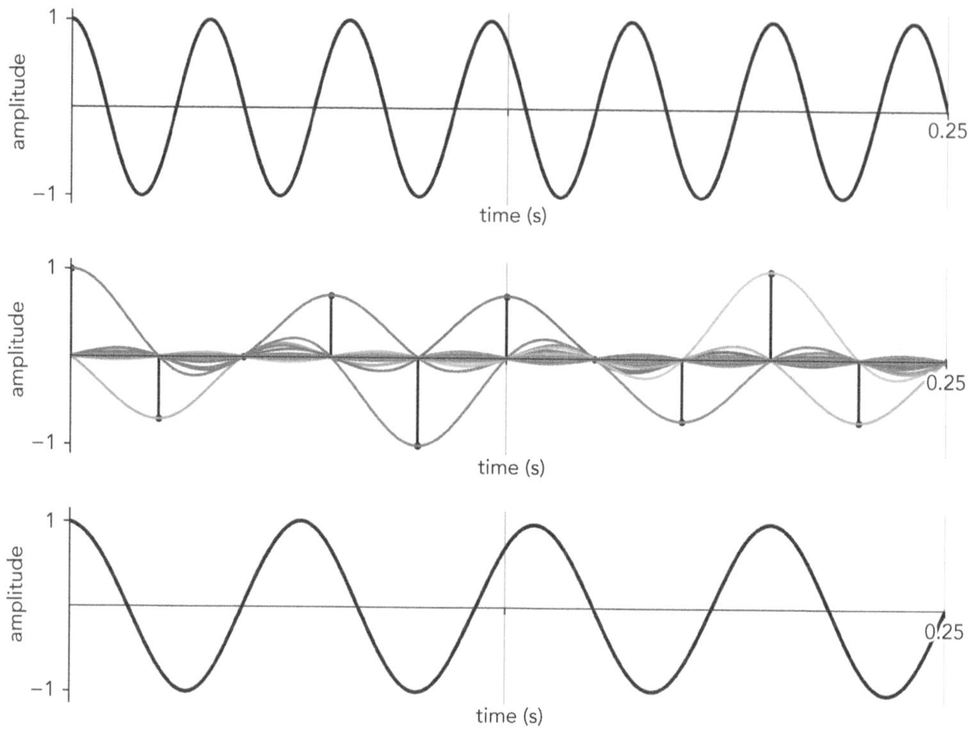

Figure 3.14
A 25 Hz sinusoid sampled (*top*), sampled with a sampling rate of 40 Hz (*middle plot stems*) and reconstructed using sinc pulses (*middle plot continuous lines*). This produces an aliased sinusoid (*bottom*) of a different frequency (15 Hz).

understanding this is that a sinusoid will need at the minimum two samples to be represented (this limit is given by a cosine wave at 1/2 sampling rate). Therefore it will not be possible to encode in a discrete-time signal any components with frequencies that are higher than that. Figure 3.14 illustrates this point, with a signal of 25 Hz sampled at 40 Hz, producing an aliased waveform at 15 Hz.

Since any components in a continuous time signal with frequencies that are outside the baseband will be represented by their corresponding aliases inside it, all of the infinite components of an ideal square wave will be recovered, but their frequencies may bear no audible relation to the original signal. Therefore we can have a jumbled-up mix of inharmonic components. This is what is causing the reconstructed wave in the second example to be very complex in shape. A simple expression can be used to determine the resulting discrete-time signal component frequency. Given a sampling frequency f_s, we have (Moore, 1990)

$$f_{out} = f_{in} - \text{int}\left(\frac{2f_{in}}{f_s}\right) f_s, \tag{3.33}$$

where f_{out} is the frequency of the encoded signal, given an input with a frequency f_{in} and int() gives the integer value of its argument. Note that each frequency in the multiple periodic images of the baseband is undistinguishable from its alias inside it. This also explains the difference between the two examples of sampled ideal square waves. In the first case, by choosing a sampling rate that is a multiple of the fundamental frequency, we ensured that the aliased frequencies coincided with the lower frequencies in the baseband, causing the signal to be a somewhat recognizable approximation of the square wave. However, it still contains aliased components; they are just not so noticeable. We can confirm this by producing a perfectly bandlimited square wave using a

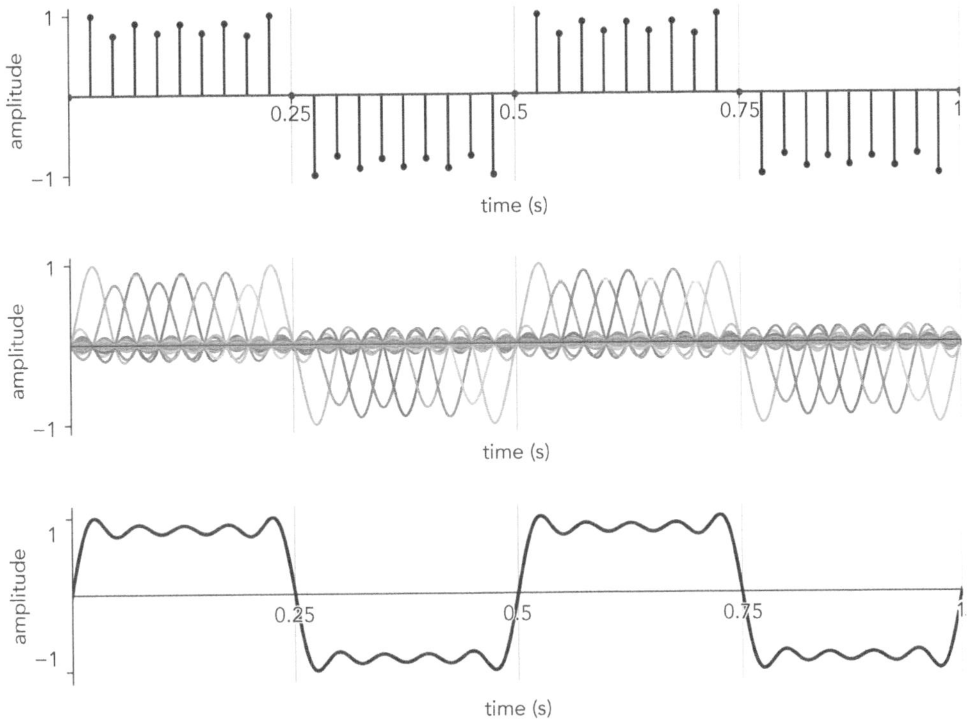

Figure 3.15
A 2 Hz bandlimited square wave sampled with a sampling rate of 40 Hz (*top*), its reconstruction with individual sinc() functions (*middle*) and the resulting continuous-time signal (*bottom*).

Fourier series additive method and comparing the result with that of Fig. 3.12. This is shown in Fig. 3.15, where we can see that the both the sampled waveform and its precise reconstruction are different from the aliased ideal square wave.

Thus, if we respect the limits of the discrete-time spectrum, we should be able to treat such signals more or less in the same way as continuous-time ones, although a number of considerations will have to made. We should assume that any signals that have been recorded into a computer system will be in the correct form, which is achieved by applying a suitable lowpass filter[15] to the input before sampling is performed. We will also need to ensure that any signal that we have generated uses a sampling rate that can accommodate all of its frequency components, or that we have taken precautions to minimize or suppress aliasing. Finally, this answers the question of how fine should a step be in the discretization process: we should sample at least at twice the highest frequency component present in the signal. If we do this, we will be able to represent and recover precisely a continuous-time waveform. In Chapter 4, we will return to the topic of sampling to give a full spectral description of the process.

Before we close the discussion, it is important to make a point on the subject of notation. So far, we have been careful so far to keep a discrete-time signal distinctively denoted by a sequence such as s_n, so that we are clear that it is not a continuous function. However, at this stage, we should be able to tell the difference, therefore we may relax this constraint and employ the usual functional notation when we mean a signal, regardless of its being continuous in time or not. In fact, it is even possible to use the variable t representing time in samples, which can be easily converted to continuous time in seconds by scaling it using the sampling rate, if necessary. At particular points, if there is the need to distinguish continuous from discrete time, we might revert to using n as the sample index, but within a function notation format.

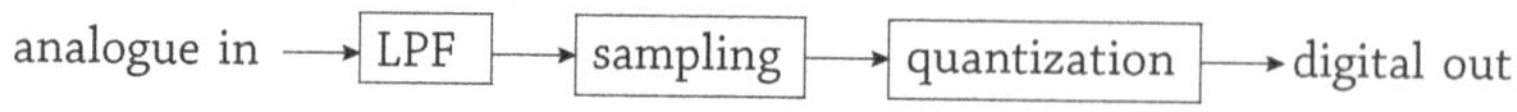

Figure 3.16 Analogue-to-digital conversion.

3.3.3 Digital Audio

Digital audio signals are not only discrete in time, but also discrete in amplitude. While there are many forms of such signals, we will be assuming in this book the most common type called *linear pulse code modulation* (PCM). These are generated by a *quantization* process that takes each sample and assigns a binary code to it representing a given amplitude linearly. The complete analogue-to-digital conversion (ADC) process is shown schematically in Fig. 3.16, where we start by removing the high-frequency components of the signal to prevent aliasing, then produce samples out of the input signal, and finally assign it discrete-amplitude values through quantization. This latter stage requires us to determine how many binary digits (bits) will be assigned to represent each quantization step, which in turns defines how precisely represented the sample will be. In general, for an integer-based digital sequence, the full signal amplitude range is divided into 2^n regions, where n is the number of bits used. The quantization error therefore will be related to how narrow or wide these are, decreasing as more bits are employed. This tends to be treated as a form of noise and determines the useful range of amplitudes in a digital signal.

Under these conditions, we can see that the error will halve for every bit that is added. Therefore the best possible signal-to-quantization-error-noise ratio (SQNR) can be determined in the dB scale as

$$\mathrm{SQNR(n)} = 20\log_{10} 2^n, \tag{3.34}$$

with n as the number of bits employed in the encoding. This is defined in relation to full-scale signals. A reduction of signal level will affect the actual value of the SQNR, causing it to increase.

For convenience and precision, we will adopt the use of *floating-point* encoding, which sets the signal amplitude range to the $[-1, 1]$ interval, and the 0dB full-scale reference to 1. This is sometimes called the *normalized* amplitude range. Floating-point numbers, which are used to represent (finitely) real numbers in a digital computer, are commonly provided in two levels of precision: 32-bit (single) and 64-bit (double). Depending on the requirements of the problem at hand, we will need to choose one or the other when programming signal processing expressions. The use of floating-point encoding has two main advantages: it provides a better SQNR if compared to the most common integer encoding types;[16] and it allows us to continue treating signals employing real-number mathematics. Therefore we can skip the full discussion related to quantization issues and assume that we are manipulating samples with sufficient precision.[17]

The opposite process to ADC, digital-to-analogue conversion (DAC), can be described using similar steps to the ones detailed in Fig. 3.16, but now applied in reverse order. It starts by producing the voltages that refer to each quantization region. It then passes this signal through a sample-and-hold mechanism, which is controlled by the sampling frequency. Finally, it reconstructs the analogue waveform by applying a lowpass filter to the sampled signal. Different actual implementations of ADC/DAC may in practice involve other steps than the ones detailed here, but they produce equivalent results.

Notes

1 Contrast this, for instance, if we were trying to track the movement of a body or a particle in space, where a single dimension, say up-down or left-right, might not be sufficient.

2 We should be careful not to interpret these outside the mathematical concept of a real number.

3 Note that if we are measuring phases in radians, then the rate of change $\omega(t)$, the frequency, is given in terms of radians per unit time, which can be turned into Hz by dividing it by 2π.

4 The voltage is the potential difference between two points in the circuit. Electric current conventionally flows from the higher voltage point to the lower voltage point.

5 Electrical power is defined as the amount of electrical energy flowing in a circuit, which for DC is just the product of the voltage and the current involved. For AC, given that the flow is periodically reversed, it depends on average measurements of the current and voltage as well as the phase difference between the alternating waves of current and voltage.

6 This is a sound that is generally concentrated in one frequency, such as 50 Hz or 60 Hz, which is due to a leakage from the AC mains power supply to the system.

7 In Chapter 1 we saw that our perception of frequency is, generally, logarithmic as well, and we noted that this is due to the fact that we perceive distances between frequencies based on their ratios rather than their differences.

8 As we discussed in Sec. 3.1.2, when we vary the frequency, we need to integrate it to get the correct phase. In fact, Fig. 3.1 demonstrates a simple frequency modulation example. See also Schottstaedt (2019).

9 Note that this is an ideal system, which will keep moving forever. If we want to introduce some form of resistance, we can include a constant d, $0 < d < 1$, to the velocity integration,

$$v_{n+1} = v_n \times d + ah.$$

10 This is an example of what we call physical modeling synthesis. See Lazzarini et al. (2016, 401) for a practical implementation of this particular case.

11 We have already encountered this fundamental tool for signal processing in Chapter 1, where it was used to describe the spectrum of an infinite long signal as well as precisely defined pulse. The Dirac delta is not a function in the usual sense, but it has some significant properties that are very useful when we want to define certain principles. We can think of it as an infinitesimally narrow pulse, one whose duration is instantaneous, precisely defined in time. Therefore it can also be defined as the limit

$$\delta(t) = \lim_{a \to 0} \frac{e^{-(t/a)^2}}{|a|\sqrt{\pi}},$$

which describes it as a "squeezing" of the normal distribution.

12 Note that we will use here the so-called *normalized* sinc function, as opposed to the original definition, $\text{sinc}(x) = \frac{\sin(x)}{x}$ for $x \neq 0$.

13 This was first proved by Shannon in his seminal article describing the sampling theorem (Shannon, 1949).

14 Confusingly, another related term, the *Nyquist rate*, which is equivalent to the sampling rate, is also used in some places. In this book, whenever we use the term, it will always be accompanied by the word *frequency* and will refer, with no exception, to $f_s/2$.

15 The sinc() reconstruction implies an ideal *brickwall* filter, whose amplitude response is a rectangle. This is difficult to achieve in practice; therefore, practical implementations of sampling attempt to approximate it as much as possible to ensure good quality of conversion.

16 The SQNR of single and double-precision signals is set at 156 and 331 dB, respectively (Lazzarini, 2017, 22).

17 For more details on quantization, see Lazzarini (2017, 20–23).

PART II

Techniques

4

Continuous and Discrete Spectra

In this chapter, we begin to explore the spectrum more properly, introducing a set of concepts and tools that will allow us to start manipulating signals in the frequency domain. The Fourier theorem, which we first introduced in Chapter 2, is the basis for a whole family of methods, which have various applications in spectral audio processing. We will begin this part of the book by exploring this theory at the ground level, that is, looking in detail at how it can be used to describe various types of signals, and introducing some key operations and concepts that compose it. This theoretical basis will help us to do two things: firstly, we will be able to return to some of the ideas we discussed in the previous chapter with a fuller understanding of their meaning; secondly, we will lay the foundations for some practical applications of Fourier techniques, such as additive synthesis. Even though our goal is to manipulate audio in the digital form introduced at the end of Chapter 3, we will start by considering the continuum in time and frequency. As the discussion unfolds, we will look at discrete representations of the spectrum so that we can fully return to digital audio processing in the next chapter.

4.1 The Fourier Series

We begin by tracking back to Sec. 2.2.5 of Chapter 2, where we looked at Fourier's original contribution, which took the form of a trigonometric series, defined by following expression[1]

$$f(t) = \frac{1}{2}a_0 + \sum_{n=1}^{\infty} a_n \cos(nt) + b_n \sin(nt) \qquad -\pi \leq t \leq \pi, \tag{4.1}$$

where the factors a_n and b_n are given as

$$a_n = \frac{1}{\pi}\int_{-\pi}^{\pi} f(t)\cos(nt)dt, \tag{4.2}$$

and

$$b_n = \frac{1}{\pi}\int_{-\pi}^{\pi} f(t)\sin(nt)dt. \tag{4.3}$$

So, as we have already seen, these equations give us a means of reconstructing a signal $f(t)$ as a sum of sine and cosine components, or partials, of different amplitudes and frequencies (Eq. 4.1, which implements a form of additive synthesis). These can be detected, or analyzed, from the signal itself (Eqs. 4.2 and 4.3).

Spectral Music Design. Victor Lazzarini, Oxford University Press. © Oxford University Press 2021.
DOI: 10.1093/oso/9780197524015.003.0004

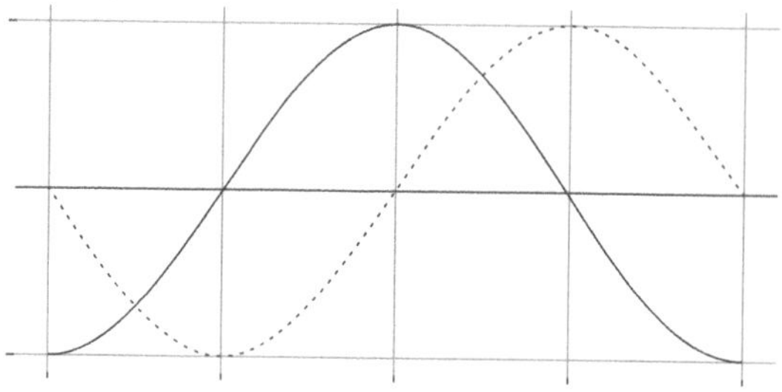

Figure 4.1
Even (cos(), solid line) and odd (sin(), dots).

4.1.1 Even and Odd Functions

The first thing we should note is that the sine and cosine terms in the sum are kept separate. Or rather, that the series is composed of these two forms of sinusoids, instead of just one or the other. This is related to an interesting property of these functions. The cosine, $f(t) = \cos(t)$, is called an *even* function, as it is symmetrical about the vertical axis ($t = 0$) since

$$\cos(t) = \cos(-t). \tag{4.4}$$

Complementarily, the $\sin(x)$ function is *odd*, since it is anti-symmetrical about the same point,

$$\sin(t) = -\sin(-t). \tag{4.5}$$

This property allows us to write a sinusoidal of any phase as sum of a sine and cosine of the same frequency[2] by appropriately scaling these,

$$\cos(\omega t + \phi) = a\cos(\omega t) + b\sin(\omega t) \tag{4.6}$$

where a and b are coefficients that are dependent on the value ϕ. We will explore this in more detail later on, but we can already intuitively see that if $\phi = 0$ then $b = 0$, and if $a = 0$ then $\phi = -\pi/2$.

We should also note the meaning of negative and positive frequencies with regard to even and odd periodic functions. In the former, a negative frequency signal cannot be distinguished from a positive, whereas in the latter case, a negative frequency implies a half-cycle phase shift. If we read a change in the sign of frequency as running the signal backward in time, then it is easy to have a mental image of this.

4.1.2 Interpreting the Fourier Formula

We can now see that it is necessary to have both sine and cosine terms in the trigonometric series in order to capture the phase of each partial. There is another reason for it, which is related to the exact mechanism that allows the Fourier coefficients a_n and b_n to be determined. Let's examine the expressions used to obtain it, given by Eqs. 4.2 and 4.3. If we break them down into separate operations we see that we have three distinct steps:

1. Multiplication of the function $f(t)$ by a cosine or a sine of a given frequency.
2. An integral sum over a given interval ($-\pi$ to π in this case).
3. Scaling by the a factor related to the interval size.

We can intuitively grasp the effect of each of these steps, and the combination of them. Firstly, we use a sinusoid to pick a component in the signal $f(t)$. If it aligns well with this component then the multiplication will be non-zero at various points. We can think of this step as using a sinusoid to tune into a component that is present in the signal. Then we sum all of these together to get the amount of the cosine or sine component that was detected in the signal. Finally, we normalize the result to keep it within the expected range of partial amplitudes.

To make it even more clear how this is happening, we can interpret the products $f(t)\cos(nt)$ and $f(t)\sin(nt)$ as ring modulation. We know the result is a mix of components containing the sums and differences of all the frequencies in the two signals (Sec. 3.2.3). Therefore, if a component in $f(t)$ has the same frequency as the sinusoid, then the resulting signal will contain a 0 Hz (DC) component.[3] We can then interpret the integration as an extreme form of lowpass filtering, which takes care of eliminating everything but DC out of that signal, and hence an amplitude at that specific frequency is detected and nothing else. We can also use a similar form of reasoning to see that if there are no components at a given frequency, we will get zero as a result.

Let's put this intuitive idea to the test, using a cosine function of period 2π, which is equivalent to the first harmonic (at the fundamental frequency). We plug $f(t) = \cos(t)$ which is such a waveform over the interval $[-\pi, \pi)$, into Eq. 4.2, since we know the cosine is an odd function, and therefore b_n coefficients will be null by definition. We do not assume anything else about the outcome, and we try to prove that a_n is one for $n = 1$ and zero elsewhere. Integrating Eq. 4.2, we get

$$\begin{aligned} a_n &= \frac{1}{\pi}\int_{-\pi}^{\pi} \cos(n)\cos(nt)dt \\ &= \frac{1}{\pi}\int_{-\pi}^{\pi} \frac{\cos(t[1-n]) + \cos(t[n+1])}{2}dt \\ &= \frac{1}{\pi}\left[\int_{-\pi}^{\pi} \frac{\cos(t[1-n])dt}{2} + \int_{-\pi}^{\pi} \frac{\cos(t[n+1])dt}{2}\right], \end{aligned} \tag{4.7}$$

that is, a sum of two integrals (related to the two ring modulation components). Of these, the integral on the right is always zero, regardless of the value of n. We can work this out, but it is easier to consider the definite integral as a sum over an interval. In this case, the interval is one fundamental period T_0, which for any sinusoid whose frequency is an integer multiple of $1/T_0$ evaluates to 0, since half the wave is positive and half is negative. We can then observe that the integral on the right is acting on such a waveform, and producing zero as a result, regardless of the value of n.

This is also the case for the integral on the left, except in the case of $n = 1$, which gives $\cos(0)$ (DC),

$$a_1 = \frac{1}{\pi}\int_{-\pi}^{\pi} \frac{1}{2}dt = 1. \tag{4.8}$$

This exercise not only proved the point but also demonstrated that our intuition was right. The integral, operating on the heterodyne signal, removes everything but the DC component.

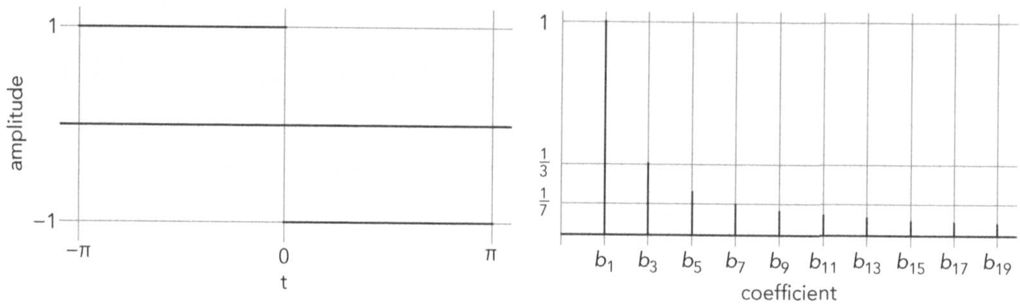

Figure 4.2
An ideal square wave (*left*) and its first 10 Fourier series coefficients.

4.1.3 The Fourier Series of a Square Wave

Now we can look at applying the equations of the Fourier theorem to a concrete case, the ideal square wave (Fig. 4.2). This can be defined as

$$\mathrm{sq}(t) = \begin{cases} 1, & -\pi \leq t < 0, \\ -1, & 0 \leq t < \pi. \end{cases} \tag{4.9}$$

We can see that this function is odd, and that it has no DC offset (i.e., it is not raised above the horizontal axis). Therefore we can assume that all a_n coefficients are zero. We can concentrate on the sine wave coefficients. Using Eq. 4.3, we have

$$\begin{aligned} b_n &= \frac{1}{\pi}\int_{-\pi}^{\pi} \mathrm{sq}(t)\sin(nt)dt = \frac{1}{\pi}\left[\int_{-\pi}^{0} \sin(nt)dt - \int_{0}^{\pi} \sin(nt)dt\right] \\ &= \frac{1}{\pi}\left[\frac{1}{n}(1-\cos(\pi n)) - \frac{1}{n}((\cos(\pi n)-1)\right] \\ &= -2\frac{\cos(\pi n)-1}{\pi n}. \end{aligned} \tag{4.10}$$

As $\cos(\pi n)$ is -1 for n odd and 1 for n even, we are left with only the odd harmonics, whose amplitude is inversely proportional to their number n, $1/n$ (Fig. 4.2). The negative sign means we have inverted sine wave partials, as the square wave is centered at $t = 0$. The Fourier series of Eq. 4.1 becomes

$$f(t) = -\frac{4}{\pi}\sum_{n=1}^{\infty}\frac{1}{2n-1}\sin([2n-1]t) \qquad -\pi \leq t \leq \pi. \tag{4.11}$$

Note that this mechanism can only work if we use cosines and sines separately to detect components. This is because an even function cannot be used to detect an odd one, and vice-versa.[4] A signal may have a mix of components with different phases at each frequency. A partial with phase in between a sine and a cosine will have a mixture of the two signals, which need to be detected separately. Therefore all along in the analysis and synthesis process, we will keep these separate.

4.1.4 Complex Representation

To make the representation of the spectrum more compact, we can bundle the cosine and sine factors together using a complex sinusoid defined by

$$e^{j\omega} = \cos(\omega) + j\sin(\omega), \tag{4.12}$$

while preserving their independence. This will allow us to formulate the Fourier theorem in more modern terms. While the Fourier series and related coefficient formulae can be used to describe the decomposition of any real function, its original form may not be not flexible enough. In particular, if we want to develop a theory to provide the foundation for a full range of spectral design applications, we need to be able to have a more general formulation. To do this, we will now proceed by rearranging the original formulae provided in the Fourier theorem. This will allow a more complete exploration of its implications for spectral audio processing.

Now we can re-write the Fourier series more compactly as

$$x(t) = \sum_{n=-\infty}^{\infty} X_n e^{j2\pi n f_0 t} \tag{4.13}$$

where[5] X_n is now a sequence of complex-valued coefficients that scale a series of complex sinusoids, which are all mixed together as before. Note that we have made some other changes to put it in terms that are better suited to the needs of spectral audio analysis:

- The sum now is over the negative and positive frequencies $[-\infty, \infty]$,
- The DC term is not summed separately, and is not scaled by 1/2,
- The result is not particularly constrained to be a single period between $-\pi$ and π, but of course we know that it still implies a periodic waveform.
- We have also included a $2\pi f_0$ factor so that the fundamental frequency f_0 can be more easily defined in Hz.

4.2 The Fourier Transform

All these modifications also mean that we need to define the coefficients X_n in a different way. For this, we introduce the *Fourier transform* (FT), which is still based on the principles of the Fourier theorem, but cast in more general terms. In particular, we will make it a full function of frequency f, a real variable, defined as (Bracewell, 2000, 6)

$$X(f) = \int_{-\infty}^{\infty} x(t) e^{-j2\pi f t} dt \qquad -\infty < f < \infty. \tag{4.14}$$

From this we can then get the coefficients n at frequencies that are multiples of a fundamental f_0, $X_n = X(nf_0)$. It is possible to recognize the relationship between the coefficient formulae of Eqs. 4.2 and 4.3 and the FT if we expand the complex exponential using Euler's formula. Note, however, that the integral is not constrained over one period of a waveform, since this has no meaning in this context. As a result of this, the scaling by the reciprocal of half a period (π) is also removed.

Note that all of these modifications did not change the practical results of the process, and more important, our intuitive understanding of it. We still have a sinusoid, although a complex one, of a given frequency that is used to capture the presence of component in the input signal $x(t)$. What we have done with it, is extend the method to apply more generally to any continuous value of frequency. In the periodic case, we can ignore any values other than the integer multiples of a fundamental frequency; they will always be zero. While we know this intuitively, we will shortly be able to determine more precisely why we can assume this.

4.2.1 The Inverse Fourier Transform

While Eq. 4.13 gives us the means of recovering the time-domain signal in the case of periodic waves, we need a more general inverse transform. This would allow us to go from waveform to spectrum and back transparently, without imposing constraints on the type of signal for which the recovery would work. This is the *inverse Fourier transform*, which takes a function of frequency $X(f)$ and gives us a function of time $f(t)$ (Bracewell, 2000, 6),

$$x(t) = \int_{-\infty}^{\infty} X(f) e^{j2\pi ft} df. \qquad -\infty < t < \infty. \tag{4.15}$$

We can interpret this as a generalized form of the Fourier series, where the frequency is a continuous variable, rather than a discrete sequence. In fact, this comparison allows us to conclude that the set of Fourier series coefficients is a discrete function of frequency, which is the spectrum of a periodic continuous-time waveform. Therefore, a periodic time-domain signal has a discrete spectrum, and we have already seen that a periodic frequency-domain function implies a discrete waveform.

4.2.2 Amplitude and Phase Spectra

The complex spectrum $X(f)$ is given by the Fourier transform in a rectangular form. We should interpret this is as separate amplitudes for cosines (real part) and sines (imaginary part). If a time-domain signal consists of a mix of components in the sine phase only, then the real part of $X(f)$ (denoted by $\Re\{X(f)\}$) is zero. Conversely, if we only have a sum of cosines in the input to the FT, then we have a purely real spectrum, $\Im\{X(f)\} = 0$. While this representation is certainly useful if we want to keep the cosines and sines separate,[6] it is not as intuitive as the polar representation.

With amplitudes A and phases θ, each component frequency ω is given as

$$s(t) = A\cos(\omega t + \theta) \tag{4.16}$$

which is far more intuitive to use.[7] Instead of speaking of cosine or sine amplitudes, we can now have component amplitudes, given by A, and phases ϕ, relative to the cosine phase. Separately, we call these the amplitude, or magnitude,[8] and phase spectra, denoted by $|X(f)|$ and $|\arg(X(f))|$, respectively. As with the cosine and sine amplitudes, each one of these is a real function of frequency.

4.2.3 The Spectra of Real Signals

As shown, the Fourier transform takes in a time-domain signal $x(t)$ and returns a complex-valued frequency-domain signal $X(t)$. The inverse operation takes the complex-valued $X(f)$ and returns $x(t)$, which we would presume at this stage is now also a complex-valued signal (since the equation involves a complex multiplication). This is actually not always the case, and will depend on the original $x(t)$. As we noted earlier, the forward-inverse transform pair is transparent; therefore, we will always recover the original signal and if it is real-valued, the result following the inverse transform will be real.

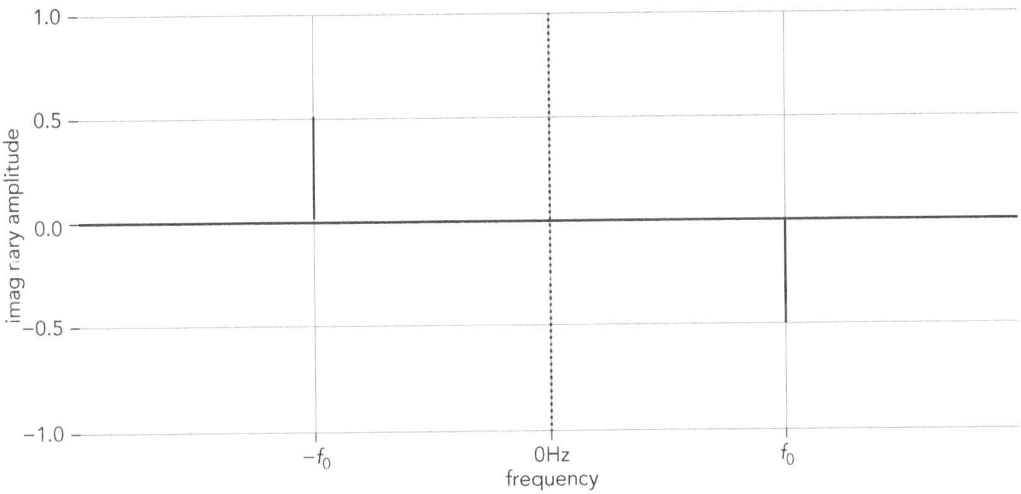

Figure 4.3
The spectrum $X(f)$ of $\sin(f_0t)$, consisting of two imaginary values at $f = -f_0$ and $f = f_0$.

A fully worked example may shine some light on how this can be the case. Let's consider a sine wave of frequency f_0, whose Fourier transform $X(f)$ will be calculated as

$$X(f) = \int_{-\infty}^{\infty} \sin(2\pi f_0 t)e^{-j2\pi ft}dt, \tag{4.17}$$

giving the result[9]

$$X(f) = \begin{cases} \frac{j}{2}, & f = -f_0 \\ -\frac{j}{2}, & f = f_0 \\ 0, & f \neq |f_0| \end{cases}. \tag{4.18}$$

So as expected, given that our signal was a sine wave, we have a purely imaginary spectrum (Fig. 4.3).[10] We should now attempt to recover it to see if we get a real-valued waveform. The inverse FT to resynthesize $x(t)$ from $X(f)$ is simply

$$\begin{aligned} x(t) &= X(-f_0)e^{-j2\pi f_0 t} + X(f_0)e^{j2\pi f_0 t} \\ &= j\frac{[\cos(2\pi f_0 t) - j\sin(2\pi f_0 t)] - [\cos(2\pi f_0 t) + j\sin(2\pi f_0 t)]}{2} \\ &= \frac{\sin(2\pi f_0 t) + \sin(2\pi f_0 t) + j[\cos(2\pi f_0 t) - \cos(2\pi f_0 t)]}{2} \\ &= \sin(2\pi f_0 t) \end{aligned} \tag{4.19}$$

While this was a more or less trivial example, where we knew that we had a single component at f_0, it has actually brought up an interesting point. Instead of finding a signal at one frequency, we actually discovered it was present in two. A complementary example may help us understand why. Let's take the FT of a cosine now,

$$X(f) = \int_{-\infty}^{\infty} \cos(2\pi f_0 t)e^{-j2\pi ft}dt, \tag{4.20}$$

which yields

$$X(f) = \begin{cases} \frac{1}{2}, & f = |f_0| \\ 0, & f \neq |f_0| \end{cases}, \tag{4.21}$$

a purely real spectrum (as expected). This also has two components, at $-f_0$ and $+f_0$. However, if we compare it with the sine case, we can see that here the value of $X(f)$ at these two frequencies is the same, whereas there they had opposite signs.

We can explain this by remembering that cosines are even functions and sines are odd functions. Therefore a cosine of a negative frequency is equivalent to one with the same, but positive, frequency. For sine waves, we need to flip the sign of the waveform, which is done by the negative amplitude. The result is that both the negative frequency and the positive frequency components contribute to the output signal. When summed up, the original amplitude of the wave at a positive (or absolute) frequency is recovered.

From another angle, we can look at the amplitude and phase spectra to determine the output signal. In the case of the sine wave, $|X(f_0)| = 0.5$ and $\arg(X(f_0)) = -\pi/2$, and the recovered signal component is

$$\frac{1}{2}\cos\left(f_0 t - \frac{\pi}{2}\right) = \frac{1}{2}\sin(f_0 t). \tag{4.22}$$

The other component at $-f_0$ has the same amplitude, but different phase, $\arg(X(f_0)) = \pi/2$,

$$\frac{1}{2}\cos\left(-f_0 t + \frac{\pi}{2}\right) = \frac{1}{2}\sin(f_0 t). \tag{4.23}$$

As we can see, the two refer to the same partial in the spectrum, except that the FT splits this into a pair of values, at negative and positive frequencies.

From these examples, we may conclude a couple of things in relation to the spectra of these waveforms:

- The spectrum is two-sided, including negative and positive frequencies,
- There is some sort of symmetry between the two sides.

Since we know that the signals we analyzed are the building blocks of other waveforms, we should by induction conclude that these points are also valid for any time-domain audio signals. In fact, as these are real functions, such characteristics are the properties of a the spectrum of real-valued signals.

What remains to be established is the type of symmetry that we have identified. We can start by noting that, in order to recover the time-domain signal as real-valued, we have to make the imaginary part vanish somehow. This can only happen if in the inverse FT sum, the negative and positive frequency sinusoid components need to have their imaginary parts with opposite signs, so they cancel each other. Looking at Eq. 4.19, we can see that this is indeed the case, as

$$\frac{j}{2}\left[e^{-j2\pi f_0 t} - e^{j2\pi f_0 t}\right] = \sin(2\pi f_0 t). \tag{4.24}$$

Complementarily, the cosine wave sum looks like this

$$\frac{1}{2}\left[e^{-j2\pi f_0 t} + e^{j2\pi f_0 t}\right] = \cos(2\pi f_0 t). \tag{4.25}$$

We have seen that two complex numbers differing only by the sign of their imaginary parts are in complex conjugation. So the symmetry we are looking for is one involving this relationship,

$$X(f) = \overline{X(-f)}, \tag{4.26}$$

which goes by the name of *Hermitian symmetry*. It also gives the following results:

- The amplitude (or magnitude) spectrum is an even function,

$$|X(f)| = |X(-f)|, \tag{4.27}$$

- A non-zero phase spectrum is odd,

$$\arg(X(f)) = -\arg(X(-f)). \tag{4.28}$$

We should also note that, since $Ae^{j\omega}$ for $\omega = 0$ is A, the DC component is always purely real. All these definitions are very important for many applications in signal processing. Assuming that the input to the FT is a real function, as indeed audio signals are, we are able to derive the negative spectrum from its positive values. We can therefore direct our attention solely to the non-negative side, which allows the number of operations to be reduced within a computational environment. This is used in real-to-complex forward and complex-to-real inverse FT transform implementations, as we will see in the next chapter.

4.2.4 The Spectra of Fundamental Signals

The spectra of cosine and sine waveforms were given with a certain degree of simplification, and with no further explanation, which was perfectly fine to demonstrate the key points about real signals. We used the intuition gained from the Fourier series coefficients formulae to describe the results of the Fourier transform. We now need to find a way to determine this more directly, which will allow us to apply the transform to arbitrary signals. In order to do this we will start from the ground up, by introducing first the fundamental building blocks, with which we will be able to derive the FT of more complex cases.

In Chapter 1, we introduced the spectrum of a signal that was perfectly defined in frequency (at 0 Hz) as represented by a Dirac δ,

$$X_{DC}(f) = \int_{-\infty}^{\infty} e^0 e^{-j2\pi ft} = \int_{-\infty}^{\infty} e^{-j2\pi ft} = \delta(f), \tag{4.29}$$

which is shown in Fig. 4.4. From this we can infer that the result of integrating a complex sinusoid of any non-zero frequency is always null. We noted before that this is also true for complete cycles of cosines or sines. This equation also confirms our intuition about the mechanism of the Fourier transform and integration as lowpass filtering.[11]

We have also seen that a signal that is perfectly defined in time can also be represented by a Dirac δ (a single infinitesimally narrow pulse). Its spectrum (Fig. 4.5) is given as

$$X_{pulse}(f) = \int_{-\infty}^{\infty} \delta(t) e^{-j2\pi ft} = 1. \tag{4.30}$$

From these two equations, we can see that the inverse Fourier transforms of $\delta(f)$ and 1 are 1 and $\delta(t)$, respectively, as indeed we stated earlier Chapter 1. These two signals make up a *transform pair*. They are the most fundamental building blocks of the Fourier theory. Armed with them, we can calculate the Fourier transform for cosines and sines more precisely as a continuous function of frequency. Let's tackle first a cosine wave of frequency[12] $\omega_0 = 2\pi f_0$. We can represent it as a sum complex exponential functions with negative and positive frequencies, as shown before. This is useful as we know that the

Figure 4.4
The spectrum of a single component at 0 Hz.

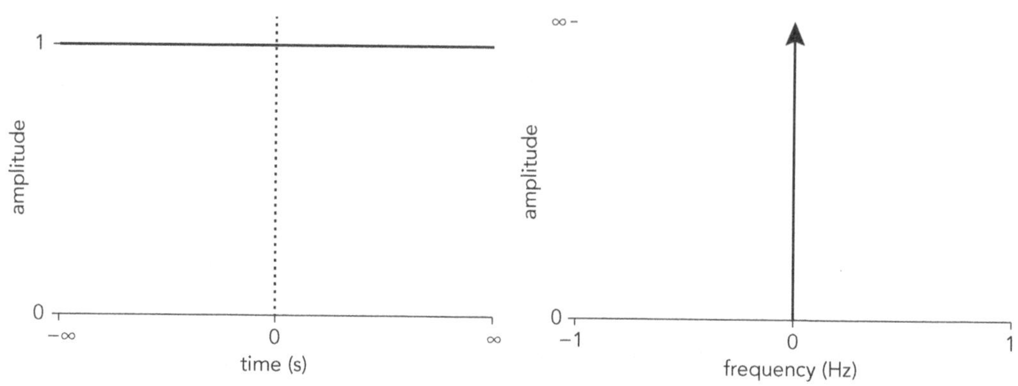

Figure 4.5
The spectrum of a Dirac δ.

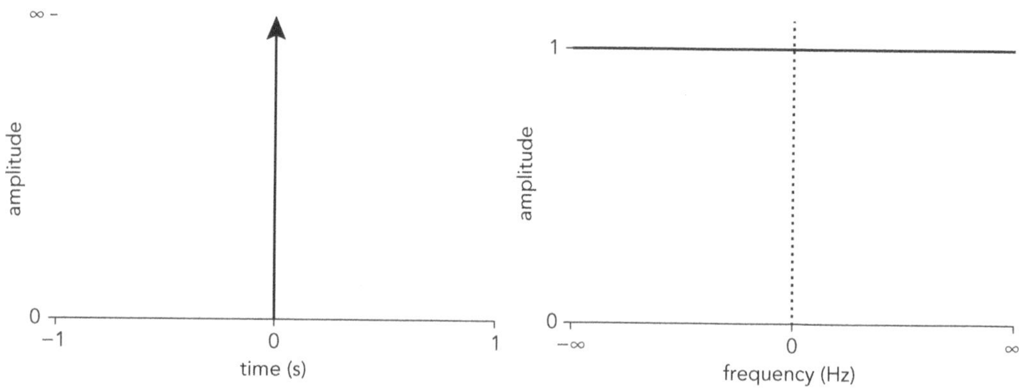

Fourier spectrum of a real signal represents a single component as a complex conjugate pair.

We plug a cosine in exponential form into the FT equation (Eq. 4.14; note that we can from now on use $\omega = 2\pi f$ as shorthand, whenever suitable). Using the fundamental definition in Eq. 4.29, we can determine $X(f)$ exactly as

$$\begin{aligned} X(f) &= \int_{-\infty}^{\infty} \left[\frac{e^{j\omega_0 t} + e^{-j\omega_0 t}}{2} \right] e^{-j\omega t} dt \\ &= \frac{1}{2} \left[\int_{-\infty}^{\infty} \left(e^{-jt[\omega+\omega_0]} + e^{-jt[\omega-\omega_0]} \right) dt \right] \\ &= \frac{1}{2} [\delta(f+f_0) + \delta(f-f_0)]. \end{aligned} \tag{4.31}$$

The interpretation of this result is that $X(f)$ is zero everywhere except at $|f_0|$, where it is defined by an infinitesimally small pulse scaled by $1/2$. What we had missed in our previous description of the cosine spectrum is the use of the Dirac *delta*, shifted to be centered at $|f_0|$. To complement this, starting with a real sine in the complex exponential

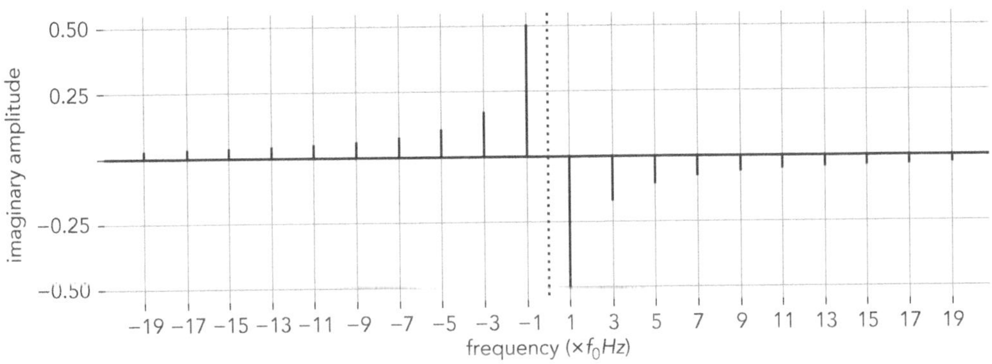

Figure 4.6
The two-sided (purely imaginary) discrete spectrum of a square wave, in the interval $[-19f_0, 19f_0]$.

form, we have

$$\begin{aligned} X(f) &= \int_{-\infty}^{\infty} \frac{(e^{j\omega_0 t} - e^{-j\omega_0 t})}{2j} e^{-j\omega t} dt \\ &= \frac{j}{2} \int_{-\infty}^{\infty} (e^{-j\omega_0 t} - e^{j\omega_0 t}) e^{-j\omega t} dt \\ &= \frac{j}{2} \int_{-\infty}^{\infty} e^{-jt[\omega+\omega_0]} - e^{-jt[\omega-\omega_0]} dt \\ &= \frac{j}{2} [\delta(f+f_0) - \delta(f-f_0)] \end{aligned} \tag{4.32}$$

which confirms our previous examples. The meaning of $j\delta(f-f_0)$ can again be extracted from previous definitions. The function[13] $\delta(f)$ relates to a cosine wave at 0 Hz. If we shift it to right side of the spectrum by f_0, we get $\delta(f-f_0)$, which is a cosine component at f_0 Hz now. The multiplication by the imaginary unit j makes it a sine component.

Using these ideas, it is possible to define the spectrum of well-known waveforms. For example, we previously calculated the Fourier coefficients of the square wave, which we determined decayed in value as $1/n$, where n is the harmonic number. With this, we can provide the ideal square wave spectrum as[14]

$$X(f) = \frac{j}{2} \sum_{n=1}^{\infty} \frac{\delta(f + (2n-1)f_0) - \delta(f - (2n-1)f_0)}{2n-1}. \tag{4.33}$$

The Dirac δ provides a mechanism to describe the discrete spectrum of the Fourier series, shown in Fig. 4.6 in the case of the square wave. It effectively allows a specific frequency point to be selected out of the continuum, to which a scaling can be applied. In the next sections, we will detail how we can use it to construct a sampling mechanism to obtain a discrete sequence out of a continuous spectrum. To support this, we first need to introduce an important new signal processing operation, *convolution*, which is fundamental for the understanding of the spectrum.

4.3 Convolution

Convolution can be described as another means of combining two functions, in addition to the usual arithmetic and algebraic operators. We can first define it for continuous signals as

$$f(x) * g(x) = \int_{-\infty}^{\infty} f(\chi) g(x - \chi) d\chi. \tag{4.34}$$

While at first look this appears quite a complicated expression, we can deconstruct it to give a better intuitive feel for it. First, as we can see, it takes the product of two signals, and then effects an integral sum of the result. The operation also shifts the signals along the domain by χ before the multiplication is done. While these steps are reasonably straightforward, there remains a difficulty in understanding the meaning of the operation as whole.

This is often related to trying to grasp the operation in terms of continuous signals, since the sum is done over the continuum, and the shift by χ is infinitesimal. This is compounded by the fact that the whole operation is defined over an infinite interval. It may be simpler to put things in terms of discrete sequences, which can be done without the risk of changing the meaning of the operation. Additionally, if there are any difficulties, we already have provided in Chapter 3 an appropriate method to move from discrete to continuous. Once we have explored discrete convolution algebraically, we will be able introduce a graphical interpretation that can also be employed in continuous domains.

4.3.1 Discrete Convolution

Redefining convolution in discrete terms allows the replacement of the integral by an ordinary sum. Therefore the shift now operates by an integral number of steps, giving the expression

$$f(n) * g(n) = \sum_{m=-\infty}^{\infty} f(m)g(n-m) \tag{4.35}$$

which is still defined over an infinite interval, but there is nothing stopping us considering it for finite sequence sizes.

So, in that spirit, let's apply this to two very short sequences, $f(n) = \{1,1,2\}$ and $g(n) = \{1,2\}$. We can think of these as infinitely long, if we want, but full of zeros besides these vales. What is the convolution $f(n) * g(n)$? That is another sequence $c(n)$, which we obtain by applying the operation. Since the signals are zero for negative indices, we can start the operation with $m = 0$ and $n = 0$:

1. For $n = 0$ have only one non-zero product ($m = 0$), $f(0)g(0) = 1$.
2. Incrementing n, we now have two products to add together ($m = [0,1]$), $f(0)g(1) + f(1) * g(0) = 3$,
3. With $n = 2$, we have $f(1)g(1) + f(2) * g(0) = 4$ ($m = [1,2]$).
4. Finally, for $n = 3$, we only have $f(2) * g(1) = 4$ ($m = 2$).

For all $n > 4$, all products are zero. This is because the convolution of a finite sequence of size N with a finite sequence of size M will have size $N + M - 1$. The result of the convolution is then $c(n) = \{1,3,4,4\}$. We can review this in terms of the shift, scale, and sum operations mentioned earlier.

- Shift: take $f(n)$ and slide it to start at the non-zero points of $g(n)$. There are only two of these, so we have $1,1,2$ and $0,1,1,2$.
- Scale: take the two shifted sequences and multiply them by their respective values in $g(n)$, $g(0)$ and $g(1)$. The two sequences are now $1,1,2$ and $0,2,2,4$.
- Sum: add the sequences together, point by point, $1+0, 1+2, 2+2, 0+4$, to get the convolution $c(n)$.

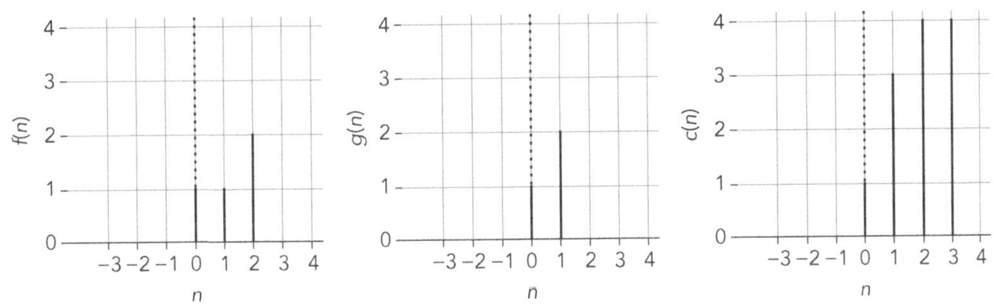

Figure 4.7 Convolution $c(n)$ (*right*) of two sequences, $f(n)$ (*left*) and $g(n)$ (*middle*).

The two sequences and their convolution are shown graphically in Fig. 4.7, where we can observe the shift-scale-sum more clearly. This is a typical example of a one-sided convolution, where we assume the signals start at $n = 0$; therefore, we do not use any negative indices. This is generally the case of time-domain convolution, where we arrange it so that time is always non-negative. Intuitively in this case, we can see that the shift in time may correspond to delayed copies of the signal, starting at a non-zero time position. This notion will be very important for the application of convolution as a means of implementing echoes and reverberation.

We have to consider, however, the cases where negative indices make sense. Of course, the first thing that should come to mind is that in the frequency domain we have been working with two-sided sequences and functions. Therefore, we should look at the behavior of convolution under these conditions. We should still expect that the shift-scale-sum approach is maintained. However there are subtleties. If we examine Eqs. 4.34 and 4.35, we may see that the 0 point is significant. The shift takes account of this point as a reference position.

This was probably not so noticeable when both sequences were one-sided. So we can use a signal $f(n)$ with 9 samples, spanning the indices $n = [-4, 4]$, $f(n) = \{0, 0, 1, 2, 3, 2, 1, 0, 0\}$. We apply this to the convolution with simple one-sided signal $f(n)$, which is zero everywhere except for $g(2) = 4/3$, which will make the process fairly evident. We can work out the result by hand, or we can just implement Eq. 4.35 to calculate it, using the Python code in Listing 4.1. It is a fairly straight translation of the mathematical expression, written to loop over two sequences of the same size and return the convolution. Note that to deal with the two-sided indexing, we set 0 at half the array size (zn) and then offset the lookup position accordingly.

Listing 4.1 Discrete Convolution function (Eq. 4.35).

```
def convol(f,g):
 n = f.size
 c = np.zeros(n)
 zn = n//2
 for i in range(0,n):
   for j in range(-zn,zn+1):
       if i-j>=0 and i-j<n and j+zn<n:
         c[i]+=f[j+zn]*g[i-j]
 return c
```

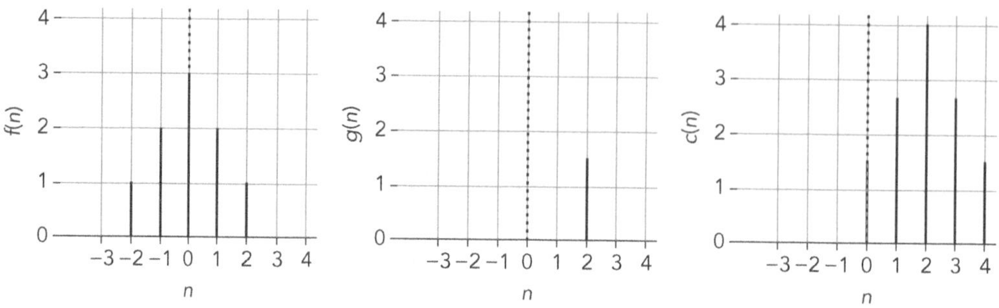

Figure 4.8 Convolution $c(n)$ (*right*) of two sequences, $f(n)$ (*left*) and $g(n)$ (*middle*), the first of which is two-sided.

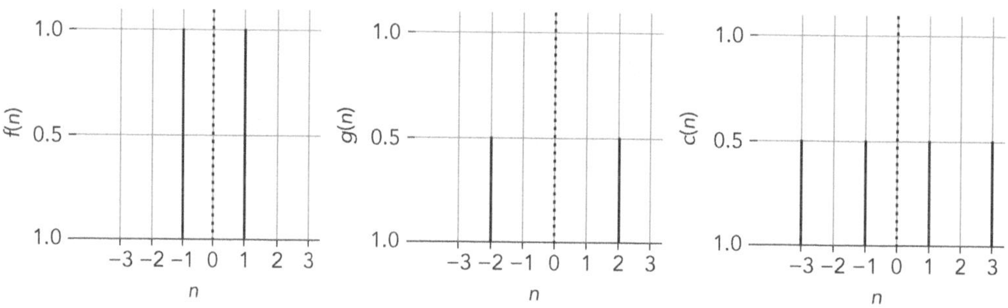

Figure 4.9 Convolution $c(n)$ (*right*) of two two-sided sequences, $f(n)$ (*left*) and $g(n)$ (*middle*).

The result of the convolution is better shown graphically, for which we can just run the code to create plots in Fig. 4.8. As expected, $f(n)$ was shifted, but perhaps surprisingly it is now *centered* at the non-zero sample of the $g(n)$. This is how the zero position reference is treated by convolution, and the sequence that was positioned there now is moved to be placed around the point with index 2. It is also scaled accordingly.

These results are very helpful to us, as we can use convolution in both the frequency and time domains (of course, it is an operation that can be applied to any real or complex functions). A third example will shine light on the nature of convolution when applied to two spectra. Consider now, two sequences that are two-sided, $f(n) = \{0,1,0,1,0\}$, and $g(n) = \{\frac{1}{2},0,0,0,\frac{1}{2}\}$, both defined in the interval $n = [-2,2]$. The result is $c(n) = \{\frac{1}{2},0,\frac{1}{2},0,\frac{1}{2},0,\frac{1}{2}\}$ (Fig..4.9).

This example is particularly important for us because the two sequences can very well represent two discrete-frequency spectra of real signals. In particular, given that each has only one positive component, they both refer to cosine waves (of different frequencies). The result is very revealing, as $c(n)$ is composed of *two* cosine waves instead of one. Where have we seen this before? If we look back at Sec. 3.2.3 of Chapter 3, we will see that this convolution represents the spectrum of ring modulation, containing components at the sums and differences of the two signal frequencies.

We may conclude that time-domain multiplication of two signals is equivalent to the convolution of their spectra,

$$f(t)g(t) = X_f(f) * X_g(f), \tag{4.36}$$

where $X_f(f)$ and $X_g(f)$ are given by the Fourier transform of $f(t)$ and $g(t)$. That is a very important result, which will also allow us to review some of the ideas we have already introduced under a new guise. Before we move on to this, we should note a few other

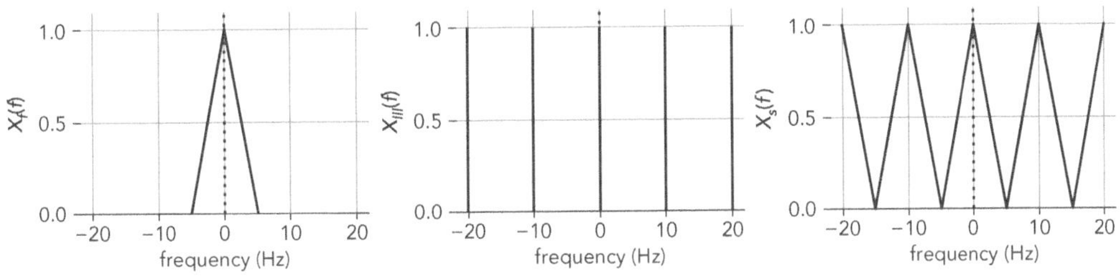

Figure 4.10 Sampling as the frequency-domain convolution $X_s(f)$ (*right*) of $X_f(f)$ (*left*) and $X_{\text{III}}(f)$ (*middle*).

things about convolution. First, that it is commutative,

$$f(n) * g(n) = g(n) * f(n). \tag{4.37}$$

Second, that it is a linear operation, which does not mean too much for what we are doing at the moment, but it is an important characteristic for later applications. Finally, everything we said in relation to the convolution of discrete sequences applies to continuous functions, and through the means of the Dirac δ, we can combine these together. In these cases, it may be easier to see what is happening from a graphical perspective, using the shift-scale-sum principle, rather than in an algebraic form. We can now return to the continuum of the Fourier transform and explore key aspects of convolution in the frequency domain.

4.4 Sampling in Time and Frequency

We can look back at the ideas related to the sampling operation through the lens of convolution, and re-apply these to the frequency domain. In Chapter 3, we saw that sampling applies an impulse train, represented by a sum of time-shifted δ functions to a continuous-time signal. This application can be in fact considered as a *product*; therefore, we have, in the frequency domain, a *convolution* of the spectrum of a Dirac comb and the spectrum of the continuous-time waveform. The former is a sum of δ functions spaced by the sampling rate f_s, and the latter is an arbitrary continuous-frequency function.

If we apply the shift-scale-sum method we developed earlier we will see that the result is a *periodic* spectrum, with period f_s, and even (which implies it is centered at nf_s, where n is an integer). This is a confirmation of what we have already noted. The continuous-time spectrum of a sampled signal repeats infinitely over frequency. We show this in Fig. 4.10, where the spectrum $X_s(f)$ of the sampled signal is given as the convolution of the spectra $X_f(f)$ and $X_{\text{III}}(f)$, of the signal $f(t)$ and the Dirac comb $\text{III}(t)$, respectively. To facilitate the illustration, we use an even function as the input, which has a real-valued spectrum.[15] This might puzzle us at first, since a sampled signal is bandlimited. However we are not completely finished with it, as the raw discrete waveform is not used directly as an analogue audio signal. It is important to remember that a continuous function needs to be reconstructed from its discrete representation before it is ready to be used.

Let's recall the reconstruction equation,

$$f(t) = \sum s_n \text{sinc}\left(\frac{t - nT_s}{T_s}\right), \tag{4.38}$$

where T_s is the sampling period. If we examine it closely, we will find that it resembles very much a type of convolution. In fact, we can actually recast it as continuous-time

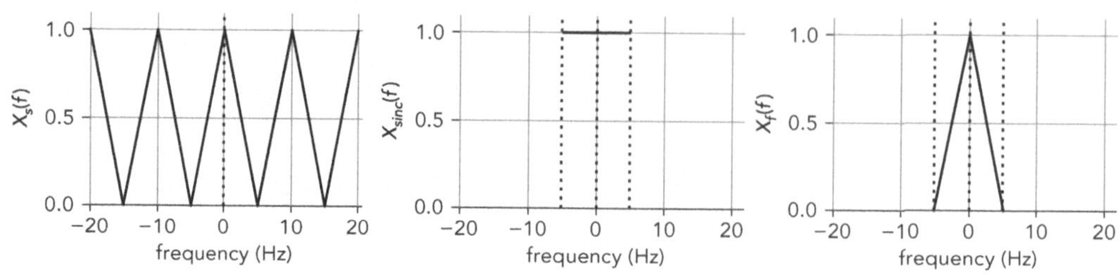

Figure 4.11 Reconstruction as the frequency-domain product $X_f(f)$ (*right*) of $X_s(f)$ (*left*) and $X_{\text{sinc}}(f)$ (*middle*).

convolution. First, we need to make the sequence s_n a continuous signal, for which, as noted earlier, we use the mechanism of the Dirac δ, to derive $s(t)$,

$$s(t) = s_n \delta(t - nT_s). \tag{4.39}$$

Then we can use the sinc() function to realize the convolution,

$$f(t) = s(t) * \text{sinc}(t). \tag{4.40}$$

Now we have the case of a time-domain convolution, what is its spectrum? Intuitively we know that since what we have is a sum of time-shifted scaled sinc() waveforms, the spectrum of this sum should be no wider than the spectrum of that signal, something in fact we assumed when first discussing reconstruction. From this intuition and what we already know about convolution, we should suspect that time-domain convolution is equivalent to frequency-domain multiplication. This is indeed the case, so we have the complementary expression to Eq. 4.36,

$$f(t) * g(t) = X_f(f)X_g(f), \tag{4.41}$$

where, again, $X_f(f)$ and $X_g(f)$ are the spectra of $f(t)$ and $g(t)$.

We know, from Chapter 3, that the spectrum of the sinc() function is bandlimited by $|f_s/2|$. Since this function is even, we have a real-valued spectrum, which is defined by (Bracewell, 2000, 138),

$$X(f) = \begin{cases} 1, & -\frac{f_s}{2} < f < \frac{f_s}{2}, \\ \frac{1}{2}, & f = |\frac{f_s}{2}|, \\ 0, & f > |\frac{f_s}{2}| \end{cases} . \tag{4.42}$$

Therefore the multiplication of the two spectra is sufficient to remove all components beyond the baseband. As this spectrum is rectangular in shape (unity inside the baseband and null elsewhere), it acts as a brickwall filter on the continuous time spectrum of $s(t)$. This is illustrated in Fig 4.11, where the real-valued spectrum $X_s(f)$ is multiplied by $X_{\text{sinc}}(f)$, which is given by the FT of a sinc waveform, resulting in the bandlimited spectrum $X_f(f)$ of the reconstructed continuous-time signal. These ideas, developed through the use of convolution, give a sound theoretical basis for sampling.

We can use these principles to understand aliasing more completely. A continuous-time signal whose bandwidth extends beyond $|f_s/2|$ will be encoded by sampling in the same way as before. However, when the convolution in the spectral domain is applied, there will be some amount of overlap between adjacent shifted copies of its spectrum. Since we will sum all of these together, the overlapped portions will be present in the resulting continuous-time spectrum of the sampled waveform (Fig. 4.12).

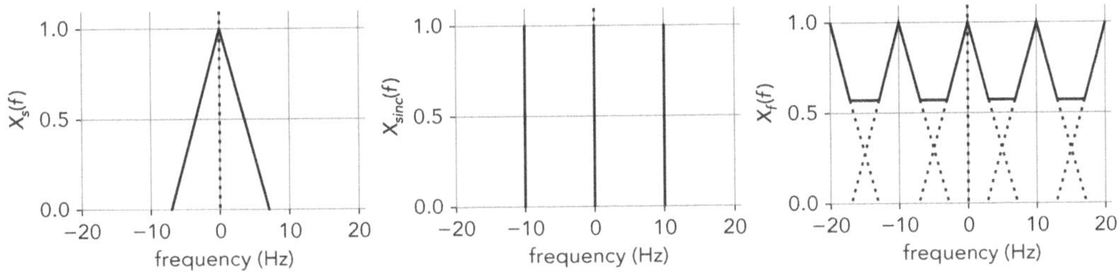

Figure 4.12 The frequency-domain convolution $X_s(f)$ (*right*) of $X_f(f)$ (*left*) and $X_{III}(f)$ (*middle*), showing an overlap of the shifted spectral images. This representation assumes a real-valued spectrum.

In Fig. 4.13, the result of aliasing is shown in the spectrum of a reconstructed signal, which is different from the one we attempted to sample (left panel of Fig. 4.12). When we reconstruct the spectrum, frequency components beyond the digital baseband will be kept as aliases of the original spectrum, which is now long gone. They will appear folded-over, but they actually belong to opposite side of a shifted copy of the spectrum. From this, we observe that aliasing from the positive side of the spectrum shows up in the negative side, and vice-versa. For cosine-phase components, this is not relevant, since they are symmetric at 0 Hz, but it is important in the case of sine components. We will always assume that, if a signal has a continuous time original, it is properly sampled with enough bandwidth to accommodate all of its components. Synthetic discrete-time signals will need to be treated with care so that they also respect the conditions of sampling. Finally, we should note that we have been deliberately employing very low sampling rates, mainly to help the discussion and to allow good visualization of time domain waveforms. In practice, audio signals will require sampling rates that are over 100 times what we are using for test signals.

So as we have seen, a sampled waveform has a periodic infinite continuous frequency spectrum. The f_s period of this spectrum is defined as $1/T_s$, where T_s is the time spacing of each sample in the signal. Conversely, we should also remember that a discretely sampled spectrum, given by the sequence of Fourier coefficients, refers to an infinite-length periodic waveform. Its period T_0 is determined by the frequency spacing of each spectral sample f_0, $T_0 = 1/f_0$. The meaning of a *sampled* spectrum is going to be developed in the next section.

4.4.1 Finite-Time Signals

We now turn to a complementary issue. What is the spectrum of a sinusoid that is finite in time, that is, it is switched on and then off? In order to determine this using the Fourier transform, it is required that this be put in terms of an infinite-time signal.

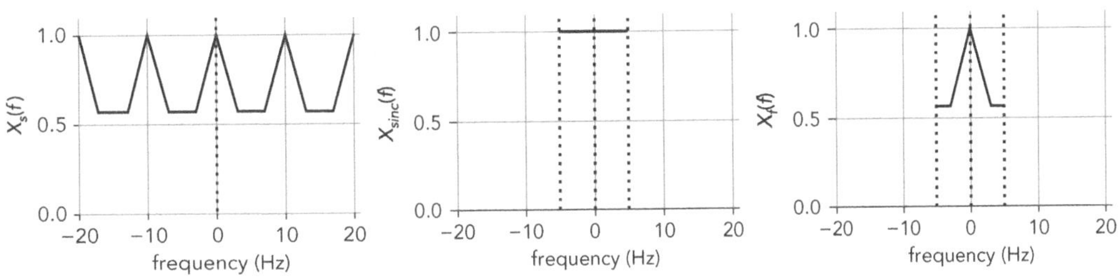

Figure 4.13 Aliasing in the reconstructed (real-valued) spectrum $X_f(f)$ (*right*), the product of $X_f(s)$ (*left*) and $X_{sinc}(f)$ (*middle*).

Following our earlier reasoning, when we discussing sampling, this can be done by multiplying the infinite-time sinusoid by another infinite time signal that is 1 over the *on* time and zero elsewhere. This is called a *rectangular window* (or rectangular function, but we prefer the other terminology).

We should know what to expect now, since a time-domain multiplication is equivalent to a frequency-domain convolution. The missing piece is the spectrum of the rectangular window, which we can easily determine with the FT. For one such signal $\text{rect}(t)$ defined over $-T/2$ and $T/2$, where T is the length of the finite-time sinusoid, we have

$$\begin{aligned} X(f) &= \int_{-\infty}^{\infty} \text{rect}(t) e^{-j2\pi ft} dt \\ &= \int_{-\frac{T}{2}}^{\frac{T}{2}} e^{-j2\pi ft} dt = \frac{e^{-j\pi fT} - e^{j\pi fT}}{-2j\pi f} \\ &= \frac{\sin(\pi Tf)}{\pi f}, \end{aligned} \tag{4.43}$$

or $T\text{sinc}(Tf)$. We should have expected this result, since the spectrum of the $\text{sinc}(t)$ function is a rectangular window.[16] These two functions form another important transform pair, and both are even and purely real functions. Note, in particular, that the width of the main lobe of the sinc() function is determined by the duration of the *on* period. A shorter duration implies a wider spectrum, whereas a longer period makes it narrower. At infinity, it shrinks to a Dirac δ. An infinitely wide spectrum with unity amplitude is also equivalent to a δ pulse in the time domain, as we have seen before. Also, since the sinc() function is scaled by the period length, as the window gets shorter, the overall amplitude of the spectrum falls (as there is less energy in the signal, which is mostly composed of zeros).

If we set the window length T to $1/f_0$, it is possible to have a finite-time signal that captures a single cycle of a sinusoid. Therefore the continuous-frequency spectrum of a single period T_0 of a real sinusoid with frequency $f_0 = 1/T_0$ is the convolution of $\text{sinc}(f/f_0)$ and the sum of two Dirac δ functions, $\delta(f + f_0) + \delta(f - f_0)$. Applying the continuous time convolution expression (Eq. 4.34), we have

$$X_{\cos}(f) * X_{\text{sinc}}(f) = \frac{T}{2}\left[\text{sinc}\left(\frac{f}{f_0} + 1\right) + \text{sinc}\left(\frac{f}{f_0} - 1\right)\right] \tag{4.44}$$

for a $\cos(2\pi f_0 t)$ waveform. This depicted in Fig. 4.14, where we show first the spectrum of a 5 Hz cosine, as we obtained before, then the spectrum of a rectangular window, and finally the convolution of the two spectra, which represents the FT of a signal consisting of a single cosine cycle (that is, it is non-repeating, and zero outside of its single period). Since both spectra are real-valued, the result of the convolution is a purely real spectrum.

It might be surprising at first that we are getting this result, since we have learned that the Fourier transform of a cosine wave is non-zero at two frequencies and zero elsewhere. What might have been missed is that we are no longer representing a 5 Hz cosine wave. We are representing a wave that for a brief period (1/5 s) contains one cycle of such a function, but is mostly zero. Its continuous-time spectrum contains many more frequencies, and its bandwidth is much wider (infinite, actually). The equivalent case for

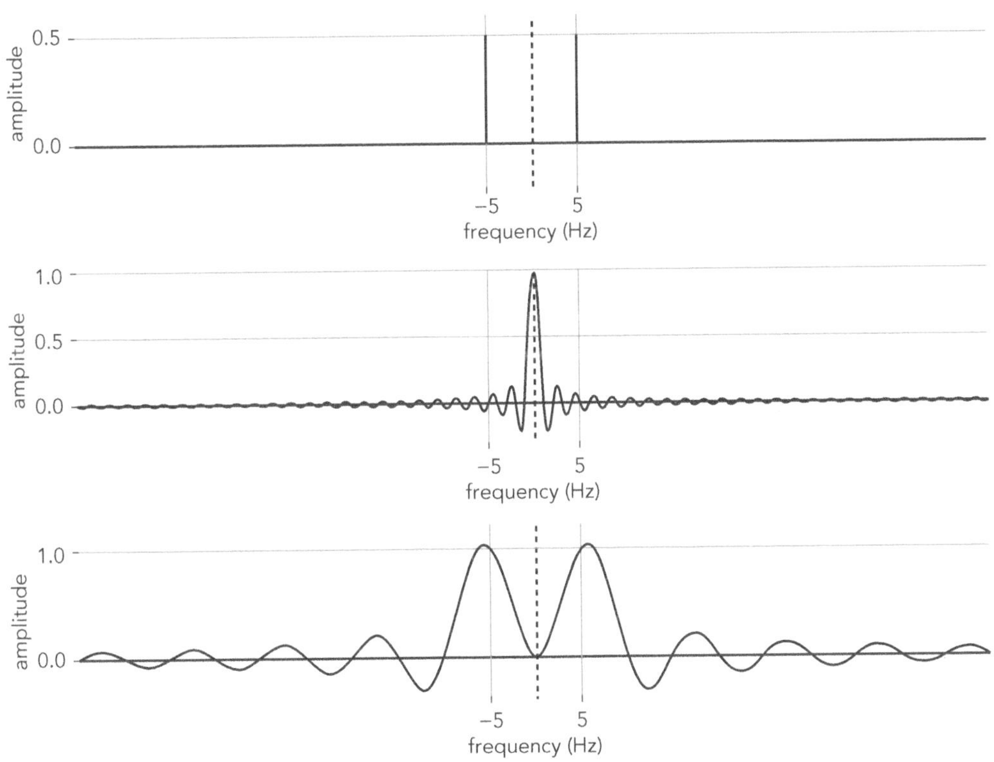

Figure 4.14
The spectrum of a finite-time cosine wave (*bottom*) as the convolution of the spectra of a 5 Hz cosine wave (top) and the sinc($f/5$) function (*middle*). These plots are scaled by $1/T$ for better visualization.

a sine wave $\sin(2\pi f_0 t)$ is very similar, and its spectrum is defined as

$$X_{\sin}(f) * X_{\text{sinc}}(f) = \frac{T}{2}\left[\text{sinc}\left(\frac{f}{f_0}+1\right) - \text{sinc}\left(\frac{f}{f_0}-1\right)\right]. \tag{4.45}$$

The example of a signal composed of single 5 Hz cycle of a sine wave is shown in Fig. 4.15. Note that in this case, the spectrum is purely imaginary.

These ideas allow us to figure out the continuous-time spectrum of finite waveforms for which we have a Fourier series description. Eqs. 4.44 and 4.45 give us a method to obtain these spectra, from where we can see that a harmonic n of frequency $f_n = nf_0$ is represented by

$$X_n(f) = \frac{\left[\text{sinc}\left(\frac{f+f_n}{f_0}\right) + \text{sinc}\left(\frac{f-f_n}{f_0}\right)\right] + j\left[\text{sinc}\left(\frac{f+f_n}{f_0}\right) - \text{sinc}\left(\frac{f-f_n}{f_0}\right)\right]}{2}. \tag{4.46}$$

For instance, using the infinite-time square wave spectrum, which we have derived before, we have

$$X_{sq}(f) = \frac{j}{2}\sum_{n=1}^{\infty}\frac{1}{2n-1}\left[\text{sinc}\left(\frac{f}{(2n-1)f_0}+2n-1\right) - \text{sinc}\left(\frac{f}{(2n-1)f_0}-2n+1\right)\right], \tag{4.47}$$

excluding any scaling to normalize the peak amplitude of the signal. The resulting square wave spectrum is shown in Fig. 4.16.

This example demonstrates that the process of extracting the Fourier series coefficients creates a sampled spectrum. The spacing samples are spaced f_0 Hz, which gives the sampling *period*, the inverse of the time duration of the rectangular window. Since the waveform produced by a given sampled spectrum is periodic, we can slice it just as we did

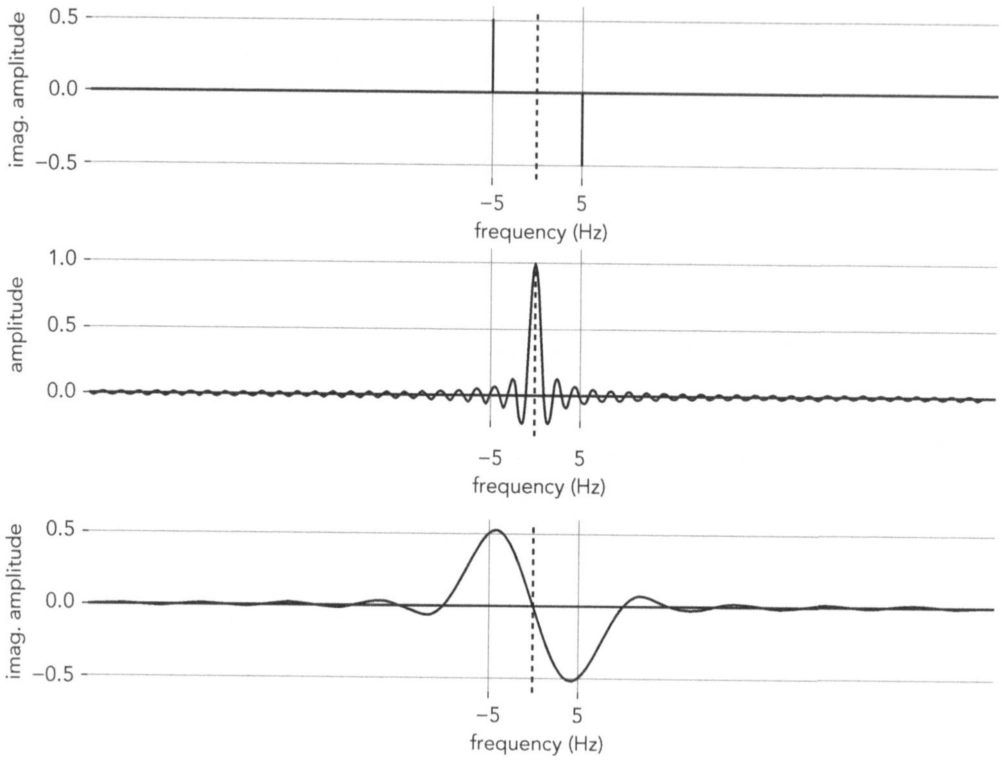

Figure 4.15 The spectrum of a finite-time sine wave (*bottom*) as the convolution of the spectra of a 5 Hz sine wave (*top*) and the sinc($f/5$) function (*middle*).

with a continuous spectrum when reconstructing a sampled waveform. The rest of the signal is just made up of repeated cycles.

4.4.2 Hard-Sync Waveforms

From this perspective, we can determine the discrete sequence of Fourier series coefficients as a result of sampling a spectrum, in a way that is similar to the process used to get a discrete-time waveform,

$$S(n) = \text{III}_{f0}X(f) \tag{4.48}$$

where $S(n)$ is the sampled spectrum, III_{f_0} is a Dirac comb with pulses spaced by f_0 Hz, and $X(f)$ is a continuous-frequency spectrum. If the comb coincides with points at which the spectrum amplitude is zero, no component is present at that frequency. The sampled spectrum obtained in this way then refers to periodic waveforms that are infinite in time.

For example, the cosine (or sine) wave magnitude spectrum dips to zero at every multiple of its fundamental frequency. Therefore if we sample it using a comb whose pulses are separated by that amount, we will have only a single component (Fig. 4.17).[17] This is equivalent to using a rectangular window that holds one cycle of the waveform. It is no different from what we have seen before, but we are looking at it from a sampling perspective.

Taking this a little further, what happens if the spacing of the samples does not line up with the dips in the spectrum? We have seen such an example, the square wave spectrum, where an infinite number of components is present. Another way to provoke this is to extend the rectangular window a little further and capture a little more of the

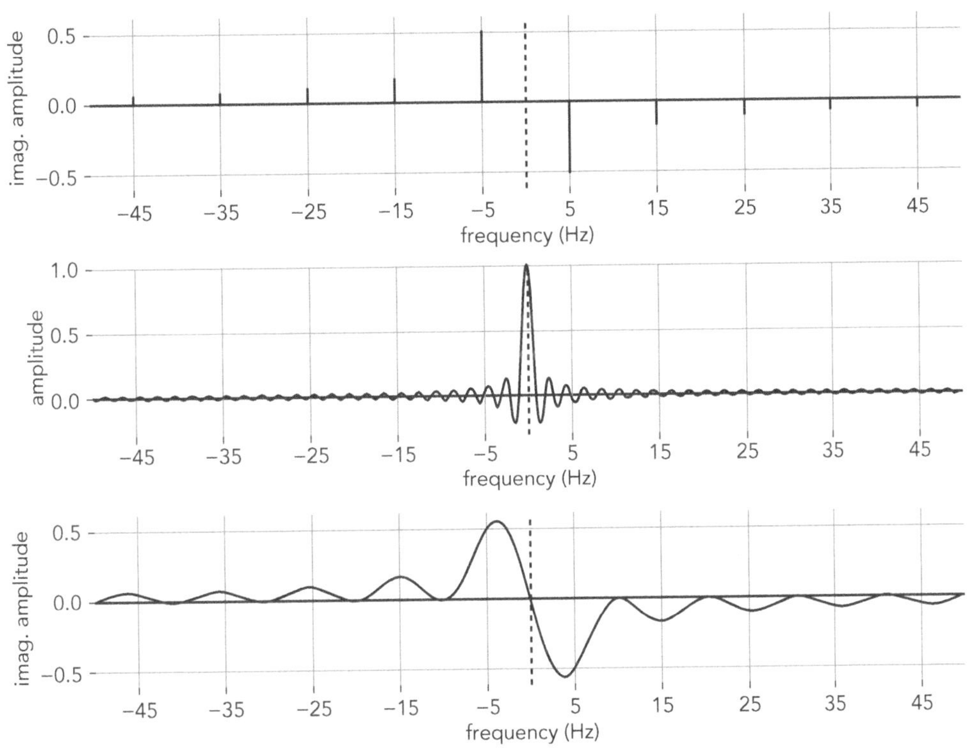

Figure 4.16 The spectrum of a finite-time square wave (*bottom*) as the convolution of the spectra of a 5 Hz square wave (*top*) and the sinc($f/5$) function (*middle*).

time-domain signal. By doing so, we will be decreasing the spacing between the samples. Alternatively, we could keep the size of the window and change the frequency of the waveform. This is a well-known synthesis effect known as *hard sync*. It is realized by resetting the start of a waveform in synchronization with a master period, which determines the length of the cycle. If the two are the same or integer multiples, the signal will not experience any timbral change. However, if the synced wave has a fractional number of cycles within one master period, we will observe a marked spectral modification.

Taking the cosine example of Fig. 4.17, let's examine its amplitude spectrum under hard sync conditions. We start by increasing its frequency slightly, keeping the window with the same duration T, which is equivalent to a master clock period $1/T$. This results in the hard sync waveform depicted in Fig. 4.18. In this case the master frequency is 4 Hz and we set the cosine at 4.6 Hz. The resulting continuous-time spectrum and its sampled version is shown in Fig. 4.19, where we also plot the corresponding Dirac comb.

Note that, as the process involves the scaling of the cosine frequency and the preservation of the window duration, it produces two related effects:

- The sinc() pulse is held at the same original width, as the window is the same size, but it is now centered at a different f_0. This changes the shape of the continuous-frequency spectrum. Comparing Figs. 4.17 and 4.19 should make this clear.
- The waveform is no longer a cosine, therefore it is not purely real anymore and components have their phase shifted accordingly.

As a result, the magnitude spectrum function does not necessarily have zero values at multiples of the fundamental frequency f_0 (the master clock frequency). Therefore, the waveform has multiple components. As we increase the frequency fractionally between

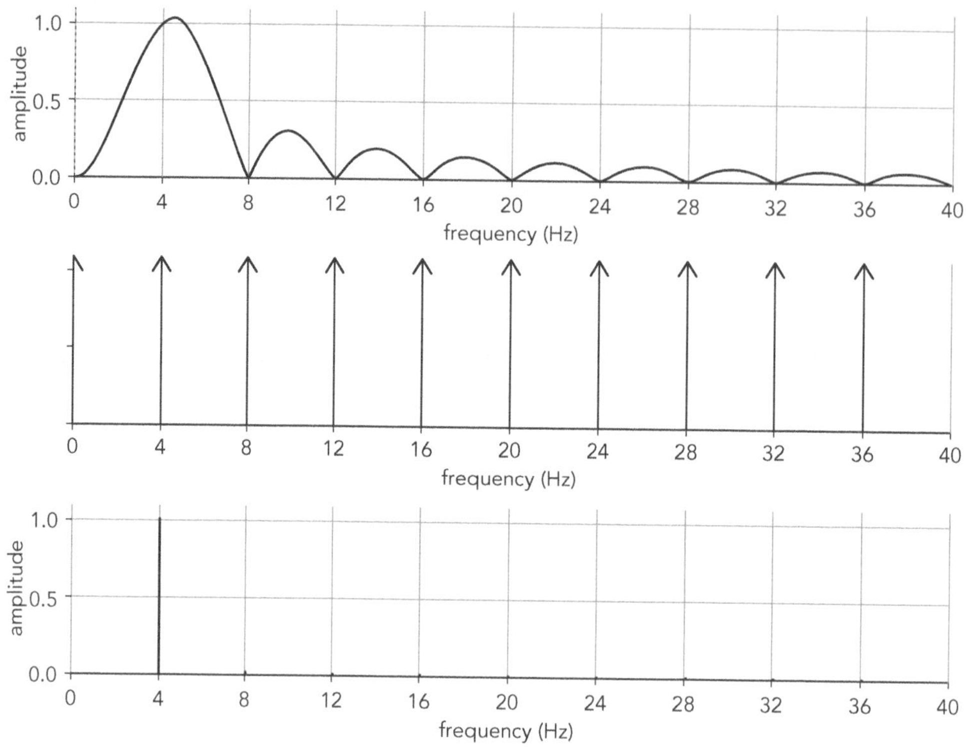

Figure 4.17 The amplitude spectrum of a finite-time cosine wave (*top*), sampled by a Dirac comb with pulses spaced at 4 Hz (*middle*), and the resulting discrete-frequency single-partial amplitude (*bottom*).

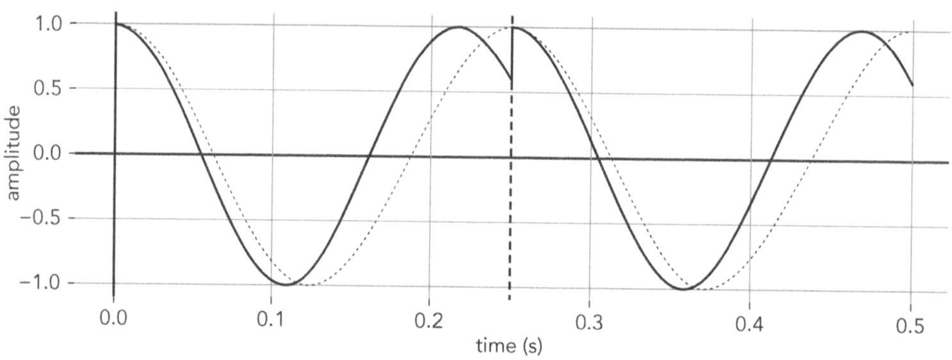

Figure 4.18 A hard-sync cosine (*solid line*) at 4.6 Hz, whose master clock period is 1/4 s (*dots*). Note that the resulting signal has a 4 Hz fundamental frequency.

multiples of the clock rate, we will see more components appearing. At halfway between 1 and 2 times, we will have the most number of partials (Fig. 4.20). As it gets closer to twice the fundamental, it starts to becomes very much like a cosine waveform again.

4.5 Classic Waveforms

The Fourier transform, together with the Dirac-δ sampling mechanism, provides a complete method to obtain the spectra of various waveforms. Since it is assumed that these are periodic, they will have a static spectrum that may be determined by defining a convenient value for f_0, for instance, 1 Hz, and then applying Eq. 4.48. We can start by looking at the classic waveforms employed in analogue synthesizers, such as the square, sawtooth, triangle, and pulse waves. In fact, as the square spectrum has already been

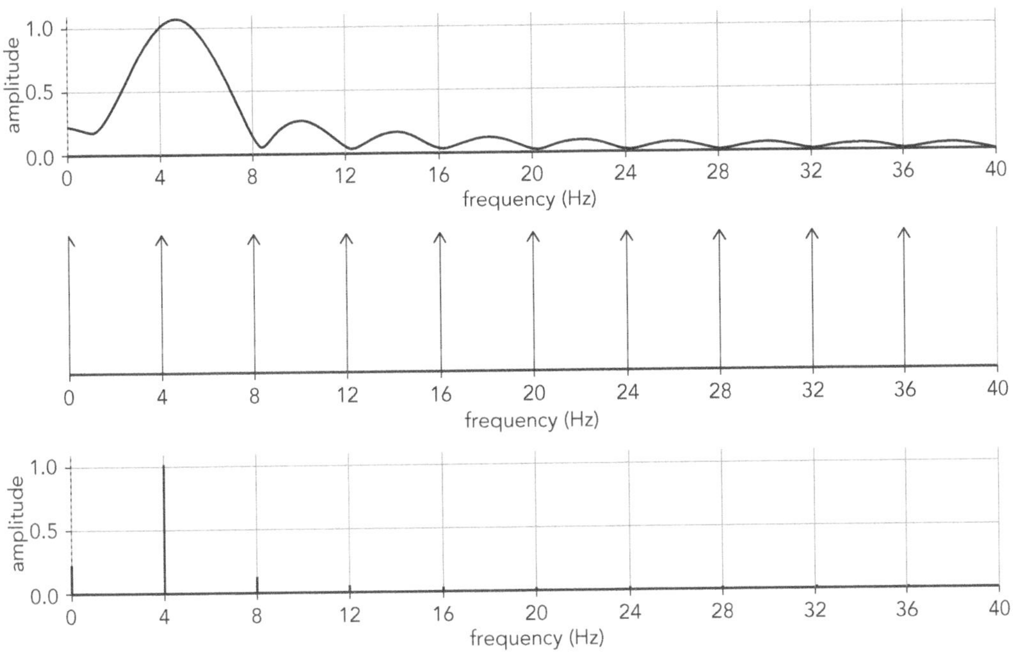

Figure 4.19
The amplitude spectrum of a hard-sync waveform, based on a 4.6 Hz cosine with an incomplete cycle (*top*), sampled by a Dirac comb with pulses spaced at 4 Hz (*middle*), and the resulting discrete-frequency partial amplitudes (*bottom*).

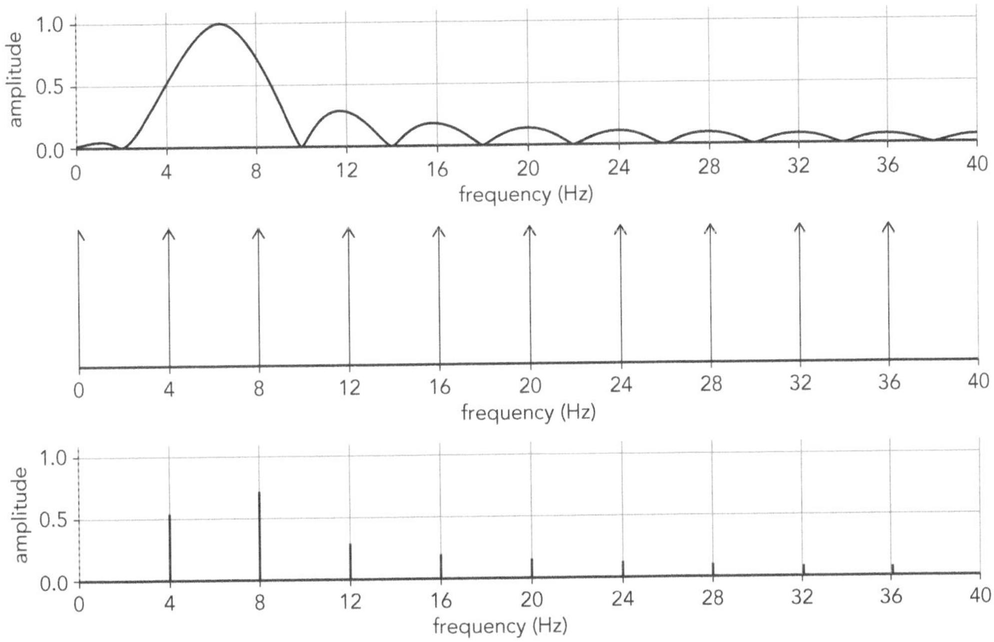

Figure 4.20
The amplitude spectrum of a hard-sync waveform, based on a 6 Hz cosine with an incomplete cycle (*top*), sampled by a Dirac comb with pulses spaced at 4 Hz (*middle*), and the resulting discrete-frequency partial amplitudes (*bottom*).

discussed before, we can use it to sketch out the process. The continuous-frequency spectrum of a square wave was given early on in Eq. 4.10 (and in Fig. 4.16, bottom panel, where it describes a single period non-repeating of a square wave). We only need to put it in the terms of a two-sided complex spectrum, define f_0 as 1 Hz, and sample it,

$$S_{sq}(n) = \text{III}_1(f)\left(j\left[\frac{\cos(\pi f) - 1}{\pi f}\right]\right) = \begin{cases} \frac{2j}{\pi n}, & |n| \quad odd \\ 0, & |n| \quad even \end{cases} \tag{4.49}$$

The resulting discrete spectrum is the one shown earlier in Fig. 4.6, whose continuous-time form is a sum of scaled and shifted Dirac functions. It refers to an infinitely repeating square-wave signal. Also note that we have based our description on the odd square wave, but a similar spectrum can also be found for the even case. In fact, the difference between the two waves is a 1/4 cycle circular rotation. A similar method can be applied to any spectrum that we can define as a 1 Hz periodic waveform. We can now look at the other classic waves.

4.5.1 The Sawtooth

The sawtooth wave is defined by a rising or falling straight line; therefore, its discontinuous point is found at the beginning and end of the cycle. It is more common to define this waveform as a descending function starting at 0, so we will look at this case. The simplest way to define a sawtooth is to extract a single cycle over the period $[0,1]$ from the function $f(t) = -2t$ by slicing this with a rectangular window $\text{rect}(t-1/2)$ that is 1 in the interval of interest and zero elsewhere. The sawtooth is an odd function; therefore, it is possible to assume all cosine components are zero. The spectrum of such a wave is then

$$\begin{aligned} X_{saw}(f) &= j\int_{-\infty}^{\infty} -2t\,\text{rect}(t-\frac{1}{2})[-\sin(2\pi ft)]dt = 2j\int_{0}^{1} t\sin(2\pi ft)dt \\ &= \frac{j}{\pi}\left[\frac{\cos(2\pi f)}{f} - \frac{\sin(2\pi f)}{2\pi f^2}\right]. \end{aligned} \tag{4.50}$$

from which we can obtain a discrete set of partials as

$$S_{saw}(n) = \text{III}_1(f)X_{saw} = \frac{j}{\pi n} \tag{4.51}$$

The plot of the sawtooth spectrum is shown in Fig. 4.21. Note that both square and sawtooth wave have $1/f$ amplitude spectra, which determines a shared tonal quality, although the lack of even harmonics in the former also gives it special characteristic, which have been described as "hollow". There has been some experimental indication that a $1/f$ spectral curve is also perceived as pleasant or naturally appealing. This spectral slope is also indicated as a -6dB/octave contour, as the amplitude falls by half every time the frequency doubles. Another interesting aspect to note is that since the sawtooth we defined is centered at $T = 1/2$, the positive side of the spectrum is now positive-valued and the negative side, negative. That is, we have imposed a circular phase shift on the usual representation, centered at 0.

4.5.2 Triangle Wave

Another classic waveform of analogue synthesis is the triangle wave. As the name indicates, it is made up of a linear rise followed by a fall, both symmetrical to each other about the center of the waveform. We can define it as a piecewise linear function in the interval $[-1/2, 1/2]$, with the negative side as $f(t) = [1+4t]$ and the positive side as $f(t) = [1-4t]$ (sliced off, as before, by a rectangular window). It is therefore an even function, and so we know that its spectrum is purely real. To get its spectrum, we can apply a similar method the one used for the square wave, where we integrate each half

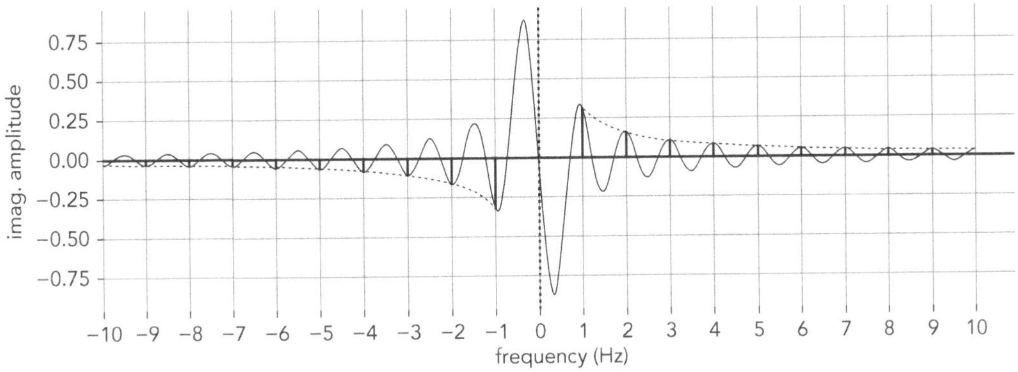

Figure 4.21 The spectrum of the sawtooth wave (*continuous line*), sampled at 1 Hz spacing (*stem plot*), and a $1/f$ curve scaled to the maximum component amplitude (*dots*), showing the −6 dB/octave spectral slope of the waveform.

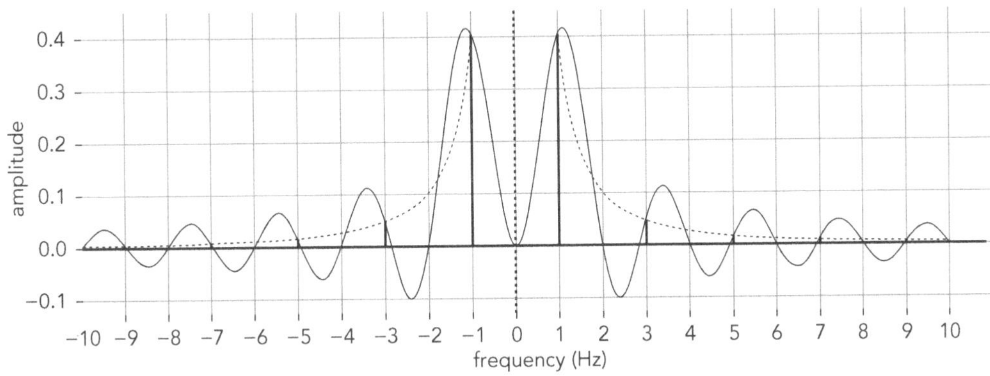

Figure 4.22 The spectrum of the triangle wave (*continuous line*), sampled at 1 Hz spacing (*stem plot*), and a $1/f^2$ curve scaled to the maximum component amplitude (dots).

separately, giving

$$\begin{aligned} X_{tri}(f) &= \int_{-\frac{1}{2}}^{0} [1+4t]\cos(2\pi ft)dt + \int_{0}^{\frac{1}{2}} [1-4t]\cos(2\pi ft)dt \\ &= 2\left[\frac{1-\cos(\pi f)}{\pi^2 f^2}\right] - \frac{\sin(\pi f)}{\pi f}. \end{aligned} \tag{4.52}$$

The amplitude of each partial can then be defined as

$$S_{tri}(n) = \mathrm{III}_1(f)X_{tri} = \begin{cases} \frac{4}{\pi n^2}, & |n| \quad odd \\ 0, & |n| \quad even \end{cases} \tag{4.53}$$

There are some interesting things to note about this waveform. Firstly, as with the square wave, all even harmonics (plus DC) are missing. Secondly, its spectral slope is steeper than in the cases we have seen before, a $1/f^2$ curve. This is shown in Fig 4.22, where in addition to the continuous and discrete 1 Hz spectra, we fit a $1/f^2$ curve to demonstrate the spectral decay.

Thus, the triangle can be considered a lowpass-filtered square wave. In fact, it can be generated from a square shape by integration, since the integral of a constant function is a linear function. In this case, the two halves of the square wave are turned into the rising and falling that characterizes the triangle shape. The integration therefore has a lowpass filtering effect with a 6dB/octave spectral magnitude slope.

4.5.3 Pulses

Finally we come to pulses, which may be the simplest of the classic waveforms in terms of their spectra. In principle, we have already discussed it, since we have seen early on that the frequency domain of an infinitesimally short pulse is unity. Therefore, if we repeat it periodically, we will have an impulse train in the spectrum of infinitesimally small pulses (or δ functions), which is a Dirac comb. However, in real-world applications, pulses will have some small duration, so they are not infinitesimally short. Still, if these are short enough, the amplitude of each partial will tend toward 1, so we have a generally flat spectrum.

As they get longer, however, the pulse spectrum will start to present a certain lowpass slope. The spectra of different types of pulses can be determined if we assume their shape to have a well-known form. It is typically defined that the pulse wave is rectangular, which simplifies the description. We know what that means in terms of spectra: some variation of the sinc() shape. The length of the pulse is therefore very important to characterize its timbral quality, and typically it can be adjusted or modulated (pulsewidth modulation, or PWM).

The pulse width is commonly described in terms of its *duty cycle*, which is the percentage of the period in which the pulse is on. A 50%-duty cycle pulse wave is therefore a DC-offset square wave, since it is 1 for half the period and 0 for the other half. We can describe this by a rectangular window that has a period of $1/2T$, where T is the waveform period. The spectrum of such a wave can be easily shown to be

$$X_{P50}(f) = \frac{\text{sinc}(f/2)}{2} \tag{4.54}$$

which represents a purely real, DC-shifted, scaled, square wave. More generally, we can put the spectra of pulses in terms of their duty cycle d, $0 < d \leq 50$, as

$$X_{pulse}(f) = \frac{d\ \text{sinc}(\frac{d}{100}f)}{100}. \tag{4.55}$$

As usual, we can sample this spectrum to get the values at the desired evenly spaced points.[18]

Plots of the spectra of waveforms with differing duty cycles are shown in Fig 4.23. Note how the spectrum is fairly wide and approaches a flat curve as the pulse becomes very narrow. Also in this case, due to the fact that each wave period is mostly made up of 0s, the amplitudes of each component is also reduced significantly. Even harmonics disappear with the 50% duty cycle as expected, and we can see that this is the result of the particular sinc() function shape, which has zeros at the even sample points. This example also demonstrates the spectral shaping possibilities of pulsewidth modulation, which is commonly used in analogue synthesis.

Note that we can generally obtain a sawtooth wave by integrating a bipolar pulse (sine rather than cosine-based), in a similar way to how we treated a square wave to produce a triangle. It is possible to see that as an ideal pulse will have an $X(f) = 1$ magnitude spectrum, multiplying this by a $1/f$ curve and shifting the phases accordingly will produce a sawtooth shape (which can then be DC offset and scaled).

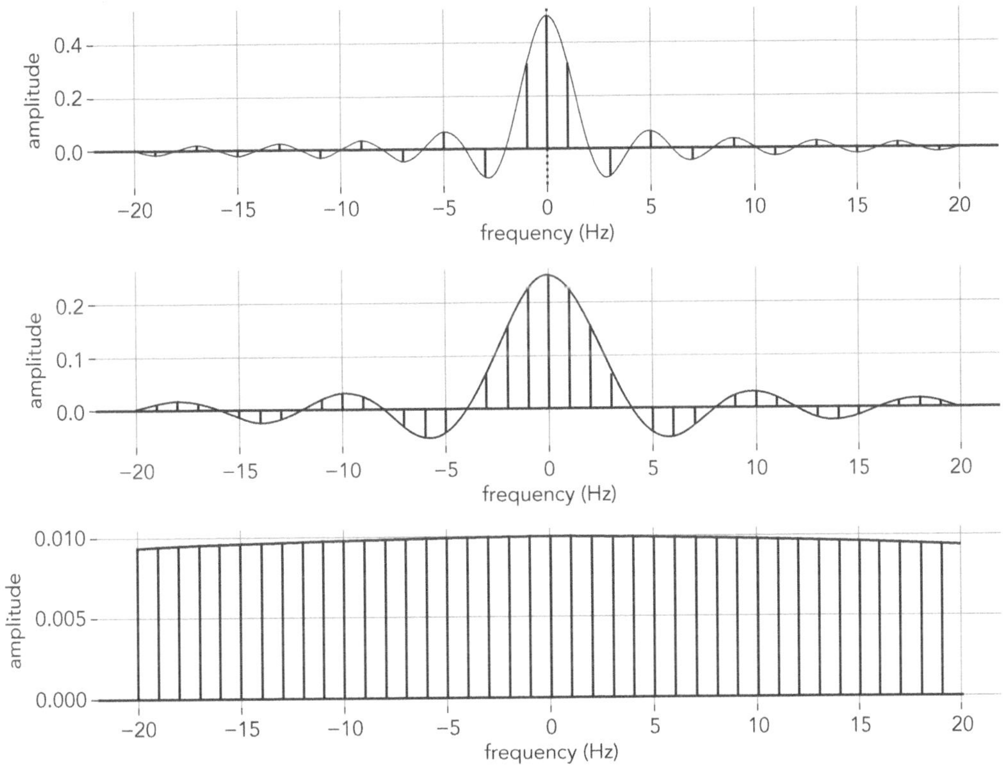

Figure 4.23 Three plots of the spectra of pulse waves with differing duty cycles: 50% (*top*), 25% (*middle*), and 1% (*bottom*).

4.5.4 Additive Synthesis

We can now introduce some practical synthesis applications of these ideas. Using Csound, we can translate these Fourier series recipes through additive synthesis routines that produce single-cycle fixed waveforms in function tables. They can then be read by oscillators. For example, the code

```
// build array of coefficients
ih[] init 10
ipi = $M_PI
ik = 1
while ik < 10 do
 // sawtooth spectrum
 ih[ik-1] = (1/ipi)*(cos(2*ipi*ik)/ik -
            sin(2*ipi*ik)/(2*ipi*ik*ik))
 ik += 1
od

// wavetable generation
isaw = ftgen(0,0,16384,10,ih)

instr 1
 kenv = linenr(p4,0.01,0.1,0.01)
 asig = oscili(kenv,p5,p6)
 out(asig)
```

```
endin

schedule(1,0,2,ampdbfs(-6),cpsmidinn(60),isaw)
```

demonstrates a very simple instrument that uses a function table containing a single period of a waveform to play a note of a given frequency and amplitude. Note that if we pass a function table to an oscillator (as the third argument), it uses it instead of a sine wave.

In this example, we use a function table generator called by `ftgen` to create bandlimited sawtooth. It takes in an optional table number (or zero), a generation time (0 for *now*), and a table size in samples (16384). The rest of the arguments are the generator routine number (10 for sine-based Fourier series) and an unlimited number of Fourier series (sine) coefficients in an array. When creating the table we must have in mind the highest fundamental frequency we will be using so that we do not provoke any aliasing. The highest component of a synthesized signal is of course Nf_0, where N is the topmost harmonic number.

This application is of course quite trivial, and it produces the type of sound captured by Helmholtz's concept of musical tone, characterized by a periodic waveform with a static spectrum. We do not need to be limited by this, as additive synthesis can do much more than only produce static waveforms. However, in order to develop a more complete method to produce time-varying spectra, we need a more developed theoretical foundation, which will be provided by the next two chapters.

4.6 The Fourier Spectrum

The Fourier transform is based on a far-ranging theory, which has many applications in several fields. In this chapter, we attempted to cut a path through it, which emphasized its use in spectral audio processing. Some particularly interesting principles were developed. We saw early on, and then confirmed later, that integration of a time-domain signal acts as a lowpass filter. This is a powerful concept that can be applied directly to understanding the operation of Fourier analysis. The process of isolating a partial in the continuous spectrum was likened to tuning into an AM radio frequency: using a sinusoid as a multiplier, a heterodyned signal is produced, to which we then apply a lowpass filter (via integration). This picks up the exact contribution of pure frequency to the overall waveform.

We have also seen that we can hop the frequency along at given steps, producing the coefficients of a Fourier series, or we can slide it continuously in infinitesimally small frequency increments, deriving a continuous spectrum. The ideas developed with the help of the FT were shown to have implications for sampled signals. Discrete-time spectra was confirmed to be periodic in frequency, which was not the general case for continuous-time signals. Conversely, we could observe that discrete-frequency spectra produced periodic waveforms.

As the theory of the Fourier transform is fundamentally based on assuming time and frequency to stretch infinitely, the question of finite-time signals was raised, since this is what we observe in practice. We were able to show the effect of switching signals on and off as a multiplication by a rectangular window, which implies a convolution in the spectral domain. A short signal has a wide spectrum and a long signal has a narrow one. In complement, we have also seen that the infinite frequency spectrum of a continuous-time signal becomes bandlimited if we sample its waveform and then reconstruct it. Since this involves a convolution in the time domain, the result is a multiplication in the frequency domain.

The original Fourier theorem was an important stepping stone in the understanding of the spectrum. However, in the form that it was first given, it tended to skew the discussion toward stable tones, static in time, which could be described by periodic waveforms. This gave rise to a classification of sounds into musical and (by exclusion) non-musical, with the justification that the latter somehow lacked order or form. This was Helmholtz's view, which, given his towering influence in the scientific study of acoustics, prevailed for many years (over a century before these ideas were more or less put to rest). This abusive use of Fourier's principles led to various kinds of criticism, some of which were perhaps just as misguided as Helmholtz's narrow description of the spectra of musical sounds.

The more modern version of the theory, as provided by the Fourier transform, however, is less conducive to a set interpretation based on periodicity. However, we still have to consider the fundamental basis given by the uncertainty principle as perhaps a limiting factor. The fact is that the Fourier transform itself does not say much about time, except that it should be infinite. Misreading the results of its application may be problematic. However, there is a lot to gain out of a proper use of the theory. Later we will see how we can reintroduce the concept of time in the analysis/synthesis methods arising from the Fourier theorem. The approach then becomes one of balancing out how much frequency response we can get out of a given time resolution.

In this chapter, we have discovered a number of important ideas that can be added to the provisional definition of spectrum we have developed in Chapter 1. They may be summarized as

> The spectrum as a complex-valued continuous function of a real variable, frequency, can be given by the Fourier transform, which is defined for an infinitely long waveform. An audio signal, as a real-valued function, has a two-sided sided spectrum that is Hermitian symmetric. The discrete-time waveform spectrum is periodic in frequency, and a discrete spectrum implies a periodic waveform.

From the Fourier transform, defined for continuous-time, continuous-frequency, we can have the Fourier series, which is a continuous-time discrete-frequency resynthesis formula. This is used to create a single cycle of a waveform, which is assumed to be periodic (and infinitely long). As we have noted a number of times before, it is not possible to process continuous time signals with a digital computer. Therefore we will need to move on to a discrete representation in both time and frequency. In fact, this has been already introduced and discussed separately, in terms of sampling, and the Fourier series, as discrete forms of time and frequency functions, respectively. The next step is to put these two together, and to this we will now turn.

Notes

1 The variable x is replaced by t since we will now mostly be speaking of time-domain functions.

2 While we should formally speak of phase angles here, since we will use the term frequency all the time, it is simpler to move on to use this more directly (and loosely). However, it is important to bear in mind the precise meanings of these terms as discussed in Chapter 3.

3 The well-known identities $\cos^2(\omega) = \frac{1+\cos(2\omega)}{2}$ and $\sin^2(\omega) = \frac{1-\cos(2\omega)}{2}$ are used to determine that.

4 Another way to describe it is to say that the sines and cosines of different frequencies form an *orthogonal* basis, from a geometric interpretation, such that they are at right angles to each other. See Steiglitz (1996 177–179) for a simple worked-out proof of this.

5 For now, we start by using subscript notation to indicate sequences; later on we will adopt functional notation to simplify the equations.

6 In particular, this form is used to add two spectra, as sums and differences treat real and imaginary parts separately and therefore are easier to compute in this way.

7 We should bear in mind that to go from rectangular to polar coordinates, starting with $a + jb$, we have

$$A = \sqrt{a^2 + b^2} \qquad \text{and} \qquad \omega = \tan^{-1}\left(\frac{b}{a}\right).$$

To go in the opposite direction we use $a = A\cos(\omega)$ and $b = A\sin(\omega)$.

8 In this form, the two terms are interchangeable. However, we will, on many occasions, employ the word *amplitude* instead, for two reasons: firstly, it is more commonly used in other audio and music applications, and therefore more directly meaningful to us; secondly, amplitude is a more generic descriptor, as it does not have the restriction of being non-negative, so we can continue to use it in cases where a bipolar value may arise from the equations. It will continue to have the same meaning, but it is not possible to exchange it by magnitude in those situations.

9 This can be obtained by the FT, but we also know that this signal has only one component, which is a sine wave with frequency f_0. Since $X(f)$ refers to the full bipolar spectrum, this component also shows up at $-f_0$.

10 A note about graphs: since we are now dealing with complex values, it can be tricky to present them in two-dimensional plots. So we will do this by keeping real and imaginary parts separate. To make it clear, we will use *imaginary amplitude* as a label when referring to the sine components of a spectrum. Conversely, if the real part is shown, then we will just use *amplitude* if the spectrum is purely real, marking it as *real* only if a distinction needs to be made. Alternatively, when suitable, we may choose to plot the magnitude (amplitude) spectrum, which again is real-valued and can be also labeled as *amplitude*, as in these cases the terms are equivalent.

11 We may wonder also why is integration a lowpass operation? One way to see this is that this makes functions smoother: integrate a function with a kink in it and it gets ironed out. Penrose talks about this property of integrals when discussing continuous functions (Penrose, 2005, Chapter 6). Since high-frequency signals are

related to fast fluctuations in a time-domain signal, when these are smoothed out, we have an equivalent operation to a type of lowpass filtering.

12 To make matters simple, we bundle up $2\pi f$ as ω. This may be called a *radian* frequency, as it includes the 2π factor.

13 We use the term *function* here (and elsewhere in this volume) in a loose way. As noted before, the Dirac δ is not a function in the usual sense, but it works for us here to think of it as such.

14 A normalization factor is needed to bring the peak waveform to 1, but its omission does not change the overall spectral distribution.

15 The amplitude spectrum of an arbitrary signal could have been equally considered, but it is simpler to visualise a real function.

16 Note that the spectrum of the sinc() function was given in a slightly different form, where the points $f = |1/2|$ where given as $1/2$, rather than 1 as implied here. However, this follows convention and we do not need in general to pay particular attention at the values of the rectangular window at these points, as discussed in Bracewell (2000, 57).

17 Here we show only the positive side of the spectrum. This is because we are plotting the amplitude spectrum, which is symmetrical at 0 Hz, and non-negative. The amplitude is also doubled, reflecting the contribution of both positive and negative components.

18 When implementing this in software, we can just get the values of $X_{pulse}(f)$ for integer values of f in order to obtain a single-cycle Fourier series recipe.

5

Discrete Time, Discrete Frequency

106

This chapter introduces a set of spectral processing tools that can be used to process audio in the computer. From here onward, we will move away from continuous signals and embrace fully their discrete counterparts. As has been noted a number of times before, if the limits of a discrete representation are observed, then it can be manipulated and interpreted in such a way that it transparently portrays the underlying reality of the continuum. Therefore, the theoretical foundation laid out in previous chapters applies to all of the discussion to this and the following topics of this volume, whether or not we are speaking of discrete or discrete frequency.

5.1 The Discrete Fourier Transform

As we noted at the end of Chapter 4, a discrete-frequency form of spectral analysis/synthesis is represented by the original Fourier series and the method to calculate its coefficients. We saw how it is possible to reinterpret this as a form of sampling in the frequency domain. In addition, in Chapter 3, the sampling theorem was introduced, providing a way of discretizing an arbitrary waveform. Now we can marry the two ideas and provide a fully discrete-time, discrete-frequency version of the continuous-time, continuous-frequency Fourier transform, the *discrete Fourier transform* (DFT). There are a couple of major modifications required to transform the FT into the DFT, namely

- To put it in terms of a sampled frequency representation, we will need to make it finite in time.
- Since we are processing discrete-time samples, the integral becomes an ordinary summation.

In other words, we will slice a signal in time (with a rectangular window), process the waveform signal sample-by-sample, and produce a finite set of spectral samples. How many of them? As many as we have taken in from the time-domain signal. The DFT of an N-sample waveform produces an N-sample spectrum. As we will see, moving from the infinity of the continuous makes many steps of the operation much more straightforward.

With these principles in mind, we can introduce the forward DFT equation as

$$X(k) = \frac{1}{N}\sum_{n=0}^{N-1} x(n)e^{-j2\pi kn/N} \qquad 0 \le k < N, \qquad (5.1)$$

Spectral Music Design. Victor Lazzarini, Oxford University Press. © Oxford University Press 2021.
DOI: 10.1093/oso/9780197524015.003.0005

where $x(n)$ is the input waveform, $X(k)$ its spectrum,[1] and N is the number of samples, and n and k are the time and frequency indices, respectively. In this definition, we have included the $1/N$ scaling, which is useful to give results in a normalized form. In some applications, we might find that it is better to leave this out of the forward transform and instead place it in the inverse formula.

It may be useful to look at the DFT in terms of real outputs, so we can split Eq. 5.1 into two parts,

$$X_{re}(k) = \frac{1}{N}\sum_{n=0}^{N-1} x(n)\cos(2\pi kn/N), \tag{5.2}$$

and

$$X_{im}(k) = \frac{1}{N}\sum_{n=0}^{N-1} x(n)\sin(-2\pi kn/N). \tag{5.3}$$

We can see that these equations look very similar to the Fourier series coefficient formulae. The main difference is that we do not have an integral, but a sum, since the signal is discrete in time. Therefore, it is expected that the mechanism we discovered for that case is also behind the operation of the DFT. That is, we are using sinusoids to pick out components in the signal. However, here, since we have a finite-length waveform, the set of sinusoids used is not infinite, but limited to the length of the analysis window (N). This is actually enough to characterize the spectrum of a finite time-domain signal, although it might not have enough frequency resolution, depending on the application (in which case we need to assess what value of N is best suited).

An indication that the analysis works whatever length is chosen is that we can recompose the time-domain signal with the inverse DFT,

$$x(n) = \sum_{n=0}^{N-1} X(k)e^{j2\pi kn/N} \qquad 0 \leq n < N, \tag{5.4}$$

in a transparent manner.[2] Also note that the inverse DFT is very similar to the trigonometric series of the original Fourier theorem. Again, the differences are mainly due to having a bandlimited spectrum, so the sum does not extend to infinity. Another thing to note about this and the forward transform is that they only differ by the sign of the exponent of the complex sinusoid. Apart from that, we can just swap one for the other to transform the signals from waveform to spectrum and backward. The sequence of N time-domain samples ($x(n)$) is often referred to as the (DFT) *analysis frame.*

5.1.1 Programming the DFT

Since everything is discrete now, we can start programming spectral analysis and synthesis operations, and use code to explore the ideas that so far have been mostly computed with pen and paper. That is an important step, which will help illustrate most of the ideas so far developed. Turning to Eq. 5.1, let's see how it can be implemented by breaking it down into its constituents. This will provide a method for translating other expressions into code. The operations in the DFT can be separated as

1. A sample-by-sample multiplication of the signal by a complex sinusoid,

$$x(n)e^{-j2\pi kn/N}$$

2. A sum of all product terms to produce a given spectrum sample.
3. Scaling of the result by $1/N$.
4. A loop over $[0, N-1]$ to produce the full frequency-domain block of samples.

If we are implementing this with a mid-level language such as C/C++, then we need to program steps 1 and 2 inside a loop, which iterates over $[0, N-1]$, the range of the sum, incrementing the value of n for every repetition. However, in Python, it is possible to apply point-by-point multiplication of two single-dimensional arrays (also known as vectors) in one go. Therefore we can eliminate the loop and, by having the signal input and a sinusoid of a given frequency as two vectors, $x(n)\exp(-j2\pi kn/N)$ can be translated directly.[3] To complement this, the NumPy library also provides a function that takes an array and sums all of its items. With these two operations, steps 1 and 2 can be taken care of in one single line.

To finalize the code, we only need to provide an empty array to take the output samples, and to implement the two remaining steps. Python also allows the product of a vector and a scalar, and we can place the scaling at the very end, which is better if it needs to be skipped. A single loop over the output sample range and it is complete. This code for the DFT is shown in Listing 5.1, and we can see that it is a straightforward translation of Eq. 5.1. It is also possible to see that Python has a transparent support for complex numbers, which can be declared using the `1j` representation.

Listing 5.1 DFT function in Python.

```
import numpy as np
pi = np.pi
# DFT
# takes real waveform x
# returns complex spectrum
def dft(x):
    N = x.size # DFT size
    X = np.zeros(N)*1j # DFT cmplx array
    n = np.arange(0, N) # sample indices
    for k in range(0, N):
        # DFT equation
        X[k] = np.sum(x*np.exp(-2j*pi*k*n/N))
    return X/N
```

Following the same approach, we can provide a separate function for the inverse DFT. As we noted, it only differs by the sign of the complex sinusoid exponent, and we could perhaps use control of flow to determine the direction of the operation and reuse the DFT code. However, for real-valued signals, it is much better to produce a real number array as the output, instead of a complex one. Therefore we implement it separately, as shown in Listing 5.2. Note that the assignment from a complex to a real variable discards the imaginary part. Since we expect this to be zero, then it is not an issue.

Listing 5.2 Inverse DFT function in Python.

```
# inverse DFT
# takes complex spectrum X
```

```
# returns real waveform
def idft(X):
   N = X.size # DFT size
   x = np.zeros(N) # DFT output array
   k = np.arange(0, N) # sample indices
   for n in range(0, N):
       # DFT equation
       x[n] = np.real(np.sum(X*np.exp(2j*pi*k*n/N)))
   return x
```


With these in hand, we can finally test the Fourier theorem in practice, to see if the results we have got so far can be confirmed. The following program creates a cosine wave, applies the DFT, and plots the real part of the spectrum:

```
import numpy as np
# sigio is a custom lib with plotting functions
import sigio
# dft is a custom lib with the dft functions
import dft
# DFT size
N = 20
# time array in samples
t = np.arange(0, N)
# sig array
s = np.zeros(N)
# cosine wave
s += np.cos(2*np.pi*t/N)
# DFT computation
S = dft.dft(s)
# plot real spectrum
sigio.realplot(S, 'stem')
```

The plot produced by this program is shown in Fig. 5.1. It does seem to confirm that the analysis is correct, there are two components, each with half amplitude. However, the plot looks odd in that the position of the stems is not where we were expecting to see them. Perhaps if we look at the spectrum of the sine wave, which is purely imaginary, we may be able to understand what is going on, at least intuitively. We can replace the cosine by a sine in the program above, and plot the imaginary spectrum, which is then placed in Fig. 5.2.

Now it is possible to guess what the result means. On the left of the picture, we have a negative value, and we know this indicates the positive side of a sine wave spectrum. At the far right, the value is positive, and that is then the negative frequency component. Intuitively, it is now clear that there is some sort of rotation, and that the plot is not centered at 0 Hz, as we have been used to seeing. Additionally, the horizontal axis is marked with sample numbers, not frequency, so it is hard to tell what it represents. Again, intuitively, since we know this the spectrum of real signal, the value of $k = 1$ must represent

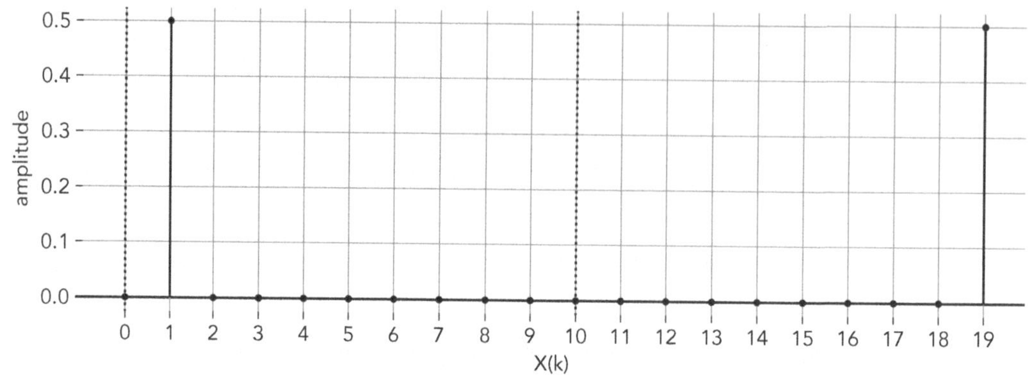

Figure 5.1 The plot of the real part of the spectrum $X(k)$ produced by the DFT of cosine wave.

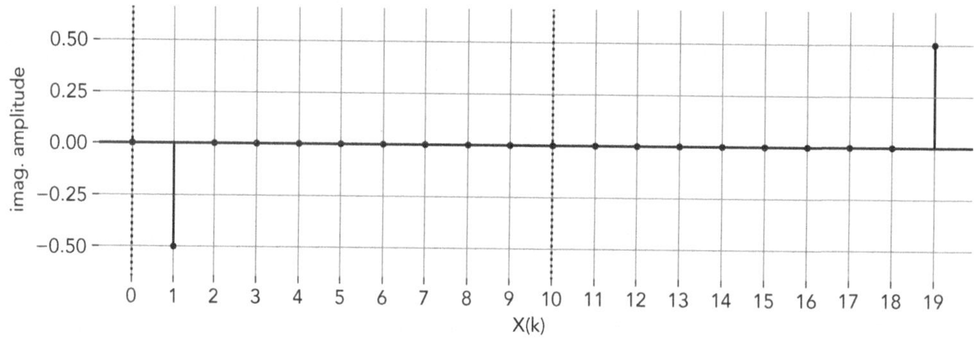

Figure 5.2 The plot of the imaginary part of the spectrum $X(k)$ produced by the DFT of sine wave.

a single-cycle sinusoid at half amplitude. Note that this since these plots are the output sequence exactly as they come out of the DFT, they may need some interpretation before we can continue discussing further what they represent.

5.1.2 Interpreting the DFT

Let's consider now a signal with more components, such as a bandlimited sawtooth wave. We can use a similar program to the above, replacing the sinusoid signal with

```
pi = np.pi
# bandlimited sawtooth
for i in t[1:N//2]:
    s += (2/pi)*np.sin(2*pi*i*t/N)/i
```

and plotting the imaginary spectrum (Fig 5.3).

Now it is possible to conclude that the two points that have zero value (0 and 10) clearly represent the DC and Nyquist frequency points, respectively. What about the others? At the moment, we can only say that they represent multiple integers of a fundamental frequency f_0, of which we do not know the value. This is the frequency of a single cycle of a waveform over N samples, and we may call it the fundamental frequency of analysis. In order to find out what it is, we need to know how to read such a signal, or, more precisely, how to reconstruct it.

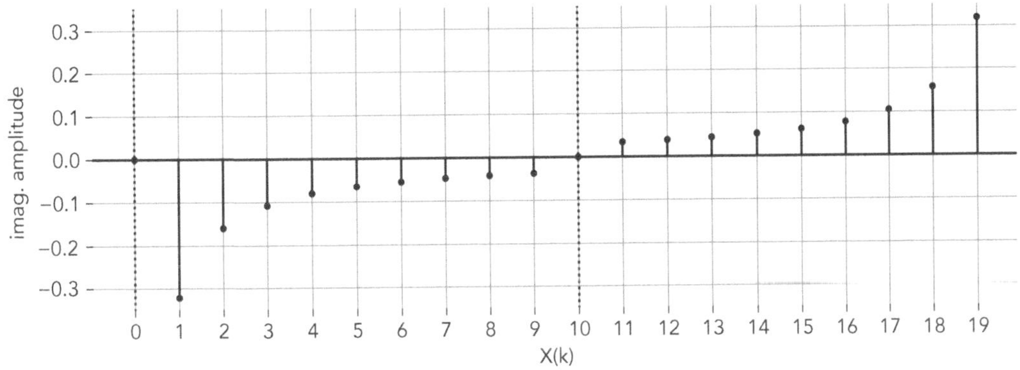

Figure 5.3 The plot of the imaginary part of the spectrum $X(k)$ produced by the DFT of a bandlimited sawtooth wave.

The missing parameter is the time between the samples, or its inverse, the sampling rate f_s. Once we know it, then we can assign each spectrum sample to a frequency. For example, if f_s is 20 Hz, then the first half of the spectrum is simply $[1,2,\ldots,10]$ Hz. If it is 40000 kHz, a much more realistic sampling rate, then with $f_0 = f_s/N = 2000$ we have $[2000,4000,\ldots,20000]$ Hz. Therefore we have

$$f_k = k\frac{f_s}{N}, \qquad 0 \le k \le \frac{N}{2}. \tag{5.5}$$

What about the other half? We can carry on using the kf_s/N expression, but we will need to remember that frequencies above f_s are aliased to the negative side of the spectrum (and we gave an expression to find these in Chapter 3). Therefore we have

$$f_k = k\frac{f_s}{N} - f_s, \qquad N/2 \le k < N. \tag{5.6}$$

These two expressions, when taken together, show that the point at $N/2$ refers to both f_s *and* $-f_s$. It is possible to represent the sampled spectrum in a circle. Starting from the right, and moving counterclockwise from 0, we have all positive frequency points. At the far left, we have $\pm$ Nyquist frequency, then in the lower side of the circle, all the negative frequencies upwards to 0 Hz. This is illustrated by Fig. 5.4, where we can see the digital baseband divided into 20 evenly spaced points in a circle, representing the sample frequencies.

If we keep going a round the circle, we will continue to encounter the same points periodically, as k increases beyond N, and similarly if we go in the opposite direction and k becomes negative. This is what is meant by saying that the spectrum of the DFT is periodic. Also, for the real-valued audio waveform, as we know that the negative spectrum is Hermitian symmetric, we should expect that the samples beyond $N/2$ are also in complex conjugation with the ones below it. Each frequency point is separated from its neighbors by f_o Hz, which determines the resolution of the spectral analysis. This can be increased at the expense of having a longer window, which reduces the time precision of the process, as the uncertainty principle makes another appearance.

To summarize, the DFT spectrum of an audio signal has the following key characteristics:

1. It is periodic, with a period of N samples, $X(k) = X(k+N)$.
2. It is Hermitian symmetric at $N/2$, with $X(k) = \overline{X(N-k)}$.

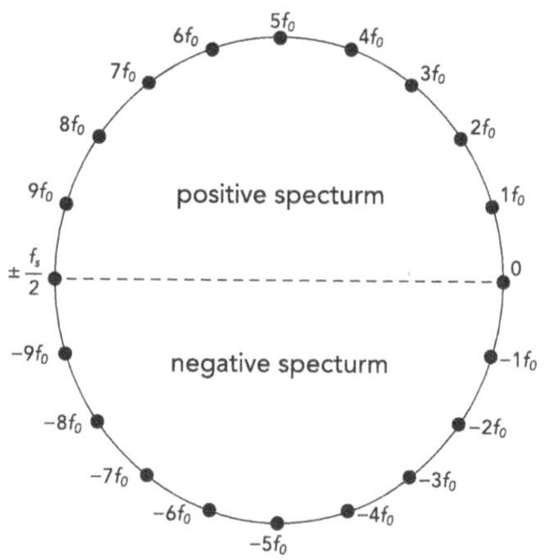

Figure 5.4 A representation of the discrete-time, discrete-frequency spectrum, where the digital based is divided into $N = 20$ equal points, separated by $f_0 = f_s/N$ Hz.

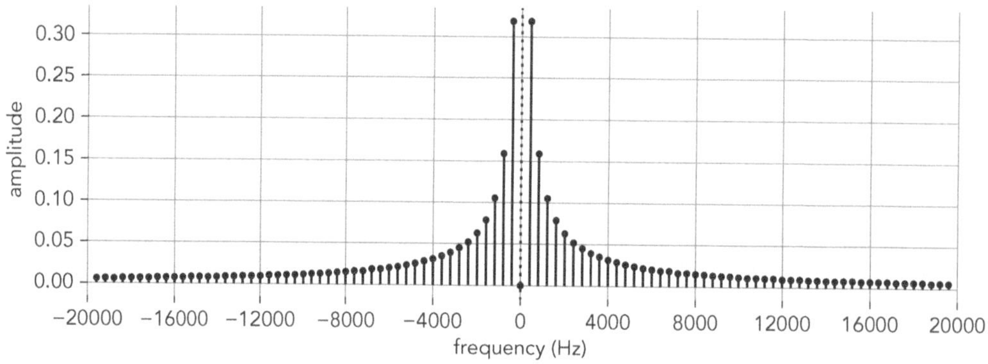

Figure 5.5 The amplitude spectrum of a 400 Hz bandlimited sawtooth wave sampled at 40 kHz, taken by a 100-point DFT.

3. It divides the f_s-wide baseband into N f_0Hz-spaced points, also called *bins*, *channels*, or *bands*.

Points 1 and 2, taken together, confirm that the symmetry exists at both 0 Hz and $f_s/2$, so in Fig. 5.4, this corresponds to between the top and bottom parts of the circle (and we can think of the horizontal line connecting these points as the axis of symmetry). To complete the discussion, we can show the sawtooth wave plot with two modifications: centered at DC, and with an interpreted horizontal axis. The analysis of a 400 Hz sawtooth wave, sampled at 40 kHz with a 100-point DFT, is shown in Fig. 5.5.

This latest plot now shows the amplitude (or magnitude) spectrum of the sawtooth wave, rather than its raw imaginary values. This is a much preferred way of representing the spectrum, as it is more general and works for spectra that are not purely imaginary or real. With it, we can see directly the distribution of amplitudes over the frequencies. Complementarily, we can have a plot of the phase spectrum, if we need to. For this particular case, as can be seen in Fig. 5.6, it is fairly predictable: components are all in the sine or inverted sine phase $(-\pi/2, \pi/2)$, except for DC and Nyquist, which are 0. However, there may be times when it is useful to look at the phase spectrum, particularly if we are combining one or more signals in one way or another.

Finally, it is important to remember that the discrete-frequency nature of the DFT implies that its inverse transform is expected to be periodic also in the time-domain,[4]

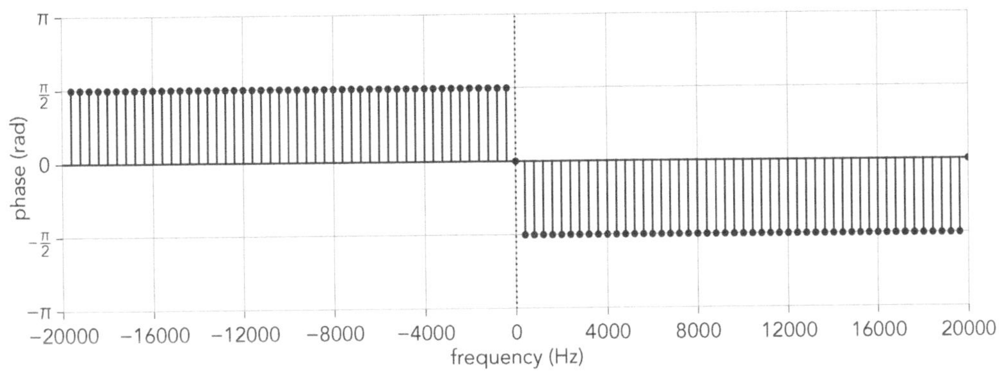

Figure 5.6
The phase spectrum of a 400 Hz bandlimited sawtooth wave sampled at 40 kHz, taken by a 100-point DFT.

$x(n) = x(n+N)$. This is the basis of the circular property of the DFT, which determines that a shift in the phase of a component rotates its waveform around the frame by a certain amount. In an intuitive way, this means that when the start of a waveform is pushed forward in a DFT analysis frame, the samples lying beyond index N reappear at the start of the sequence. Such a rotation produces no change in the amplitude spectrum, only in the phase spectrum.

5.1.3 Analysis Windows

We have been very careful with the examples given so far, as they are all representing complete cycles of waveforms. In practice, it will be hard to tune our analysis window size to a given f_0, and besides we want to be to deal with sounds that may not have a reasonable value for a fundamental frequency, or that are not periodic at all. Therefore, it is important to start considering the more difficult cases, and the meaning of the DFT analysis in different situations.

In order to assess the impact of these cases, we can proceed as before: look at how sinusoids behave and then extrapolate it to signals with any number of components. A good place to start is to look at what happens if we have incomplete waveform cycles within one analysis window (or frame). For example, what happens if we analyze a signal such as

```
t = np.arange(0, N)
x = np.cos(2*np.pi*1.15*t/N)
```

which completes a cycle, starts another and then it is cut short. Its magnitude DFT plot with $N = 50$ is shown in Fig. 5.7. Unsurprisingly, there are many components in the spectrum, even though we started off with a cosine wave. As has been discussed in Chapter 4, we are not analyzing a sinusoid anymore; therefore, we should expect that there will be more than one partial in the spectrum.

As before, in the case of the hard sync spectrum, the result is due to the application of the rectangular window. Therefore we have the convolution of a sinc() function with the cosine, which is then sampled at f_0 frequency intervals. Because the rectangular window spectrum is very wide, it smears the analysis over a number of frequencies. That is no good if we want to isolate the amplitude peaks that indicate the presence of a partial. It

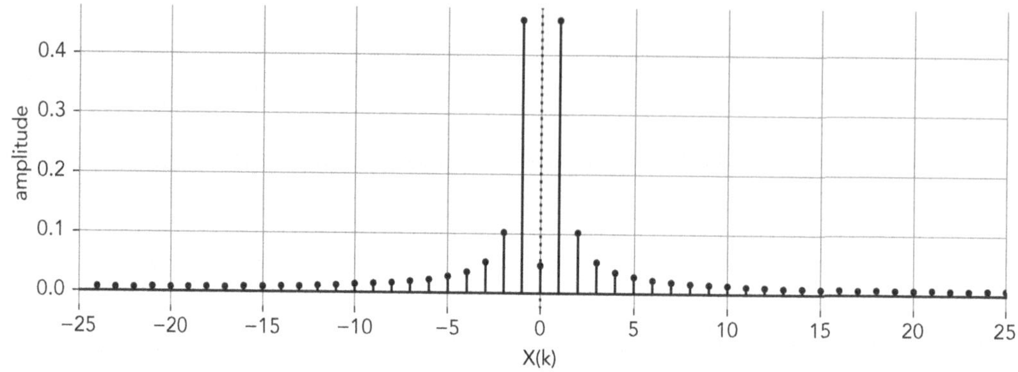

Figure 5.7
The magnitude spectrum of 1.15 cycles of a cosine wave, taken by a 50-point DFT and rotated to place 0 Hz at the center.

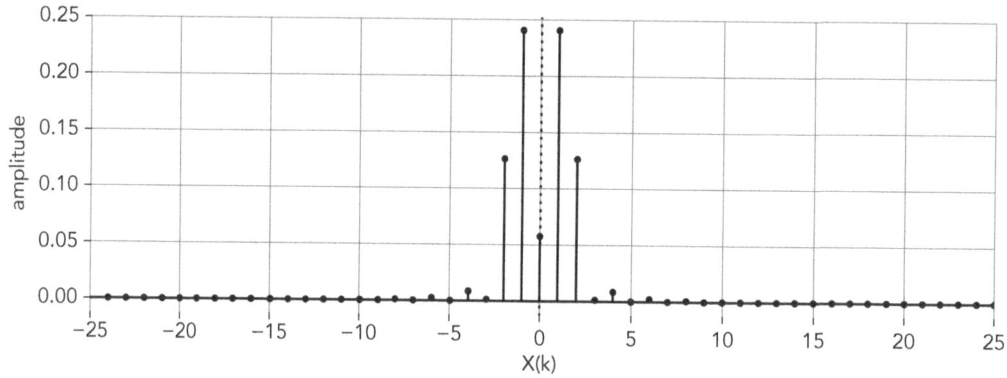

Figure 5.8
The magnitude spectrum of 1.15 cycles of a cosine wave, taken with a Bartlett window.

is possible to see that the peak is located somewhere above $X(1)$, as we would expect, but there is a lot of smearing that could be problematic in more complicated cases.

We should not expect much from just slicing off a piece of wave with any discontinuities that it may have at one or both ends of the frame. If we are using the DFT as a tool to analyze arbitrary signals, we need to improve this. The solution is to employ a better window, which is tapered at its ends. If the signal is sampled correctly in time, then we should not expect discontinuities anywhere else in a frame; therefore, just taking care of the beginning and the end of the window should make the analysis input become smoother.[5]

Let's make a first attempt, and employ a triangle-shaped window, also known as the *Bartlett* window. This is a good first step, as we know that this is the integral of a rectangular function, and we also know about its spectrum, which is less wide, with a $1/f^2$ rolloff. The code line becomes

```
x = np.cos(2*np.pi*1.15*t/N)*np.bartlett(N)
```

and the result is shown in Fig. 5.8. This is definitely an improvement, so we are moving in the right direction.

We can do better by looking for a window with two particular characteristics in relation to its spectrum:

1. It should have a narrow bandwidth.

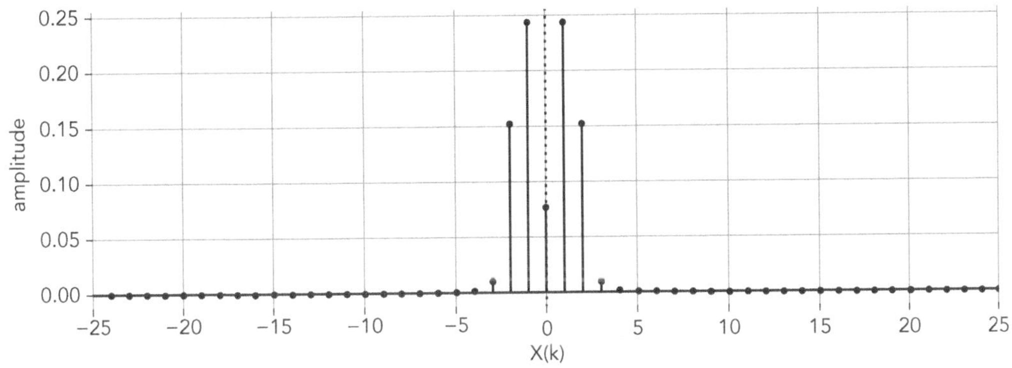

Figure 5.9
The magnitude spectrum of 1.15 cycles of a cosine wave, taken with a hanning window.

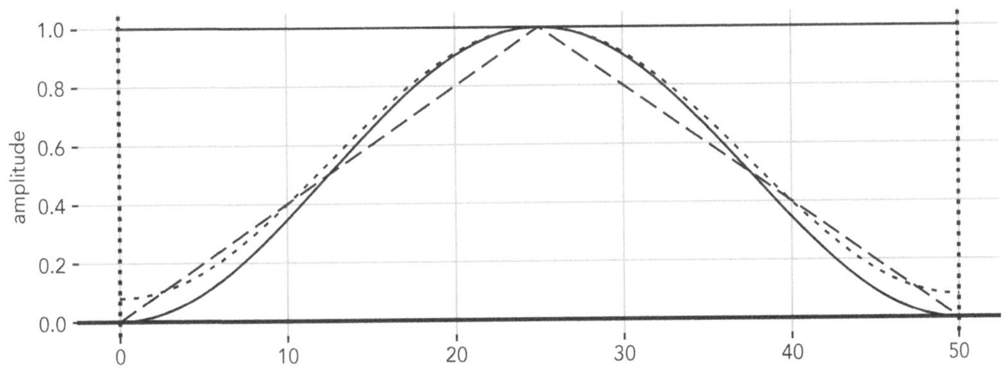

Figure 5.10
Rectangular (*solid*), hanning (*solid*), Hamming (*dots*), and Bartlett (*dashes*) windows.

2. It should be easily described.

If we want to find a finite narrowband spectrum with a simple description, then the solution points to designing the window out of sinusoids. So let's have a shape based on a sinusoid that is 1 in the middle and 0 at the ends,

```
w = 0.5 - 0.5*np.cos(2*np.pi*t/(N-1))
```

called a *Von Hann*, *Hann*, or *hanning* window. It is applied to the signal analysis shown in Fig. 5.9. It produces a marginally better result than the Bartlett case, but it has the advantage of meeting both requirements above. A variation on this function is the *Hamming* window, $0.54 - 0.46\cos(2\pi * n/(N-1))$, which has the characteristic of being non-zero in its entirety. A plot of these four windows (rectangular, hanning, Hamming, and Bartlett) is shown in Fig. 5.10.

It is useful to examine the hanning window case in a little more detail. We can easily guess what its spectrum is: there is a component at 0 Hz (easily seen with the DC offset of 0.5) and two others, at each side, with amplitude 0.25. The discrete magnitude and continuous-frequency magnitude plots are shown in Fig. 5.11, together with the continuous-frequency spectrum of the rectangular window, for reference. We can see that its bandwidth extends further than the central lobe of the sinc() function, but then it goes to zero. The effective bandwidth of the hanning window is $4f_s/N$.

Remembering that the multiplication by a window applies a convolution of its spectrum and the signal partials, we should expect the following results:

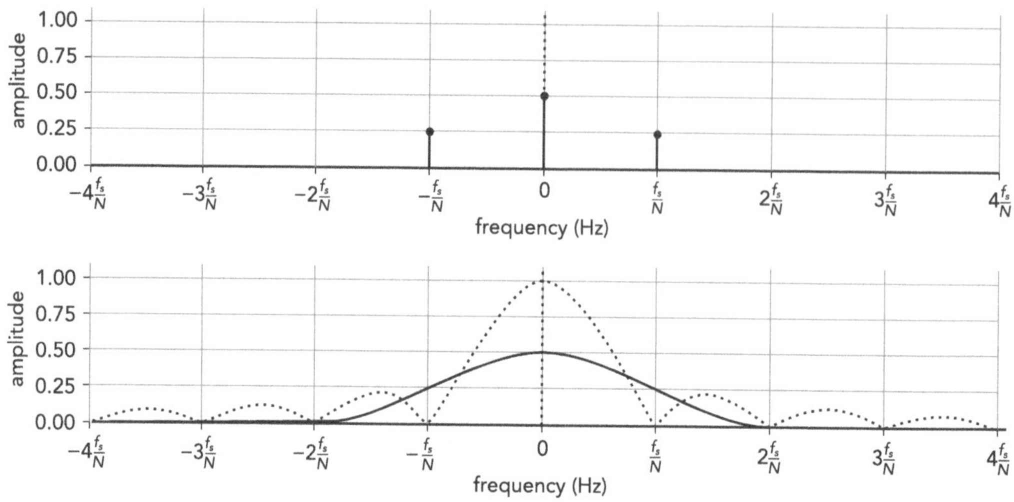

Figure 5.11 The DFT magnitude spectrum of a hanning window (*top*), and its continuous-frequency form (*bottom*), with the rectangular window spectrum as reference (*dots*).

- A single sinusoidal component at one of the frequency points will appear as three components on each side of the spectrum. A central one with half the original amplitude and two others on each side, with quarter amplitude.
- The bandwidth of frequency point (or bin) will be $4f_s/N$, which will lead to a certain degree of overlap between consecutive bands.
- If a signal component is not placed at a bin center frequency, it will appear as four components (each side). This can be clearly seen in Fig. 5.9, where the sinusoid shows up in points 0,1,2, and 3 (as well as $0, -1, -2$, and -3).

Therefore, a component anywhere in the spectrum appears as a peak with some sidebands. Note that, compared to the rectangular window, it is much better in the case that components are not multiples of f_0 (which is the general case), but on the other hand, it is not perfect if they are. On the whole, the hanning window will be very useful, with many applications in the methods we are going to explore. A major advantage is that we know what its spectrum looks like in an analytic way, and we can even apply it directly as a convolution in the frequency domain.

For this, it is worth considering its phase spectrum. The plots we have used so far are all of its magnitude spectrum, so they only tell us half of the story. Looking at the time-domain expression, we should guess the component phases are 0 and $-\pi$, since we only have DC and an inverted cosine. Let's check it. Figure 5.12 plots the phases from a 50-point DFT of the window. The result may look surprising at first, in that the values at -1 and 1 are not $-\pi$: they are not exact inverted cosines. An inspection of the hanning window expression shows that the period of the cosine is slightly shorter than the window by a factor of 1/N. However, as the window gets larger, this factor becomes very small and phases each side of DC approximate $-\pi$. The plot of the phase spectrum of windowed sine wave (with frequency $4f_0$) is shown on Fig 5.13, using a 1,000-point DFT. It confirms that the components each side of the peak have opposing phases (a $-\pi$ shift) compared to the central one. This is the effect of the convolution in the phase spectrum.

There are many other types of analysis windows beyond the cosine-based, the triangle, and the rectangular. Each one is optimized for a particular application. In the

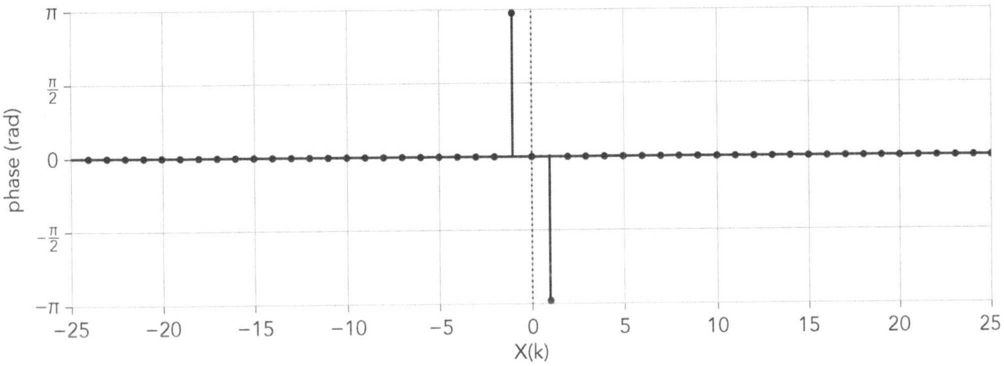

Figure 5.12
The DFT phase spectrum of a hanning window.

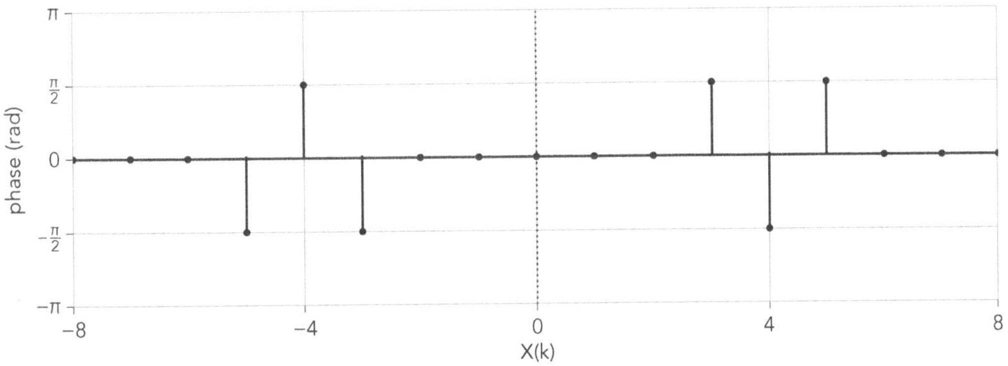

Figure 5.13
The DFT phase spectrum of a sine wave multiplied by a hanning window.

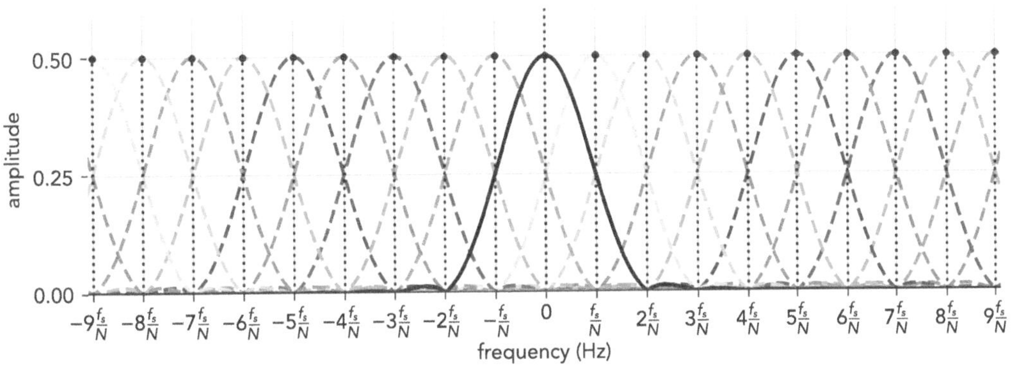

Figure 5.14
The windowed DFT analysis as a bank of bandpass filters.

methods of spectral processing we will see, the hanning window is by far the most commonly used. In general we should always bear in mind that the effect of windowing is a convolution in the spectral domain, and the net result to the DFT is to create an analysis band around each frequency point, as shown in Fig 5.14 for the hanning window case. This in fact can lead us to interpret the DFT as a bank of N parallel bandpass filters with equal-sized bandwidths across the full digital baseband. We will continue to develop this conclusion, which is very useful, in intuitive terms until it can be confirmed more formally.

5.2 The Fast Fourier Transform

Now that we have looked at the major characteristics of the DFT, we can turn to its computation. Although we have already presented a straight translation of the formula into a Python function, this is not the most common way to implement the operation. As we have already seen in Chapter 2, in 1965, Cooley and Tukey introduced the idea that the complex Fourier series can be calculated efficiently if the number of coefficients is a highly composite number.[6] They noted in particular that powers of 2, 3, and 4 could lead to the most gains in computation.

5.2.1 Radix-2 FFT

Given the special status of the number 2 as the basis of the binary system employed by digital computers, Cooley and Tukey concluded that it was the most optimal value for the radix of a fast algorithm for the DFT.[7] In this section, we will explore this algorithm, known as the *radix-2 fast Fourier transform*, which is the most commonly used method of computing the DFT in audio applications.

For this, then, we start by fixing the size of the DFT, N, to a power-of-2 number, and exploit the factorization possibilities provided by this. It is best to begin with an intuitive overview of the process (Mulgrew et al., 1999, Steiglitz, 1996). Consider the smallest transform size, with $N = 1$. This of course, is quite trivial, as $X_1(0) = x(0)$, an identity. Now consider the next size up, $N = 2$, where we have

$$\begin{aligned} X_2(0) &= x(0)e^0 + x(1)e^0 \\ X_2(1) &= x(0)e^0 + x(1)e^{-j\pi}. \end{aligned} \tag{5.7}$$

We can now reinterpret this as a combination of two half-size transforms (size 1),

$$\begin{aligned} X_2(0) &= X_1(0) + X_1(1) \\ X_2(1) &= X_1(0) - X_1(1). \end{aligned} \tag{5.8}$$

This shows we can build a DFT size 2 as two DFTs size 1 of the even and odd samples (indices 0 and 1) of the input signal x. Can we go upward from here? It is possible to show that we can obtain a 4-point DFT out of two size 2 DFTs of the even and odd samples (indices 0,2 and 1,3) of the input. And so on, we can get DFTs with N = 8, 16, etc.. The principle extends to all power-of-two sizes. This is the basis of the factorization process.

Now that we have an intuitive idea of what is going to be done, let's try to put it in more formal terms. We will derive the algorithm for the radix-2 FFT from the top down, even though its implementation proceeds as outlined above, from a base case of 2-point DFTs. The first step is to show how we can divide the DFT formula into two parts. Starting with a N-size DFT (N a power-of-two), we have[8]

$$\begin{aligned} X(k) &= \sum_{n=0}^{N-1} x(n)e^{-2j\pi k\frac{n}{N}} \\ &= \sum_{m=0}^{N/2-1} x(m)e^{-2j\pi k\frac{2m}{N}} + \sum_{m=0}^{N/2-1} x(2m+1)e^{-2j\pi k\frac{2m+1}{N}} \\ &= \sum_{m=0}^{N/2-1} x(m)e^{-2j\pi k\frac{m}{N/2}} + e^{-2j\pi\frac{k}{N}} \sum_{m=0}^{N/2-1} x(2m+1)e^{-2j\pi k\frac{m}{N/2}}. \end{aligned} \tag{5.9}$$

As before, we are splitting a DFT into two half-size transforms of the even- and odd-indexed sequences. We can see that they only only differ by a complex exponential factor. Calling these E and O, for the even and odd parts, respectively, and setting $e^{-2j\pi\frac{k}{N}}$ to ω^{-k}, we have

$$X(k) = E(k) + \omega^{-k}O(k). \tag{5.10}$$

However, there are some issues that need to be resolved before we go any further. The transform is N-sized, therefore k should range from 0 to $N-1$, but $E(k)$ and $O(k)$ are only of size $N/2$. What happens then when k exceeds that length? This should be no problem, as the DFT spectrum is periodic. Therefore, for these half-size transforms we have

$$\begin{aligned} E\left(k+\frac{N}{2}\right) &= E(k) \\ E\left(k+\frac{N}{2}\right) &= O(k). \end{aligned} \tag{5.11}$$

Now we can rewrite Eq. 5.10 fully worked out as

$$X(k) = \begin{cases} E(k) + \omega^{-k}O(k), & 0 \le k < \frac{N}{2} \\ E(k-\frac{N}{2}) + \omega^{-k}O(k-\frac{N}{2}), & \frac{N}{2} \le k < N \end{cases} \tag{5.12}$$

so we can get the results directly from the half-size transform outputs. This is almost what we need, but we have one more step to go. Note that the two halves of $X(k)$ can be computed independently. Taking advantage of the relationship,

$$\omega^{-(k+N/2)} = e^{-2j\pi(k+N/2)/N} = e^{-2j\pi k/N}e^{-j\pi} = -e^{-2j\pi k/N}, \tag{5.13}$$

we can derive two expressions, operating over half the transform size ($0 \le k < N/2$), which can be used to merge the even and odd parts to give the DFT. Each one of these computes half of the spectrum,

$$\begin{aligned} X(k) &= E(k) + \omega^{-k}O(k) \\ X\left(k+\frac{N}{2}\right) &= E(k) - \omega^{-k}O(k). \end{aligned} \tag{5.14}$$

This gives us the means to calculate the DFT in two separate parts. The next step is to apply the same process to the half-size transforms $X(k)$ and $X(k+N/2)$. These have the same characteristics as the full transform, and therefore we can split them into even and odd parts again and produce four quarter-size transforms. We repeat this again until the transform size is two, and each half is a single point. At that stage, we have $N/2$ transforms to compute. These have the form we already derived in Eq. 5.8, which is the same as Eq. 5.14.

The algorithm is implemented by starting with the 2-size transforms, and then upward, doubling the size of each transform and halving their number. At each step, we employ Eq. 5.14 to combine the odd and even elements and produce a partial result, until we reach the size of the complete DFT. For a transform of size N, we will have $\log_2 N$ steps to complete the full DFT. For large values of N, this represents a considerable saving in computation time, as the standard expression has N^2 complexity as opposed to $N \log_2 N$ in the case of the FFT.

The process of splitting transforms into even and odd parts has some effect on the sequence indices. At the end of it, all even indices will be on the left side and all odd ones will be on the right. Let's illustrate this point with an example based on a 8-point DFT,

1. The starting sequence for $N = 8$ is

$$\{X_8(0), X_8(1), X_8(2), X_8(3), X_8(4), X_8(5), X_8(6), X_8(7)\}$$

2. Splitting it once into even and odd parts gives us two $N = 4$ transforms,

$$\{X_4(0), X_4(2), X_4(4), X_4(6)\} \text{ and } \{X_4(1), X_4(3), X_4(5), X_4(7)\}$$

3. Applying the process once more to get four 2-point DFTs, we get

$$\{X_2(0), X_2(4)\}, \{X_2(2), X_2(6)\}, \{X_2(1), X_2(5)\}, \text{ and } \{X_2(3), X_2(7)\}$$

This shows that in order to get the correct result at the end, some reordering of the input is required. Cooley and Tukey noted in their paper that the sequence indices are bit-reversed. If we translate each base-10 index value into base-2, we will see that the orderings at end of the FFT process are related by reversing the bit order of indices. For $N = 8$, each index can be represented by 3 bits, therefore some indices (such as 000 and 111) will not change, while others (e.g., 110) will. Before a signal is sent to the first stage of the FFT, it needs to be reordered by applying bit-reversal of its indices.

We can do this by preparing a bit-reverse array with a simple algorithm,[9] and then using it to change the position of the samples in the input array. The Python function in Listing 5.3 implements this operation.

Listing 5.3 Data reordering function.

```python
# Reorder N complex samples
# in s array
# bit-reversing indices
# return reordered array
def reord(s,N):
   # initialise bitrev array
  bp = np.arange(0,N)
  # prepare it
  i, n = 1, N//2
  while i < N:
   for j in range(0,i): bp[i+j] = bp[j] + n
   i, n = i*2, n//2
  # create complex out array
  out = np.zeros(N).astype(complex)
  # reorder input
  for i in range(0,N): out[i] = s[bp[i]]
  return out
```

Applying this function to a sequence of indices $[0, 1, 2, \ldots, 7]$ produces $[0, 4, 2, 6, 1, 5, 3, 7]$, which is the correct order of the input, as seen in the example above, for $N = 8$.

We have everything at hand to attempt an implementation of the FFT. A diagram of the operations at each step is shown in Fig 5.15, for $N = 8$, where we can see that the

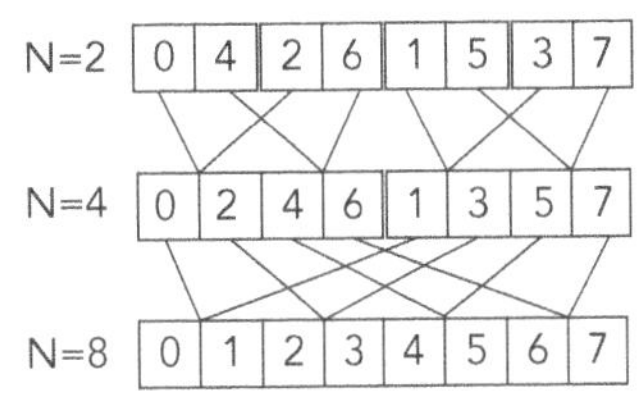

Figure 5.15 A diagram of the Radix-2 FFT operations, for $N = 8$

algorithm combines pairs of numbers to produce the result, first every other, then every two, and finally every four. As we can observe, the process involved in the FFT algorithm is recursive, the top-down division of transforms into even and odd parts is reapplied until we reach a base point. The code should therefore represent this as faithfully as possible.

There are actually two ways in which we could program the FFT, using loops to iterate over the $\log_2 N$ steps to produce the final result, or use self-recursion, where a function can call itself a number of times. The advantage of the former is a more efficient use of computing resources, but the latter may provide simpler and more compact code.[10] Since the main focus of the discussion is the algorithm, we will attempt to present in its clearest form, as a recursive function.

We have a base case, which is the size-2 DFT, and a pair of expressions to compute each step. So we can try to translate this into Python code and then call it recursively. Eq. 5.14 takes the even and odd parts and merges them into the output spectrum. This can be programmed *in place*, that is, the new values replacing the old in the input array. We will do that, but to make the exposition clearer, we will use two temporary arrays to hold the even and odd parts,

```
e,o = s[k:kn],w[::N//n2]*s[kn:kn2]
s[k:kn],s[kn:kn2] = e+o,e-o
```

Note that we are using slices to process only the parts of the array that interest us. The variables k, kn, and kn2 are counters that control the slicing. The n2 variable holds the transform size for a given step, and N is the full DFT size. The expression requires an array w, which holds the values of the complex exponential ω^{-k}. So, the first line in the code above extracts the even and odd parts of the transform, and the second combines them to produce the result as determined by Eq. 5.14.

Around this expression is a scaffolding of counting variables that controls the process, completing the function code. This is shown in Listing 5.4.[11]

Listing 5.4 A recursive radix-2 FFT function in Python.

```
# Recursive N-point Radix-2 FFT
# input array s (size N)
# cmplx exp array w (size N/2)
# and transform size N
# returns the DFT spectrum
def fft(s,w,N,n=1,k=0):
  n2 = n*2
  kn,kn2 = k+n,k+n2
  e,o = s[k:kn],w[::N//n2]*s[kn:kn2]
```

```
    s[k:kn],s[kn:kn2] = e+o,e-o
    if kn2 < N: return fft(s,w,N,n,kn2)
    if n2 < N: return fft(s,w,N,n2)
    return s
```

In this function, recursion happens in two directions: first we proceed by applying the N/n2 transforms to the input array, hopping along in n2 increments each recursive call. Once the partial result for a step is completed (kn2 becomes N), we increment the transform half size (n) and make another call to start again, resetting k to zero. This process goes until n2 becomes the DFT size, then the function returns the spectrum s to its caller.

The FFT code needs a driver, which will create the w array, call the reordering and the fft() function. This is implemented by

```
# DFT (Radix-2 FFT)
# takes waveform s
# returns spectrum
def dft(s):
  N = s.size
  n = np.arange(0,N//2)
  w = np.exp(-2j*np.pi*n/N)
  return fft(reord(s,N),w,N)/N
```

Note that we can implement the inverse DFT with the same FFT code given above. All we need is to change the sign of the complex exponential,

```
# inverse DFT (Radix-2 FFT)
# takes spectrum s
# returns waveform
def idft(s):
  N = s.size
  n = np.arange(0,N//2)
  w = np.exp(2j*np.pi*n/N)
  return fft(reord(s,N),w,N)
```

A magnitude plot of a 16-point DFT of a single cycle of a cosine wave is shown in Fig. 5.16, which also includes plots of the partial transforms produced in the each stage of the FFT. Note how at each step, the results of the previous transforms are merged and the output gets increasingly closer to the full spectrum.

5.2.2 Real-to-Complex and Complex-to-Real Transforms

The DFT functions we have developed so far, both using the direct formula implementation and the FFT, can take a real waveform input and produce a complex spectrum. This is because we are mostly interested in real time-domain data, audio signals, and thus it is easier if we get these in the right form, although the DFT itself can use complex inputs if needed. We are not doing anything special, effectively just casting from one type to another.

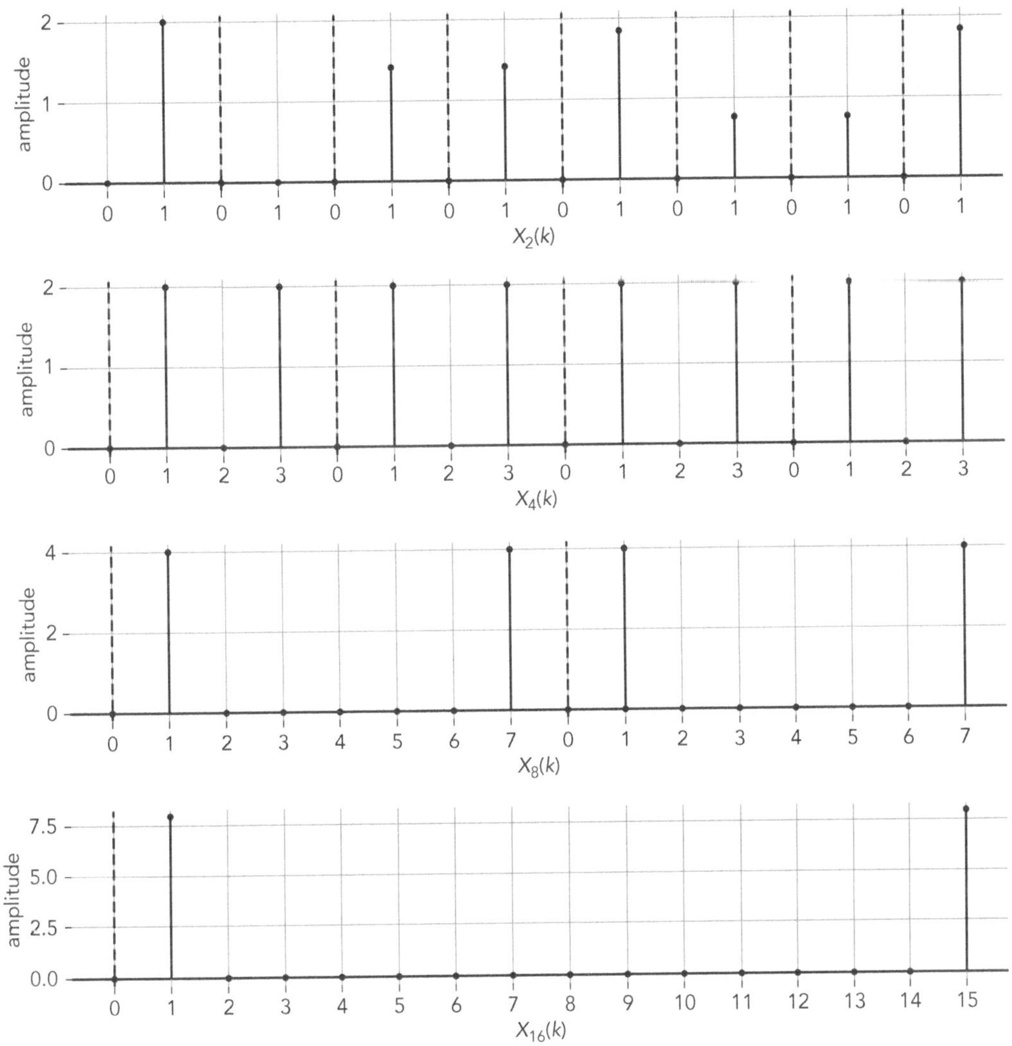

Figure 5.16
Four magnitude plots showing the partial transforms ($X_n(k)$, for $N = \{2,4,8\}$), and the complete 16-point DFT of a single cycle of a cosine wave. The dashed vertical lines separate the transforms at each stage.

Two-sided spectra seem a bit wasteful, as effectively the DFT produces double the data we need. If we were to modify (process) the spectral data to realize some effect, we would need to make sure the modification was applied properly to both sides, so that the Hermitian symmetry was preserved. If that is not the case, we will end up with non-zero imaginary parts following the inverse transform. Therefore having two sides is not only a lot of redundancy, but may also pose some headaches in future applications. It would be much better if we were only presented with the positive spectrum.

What we need to do is to find a way to produce $N/2+1$ bins, the non-negative side of the spectrum, from N input samples. If we do that, we should be able to cut down the redundancy by a certain amount. To do this, we can again use some intuition to imagine the method, and then work out the algorithm. Let's first consider what the non-negative spectrum is, from the perspective of an arbitrary input to the DFT. Taking the half-size, $M = N/2$, transform as the basis, if the input were $N/2$ real-valued samples, the DFT will be symmetrical in two halves. That is not the general case of a set of M spectral samples representing the positive spectrum: there is no guarantee of any symmetry between the first and the second block of $M/2$ data. Therefore the sequence of M samples could just as well be produced by a complex-valued input.

This insight leads us to propose that the M-sized DFT representing the positive spectrum of an N-sized DFT can be developed as the transform of a complex-valued waveform. To develop this idea further, a reinterpretation of the input data is required: if we consider the N real-valued samples of an audio signal as M complex-valued samples, we can then apply the transform to it. Since the input is not actually complex, it is necessary to make some adjustments to derive the single-sided spectrum from the raw results out of the DFT. This will involve some extra amount of computation, but the net result is that this allows us to use a half-size transform with real inputs.

Now we can proceed more formally to develop the algorithm (Chu and George, 1999). The first step is to take the N-size waveform $x(n)$ and split it into M even and odd samples,

$$x(n) = \begin{cases} a(m), & n \text{ odd}, 0 \leq m < M \\ b(m), & n \text{ even}, 0 \leq m < M \end{cases}. \tag{5.15}$$

With this done, we can fictitiously make a complex signal $y(m) = a(m) + jb(m)$, to which an M-size DFT can be applied,

$$Y(k) = \sum_{m=0}^{M-1} y(m)e^{-2j\pi km/M} \qquad 0 \leq k < M. \tag{5.16}$$

Now we have $Y(k)$, but we want to get the non-negative spectrum of $x(n)$, $X(k)$ ($0 \leq k \leq M$). The direction of travel now is to derive it from $Y(k)$. It is possible to split its DFT equation in two, corresponding to the real and imaginary parts,

$$\begin{aligned} Y(k) &= \sum_{m=0}^{M-1} \Re\{y(m)\}e^{-2j\pi km/M} + j\sum_{m=0}^{M-1} \Im\{y(m)\}e^{-2j\pi km/M} \\ &= R(k) + jI(k) \end{aligned} \tag{5.17}$$

The spectra $R(k)$ and $I(k)$ are the DFTs of the imaginary and real parts of $y(m)$. Since these are real-valued functions (purely real and purely imaginary, respectively), their DFT spectra will be Hermitian-symmetric,

$$\begin{aligned} R(k) &= \overline{R(M-k)} \\ I(k) &= \overline{I(M-k)}. \end{aligned} \tag{5.18}$$

Therefore, if $Y(k) = R(k) + jI(k)$, then

$$\overline{Y(M-k)} = \overline{R(M-k)} + \overline{jI(M-k)} = R(k) - jI(k). \tag{5.19}$$

Now we have enough relationships to allow us to find $R(k)$ and $I(k)$ from $Y(k)$,

$$\begin{aligned} R(k) &= \frac{1}{2}\left[Y(k) + \overline{Y(M-k)}\right] \\ I(k) &= \frac{j}{2}\left[\overline{Y(M-k)} - Y(k)\right] \end{aligned} \tag{5.20}$$

This is a significant step, since $R(k)$ and $I(k)$ are transforms of the even and odd samples of $x(n)$, $a(m)$, and $b(m)$. Using the same method employed to split the DFT in two for fast

Fourier transform, we have,

$$
\begin{aligned}
X(k) &= \sum_{n=0}^{N-1} x(n)e^{-2j\pi k\frac{n}{N}} \\
&= \sum_{m=0}^{M-1} a(m)e^{-2j\pi k\frac{m}{M}} + e^{-2j\pi\frac{k}{N}} \sum_{m=0}^{M-1} b(m)e^{-2j\pi k\frac{m}{M}} \\
&= R(k) + \omega^{-k} I(k).
\end{aligned}
\tag{5.21}
$$

for $0 \leq k \leq M$, and with $\omega^{-k} = e^{-j\pi\frac{k}{M}}$. This allows us to get $X(k)$ from $Y(k)$, and therefore we have a complete method.[12] The points at $k = M$ (the Nyquist frequency) and $k = 0$ (DC) are special. Since they are purely real, we can treat them separately, using the following expressions:

$$
\begin{aligned}
X(0) &= \Re\{Y(0)\} + \Im\{Y(0)\} \\
X(M) &= \Re\{Y(0)\} - \Im\{Y(0)\}
\end{aligned}
\tag{5.22}
$$

In many practical implementations, to save space, these two points are stored as a pair in place of the real and imaginary values for $k = 0$. This allows us to pack up $N/2$ complex numbers in the same storage space as N real numbers. It is a common approach to provide in-place real-to-complex transforms. While this is a good convention, it is not as useful if we are employing vectorial operations (e.g., multiplication of two arrays, etc.) in Python, therefore it is better to produce N/2 + 1 spectral samples, with the first and last point having a zero imaginary part. In summary, these are the steps for a real-input DFT:

1. Re-interpret the N real samples in $x(n)$ as M complex values $y(m)$.
2. Take a DFT of the complex input, getting $Y(k)$
3. Get the real and imaginary parts, $R(k)$ and $I(k)$ using Eq. 5.18.
4. Compute $X(k)$ by merging $R(k)$ and $I(k)$ (Eq. 5.21).

The code is implemented in Listing 5.5. It calls the previously defined DFT driver function, which then employs the recursive FFT. With this we have a complete algorithm for efficient spectral analysis of audio signals.

Listing 5.5 A real-to-complex DFT in Python.

```
# real-to-complex DFT (Radix-2 FFT)
# takes real waveform s
# returns complex non-negative spectrum
def rdft(s):
  N = len(s)
  M = N//2
  s = dft(np.array(s[:N:2] + s[1:N:2]*1j))
  n = np.arange(0,M)
  w = np.exp(-1j*np.pi*n/M)
  cjs = np.flip(s[1:],0).conjugate()
  R = .5*(s[1:] + cjs)
  I = .5j*(cjs - s[1:])
  r,i = s[0].real*.5,s[0].imag*.5
```

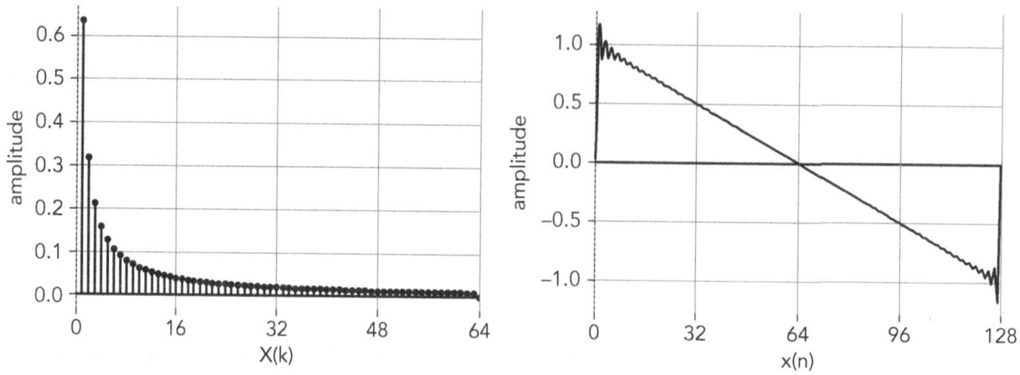

Figure 5.17 The real-to-complex DFT amplitude spectrum (*left*) and the corresponding complex-to-real inverse DFT waveform (*right*), using a bandlimited sawtooth wave as an input.

```
    s = np.zeros(M+1).astype(complex)
    s[0], s[M] = r+i + 0j, r-i + 0j
    s[1:M] = R[:] + w[1:]*I
    return s
```

To complement this code, we present the inverse operation in Listing 5.6. It follows the same principles, the only difference being that we start from $X(k)$, derive $Y(k)$ from it, apply the inverse DFT, and then reinterpret the complex output as a real sequence of samples. Note that in order not to modify the input array (causing a side effect), the code first makes a local copy of the data before operating on it. This is needed because the complex DFT is only called at the end (the DFT itself also operates on a copy of the input, which is made at the reordering stage). This pair of transforms can then be used in spectral analysis-resynthesis of audio signals. The plots in Fig. 5.17 show the real-to-complex DFT and corresponding the inverse transform outputs, using a 128-point bandlimited sawtooth as input. Note that the transform produces $N/2+1$ points covering the $[0,f_s/2]$ frequency range. The transform size is conventionally set to N points, even though the DFT is only run on half the number of (complex) samples.

Listing 5.6 A complex-to-real inverse DFT in Python.

```
# complex-to-real DFT (Radix-2 FFT)
# takes complex non-negative spectrum
# returns real waveform
def irdft(s):
    M = len(s) - 1
    N = 2*M
    zro,nyq = s[0].real,s[M].real
    s = np.array(s[:M])
    s[0] = complex(zro+nyq,zro-nyq)
    n = np.arange(0,M)
    w = np.exp(1j*np.pi*n/M)
    cjs = np.flip(s[1:],0).conjugate()
    R = .5*(s[1:] + cjs)
    I = .5j*(s[1:] - cjs)
    s[1:] = R + w[1:]*I
```

```
    s = idft(s)
    out = np.zeros(N)
    out[0::2], out[1::2] = s.real, s.imag
    return out
```

The Python code to produce Fig. 5.17 is as follows:

```
import numpy as np
import sigio
import fft

N = 128
t = np.arange(0,N)
s = np.zeros(N)
for i in t[1:N//2]:
  s += (2/np.pi)*np.sin(2*np.pi*t*i/N)/i

ss = fft.rdft(s)
z = fft.irdft(ss)
sigio.sigplot((ss,z),('stem','line'), gridstep=(16,32))
```

where the module `fft` contains the real-to-complex and complex-to-real DFTs as shown above and `sigio` has various custom plotting functions.

The FFT algorithm can also be efficiently computed in parallel, since the operations in each iteration are independent of each other. A full discussion of parallel implementations is beyond the scope of this book, but the interested reader is directed to Lazzarini (2020) for a complete description of the process. Parallel code running in dedicated hardware such as general-purpose graphical processing units can be particularly useful for large transform sizes.

5.2.3 Other Radices

We may wonder if the method employed to obtain the radix-2 FFT can be redeployed to derive an algorithm for other radices. This question is interesting, since Cooley and Tukey had noted that, for instance, the radix-3 factorisation is also very efficient. Since the road has already been traveled, we may as well try to apply the process to powers of 3. As before, we split the DFT equation, but now in three parts,

$$
\begin{aligned}
X(k) &= \sum_{n=0}^{N-1} x(n)e^{-2j\pi k\frac{n}{N}} \\
&= \sum_{n=0}^{N/3-1} x(3m)e^{-2j\pi k\frac{3m}{N}} + \sum_{n=0}^{N/3-1} x(3m+1)e^{-2j\pi k\frac{3m+1}{N}} + \sum_{n=0}^{N/3-1} x(3m+2)e^{-2k\pi k\frac{3m+2}{N}} \\
&= \sum_{n=0}^{M-1} x(3m)e^{-2j\pi k\frac{m}{M}} + e^{-2j\pi\frac{k}{n}} \sum_{n=0}^{M-1} x(3m+1)e^{-2j\pi k\frac{m}{M}} + e^{-4j\pi\frac{k}{n}} \sum_{n=0}^{M-1} x(3m+2)e^{-2j\pi k\frac{m}{M}} \\
&= A(k) + \omega^{-k}B(k) + \omega^{-2k}C(k),
\end{aligned}
\tag{5.23}
$$

where $M = N/3$ and, as before, $\omega^{-k} = e^{-j2\pi k/N}$. It is possible to make the same assumptions as we did before about the nature of $A(k)$, $B(k)$, and $C(k)$. Therefore we can compute $X(k)$ in the interval $[0, M)$ using

$$\begin{aligned} X(k) &= A(k) + \omega^{-k}B(k) + \omega^{-2k}C(k) \\ X(k+M) &= A(k) + \omega^{-(k+M)}B(k) + \omega^{-2(k+M)}C(k) \\ X(k+2M) &= A(k) + \omega^{-(k+2M)}B(k) + \omega^{-2(k+2M)}C(k). \end{aligned} \tag{5.24}$$

Now we have a triplet of equations, which can be computed recursively from the base case of $N = 3$ upward to any power-of-three. As before, it should be expected that the data will come out scrambled at the end, if it is not reordered to start with. Since in radix-2 case the base-2 digits of each index need to be reversed in order, it should follow that here the base-3 (ternary) digits are to be reversed.[13] A similar reordering algorithm can be developed, now adapted to work in base-3.

We can put this all together in Python code to provide a radix-3 recursive FFT implementation. This is shown in Listing 5.7. As before, driver functions are provided to launch the FFT computation. Note that, in the absence of any obvious identities, we employ a full cycle of the complex exponential ω and offset the reading as required by Eq. 5.24.

Listing 5.7 Radix-3 FFT implementation in Python.

```
# Reorder N complex samples
# in s array
# tit-reversing indices
# return reordered array
def reord3(s, N):
 bp = np.arange(0,N)
 i, n = 1, N//3
 while i < N:
   for j in range(0,i+N//3-1):
        bp[i+j] = bp[j] + n
        bp[i+j+1] = bp[j+1] + n
   i, n = i*3, n//3
 out = np.zeros(N).astype(complex)
 for i in range(0,N): out[i] = s[bp[i]]
 return out

# Recursive N-point Radix-3 FFT
# input array s (size N)
# cmplx exp array w (size N)
# and transform size N
# returns the DFT spectrum
def fft3(s,w,N,n=1,k=0):
   n3 = n*3
   kn1,kn2,kn3 = k+n,k+2*n,k+n3
   w0,w1,w2 = w[:N//3:N//n3],w[N//3:2*N//3:N//n3],
              w[2*N//3::N//n3]
```

```
    x0,x1,x2 = s[k:kn1],s[kn1:kn2],s[kn2:kn3]
    s[k:kn1],s[kn1:kn2],s[kn2:kn3] = (x0+x1*w0+x2*w0**2,
                                       x0+x1*w1+x2*w1**2,
                                       x0+x1*w2+x2*w2**2)
    if kn3 < N: return fft3(s,w,N,n,kn3)
    if n3 < N: return fft3(s,w,N,n3)
    return s

# complex DFT (Radix-3 FFT)
# takes complex waveform s
# returns complex spectrum
def dft3(s):
  N = s.size
  n = np.arange(0,N)
  w = np.exp(-2j*np.pi*n/N)
  return fft3(reord3(s,N),w,N)/N

# complex inverse DFT (Radix-3 FFT)
# takes complex spectrum s
# returns complex waveform
def idft3(s):
  N = s.size
  n = np.arange(0,N)
  w = np.exp(2j*np.pi*n/N)
  return fft3(reord3(s,N),w,N)
```


The following minimal example demonstrates the use of this code

```
import numpy as np
import sigio
import fft

N = 81
t = np.arange(0,N)
s = np.zeros(N)
for i in t[1:N//2+1]:
  s += (2/np.pi)*np.sin(2*np.pi*t*i/N)/i

X = fft.dft3(s)
x = np.real(fft.idft3(X))
sigio.sigplot((X,x),('stem','line'),gridstep=(9,9))
```

It runs an 81-point DFT using the radix-3 FFT algorithm on a bandlimited sawtooth wave. Note that since this is an odd-sized transform, the Nyquist frequency is not a multiple of the analysis f_0 and it lies between two bins in the spectrum. The output of this program is shown in Fig. 5.18.

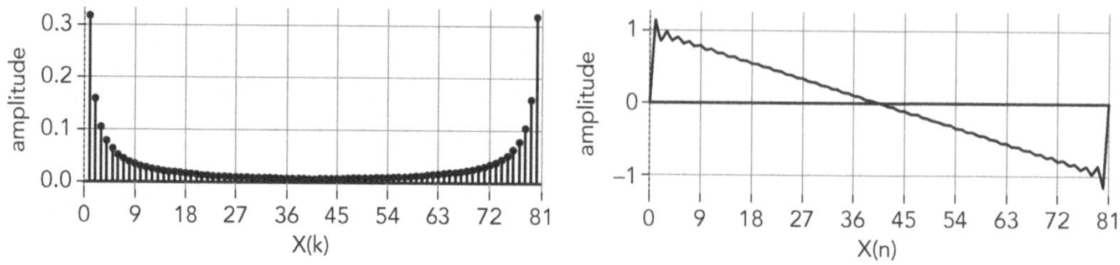

Figure 5.18 An 81-point DFT amplitude spectrum of a bandlimited sawtooth wave (*left*) and its inverse transform output (*right*).

This indicates that once the first, simplest, FFT methods are derived, other factorizations of the DFT may follow. With these, it should be possible to provide implementations that would cover a large variety of N sizes. While the methods discussed here work with exact powers of a given number, it is possible to have mixed radices, and vary the split at different stages to cover other composite sizes. Of course, all of these involve more computation than the basic radix-2, and therefore we should prefer to employ this as the standard method of DFT analysis and synthesis.

5.3 Discrete-Time Convolution

We can now turn to some audio processing applications of the DFT. For this, we will start by reviewing some ideas introduced in Chapter 4 (Sec. 4.3). There, the operation of convolution was presented for both the continuous and discrete cases. The latter was shown to be described by

$$f(n) * g(n) = \sum_{m=-\infty}^{\infty} f(m)g(n-m) \tag{5.25}$$

where $f(n)$ and $g(n)$ are two arbitrary functions. We were able to employ the discrete form of convolution in the spectral domain to describe different types of waveforms. Now we are interested in its time-domain applications and how we can implement this equation for the cases where signals are 0 for $t < 0$, and finite in length.

Let's develop this idea a little further. Consider an input $x(n)$ that is put through a device called a *delay line*. This can be easily implemented in a computer by storing the samples somewhere and then retrieving them later. The delayed signal can be represented as

$$y(n) = x(n-m), \tag{5.26}$$

where we see that the signal at time n is the same as the input at time $n-m$, a delay of m samples. Using the sampling rate f_s, we can interpret this as a delay of m/f_s seconds. Now it is possible to look at Eq. 5.25 and see clearly that convolution involves the delay of one of the signals, which has been described before as a time shift.

The next step is to consider a case where we would like to combine the original input with the delay, to make an echo effect. A delayed signal on its own is only a copy of a signal, if we do not have a reference, it is hard to notice that there is any effect going on. Therefore we can add the two signals together as in

$$y(n) = x(n) + x(n-m) \tag{5.27}$$

and the result is as expected, an echo. Now consider a signal $h(n) = \{1,0,\ldots,1\}$, with $m-1$ samples between the first and the last ($h(0) = 1$ and $h(m) = 1$). We can rewrite Eq. 5.27 as

$$y(n) = \sum_{k=0}^{m} h(k)x(n-k), \tag{5.28}$$

so that when $k=0$ we have $h(0)x(n)$ and when $k=m$, $h(m)x(n-m)$. This equation is very similar to Eq. 5.25, except for the fact that it is defined for a finite, positive-only, m. This is what we use to implement time-domain convolution. The function $h(n)$ is called an *impulse response* (IR), which is equivalent to the output of a system given an impulse (i.e., a sequence $\delta_{0n} = \{1,0,0,\ldots\}$, called a Kronecker delta)[14] as its input. In this particular case, we could interpret the IR as the sum of the response of a surface standing a certain distance away and the impulse itself (in an ideal situation where there is total reflection and no absorption).

Discrete-time convolution can be used in many applications to impart the characteristics of a system onto a signal. The *system* in question can be anything, such as a physical object or space, the model of a filter, or even another audio signal. For instance, a typical example is reverberation, where we can measure the IR of a space and then process a dry input sound using the convolution of the two signals. Each reflection in the room is recorded in the measured IR, and then when applied to the signal will simulate a sound being played back in that space. This is very simple conceptually, but can have many creative uses.

5.3.1 Direct Convolution

We will first look at a time-domain approach to implementing discrete-time convolution, even though our destination is eventually to use the DFT to compute it. This will allow us to discuss some important details that have great significance for the various types of spectral processing we will be exploring in this and subsequent chapters. Generally speaking, in practical implementations of convolution, we have a finite impulse response that is applied to an input of arbitrary length. Therefore the IR length defines the maximum delay that will be applied, and that determines how much storage we will need to hold the input signal (the delay line size). Once that is defined, we can implement Eq. 5.28 as shown in Listing 5.8. It depends on externally defined arrays for IR and delay, as well as read/write position variable. It takes a sample of the input signal and returns the convolution output for time n and the updated delay line position.[15]

Listing 5.8 Direct convolution function in Python.

```
# Convolution
# takes a sample s
# an impulse response array h
# a delay buffer array d
# and a read pointer p
# returns the updated read pos
# and the convolution sample
def convolution(s,h,delay,p):
  out, size = 0., h.size
```

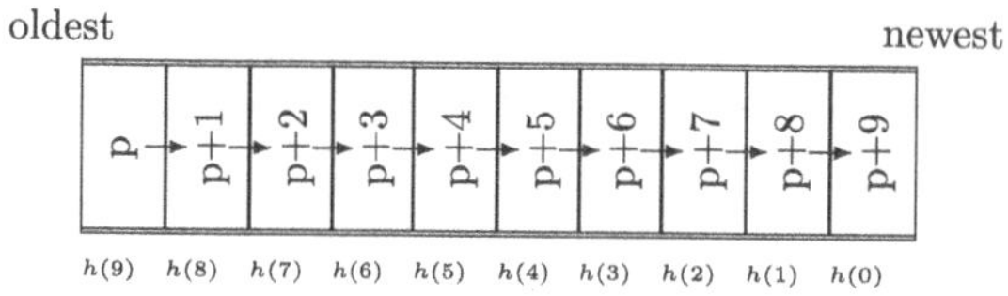

Figure 5.19 A schematic view of direct convolution.

```
  delay[p] = s
  p += 1
  for k in range(0, size):
    out += h[size-k-1]*delay[(p+k)%size]
  return p%size, out
```

This very literal implementation of the convolution operation is often called *direct convolution*. It involves placing a signal in a delay line, then retrieving the convolution output as the sum of all delayed samples multiplied by their corresponding impulse response values,

```
out += h[size-k-1]*d[(p+k)%size]
```

Note that the delay line is read circularly, wrapping around the end of the array, back to beginning (using the modulo operator, `%size`). For each single output sample we need as many multiplications and additions as size of $h(n)$, which may be quite a considerable computation load if the IR is large. The implementation reads the `h` array backward, since the delay position moves forward every sample, replacing the oldest sample with the latest in `d`. When the convolution is computed, the position `p` points to the m-th delayed sample, which corresponds to $h(m)$ for an IR with length $m+1$. If we know the size l of the input signal, then the convolution is $l+m-1$ samples long. Figure 5.19 illustrates the direct convolution algorithm, showing a 9-sample circular delay line from the oldest and newest sample, with the respective impulse response samples for each position.

To develop a feel for how the process works, we can put some test signals and IRs through this and plot the output. The basic program for this is

```
# very simple IR 8-sample IR
h = np.array([1, 0, 0, 0, 0, 0, 0, .5])
# delay line
delay = np.zeros(h.size)
# 7-sample triangle signal
s = np.array([0,1,2,3,2,1,0])
# ouput array
out = np.zeros(s.size+h.size-1)
# convolution of s and h
p = 0
j = 0
for i in range(0, out.size):
  if i < s.size: ss = s[i]
  else: ss = 0.
```

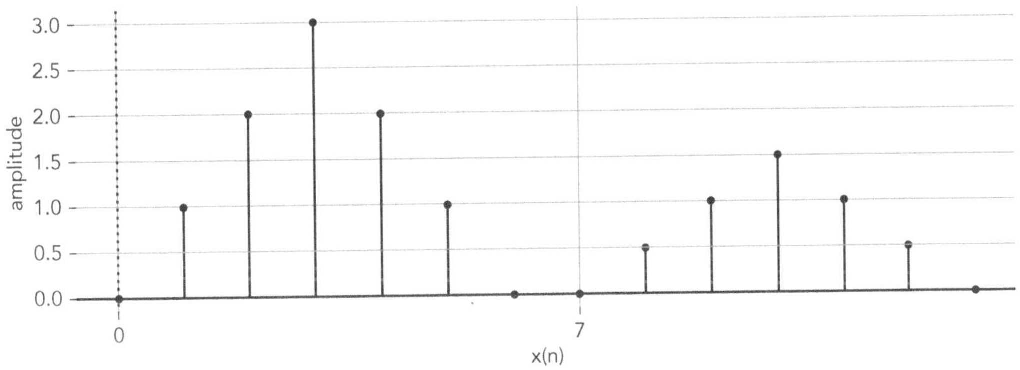

Figure 5.20
A simple convolution example.

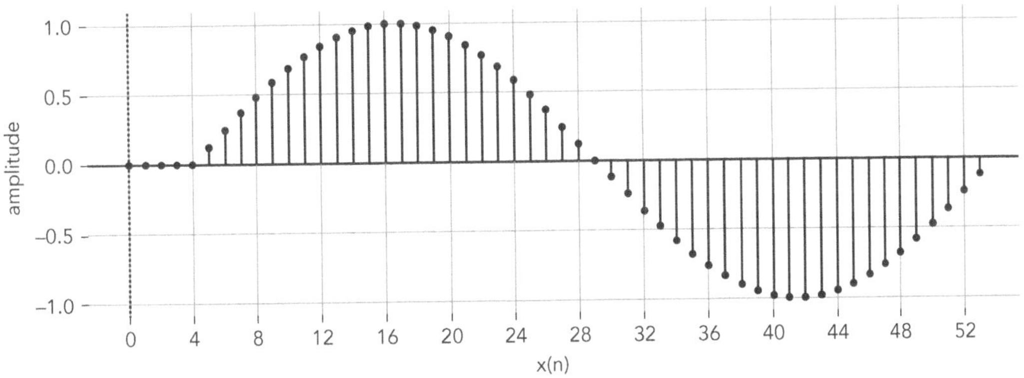

Figure 5.21
Sine wave delayed by 5 samples through convolution.

```
    p,out[j] = convolution(ss,h,delay,p)
    j += 1
# plot output
sigio.waveplot(out,gridstep=h.size-1,midmark=False)
```

which, in this case, uses a simple 8-sample IR that is similar to the echo example discussed, except that the second non-zero sample is half amplitude. The result is plotted in Fig. 5.20. We can see how the triangle is echoed after seven samples as the final sample at $h(7)$ is the only other non-zero value in the IR.

We should expect that this works for any signal. Let's try a sine wave, delayed by 5 samples,

```
# delay by 5 samples
h = np.array([0, 0, 0, 0, 1])
# sine sig
N=50
t = np.arange(0,N)
s = np.sin(2*np.pi*t/N)
```

which is shown in Fig. 5.21. This effect can be called a *pure delay*, and an IR of this kind has no other effect on the signal.

In these examples, in order to show clearly the effect of delayed copies of the signal, care was taken not to use IRs that would lead to the overlapping of the input. However,

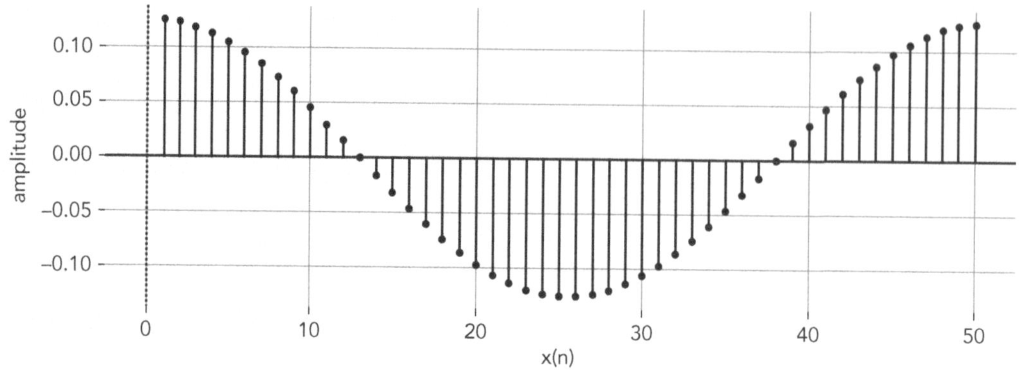

Figure 5.22 Convolution of sine wave and $h(n) = \{1, -1\}$.

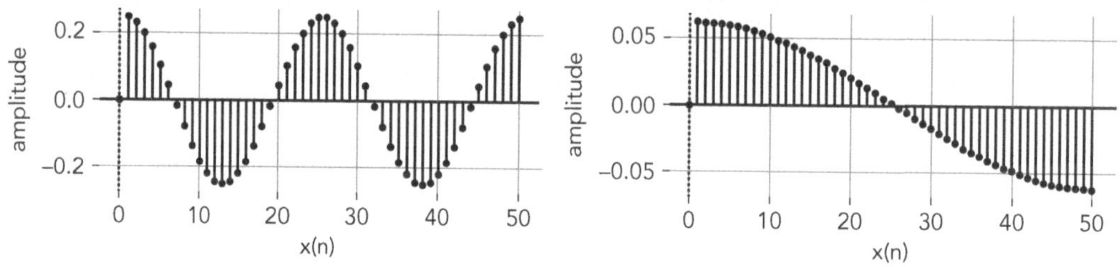

Figure 5.23 Convolution of sine waves and $h(n) = \{1, -1\}$ with $f_0 = f_s/100$ Hz *(left)* and $f_0 = f_s/25$ Hz *(right)*.

this is actually one of the important things about convolution, the fact that various repetitions are mixed together at different levels to create the final result. Let's explore this idea a little further. Consider a sine wave as before, but now an IR that is $h(n) = \{1, -1\}$ (yes, they can be bipolar). The result is plotted in Fig. 5.22.

It may at first be surprising that we have come out with a cosine wave, from a sine input. However, if we examine it closely, it should be clear why this is the case. The IR mixes two copies of the input, spaced by one sample, but with inverse phases. Another way to look at this is that we are taking the difference between two samples, and we know that, for continuous signals,

$$\frac{d\sin(t)}{dt} = \cos(t), \tag{5.29}$$

so therefore with this IR, convolution implements discrete-time differentiation. That is one of the conclusions we can reach. The other is to look at the effect of the convolution in the signal amplitude. The signal going in was peaking at 1, but what comes out has a maximum amplitude of around 0.12. So the IR is also acting as a filter by changing the signal amplitude. What kind of filter? We can find out by guessing that is either a lowpass or a highpass, and putting two signals through, one with a higher frequency and another with a lower frequency. This is shown in Fig. 5.23, where we can see that the amplitude of the higher-frequency wave is reduced by a lesser amount if compared to the lower wave. Therefore it is a highpass (or lowcut) filter.

Now consider the case of $h(n) = \{0.5, 0.5\}$, where the convolution will produce the average of every two samples. From what we have learned so far, we should expect this to be some sort of filter. Since we have a spectrum analysis tool in the form of the DFT, we can use it to explore the results of the convolution. Using a sawtooth wave, we run the convolution program and show the results in Fig. 5.24. The top two graphs show the

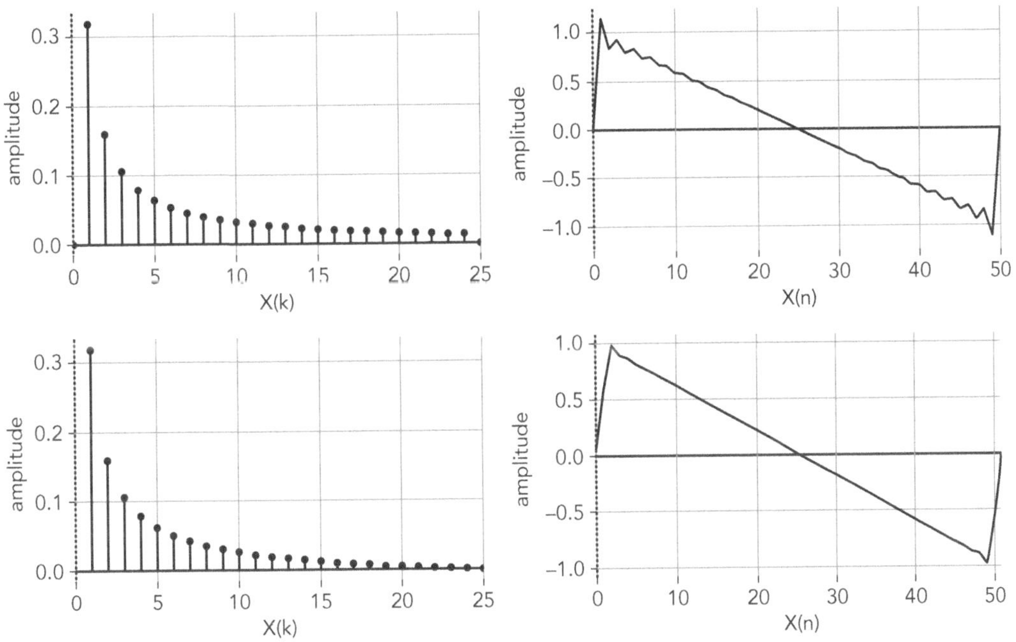

Figure 5.24
A bandlimited sawtooth and its magnitude spectrum (*top*), and the result of its convolution with a $h(n) = \{0.5, 0.5\}$ IR (*bottom*).

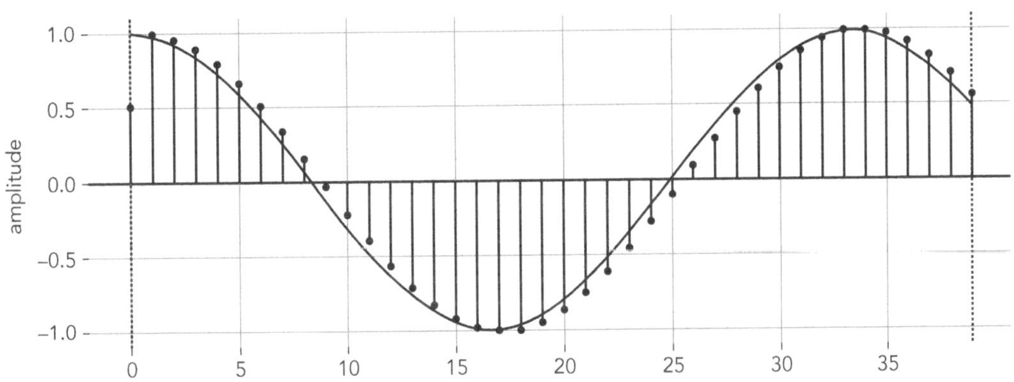

Figure 5.25
Comparison of a cosine wave (*solid line*) and the result of its convolution by the lowpass IR $h(n) = \{0.5, 0.5\}$ (*stem plot*), demonstrating the half-sample delay imposed by the process. The two signals are normalized to peak amplitude of 1.

bandlimited sawtooth and its spectrum, and below these, the results of the convolution. As we can see, the filter has had a lowpass effect. Averaging in general should be expected to have a smoothing effect, and that is what we observe in this case.

We should expect that filters such as these have some effect on both the amplitudes[16] and the phases of the signal.[17] We can see in the plots of the highpass filter outputs, that the signals are delayed by a certain amount, since the input was in the sine phase and the output came out in what looks to be the cosine phase. This is due to the filter effect on the phase spectrum, delaying components by a certain number of samples. We can also demonstrate this in the case of the simple averaging lowpass filter, which imposes a half-sample delay on its input, as shown in Fig. 5.25. A linear-phase behavior with a delay of $(N-1)/2$ samples across all the spectrum is achieved when the values of the IR are symmetrical about their middle point.[18] This is the case, for instance, of the averaging lowpass filter.

These examples of convolution taught us many important points about the process, namely:

- The impulse response imparts an effect on an input signal, modifying its spectrum.

- The operation can be characterized as a form of filtering.
- The effects may be described through an amplitude response (how it affects the amplitude spectrum of the input), and a phase response (the effect on the phase spectrum).
- An impulse response can be constructed to have a linear phase effect (a $(N-1)/2$ delay) if its sample sequence is mid-point symmetric.

Note that all these points stem from the fact that the convolution in the time domain is equivalent to multiplication in the frequency domain, as we have seen before. This connection allows us to apply here the ideas we have been developing about the spectrum. Time-domain convolution as discussed here is a linear time-invariant (LTI) process. That is, it does not cause non-linear distortion and it does not change over time (as the IR is fixed). The method of direct convolution, however, is fairly expensive in computational terms, and is not practical for realtime implementations of processes requiring longer IRs.

5.3.2 Fast Convolution

We can now return to the DFT and apply it to implement discrete-time convolution taking advantage of the fast Fourier transform algorithm. This allows the order of complexity to be reduced from N^2 to $N \log_2 N$ operations (in the case of the radix-2 FFT), and we should be able to save computational resources. Particularly in the case of large IRs, this makes convolution more practical. We should proceed as follows to compute the convolution $y(n)$ of a signal $x(n)$ of length N_x with an impulse response $h(n)$ of length N_h:

- Set the DFT size to a power of two size N, $N \geq N_x + N_n - 1$.
- Pad $x(n)$ with $N_h - 1$ zeros and $h(n)$ with $N_x - 1$ zeros.
- Compute their spectra using the real-to-complex FFT, obtaining $X(k)$ and $H(k)$.
- Take the product of the two spectra $(Y()k) = X(k)H(k))$.
- Computer the result $y(n)$ using the complex-to-real FFT of $Y(k)$.
- Discard any samples for $n \geq N_x + N_n - 1$.

The process relies on being able to pad a signal with zeros, and take its DFT, take its inverse DFT and then remove the padded zeros transparently. There is nothing preventing us from doing it, it is perfectly legitimate. With the basic algorithm for fast convolution laid out, we can code it as shown in Listing 5.9.[19]

Listing 5.9 Fast convolution function in Python.

```
# Fast convolution
# takes a signal x
# and an impulse response h
# returns the convolution signal
def fast_convolution(x,h):
    L = x.size + h.size - 1
    N = 1
    while N < L: N *= 2
    xx, hh = np.zeros(N), np.zeros(N)
```

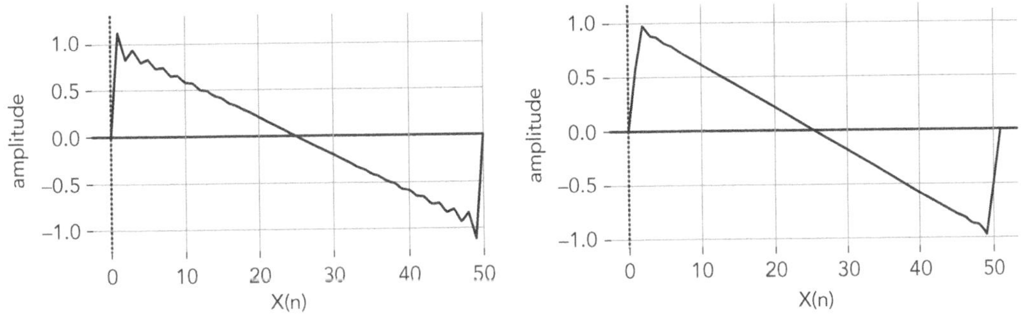

Figure 5.26 A bandlimited sawtooth (*left*), and the result of its convolution with a $h(n) = \{0.5, 0.5\}$ IR (*right*), using the fast convolution algorithm.


```
    xx[0:x.size], hh[0:h.size] = x, h
    Y = fft.rdft2(xx)*fft.rdft2(hh)
    return fft.irdft2(Y)[:L]
```

As before, to check that the algorithm is correct, we can put through some simple test signals. If we run the sawtooth example of Fig. 5.24, which uses the simple averaging filter, we will have a reference to check the result against. The code for this is

```
N=50
h = np.array([0.5,0.5])
t = np.arange(0,N)
s = np.zeros(N)
for i in t[1:N//2]:
   s += (2/np.pi)*np.sin(2*np.pi*t*i/N)/i
y - fast_convolution(s,h)
sigio.waveplotM((s,y), ('line','line'),gridstep=10)
```

and the output is plotted in Fig. 5.26. As we can see, the process works as expected.

Since this is a fully spectral process, we may try to design filters directly as a function of frequency, and only take one DFT of the input signal and one inverse DFT of the product. In this case, we use half of the algorithm given above, with no zero padding. Let's outline it:

- Set the DFT size to a power of two size $N = 2L$, where L is the desired impulse response size.
- Construct a (non-negative) filter spectrum $H(k)$ of size L from any desired frequency response.
- Compute the input spectra using the real-to-complex FFT, obtaining $X(k)$.
- Take the product of the two spectra $(Y()k) = X(k)H(k))$.
- Computer the result $y(n)$ using the complex-to-real FFT of $Y(k)$.

The following program implements this process, using a $1/f$ linear-phase frequency response, defined by

$$H(k) = \frac{1}{k}e^{-j\pi k}, \tag{5.30}$$

and feeding a pulse waveform into it,

```
L = 64
```

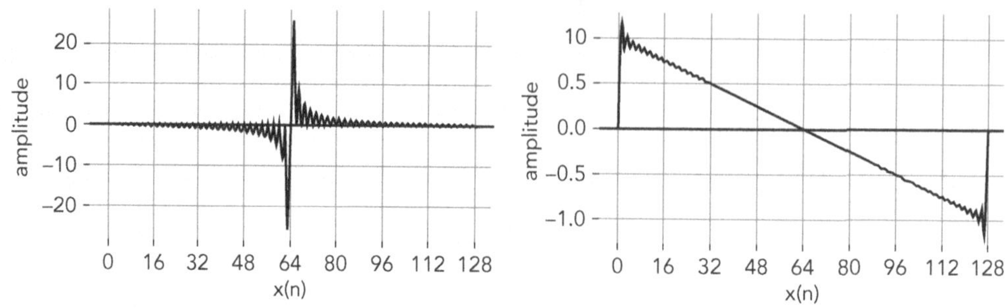

Figure 5.27 A bandlimited pulse (*left*), and the result of its convolution with a 1/f linear-phase frequency response (*right*).


```
N = 2*L
# filter spectrum
H = np.zeros(L+1).astype(complex)
for k in range(1,L):
    H[k] = (1/k)*np.exp(-1j*np.pi*k)
# bandlimited pulse (centred)
n = np.arange(0,N)
x = np.zeros(N)
for i in n[1:L]:
  x += (2/np.pi)*np.sin(2*np.pi*n*i/N + np.pi*i)
y = fft.irdft2(fft.rdft2(x)*H)
sigio.waveplotM((x,y), ('line','line'),gridstep=16)
```

We can see the output of this program in Fig. 5.27. Starting with a pulse wave and applying a $1/f$ (−6 dB/octave rolloff) filter gives a sawtooth wave. This could have been predicted, since we know from Chapter 4 that if we integrate a pulse we get a sawtooth, as its magnitude spectrum is equal to one at all frequency points. We can also see that the output has been shifted around by about half waveform a period, since we started with a centred pulse and got a 0-based sawtooth. This is due to the filter phase response (πk).

This example actually raises the question about convolution sizes and the exact nature of this spectral filtering process. We have said that the convolution of two sequences produces a sequence that is the length of the two added together minus one. However, now we got a result from multiplying two N-sized spectra, which is N-size instead of $2N - 1$, what is going on? The answer is that the DFT shifts samples circularly, so any delays caused by convolution will be moved around the DFT frame, if there is no zero padding. Consider the following code:

```
N = 128
h = np.zeros(N)
h[N//4] = 1
H = fft.rdft2(h)
n = np.arange(0,N)
x = np.zeros(N)
for i in n[1:N//2]:
  x += 4/((N/2)*np.pi)*np.sin(2*np.pi*n*i/N + np.pi*i)
y = fft.irdft2(fft.rdft2(x)*H)
```

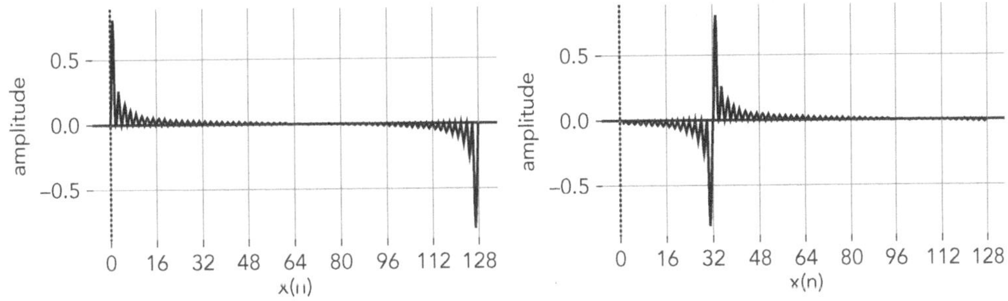

Figure 5.28
A bandlimited pulse waveform shifted circularly by 1/4 frame.

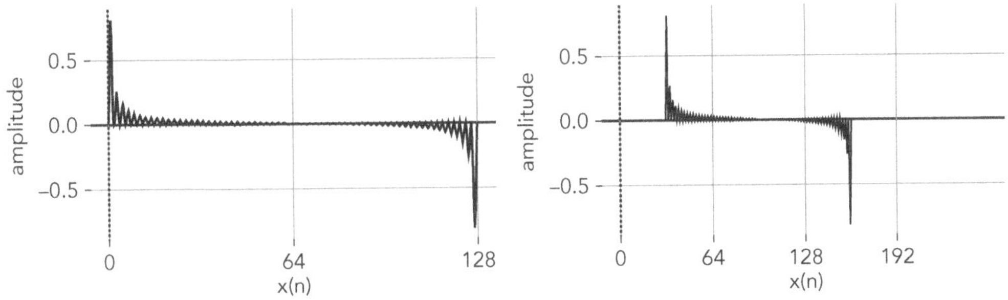

Figure 5.29
The convolution of a bandlimited pulse and an impulse response consisting of single non-zero sample at the 1/4-size position.


```
sigio.waveplotM((x,y), ('line','line'),gridstep=16)
```

where we multiply a signal consisting of one non-zero sample at $N/4$ by a pulse waveform. The result is plotted in Fig. 5.28. This yields a circular shift of 1/4 frame, due to the periodic nature of the DFT analysis ($y(n) = y(n+N)$).

Now consider the same two signals, correctly zero-padded to a $2N$-size DFT, that is, an ordinary (fast) convolution. The result is shown in Fig. 5.29, which is what we should expect, a shift forward, rather than circular. This is the normal result of convolution, whereas the pure spectral product implemented above is actually a different process.[20] As is, this only works in practice for single-shot applications, such as the one described here. If we want a streaming process, where an audio signal is constantly being filtered, we will need to use one of the forms of partitioned convolution with a time-domain IR. However, we can use purely spectral methods if we reintroduce time into the process, and make use of (non-rectangular) windowing. This will be explored later in Chapter 6.

5.3.3 Partitioned Convolution

The fast convolution algorithm is a good step forward from direct convolution. However, it may not be very practical at all. We might have noticed that in all examples we required the complete impulse response and input signals, since we needed to know beforehand the convolution length. That is a major limitation, as we cannot implement this for an input signal whose length we do not know in advance. Therefore any realtime implementations are out of question.

The solution for this is to implement *partitioned* convolution, a revised version of the algorithm that sets the DFT size according to the impulse response length.[21] The input signal then can be streamed in separate partitions, which allows us to run the process in realtime just by having the complete IR. The key to make this work is to be able to reconstruct the output signal from the individual partitions. Since we know that the convolution length is set to $2L-1$, L given as the IR size, it is possible to see that the final $L-1$ samples overlap with the first $L-1$ samples of the next partition.

The algorithm can be given as:

1. Set the DFT size to a power of two size $N \geq 2L-1$, where L is the desired impulse response size.
2. Pad $h(n)$ with $N-L-1$ zeros, take its DFT to get $H(k)$.
3. Take a size L rectangular window of the input, pad it with $N-L-1$ zeros.
4. Compute its spectrum, obtaining $X(k)$.
5. Take the product of the two spectra ($Y(k) = X(k)H(k)$).
6. Compute the result $y_p(m)$ using the complex-to-real FFT of $Y(k)$.
7. Discard any samples for $n \geq 2L-1$.
8. Mix $y_p(m)$ with the L$-$1 final samples of the previous partition, $y_{p-1}(m+L)$ to produce $y(n)$.

This algorithm is implemented as in a Python function in Listing 5.10. It takes an input and an impulse array and returns the convolution. However, in the current form, it does not represent any advance, as we still have to send in the complete signal. It allows us to see that it operates and to step through the algorithm, and it works as a drop-in replacement for the basic fast convolution function.

Listing 5.10 Partitioned convolution function in Python.

```
# partitioned convolution
# takes a signal x
# and an IR h
# returns the convolution
def part_convolution(x,h):
 n, l = h.size, x.size
 N = 1
 while N <= (2*n-1): N *= 2
 xx, hh = np.zeros(N), np.zeros(N)
 y = np.zeros(n+l-1)
 hh[:n] = h
 H = fft.rdft2(hh)
 for p in range(0,l,n):
  if p+n > l:
    M = l - p
    xx[n:] = np.zeros(N-n)
  xx[:n] = x[p:p+n]
  y[p:p+2*n-1] += fft.irdft2(fft.rdft2(xx)*H)[:2*n-1]
 return y
```

It is possible to implement partitioned convolution in a less simplistic way, to allow for *streaming* spectral processing, where input data is provided in small buffers, processed, and a buffer is returned. With this approach, we have the bases for a realtime implementation of the algorithm. For this to work, it is necessary to store part of the convolution output so it can be overlap-added to the next block. In other words, there will be some *state* to be kept between calls to the convolution function.

We have faced this situation before, as the direct convolution function implemented in Listing 5.8 had to keep track of the delay read/write position. This and the delay line memory constituted the state of the operation. In fact, the direct convolution function was implemented in a streaming way, taking and producing single samples. In that code, we simply provided the delay as an external array object, and returned the updated read-write position, which was then reused as an argument. In the present case, it might not be a great idea to continue doing this, as it does not scale well and can lead to all sorts of tangles later on.

What we need is a convolution object, which will hold and manage its own state, separate from the rest of the program. For this, we need to design a *class*: the type describing the object. Once this is done, we can create as many object variables as we need. At this stage, there is no need for overcomplex code, as we can implement the algorithm more or less as before. One modification we will do to the previous implementation is to enforce the partition size L to be the same as $N/2$, so that the windowing of the input is simply half the transform size. This simplifies the overlap-add code and makes no difference to the result, since the IR is going to be padded with zeros anyway if it has fewer samples than $N/2$.

The partitioned convolution class is shown in Listing 5.11. It consists of four functions (or methods): an initializer (or constructor, `__init__()`), an impulse response setter (`impulse()`), and the actual convolution processing code (`ola()`). This takes in an array with L samples and returns an array with L samples each time it is called, so we can continue to stream data in and out for as long as it is required.

Listing 5.11 Partitioned convolution class in Python.

```
# partitioned convolution class
# initialised with IR size L
class pconv:
  def __init__(self, L):
   self.L = L
   self.bufin = np.zeros(2*L)
   self.bufout = np.zeros(2*L)

  # sets the impulse response
  def impulse(self, h):
      hh = np.zeros(self.L*2)
      hh[:self.L] = h[:self.L]
      self.H = fft.rdft2(hh)

  # convolution overlap-add process
  # takes an L-size input array
  # returns an L-size output array
  def ola(self, inp):
    l = self.L
```

```
        self.bufin[:l] = inp
        y = fft.irdft2(fft.rdft2(self.bufin)*self.H)
        self.bufout[:l] = y[:l] + self.bufout[l:]
        self.bufout[l:] = y[l:]
        return self.bufout[:l]
```

With the convolution class designed, we can go ahead and use it. The programming approach is slightly different from what we have seen in previous examples. It involves instantiating the class to create an object, then calling the methods on this to do the work. The following code excerpt outlines this:


```
# create a convolution object
cvl = pconv(L)
# set the impulse response
cvl.impulse(h)
# process data
while on is True:
    ...
    out = cvl.ola(s)
    ...
```

where the data input `s` can come from any source and the output buffer `out` can be sent to any destination. Implementing partitioned convolution as a class is a great idea. We will employ it again in situations similar to this one, where an algorithm requires some state to be managed. This approach is called *object-oriented programming*. It is a very important paradigm for sound and music computing, and it is seen in many systems. For example, all opcodes in Csound are classes from which anonymous objects are instantiated, and Csound instruments are themselves classes, invoked via the scheduling mechanism.

5.3.4 Multiple Partitions

While the overlap-add partitioned convolution method works well with any input, it still has some limitations when it comes to realtime audio processing. As it may have been noticed, it requires a buffer of the size of the impulse response size as input. This means that the sound will only begin to come out after the buffer is filled with data from the source. This will impose a certain amount of latency between the input and the output. If the IR is long, for instance, the recording of a room response for reverberation applications, a noticeable delay will be present.

To solve this issue, a variation on the method can be developed to partition the input data as well as the impulse response, enabling a shorter delay, as the partition size can be determined independently of the IR size.[22] This is equivalent to having multiple partitions of the impulse response and a spectral delay line to place the input data blocks. For N partitions of L/N samples, we would then have a N-length delay line, and N products of impulse response and input spectra. It is not unlike the direct convolution approach, except that instead of single samples, we have whole frames of frequency-domain data (as shown schematically in Fig. 5.30). At one extreme where there are $N = L$ partitions of

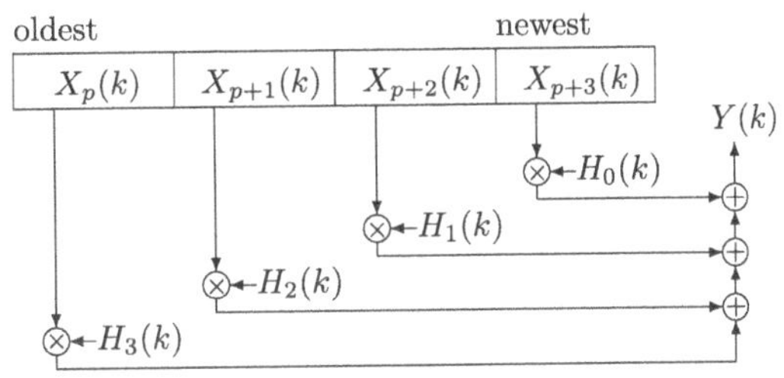

Figure 5.30
A schematic view of multi-partition convolution (4 partitions).

1 sample we have exactly the case of direct convolution. Generally, partitions are going to be much bigger than that. At the other extreme $N = 1$, we have one partition that is the same size as a whole impulse response, which is the case of the single-partition convolution. The size should be determined by balancing the amount of tolerable latency and the efficiency of computation.

The multi-partition overlap-add (OLA) algorithm is structured as follows:

1. Set the total convolution size N to the appropriate power of two to hold $2L - 1$ samples.
2. Decide on the size M of the partition (also a power of two) to give $P = N/M$ blocks of M samples.
3. Compute each spectrum $H_p(k)$ from the P partitions.
4. For each M samples of the input, compute the spectrum $X_p(k)$ of the $2M$-size padded frame and place it in a spectral delay line.
5. Sum the products of each input partition and its corresponding IR partition.
6. Take the inverse DFT of the result.
7. Overlap-add as in the single-partition case (M samples), saving the final M samples for the next iteration.

As can be seen, this is effectively just an extension of the original partitioned convolution algorithm (Listing 5.12). The modifications are creating a spectral delay line, placing the IR in partitions, and when processing, loop over the delay to take the products and accumulate the results. However, the scaffolding provided by the single partition case can be reused considerably. In particular, the basic OLA mechanism is retained without any changes.

Listing 5.12 Overlap-add multi partition convolution class in Python.

```
# multi-partition convolution class
# initialised with IR size L and
# partition size M
class mpconv:
  def __init__(self, L,M):
   self.L = L
   self.M = M
   self.P = L//M
   self.dline = np.zeros((self.P,M+1)).astype(complex)
   self.bufin = np.zeros(2*M)
   self.bufout = np.zeros(2*M)
```

```
        self.p = 0

    def impulse(self, h):
        M = self.M
        hh = np.zeros(M*2)
        self.H = np.zeros((self.P,M+1)).astype(complex)
        for i in range(0,self.P):
          hh[:M] = h[i*M:(i+1)*M]
          self.H[i] = fft.rdft2(hh)

    def ola(self, inp):
      m = self.M
      P = self.P
      self.bufin[:m] = inp
      self.dline[self.p] = fft.rdft2(self.bufin)
      self.p = (self.p+1)%P
      j = self.p
      Y = np.zeros(m+1).astype(complex)
      for i in range(0,P):
         Y += self.dline[(j+i)%P]*self.H[P-i-1]
      y = fft.irdft2(Y)
      self.bufout[:m] = y[:m] + self.bufout[m:]
      self.bufout[m:] = y[m:]
      return self.bufout[:m]
```

An alternative to OLA is provided by the overlap-save (OLS) algorithm. The approach here is to perform the overlapping at the input stage rather than at the output. In this case, we place the current output data in the second half of DFT buffer, which was filled with zeros in the OLA method. After the spectrum is computed, the input is copied (saved) into the first half of the input array, ready for the next iteration. The multi-partition convolution is performed as before, but the second half of the output buffer will now contain the overlapped data, so we can just output it directly.

We might wonder how the convolution output happened to be magically overlapped with no mixing in place. What is happening is that a DFT is taken of two P-size (where P is to the partition size) blocks of input data at a time, no zero padding involved. The first half of the array holds the previous partition and the second half the current one. When the convolution is performed, the current block's tail (final $P-1$ samples) is shifted around to the first half. This is due to the periodic/circular property of the DFT (as shown in Fig. 5.28). With this, the convolution tail of one block (first half) ends up mixed with the next block convolution start. It is therefore slightly more efficient than OLA, since we do not need to perform the mixing and there is no wasting of half of the transform array with zero-padding. In fact, the DFT size may be longer than twice the partition size, making the process even more optimized for computation.

The OLS method can be implemented within an existing multi-partition class, since it uses most of its components in a slightly different way. The following code adds a function to the `mpconv` class to support OLS:

```
def ols(self, inp):
  m = self.M
  P = self.P
  self.bufin[m:] = inp
  self.dline[self.p] = fft.rdft2(self.bufin)
  self.bufin[:m] = inp
  self.p = (self.p+1)%P
  j = self.p
  Y = np.zeros(m+1).astype(complex)
  for i in range(0,P):
     Y += self.dline[(j+i)%P]*self.H[P-i-1]
  return fft.irdft2(Y)[m:]
```


It is written in such a way that it can be used as a drop-in replacement for the OLA function. Note that since we do not need to save the output in the OLS method, we can just output the inverse DFT data directly. Multi-partition convolution is the standard method for fast convolution in modern applications. Careful choice of partition size, balancing efficiency of computation and initial delay, is essential in realtime implementations.

5.3.5 Spectral Design Applications

Csound provides both OLA and OLS multi-partitioned convolution implementations. The opcode `ftconv` implements the former method. Similarly to the direct convolution case, it uses a function table to retrieve an impulse response. The partition size can then be given as one of input parameters,

```
aout = ftconv(asig,ifn,iplen)
```

It can also be used with multichannel IRs, producing more than one output. OLS is implemented by `pconvolve`, which takes an impulse response directly from a soundfile instead,

```
aout = pconvolve(ain,Sfile,iplen)
```

Similarly, multichannel files can be used as IRs and the opcode can then produce more than one input. A simple stereo reverb, using an impulse response recorded from a church ("church.wav"), may be implemented as

```
nchnls = 2
iR = ftgen(0,0,0,1,"church.wav",0,0,0)
instr 1
 asig = diskin:a("fox.wav")*0.05
 aL,aR  ftconv asig,p4,64
 out(aL, aR)
```

```
endin
schedule(1,0,5,iR)
```

where the stereo file samples are copied onto a table (using GEN1) and a table reference is passed to the convolution opcode. This is fed with a mono signal coming from a soundfile, producing a stereo output. Csound is now running with two output channels (`nchnls=2`, the default is mono).

The key aspect in convolution is the choice of IR. In the example above, the recorded response is very resonant and to avoid going beyond the maximum amplitude, it is necessary to scale the input significantly. A different IR may have other characteristics. The result is very much dependent on it. Note that this can be anything, not only the recording the response of a space, but literally any sequence of samples. We have seen in the direct convolution example how a filter IR can be designed, but there are many other possibilities

By using another sound as an IR, the convolution operates a form of cross-synthesis, where the common components of the two inputs. For example, employing a piano C2 tone ("pianoc2.wav") instead of a reverb IR allows the input to excite the harmonics of that sound, creating something that resembles a piano-string resonance chamber:

```
iR = ftgen(0,0,0,1,"pianoc2.wav",0,0,0)
instr 1
 asig = diskin:a("fox.wav",1,0,1)*0.005
 aL = ftconv:a(asig,p4,64)
 out(aL, aL)
endin
schedule(1,0,30,iR)
```

There are many opportunities for interesting non-standard spectral combinations. A careful choice of inputs is required. Note that convolution smears the two sounds in time, so if a continuous sound is used as an IR, a short percussive audio input will create long reverb tails, as each impulse is actually making the IR play from the beginning. For instance, keeping the piano IR and using an impulse generator (`mpulse`) producing an impulse every second, the result is a regular series of piano strikes,

```
iR = ftgen(0,0,0,1,"pianoc2.wav",0,0,0)
instr 1
 asig mpulse 0dbfs/2, 1
 aL = ftconv:a(asig,p4,64)
 out(aL, aL)
endin
schedule(1,0,30,iR)
```

and therefore the whole IR can be "performed" through short impulses in the input. This has interesting applications in composition and realtime computer music performance.

While the multi-partitioned approach can reduce input/output latency considerably, we cannot make it zero unless direct convolution is used. An approach combining more than one convolution opcodes can, however, produce an efficient zero-latency implementation. The idea is as follows:[23]

- The first few samples of the IR use direct convolution.
- The next ones use a small partition of multi-partition convolution.
- The IR tail uses a large partition.

The idea is that direct convolution allows sound to start streaming straight away, covering the beginning of the process. Following this, and offset, partitioned convolution kicks in, and we can have multiple partition sizes coming in later on. The key to this is to make sure the IRs used by each opcode start at the correct time, and for this we need to skip the portions that are covered by the earlier partitions. We also need to limit the length of the earlier partitions so that they do not overlap with the later stages. The `ftconv` opcode allows us to skip the beginning of an IR and to set the IR length required through its optional parameters,


```
aout = ftconv(asig,ifn,iplen,iskips,irlen)
```

which we can use to offset the IR start for the later parts of the convolution, and limit the convolution size in the case of the earlier parts.

An example with three stages, using `dconv` to cover the first `ipart` samples, followed by two instances of `ftconv`, covering the middle (up to `ipart*7`) and the tail of the IR, is given by the following user-defined opcode (UDO),

```
opcode ZConv,a,aii
 asig,ifn,iprt xin
 a1 = dconv(asig,iprt,ifn)
 a2 = ftconv:a(asig,ifn,iprt,iprt,iprt*7)
 a3 = ftconv:a(asig,ifn,iprt*8,iprt*8)
 xout(a1+a2+a3)
endop
```

which can be used as a drop-in replacement for the (mono version) `ftconv` opcode,

```
instr 1
 asig = diskin:a("fox.wav",1,0,1)*0.05
 aL = ZConv(asig,p4,64)
 out(aL, aL)
endin
schedule(1,0,30,iR)
```

In this case, the first 64 samples are covered by direct convolution. The middle part will contain seven partitions of 64 samples, and the final part will use a 512-size partition, covering the tail of the IR (Fig. 5.31). It is of course possible to use more convolution stages in the process, but number of partition sizes used needs to be determined by the computational requirements of the implementation.[24]

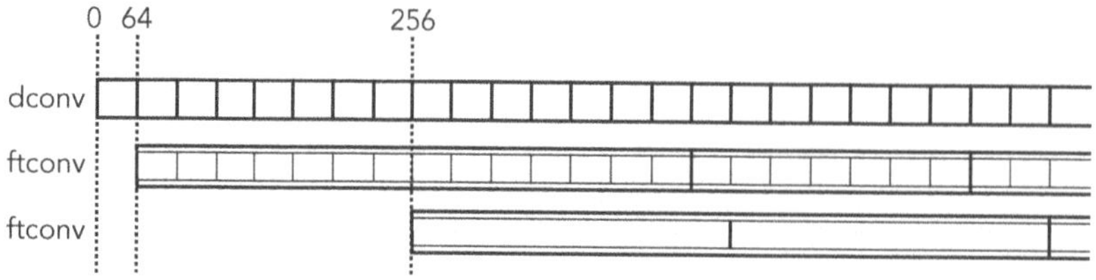

Figure 5.31 Convolution with three partition sizes.

5.4 Time-Varying Convolution

At this stage, we probably have had enough results out of focusing solely on the spectrum and ignoring temporal aspects of signal processing. Although with the streaming approach provided by partitioned convolution we were now able to process arbitrary time-domain signals, whose spectra may of course vary in time, everything we have done so far is linear time invariant. That means that the IRs are fixed, and their application to an input signal does not vary in time. In fact, as we have seen, what is observed is a smearing of the temporal elements of each input.

We can begin now to reintroduce time as a parameter of spectral processing. The first step is to make convolution time varying. This may appear non-standard at first, but it will allow some very interesting applications. The method is fairly simple and intuitive (Brandtsegg et al., 2018):

- Choose an IR length L and divide the full convolution length $2L-1$ into power-of-two partitions as before.
- Instead of pre-storing the IR as fixed spectral frames ready for the convolution, stream in a signal at the same time as the input. New samples replace old samples, and what was the IR becomes continuously renewed.
- Apply the OLA method.

From a spectral perspective, the following operation takes place,

$$Y(t,k) = S(t,k)X(t,k) \tag{5.31}$$

where the input spectra S and X are now functions of two variables, time t and frequency k. Time advances partition by partition, and the output $Y(t,k)$ then needs to inverse-transformed and overlap-added as before.

Making the spectra a function of both time and frequency reintroduces time into the process and makes it linear time varying. Because the IR length is fixed, this can be also characterized as a periodic time-varying filter. In fact, speaking of impulse responses at this point is somewhat unhelpful, as the two signals are streamed in the same way and there is no function difference between $S(t,k)$ and $X(t,k)$. So it is better to start referring to L as a *filter length* (for which the convolution length is $2L-1$).

In a practical implementation, it is better to round up the filter length to a power-of-two size to avoid having to deal with a convolution tail that does not quite fit a DFT frame. Of course, direct convolution implementations would have no such problems, but there are practical realtime computation limits to this. Also, we should note that the OLA method is required here, since the OLS method assumes that the IR is not going be changing in time (this is required for the overlapping to be done at input stage by the DFT).

5.4.1 Implementation

As noted, time-varying convolution is fairly straightforward once we have the OLA implementation in place. It only requires another input and a couple of lines to be added to the original time-invariant code:

```
def tvf(self, inp, inp2):
  m = self.M
  P = self.P
  self.bufin[:m] = inp
  self.dline[self.p] = fft.rdft2(self.bufin)
  # transform the second input
  # place it in the H delay line
  self.bufin[:m] = inp2
  self.H[self.p] = fft.rdft2(self.bufin)
  self.p = (self.p+1)%P
  j = self.p
  Y = np.zeros(m+1).astype(complex)
  for i in range(0,P):
     Y += self.dline[(j+i)%P]*self.H[P-i-1]
  y = fft.irdft2(Y)
  self.bufout[:m] = y[:m] + self.bufout[m:]
  self.bufout[m:] = y[m:]
  return self.bufout[:m]
```

so that the `self.H` buffer now becomes effectively a second delay line. The signals are processed as before, that is, the read/write pointer in this second buffer, moves in the opposite direction from the one used for the other delay line.

5.4.2 Spectral Design Applications

Csound implements time-varying convolution in two opcodes, `liveconv` and `tvconv`, which approach the process from two different directions. The first one, implementing what is called *live convolution*, is effectively a modification of the fixed-IR `ftconv`, which we have studied earlier. It takes an impulse response from a function table, but allows this to be updated at any time (or even continuously),

```
asig = liveconv(ain,iftn,iplen,kupdate,kclear)
```

where parameters are as in `ftconv`, with the added `kupdate` switch, which accepts two values (1 and −1) for loading and unloading the IR. So if this is set to 1, the IR is loaded partition by partition from the function table, whereas setting this to −1 unloads the convolution partition by partition. The `kclear` parameter works like a *panic* switch, clearing all delay line buffers immediately.

The following code example demonstrates the operation of live convolution. It uses three instruments; the first writes a signal to the IR function table, the next one can be used to switch convolution off (by unloading the IR), and the final instrument is the actual convolution engine. We schedule instrument 1 to regularly update the table, and the live convolution is constantly reading from this, therefore replacing the old IR if a new one is available. The convolution is between two soundfiles, "beats.wav"[25] and "fox.wav" (which have used before). The output will contain a rhythmic loop of the words read in the second file, defined by the beat patterns in the first:

```
// IR load switch
giload = 1
// IR ftable
isize = 2^18
print(isize)
iir =   ftgen(1,0,isize,2,0)
// IR duration
idur = ftlen(iir)/sr

// instr 1 writes to ftable
instr 1
 giload = 1
 iskp = p5
 asig = diskin:a("fox.wav",1,iskp)
 amp = linen(1,0.01,p3,0.01)
 andx = phasor(1/p3)
 tablew(asig/0dbfs,andx,p4,1)
 out(tablei:a(andx,p4,1)*0dbfs/8)
endin

// instr 2 switches
// convolution off, unloading
instr 2
giload = -1
endin

// instr 3 is the
// convolution engine
instr 3
 kup = giload
 asig = diskin:a("beats.wav",1,0,1)
 aconv = liveconv(asig,p4,2048,kup,0)
 out(aconv*.005)
endin

// run conv engine for 50s
```

```
schedule(3,0,50,iir)
// load IR with soundfile
schedule(1,0,idur,iir,0)
// stop convolution
schedule(2,2,0)
// start again
schedule(1,6,idur,iir,0.4)
// replace IR at 16,26,36 secs
schedule(1,16,idur,iir,0.7)
schedule(1,26,idur,iir,1)
schedule(1,36,idur,iir,2)
// stop convolution
schedule(2,46,0)
```


The live convolution approach is designed for applications where we would like to replace the IR periodically, but effectively reuse for a certain amount of time. In the example above, because the function table samples are only renewed about every 10 seconds, the convolution is fixed for a while at each stretch of time.

The approach taken by `tvconv` is to do away with the function table altogether and implement convolution as a periodic time-varying filter, as in the Python code example above. In this case, we have two inputs, partition, and filter sizes,

```
asig = tvconv(asig1,asig2,xfreez1,xfreez2,iparts,ifils)
```

and the two other parameters `xfreez1` and `xfreez2` work as *freezing* switches, which stop the signal going into the delay lines if they are set to 0.

As the opcode implements time-varying convolution in the full, straightforward, fashion described earlier, it can lead to more radical results. The example below shows an instrument that uses `tvconv` and the same two soundfiles used in the live convolution example. The instrument uses two gating control signals[26] (based on a square wave) that can be used to freeze periodically each input. If the gating frequency is 0, then the gate is always high (1) and there is no freezing. These signals have slight different frequencies which make them drift slightly and sometimes freeze one input, while the other is running, and vice-versa. The gating function is constructed using GEN7, which creates a table with piecewise linear sections.

```
// gating function (square)
igt = ftgen(0,0,1024,7,1,512,1,1,0,511,0)

instr 1
 // inputs
 asig1 = diskin:a("fox.wav",1,0,1)*p4
 asig2 = diskin:a("beats.wav",1,0,1)*p4
 // HP filter to attenuate low drums
 asig2 = buthp(asig2,1000)
 // gating signals with 10% drift
 afrz1 = oscili(1, p5*0.95,p6)
```

```
  afrz2 = oscili(1, p5*1.05,p6)
  // time-varying convolution
  asig = tvconv(asig1,asig2,afrz1,afrz2,p7,p8)
    out(asig)
endin

// short filter, no gating
schedule(1,0,10,.5,0,igt,256,1024)
// short filter, .5~Hz gating
schedule(1,10,20,.5,.5,igt,256,1024)
// longer filter (>1s), no gating
schedule(1,30,10,.3,0,igt,256,sr)
// longer filter (>1s), .1HZ gating
schedule(1,40,40,.3,.1,igt,256,sr)
```


This example demonstrates also the different results obtained by varying the filter length. Short filters enable very snappy time responses, but the cross-synthesis blending of the two spectra is less pronounced. Also in this case, freezing the signals can lead to buzzing artifacts. Longer filters produce a smoother result, with more smearing of the details in the sounds (sometimes also sounding close to reverberation or resonances), but achieving more pronounced cross-synthesis. Freezing the inputs can be used to create looping repetitions. Time-varying convolution presents great creative potential, especially for live performance applications, where the blending and interactive characteristics of the method can be exploited in many interesting ways.

By reintroducing time as a variable in spectral processing, we may be able to balance a reasonable time response with the desired frequency-domain effect. More possibilities are opened out as we move away from convolution-based method to consider other streaming approaches. In particular, the OLA mechanism, in a slightly different form from the one used here, will be fundamental to support the development of a series of time-frequency algorithms. Therefore, we should consider the ideas explored in discrete-time convolution also as stepping stones toward a wider range of possibilities for spectral design.

5.5 The Discrete Spectrum

This chapter opened the way for computational methods of spectral design. We were able to begin applying the fundamental theory developed for the continuous cases to discrete-time, discrete-frequency transforms. The DFT was first cast as a version of the formulae for the computation of coefficients for the Fourier series. This analogy was useful because it provided support for half of the work, since the FS implies a discrete-frequency spectrum. From there, all it required was to move from continuous-time to sampled input signals. Since we have already explored their spectra, the transform method was straightforward to derive.

In the spirit of advancing the definition of spectrum, to which we have been adding at each chapter, the ideas built into the DFT tell us that

The discrete-frequency spectrum of a discrete-time signal is finite sequence with the same length as its time-domain counterpart, which can be obtained by the application of the discrete Fourier transform. It is also periodic in frequency and describes a waveform that is periodic, both periods being equivalent to the length of the sequence. As in the continuous-time case, an inverse transform can be used transparently to recover the original time-domain input. The spectrum as given by the transform is ordered in such way that it contains the non-negative frequency samples first, followed by the negative frequency samples, in ascending order. Therefore, for audio signals, it is Hermitian-symmetric about the center of the sequence.

Refining our computational methods, an efficient form of the DFT for highly composite frame sizes, the FFT, was introduced and derived. This led to the development of a complete spectral analysis and resynthesis algorithm for real-valued inputs, which took advantage of the specific characteristics of audio signals. We were able to show practical implementations of this in Python. This led to the introduction of a practical audio processing application of DFT, discrete-time convolution. We explored this first as a linear-time invariant method, with which different types of filtering could be implemented, as well as reverberation and cross-synthesis. The discussion was completed by reintroducing time as a parameter of spectral processing in time-varying convolution applications. This pointed the way to the development of streaming spectral processing algorithms, which are based on functions of both frequency and time. This is the direction we will head toward in the next chapter.

Notes

1 As we had anticipated, we now simplify the notation of sequences, replacing subscripts with a straightforward functional form.

2 Since we have included the normalization factor in the forward transform, we omit it here.

3 Note that we can represent the complex sinusoid either as an exponential expression or through the exp() function.

4 In Chapter 4, it was shown that a spectrum sampled at f-spaced points defines a waveform with period $1/f$.

5 This approach was pioneered by Gabor in his seminal article (Gabor, 1940), as a means of taking good localized analyses. As we will see in Chapter 6, windowing is an essential element of time-frequency analysis and synthesis. In his work, Gabor employed a Gaussian window, and the FT performed on such windowed signals is sometimes called the *Gabor transform*.

6 While the full application to the Fourier series was first introduced in Cooley and Tukey (1965), the mathematical bases for these methods have a long history, which can be traced back to Gauss (Heideman et al., 1985).

7 Their results actually showed that a radix-3 algorithm would be marginally faster (by around 6%), but it would not have the advantages brought by an implementation in binary computers.

8 We are excluding here the $1/N$ scaling term for sake of simplicity.

9 Elster (1998) discusses this and related algorithms.

10 Although this is not the case in Python, tail-recursive functions such as the one we will develop here can be optimized in some languages by the compiler/interpreter, putting these on a par with iterative code as far as performance is concerned.

11 This is a tail-recursive function, since the self-reference call is the last operation in each case (except for the final return to the caller, of course).

12 Note that we could also take advantage of the relationship

$$X(M-k) = \overline{R(k) - I(k)},$$

implied by Eq. 5.19, to calculate the second set of $M/2$ points from $M/2 \leq k < M$ instead of just using Eq. 5.21 for the whole spectrum. This may be a better approach if we want to save memory and process the samples in place, particularly if we are using a loop to iterate over the data.

13 A tit-reverse operation?

14 Also called a unit-sample impulse, this is the discrete-time counterpart of the Dirac delta function.

15 A discrete convolution function was already shown in Chapter 4, Listing 4.1. However, that code was more general (it allowed two-sided convolution, for instance), and was not designed for streaming audio applications.

16 If they are not allpass as in the case of the pure delay.

17 The combined effect on the amplitude and phase of a signal is known as the *frequency response*. Often this term is used to mean solely the amplitude response of a system, but that is not formally correct.

18 N is the IR size. We are giving this here with no further explanation or derivation, except for some empirical backing. However, we will be able to come back to the subject later on and give a better justification for this assertion.

19 This employs the functions `rdft2()` and `irdft2()` from the custom `fft` Python library we have been developing. These implement slight variations on the code discussed earlier (for `rdft()` and `irdft()`), placing the scaling by $1/N$ at the inverse transform stage. For numerical reasons, this suits better the computation of fast convolution, since the product will not include an extra $1/N$ factor. Another difference is that they employ an iterative implementation of the FFT,

```
def stage(s,w,N,n,k):
  n2 = n*2
  kn,kn2 = k+n,k+n2
  e,o = s[k:kn],w[::N//n2]*s[kn:kn2]
  s[k:kn],s[kn:kn2] = e+o,e-o

def fft2(s,fwd=1):
  N = s.size
  n = np.arange(0,N//2)
  w = np.exp(-fwd*2j*np.pi*n/N)
  s = reord(s,N)
  k, n = 0, 1
  while n < N:
   while k < N:
    stage(s,w,N,n,k)
    k += n*2
   n *= 2
   k = 0
  return s
```

which is faster and allows for larger DFT lengths, since with the recursive implementation, sizes $> 1,024$ break the maximum recursion depth in Python.

20 This can also be called periodic convolution, as opposed to the aperiodic form, for which we need to zero-pad inputs (Stockham, 1966).

21 This algorithm was first discussed in the computer music literature by F. R. Moore (1990). However, the principles of fast convolution, including partition segmentation, were introduced in Stockham (1966), not very long after Cooley and Tukey published their FFT paper.

22 This idea was first introduced in Soo and Pang (1990) for the computation of the frequency-domain least-square mean algorithm. It was then used in time-domain convolution for surround sound applications in Torger and Farina (2001). However, in that paper, the method does not employ a spectral delay line, but does the mixing of the output blocks in the time domain.

23 This method was first discussed in Gardner (1995), albeit without the multi-partition approach used here. In that paper, multiple-size partitions were employed but as separate single-partition convolution streams mixed together.

24 In Lazzarini et al. (2016, 305) a multi-size partition opcodes is shown, where the number and growth ratio of partition sizes can be defined for each instance.

25 This soundfile is also distributed with the Csound reference manual and is a recording of drums playing a short beat pattern.

26 A gate is a signal that is either 0 or 1, such as, for instance, a non-bandlimited square wave, and can be used as synchronizing control, or to switch on/off certain parameters.

6

Time-Frequency Processing

Readmitting time as a parameter for our processes is a fundamental step. After all, music and sound design are time-dependent forms of art, and the removal of most temporal considerations was never completely resolved in the discussions of the previous two chapters. Here we will embark in the development of ideas that fully embrace time and frequency in equal balance. This brings to the fore a somewhat novel idea for us, that of *spectral signals*.

So far we have reserved the use of the word signals for time-domain functions, which in a digital environment means a sequence samples, one for each time point. However, now we begin to see a series composed of whole spectral frames, instead of single samples. That is what goes on in time-varying convolution, and it becomes even more explicit as we develop a full method for streaming spectral processing. Therefore, we may now describe these functions of time and frequency as spectral signals. This approach allows us to develop various ways to modify sounds, and to be able to manipulate their time and frequency dimensions in an independent manner.

6.1 Sub-band Signals

The first step in this discussion is to consider time-domain signals at *sub-band level*, that is, removing part of the frequency spectrum and concentrating on a particular range of frequencies. From what we have studied so far, this can be achieved by employing some sort of filtering, which should be of a bandpass type, if we are speaking of a region in the middle of the spectrum.

6.1.1 Designing a Bandpass Filter

Let's start by designing such a filter. We should have everything we need for the task. Remembering the principles of convolution in the frequency domain, if we have a low-pass shape we are happy with, centered at 0 Hz, we can then move it around the spectrum by multiplying its corresponding impulse response (or its inverse transform) by a sinusoid. This results in the filter being centered at the sinusoid frequency. With such an impulse response $h(n)$, we can then apply convolution in the time-domain with an input waveform $x(n)$ to produce a sub-band signal centered at frequency f,

$$s_f(n) = x(n) * h(n) \tag{6.1}$$

Spectral Music Design. Victor Lazzarini, Oxford University Press. © Oxford University Press 2021.
DOI: 10.1093/oso/9780197524015.003.0006

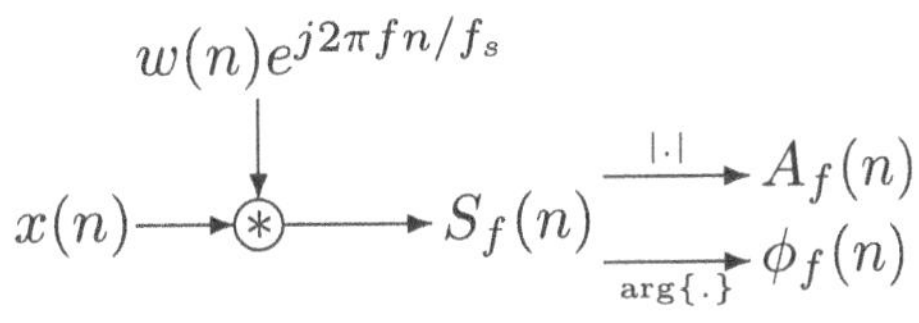

Figure 6.1
Sub-band analysis signal flowchart.

Therefore, the first consideration is what lowpass shape to use. Looking back at Chapter 5, we can see that all the various window shapes employed have lowpass curves. The rectangular window is problematic in that it may not be selective enough as its spectrum has quite substantial side lobes, but the triangle and hanning windows are good options. We may not need to decide on this now, but it is sufficient to say that the lowpass shape is a window of some description with good suppression characteristics. This leads us to derive an IR for a digital bandpass filter centered at frequency f, as

$$h(n) = w(n)e^{j2\pi fn/f_s} \tag{6.2}$$

where f_s is the sampling rate (time in seconds is of course given by n/f_s), and $w(n)$ is whatever window shape we chose. Note that we are using a complex exponential as it is the most general form of a sinusoid. We should be used to this approach by now.

With the design of the filter ready, we can apply convolution to get $S_f(n)$,

$$\begin{aligned} S_f(n) = x(n) * h(n) &= \sum_{m=-\infty}^{\infty} x(m)h(n-m) \\ &= \sum_{m=-\infty}^{\infty} x(m)w(n-m)e^{j2\pi f[n-m]/f_s} \\ &= e^{j2\pi fn/f_s} \sum_{m=-\infty}^{\infty} x(m)w(n-m)e^{-j2\pi fm/f_s} \end{aligned} \tag{6.3}$$

We have done very well so far deriving the sub-band signal, but what can we do with it? First it should be noted that $S_f(n)$ is complex and can be put in a rectangular form as

$$S_f(n) = a(n) + jb(n) \tag{6.4}$$

therefore we can easily derive the amount of energy $A_f(n)$ in the sub-band by taking its magnitude,

$$A_f(n) = |S_f(n)| = \sqrt{a(n)^2 + b(n)^2}. \tag{6.5}$$

Moreover, we can also extract its sample-by-sample phase,

$$\phi_f(n) = \arg\{S_f(n)\} = \tan^{-1}\left(\frac{b(n)}{a(n)}\right). \tag{6.6}$$

This may be described as an amplitude modulation/phase modulation decomposition performed at a sub-band level. From this equation, we can go one step further and obtain the frequency as the derivative of the phase.[1] The flowchart for this sub-band analysis filter is shown in Fig. 6.1.

6.1.2 The Phase Vocoder

Now consider a bank of such filters, centered at different (but contiguous) frequencies; from each one we can derive an amplitude and frequency signal (Fig. 6.2). With

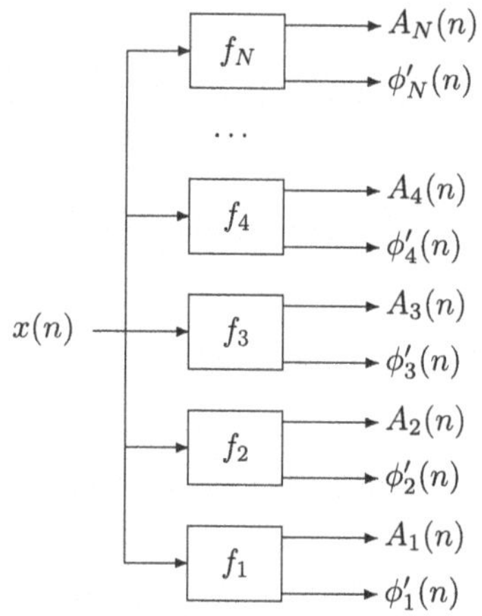

Figure 6.2
The phase vocoder filterbank.

these, we have a complete streaming analysis system from which we can derive signals to drive a bank of oscillators for additive synthesis. Furthermore, and this is the real gain, we can perform all sorts of standard and non-standard modifications to these signals. Suppressing or boosting the amplitude signals in different bands can be used to create time-varying filters. Frequencies may be scaled in order to change the pitch of a tone, independently of time considerations. Interpolating the amplitude and frequency signals, or throwing away samples from these, allows us to apply timescale modifications. Many other operations are possible.

This sub-band method is called the *phase vocoder* (PV) and was first introduced by Golden and Flanagan (1966). The name was given due to the fact that, unlike the original vocoder, it also carried the phase signals alongside the amplitudes in each sub-band, although these are normally translated into frequencies in most implementations.[2] It was a seminal idea, but it took a little while to be adopted more widely, as the originally proposed standard implementation required significant amounts of computation. However, almost in synchrony, Cooley and Tukey brought out the fast Fourier transform algorithm, which was then later applied to the computation of PV signals,[3] allowing practical implementations of the phase vocoder filterbank analysis method to be used in musical applications.[4]

6.2 The Short-Time Fourier Transform

The link between the phase vocoder and the DFT needs to be established for us to be able to take advantage of the FFT in its implementation. While this connection may not be apparent at first, the fact that we have been employing windows, complex exponentials, and convolution possibly gives a clue that we could translate the PV formula as a spectral transform. If we track back to Eq. 6.3,

$$\begin{aligned} S_f(n) &= e^{j2\pi fn/f_s} \sum_{m=-\infty}^{\infty} x(m)w(n-m)e^{-j2\pi fm/f_s} \\ &= e^{j2\pi fn/f_s} F_f(n) \end{aligned} \tag{6.7}$$

we may note that the inside sum (represented by $F_f(n)$) is very familiar. It looks like some sort of variation of the Fourier transform of a signal $x(m)w(n-m)$, where the time is discrete, but the frequency is not (we can plug any value of f there). We can concentrate on $F_f(n)$, and the complex sinusoid may be ignored for now, since it only shifts the spectrum of a signal to be centered on f. By removing it, the spectrum of the sub-band becomes centered on 0 Hz.

If the perspective is changed from considering this as a time-domain signal $F_f(n)$ to looking at it as a function of frequency[5] $F_n(f)$, we find the answer to our initial impression. From this point of view, there is actually a name for this expression; it is called the (discrete-time) *short-time Fourier transform* (STFT),[6]

$$X_n(f) = \sum_{m=-\infty}^{\infty} x(m)w(n-m)e^{-j2\pi fm/f_s}. \tag{6.8}$$

The *short time* of the title refers to the fact that the window is expected to have a finite non-zero portion, so the signal is only observed over a limited interval. Now we are definitely in familiar territory as we have already discussed the use of windows with the DFT and this is not too different from what we have done before. The main obstacle now is to move this expression from an infinite sum to the more familiar interval $[0, N-1]$, and as a result obtain a discrete-frequency formula. Once we have done this, the FFT then can take over the computation duties.

We will proceed in stages.[7] The first step is to make the substitution $m = m + n$ in Eq. 6.8, so that we can make the window function depend on m only,

$$\begin{aligned} X_n(f) &= \sum_{m=-\infty}^{\infty} x(m+n)w(-m)e^{-j2\pi f[m+n]/f_s} \\ &= e^{-j2\pi fn/f_s} \sum_{m=-\infty}^{\infty} x(m+n)w(-m)e^{-j2\pi fm/f_s}, \end{aligned} \tag{6.9}$$

which brings out the complex scaling factor $e^{-j2\pi fn/f_s}$. This form of the STFT allows us to place the sum in a finite form, since the window $w(n)$ has a finite non-zero length N, in the interval $m = [0, N-1]$,

$$X_n(f) = e^{-j2\pi fn/f_s} \sum_{m=0}^{N-1} x(m+n)w(-m)e^{-j2\pi fm/f_s}. \tag{6.10}$$

Since the window is supposed to be symmetric, the fact that it is being used backward does not have any bearing on the result.

We need to decide now how many analysis bands we will employ. It makes sense to make this equal to N, since we have that number of input samples. Therefore, if we have N frequency bands, which are multiples of f_s/N, the following relationship applies:

$$f_k = k\frac{f_s}{N}, \tag{6.11}$$

and the sum becomes

$$X_n(k) = e^{-j2\pi kn/N} \sum_{m=0}^{N-1} x(m+n)w(-m)e^{-j2\pi km/N}. \tag{6.12}$$

So we have succeeded in finding a way to put the expression in terms that can be computed with an FFT. The expression in Eq. 6.12 is a frame of DFT data scaled by a complex

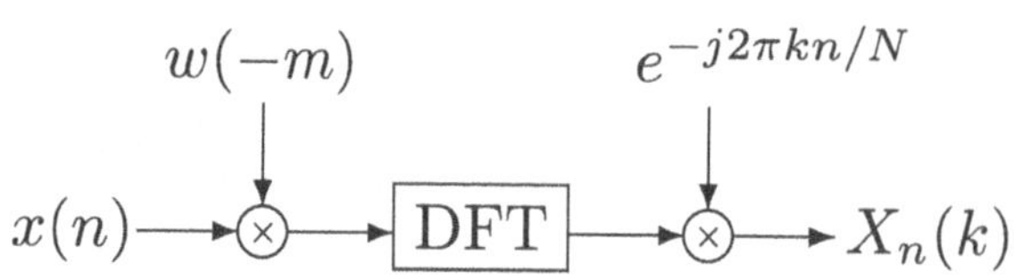

Figure 6.3 Spectral analysis flowchart.

factor, which operates on N samples of the input starting at time n. The STFT operates on each sample of the input, sliding the window forward on a sample-by-sample basis, picking up N samples at a time, and producing N pairs of amplitude and frequency data (one per sample). When viewed from the perspective of spectral analysis, we have a frequency-domain signal, a time sequence of DFT frames. We can also look at this from a different standpoint: at each frequency there is a separate sub-band time-domain signal. The flowchart in Fig. 6.3 provides a schematic view of this STFT-based streaming spectral analysis process.

6.2.1 Analysis Frame Rate

As it stands, we are moving the analysis window on a sample-by-sample basis. If we were to implement this sequence of DFTs at a normal sampling rate (say, a new frame 44,100 times per second), even with the FFT it would be a struggle. Therefore, the current method is not practical unless we can reduce the *analysis frame rate*, or how many times a second we produce new output data. This will be directly linked to what we can get away with when trying to reconstruct the input with an inverse transform (or additive synthesis via an oscillator bank).

Intuitively we should expect that with a rectangular window, we could safely jump N samples and still be able to resynthesize the results with no problems; after all, the inverse DFT will be reproducing the data that is cut from the original and then it is just a matter of concatenating the outputs. This is correct, but not particularly helpful, since the goal is to be able to change the spectral data in some way, and once that is done, the simple concatenation will break down. This is because we will not be able to ensure smooth continuity between adjacent frames, and there will likely be jumps at the frame boundaries causing clicks at the output.

Since rectangular windows are a bad idea, we are now looking at the usual suspects, shapes that decay toward zeros at each end. Of these, the hanning window is the most commonly used and it is a good idea to stick with it, since it has a simple well-described spectrum and time-domain waveform. However, intuitively again, we should expect that we cannot hop one full frame of N samples ahead and expect the input to be reconstituted perfectly for the very fact that the samples at the edge of the frame are zero. The result will be an amplitude-modulated signal. The question now is what is the fewest number of samples we can jump ahead and still have a well-reconstructed output? That will determine the minimum frame rate.

To answer this question, it is best to look at the issue from a sub-band perspective. Remembering that each band is a filter whose spectral shape and width are determined by the analysis window, it is possible to know the minimum sampling rate required to precisely encode the amplitude and frequency signals produced by the analysis. Note that since at the very start we removed the band-centering complex factor, the signal on each

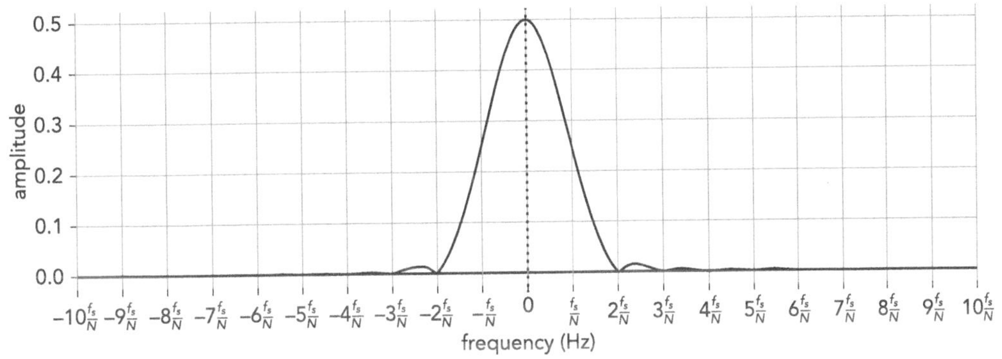

Figure 6.4
Magnitude spectrum of the hanning window, showing its effective bandwidth of $4\frac{f_s}{N}$.

band is a low-frequency waveform close to 0 Hz. This should give us enough support to look for a lower sampling rate for the analysis.

As we know very well at this stage, the digital baseband extends from $-f_s$ to $+f_s$, therefore the minimum rate f_s for the signal will be determined by the extent of the filtering bandwidth. This is because the window acting as a filter guarantees that the signal in the sub-band does not contain any components beyond a given band. For the hanning window, from Chapter 5 we know that its effective bandwidth B in Hz is

$$B = 4\frac{f_s}{N}, \tag{6.13}$$

as shown in Fig. 6.4.

This can be translated into an analysis rate of $4f_s/N$ frames per second. To realize this, each window needs to be overlapped by 1/4 length. Therefore if we hop by at most N/4 we will be able to resynthesize the signal without aliasing. For a hopsize $D = N/O$, where O is the number of overlapping frames, the sub-band sampling rate is f_s/D. Since the analysis signals are low-frequency (slow-varying), it is expected that we can downsample them at a fraction of the input signal sampling rate, f_s. The exact frame rate and the maximum hopsize D are properties of the particular analysis window employed.

6.2.2 Phase Alignment

We have until now ignored the complex exponential factor $e^{-j2\pi kn/N}$ that scales the DFT in each analysis frame, but it is important to explore its meaning. The first thing that should be noted is that it has no effect whatsoever on the amplitude spectrum (it has unity magnitude). Also, if we were hoping the analysis by N samples, it would not have any effect on the phases spectrum either. It is only when the analysis time point n is moved between multiples of N that it does anything. We can interpret it as a phase alignment operator, which makes sure that the phase of successive analysis is captured accurately in relation to each other.

In Chapter 5 we have already seen that the DFT is periodic in time, and any linear phase shifts such as this one are circular. This may be expressed as

$$e^{-j2\pi kn/N}X(k) = \sum_{m=0}^{N-1} \mathrm{R}_N^{-n}[x(m+n)w(-m)]e^{-j2\pi km/N}, \tag{6.14}$$

where the circular rotation operator $\mathrm{R}_N^n[]$ is defined as

$$\mathrm{R}_N^n[s(m)] = s([m+n] \bmod N). \tag{6.15}$$

Since the window hops forward by D samples, each analysis has a D/N-frame backward circular rotation relative to its predecessor. We can either effect the scaling in the frequency domain or apply the circular shift in the time domain. If the output is already in polar form, the multiplication becomes an addition, in which case it may be simpler to do the latter operation. However, if the analysis system includes the correct rotation as part of the algorithm, the complex factor is automatically taken into account. This may be the most convenient way to implement a streaming frequency analysis.[8]

6.2.3 Resynthesis

Since we are analyzing the input at a frame rate of f_s/D, at each band output signals are being updated at a rate that is a fraction of the synthesis sampling rate. At the analysis stage we have *decimated* the input, that is, discarded the intermediate values between points. Therefore when we come to resynthesize it, for each sample of the analysis signal, there are $D-1$ others we do not have any data for. Some sort of interpolation is required to up-sample the analysis signal back to the sampling rate.

It is possible to consider this process as smoothing the transition between the successive samples of spectral analysis. By now we should already know that such an operation requires some sort of lowpass filtering. So, to implement this we can apply a convolution of the analysis signals and an appropriate filter impulse response. A good candidate for the job is the same window we used in the analysis, since we have already demonstrated that it has the correct spectral shape and bandwidth. It only then needs to be centered at the band frequency, for which we can multiply by a sinusoid. We can proceed using the following steps. For each band,

1. Take the sub-band amplitude $A_k(i)$ and phase $\phi_k(i)$ (integrated from the frequency $\phi_k'(i)$) signals, where i is the frame index, which are sampled at f_s/D Hz.
2. Re-sample these at f_s Hz, by selecting their values at $n = iD$, and filling the intermediate samples with zeros, producing $A_k(n)$ and $\phi_k(n)$.
3. Construct a filter impulse response composed of the window function $w(n)$ and a sinusoid at the band center frequency,

$$w(n)e^{j2\pi f_k n/f_s}. \tag{6.16}$$

4. Apply the convolution to generate the output signals, taking the real part of it (as this is a proper audio signal), and disregarding the STFT phase rotation,

$$s_{f_k}(n) = \Re\left\{\sum_{m=-\infty}^{\infty} A_k(m)w(n-m)e^{j(2\pi f_k[n-m]/f_s+\phi_k(m))}\right\}. \tag{6.17}$$

The complete output signal is the mix of all k bands,

$$x(n) = \sum_{k=0}^{N-1} s_{f_k}(n). \tag{6.18}$$

This gives us the general lines of a resynthesis method. What we need to do now is to cast this in terms of a practical synthesis method so that we can implement Eqs. 6.17 and 6.18 in a practical way. We may re-interpret the time-domain convolution of Eq. 6.17

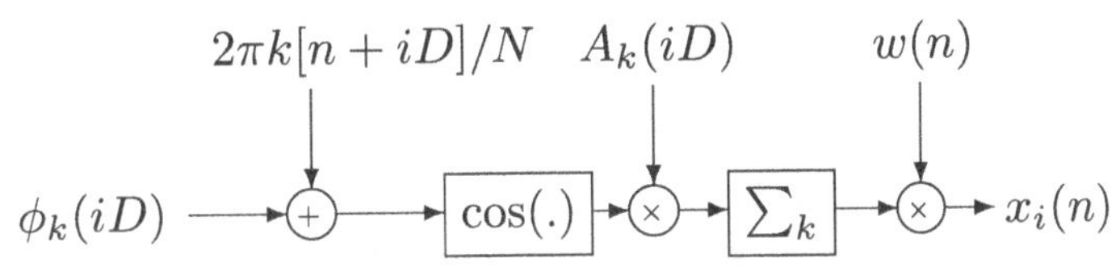

Figure 6.5 Additive frame resynthesis flowchart.

as an overlap-add (OLA) operation. Viewed from the perspective of each analysis time iD, the output is an N-size window $w(m)$ applied to the time-domain synthesis frame. Considering that for $m \neq iD$, i integer, $A_k(m) = 0$, then Eq. 6.17 can be re-written as

$$s_k(n) = \Re\left\{ e^{j2\pi kn/N} \sum_{i=-\infty}^{\infty} w(n - iD) A_k(iD) e^{j[\phi_k(iD) - 2\pi kiD/N]} \right\}, \tag{6.19}$$

where $f_k = kf_s/N$. This represents an OLA operation, where each frame i is[9]

$$s_k(i,n) = w(n) A_k(iD) \cos\left(2\pi k \frac{n+iD}{N} + \phi_k(iD)\right) \qquad 0 \leq n \leq N-1. \tag{6.20}$$

To produce the sub-band signal, we need to overlap-add a sequence of these, each starting at a consecutive analysis time point, as per 6.19. So if the spacing between these is D samples, there will be N/D frames contributing to every single sample of the output. Thus, the OLA operation is equivalent to the shift, scale, and mix process of convolution as first described in Chapter 4. The window is shifted to the start of each new synthesis frame, scaling it, and the result is summed together.

The method therefore produces as an output a mix of successive synthesis frames, but we have only outlined the form of these. If we are employing an additive synthesis method, then Eqs. 6.18 and 6.20 can be put together, giving[10]

$$x_i(n) = w(n) \sum_{k=0}^{N-1} A_k(iD) \cos\left(2\pi k \frac{n+iD}{N} + \phi_k(iD)\right) \tag{6.21}$$

for $0 \leq n \leq N-1$. This method has the advantage of being conceptually simple (Fig. 6.6). It is also fairly flexible for applying modifications to the amplitude and frequency data, and for omitting bands if we want to.

Its disadvantage of course is the computation complexity of using one oscillator per band. As an alternative, we should expect that an inverse STFT method computed by the FFT algorithm is also possible. Starting with a spectral-domain frame at analysis point i, $X_i(k)$, we have

$$x_i(n) = w(n) \sum_{k=0}^{N-1} X_i(k) e^{j2\pi k[n+iD]/N} \qquad 0 \leq n \leq N-1, \tag{6.22}$$

which is equivalent to an inverse DFT rotated circularly at time points iD (Fig. 6.6).[11] This is much more efficient, but also more restrictive, as the input to resynthesis has to be in the correct format as expected for a DFT, and spectral manipulations may need to be more carefully planned (in particular the ones applied to the frequency data). Whichever synthesis method is chosen, each frame $x_i(m)$ is overlap-added to compose the output signal.

Finally, we should consider the question of DFT normalization. This is not explicitly given in the equations above, but could be present as a factor in the windowing. In

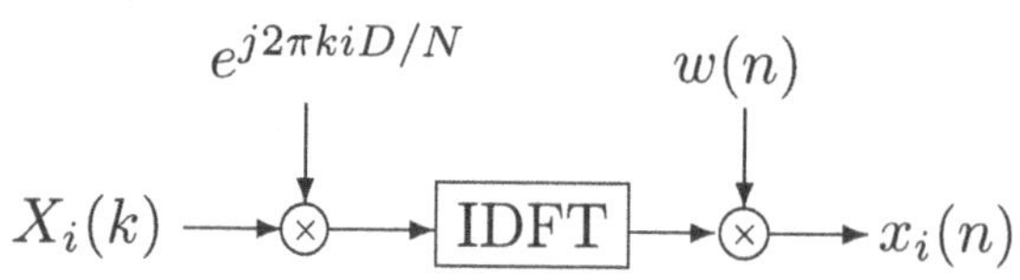

Figure 6.6 DFT-based frame resynthesis flowchart.

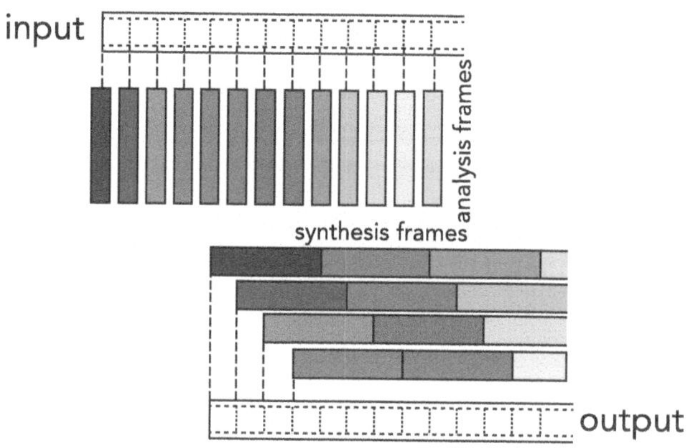

Figure 6.7 Spectral analysis-synthesis schematics.

Chapter 5, we first applied scaling by $1/N$ to the forward transform. The advantage of this is that all amplitudes are given then as true values, without the presence of an N factor, making a reading of the spectrum much cleaner. Later on, we moved this to the inverse transform, to avoid any numeric issues when taking the product of two spectra. Whatever we decide to do, it is always important to make sure the scaling is applied at one or the other stage, or otherwise the output will most likely be completely out of range.

6.3 Spectral Analysis-Synthesis

The mechanism provided by the discrete short-time Fourier transform and its inverse is the basis for the complete spectral analysis-synthesis system we are seeking to implement in this chapter. It takes in an input signal and analyzes it at evenly spaced points determined by the hopsize D. This constitutes a spectral signal, that is, a sequence of DFT frames, or from another perspective, parallel sub-band complex time-domain signals, downsampled to f_s/D Hz. This latter view considers the DFT of a windowed signal as a bank of overlapping filters, as already discussed in Chapter 5. The time-domain signal can be reconstituted by employing the inverse DFT, windowing, and overlap-adding the successive time-domain synthesis frames. Between analysis and resynthesis, the spectral data can undergo various types of modifications, which may be standard as well as non-standard.[12] A schematic diagram of this spectral analysis-synthesis system is shown in Fig. 6.7.

We have a few details to clear up before we can proceed to develop an implementation. Firstly, since the discussion so far has been approached from a general perspective, the principles of the real-to-complex transforms developed earlier have not been applied. As we progress closer to practical applications, we should re-introduce these. Therefore, now it should be determined that, for a window of size N, the output of the analysis

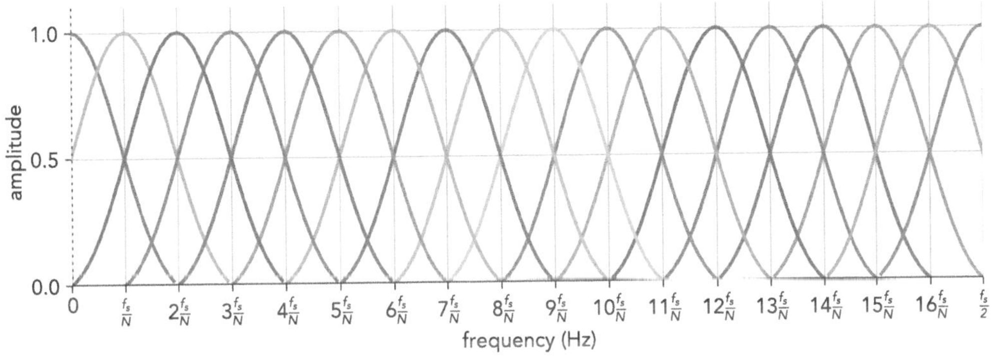

Figure 6.8 Overlapping equal-bandwidth spectral analysis bands for $N = 32$, using a hanning window.

step will be composed of $N/2 + 1$ bands (or bins), covering the non-negative spectrum (Fig. 6.8). As we employ a real-to-complex and complex-to-real transforms, the negative spectrum is treated silently and properly.

At the analysis output, each bin will be f_s/N Hz wide. Since this is fixed across the spectrum, the analysis becomes more perceptually selective as the frequency increases. This means that lower-order bins encompass wider musical intervals than the higher-order ones. For example, there is an octave spanning 100-200 Hz, and another between 200 and 400 Hz; therefore, if bands are equal-sized, the first octave will be covered by half the number of bins than the second. Another way to put this is to introduce the filter selectivity factor Q, defined as

$$Q = \frac{f}{B} \tag{6.23}$$

where f is the bin frequency and B its bandwidth.[13] The DFT-based analysis method can then be defined as equal-bandwidth, but not equal-Q. Ideally, we would prefer the latter, but that is not achievable directly. All the conveniences provided by the straight FFT output against the complexities of computing a log-based transform justify the use of a model that is imperfect from a perceptual point of view.

The resolution of the analysis is therefore fixed, as each bin is able to handle one component and provide its amplitude and frequency (or phase). The precision in frequency can be increased at the expense of time, as determined by the uncertainty principle. If we double the size of the frame, the spectrum will be split into more finely separated bins, but at the same time the analysis will be averaging the waveform over a longer period. A question may be raised as to how does the analysis cope with more than one component falling within each bin. In that case, both amplitude and frequency will refer to the interactions between these, with average values being produced, which will possibly also vary on a frame-by-frame basis. Therefore if the analysis cannot resolve to a single component, a composite AM/FM signal is produced. This can still be manipulated, modified, etc., as if it were a single component. Therefore a spectral analysis-synthesis system can deal with arbitrary input signals, as long as we understand the nature of its output and do not expect to look at a given bin for more than one component to be cleanly detected.

The format of the analysis output is rectangular complex, that is, and it is given as a real/imaginary pair for each bin. Again, we can interpret this from a spectral frame perspective, where we have $N/2 + 1$ pairs of values $a_i(k) + jb_i(k)$, where i is the frame index and k the bin number. Alternatively, from a sub-band view, there are $N/2 + 1$

pairs of time-domain signals $a_k(i) + jb_k(i)$. Earlier on, in Sec. 6.1 and then later again, in Sec. 6.2.3, we considered the output of the analysis in a different from, as amplitude and phase signals $A_k(i)$ and $\phi_k(i)$, respectively. From a spectral perspective, this implies

$$X_i(k) = A_i(k)e^{\phi_i(k)}, \tag{6.24}$$

where the usual relationships $A_i(k) = |X_i(k)|$ and $\phi_i(k) = \arg\{X_i(k)\}$ apply (as in Eqs. 6.5 and 6.6).

We have also mentioned that it is more common to keep the analysis data in an amplitude-frequency rather than amplitude-phase format. The reason for this is that raw phases are not amenable to manipulation, and it is far easier to deal with their derivatives. While obtaining the amplitudes from the rectangular-form output is fairly straightforward (the square root of the squared real and imaginary parts), phases and frequencies require a little more thought. There are a number of approaches to compute these, so it is worth taking time to explore some of the different methods.

6.3.1 Phase Difference Method

The simplest way to obtain a phase function from the rectangular data is, as shown before, to compute the arc tangent of the quotient of the imaginary and real parts,

$$\phi_i(k) = \tan^{-1}\left(\frac{b_i(k)}{a_i(k)}\right), \tag{6.25}$$

where $X_i(k) = a_i(k) + jb_i(k)$. This needs to be computed with special care for the cases where the real part is zero, so many computer languages include a library function called `atan2(x,y)`, which takes separate arguments for the numerator and denominator and provides a result in the $[-\pi, \pi)$ interval.

There are occasions, however, that may call for different methods of obtaining the phase. For example, it is possible to have it in complex exponential form as

$$e^{j\phi_i(k)} = \frac{X_i(k)}{|X_i(k)|}, \tag{6.26}$$

simply dividing the rectangular form by its magnitude (the amplitude, or absolute value). From there, we can get the actual phase function via its logarithm,

$$\phi_i(k) = \frac{1}{j}\log\left(\frac{X_i(k)}{|X_i(k)|}\right). \tag{6.27}$$

A similar approach can give us yet another method to obtain the phases. Taking the logarithm of the spectrum,

$$\log(X_i(k)) = \log(A_i(k)e^{j\phi_i(k)}) = \log(A_i(k)) + j\phi_i(k), \tag{6.28}$$

separates the (log) amplitude and phase in real and imaginary pairs. From this we then have

$$\phi_i(k) = \Im\left\{\log(X_i(k))\right\}. \tag{6.29}$$

Moving on to obtaining a frequency signal from the analysis phases, again the simplest method is to compute the derivative, which in discrete-time is the difference of consecutive values in the sequence, defined by the $\Delta_i[]$ operator

$$\Delta_i[\phi(k)] = \phi_{i+1}(k) - \phi_i(k). \tag{6.30}$$

Note that this formula has a rectangular form, given by

$$\Delta_i[\phi(k)] = \arg\left\{\frac{X_{i+1}(k)}{X_i(k)}\right\}. \tag{6.31}$$

It is also possible to arrive at the frequencies directly from the real and imaginary parts. We can first develop it considering a continuous time derivative of a phase function $\phi_k(t)$,[14]

$$\begin{aligned}\omega_k(t) &= \frac{d\phi_k(t)}{dt} = \frac{d}{dt}\tan^{-1}\left(\frac{b_k(t)}{{}_k(t)}\right)\\ &= \frac{1}{1+[b_k(t)/a_k(t)]^2}\frac{d[b_k(t)/a_k(t)]}{dt}\\ &= \frac{a_k(t)b_k'(t) - b_k(t)a_k'(t)}{a_k(t)^2 + b_k(t)^2}.\end{aligned} \tag{6.32}$$

Going back to discrete time we have

$$\Delta_i\phi(k) = \frac{a_i(k)\Delta_i[b(k)] - b_i(k)\Delta_i[a(k)]}{a_i(k)^2 + b_i(k)^2}, \tag{6.33}$$

which is the form used by Golden and Flanagan in the original phase vocoder paper.[15]

The frequency signal $\Delta_i[\phi(k)]$ has a few characteristics worth noting:

- As discussed before, it is downsampled at f_s/D.
- It is given in radians per hopsize samples, that is, how many 2π units in one sample increment.
- It refers to a frequency deviation amount, to which we need to add the band center frequency kf_s/N to recover the bin frequency.[16]

Therefore, if we want to convert $\Delta_i[\phi(k)]$ into a frequency signal $f_i(k)$ in Hz at each band k, we need to apply the following expression:

$$f_i(k) = f_s\left[\frac{k}{N} + \frac{\Delta_i[\phi(k)]}{2\pi D}\right] \tag{6.34}$$

where D is of course the hopsize and N the transform size. The method of taking direct phase differences from consecutive analysis is probably the most common approach to estimate the frequency at each bin. It is simple both conceptually and computationally, and yields good results for most practical applications. If phases are computed in the principal value range $[-\pi, \pi)$, then we will need to be careful to unwrap the differences back into that range. This is necessary if successive phases are given at the extremes of the range (as the phase wraps around it) and the differences then are computed as a multiple of 2π. However, this should not cause any major difficulties in implementation.

6.3.2 Instantaneous Frequencies

If we approach the problem from a different perspective, it is possible to develop a method that does not rely on past phase values, producing the instantaneous frequency directly for a single time point. This idea was first introduced by Friedman (1985) as a means of improving the frequency resolution of the STFT, but of course it can be applied to provide frequency signals for each sub-band in an analysis-synthesis system.

To develop it, we need to revert again to a continuous-time representation. Once the method is obtained, then it can be implemented with a discrete-time discrete-frequency transform.

The continuous-time, continuous-frequency short-time Fourier transform can be given as

$$X(t,f) = \int_{-\infty}^{\infty} x(t+\mu)w(\mu)e^{-j2\pi f\mu}d\mu \tag{6.35}$$

which is a straight recast of Eq. 6.9 into a continuous form,[17] with an integral replacing the ordinary sum.

Now we can proceed by finding the instantaneous frequency $\phi(t,f)$ as the derivative of the phase. From Eq. 6.29, we have

$$\phi(t,f) = \Im\{\log(X(t,f))\}. \tag{6.36}$$

and so,

$$\begin{aligned}\frac{\partial}{\partial t}\phi(t,f) &= \frac{\partial}{\partial t}\Im\{\log(X(t,f))\}\\ &= \Im\left\{\frac{1}{\log(X(t,f))}\frac{\partial}{\partial t}\log(X(t,f))\right\}.\end{aligned} \tag{6.37}$$

The solution to this is down to finding the partial derivative of $X(t,f)$ with respect to time. We proceed by re-writing the STFT as

$$X(t,f) = e^{j2\pi ft}\int_{-\infty}^{\infty} x(\tau)w(t-\tau)e^{-j2\pi f\tau}d\tau \tag{6.38}$$

by substituting $\mu = t-\tau$. This allows us to isolate the time variable, making the window depend on it. We can now differentiate it directly,

$$\begin{aligned}\frac{\partial}{\partial t}\log(X(t,f)) &= \frac{\partial}{\partial t}e^{j2\pi ft}\int_{-\infty}^{\infty} x(\tau)w(t-\tau)e^{-j2\pi f\tau}d\tau\\ &= \frac{\partial}{\partial t}e^{j2\pi ft}\int_{-\infty}^{\infty} x(\tau)w(t-\tau)e^{-j2\pi f\tau}d\tau + e^{j2\pi ft}\frac{\partial}{\partial t}\int_{-\infty}^{\infty} x(\tau)w(t-\tau)e^{-j2\pi f\tau}d\tau\\ &= j2\pi f\int_{-\infty}^{\infty} x(\tau)w(t-\tau)e^{-j2\pi f\tau}d\tau + e^{j2\pi ft}\int_{-\infty}^{\infty} x(\tau)\frac{\partial}{\partial t}w(t-\tau)e^{-j2\pi f\tau}d\tau\\ &= j2\pi fX(t,f) + X_{dw}(t,f),\end{aligned} \tag{6.39}$$

where $X_{dw}(t,f)$ is the STFT taken with a time derivative of the window $w(t)$.

Substituting it back in Eq. 6.37, we have the following expression for the instantaneous frequency,

$$\begin{aligned}\frac{\partial}{\partial t}\phi(t,f) &= \Im\left\{\frac{j2\pi fX(t,f)+X_{dw}(t,f)}{X(t,f)}\right\}\\ &= 2\pi f + \Im\left\{\frac{X_{dw}(t,f)}{X(t,f)}\right\}.\end{aligned} \tag{6.40}$$

This can be computed by taking two windowed transforms of the same input data, one employing the derivative of the window function. If this is a discrete-time function, we can obtain its derivative by taking the first-order (or sample-by-sample) differences of whatever window is employed. When a hanning type is used, it is probably simpler to obtain this result analytically through the usual trigonometric identity, as[18]

$$w'(n) = \frac{\pi}{N-1} \sin\left(\frac{2\pi n}{N-1}\right). \tag{6.41}$$

The quotient of transforms in Eq. 6.40 can also be expressed as complex multiplication using the complex conjugate of the denominator. Reworking the expression in this form, and then putting it in discrete terms, we have

$$f_i(k) = f_s \left[\frac{k}{N} + \frac{\Im\{\overline{X_i(k)}X_i^{dh}(k)\}}{2\pi|X_i(k)|^2}\right], \tag{6.42}$$

where $X_i^{dh}(k)$ and $X_i(k)$ are the spectra of the ith input frame, the former being computed using the time derivative of the analysis window. This is an expression that can be implemented as an alternative to the direct difference method, as it provides the instantaneous frequencies in Hz of each bin for a given time point.

Friedman described this method as the *instantaneous frequency density* or distribution (IFD). It was later independently developed by Abe, Kobaiashi, and Imai as the IF spectrogram (Abe et al., 1996). An interesting aspect of their work was to make the identity

$$\frac{a_i(k)\Delta_i[b(k)] - b_i(k)\Delta_i[a(k)]}{a_i(k)^2 + b_i(k)^2} = \frac{\Im\{\overline{X_i(k)}X_i^{dh}(k)\}}{|X_i(k)|^2} \tag{6.43}$$

which connects the original phase vocoder method with the instantaneous frequency approach. Finally, Friedman's method is related to the concept of *frequency reassignment* (Auger and Flandrin, 1994), which is a fancy way to say that the bin center frequencies are shifted to match the estimated instantaneous frequencies.

6.3.3 One-Sample Hopsize

The method pioneered by Friedman works effectively by computing an infinitesimal shift in the transform window and with this obtaining the phase derivative. We can note, for instance, the similarities between the IFD and the phase difference equation in rectangular form (Eq.6.31) to develop the intuition that the infinitesimal step ahead can be effected by employing a derivative window. A discrete implementation of course determines that the minimum shift for the difference method should be one sample. Following the principles given by the IFD, it should possible to reach an equivalent result by having transforms that are spaced by a single sample.

In fact, we can employ a method to achieve this without actually having to compute a second transform (Charpentier, 1986). The idea is based on shifting the analysis window one sample backward over the same analysis data, then computing the phase difference of the two spectra in rectangular form. For this to work, we need to apply the window as a convolution in the frequency domain. As we have seen in Chapter 5, the hanning window discrete spectrum $W(k)$ is composed of three points $W(-1) = -1/2$, $W(0) = 1$, and $W(1) = 1/2$. Thus, its convolution with a spectrum $X(k)$, can be calculated as

$$X^w(k) = X(k) - \frac{X(k-1) + X(k+1)}{2} \tag{6.44}$$

Starting with an $X(k)$ computed with just a rectangular window at time n,

$$X_n(k) = \sum_{m=0}^{N-1} x(n+m)e^{-j2\pi km/N}, \tag{6.45}$$

we can apply Eq. 6.44 to obtain $X_n^w(k)$. Now we derive the spectrum with the window shifted back to $n-1$ using

$$X_{n-1}^w(k) = e^{-2j\pi k/N}\left[X_n^w(k) - \frac{e^{j2\pi/N}X_n^w(k-1) + e^{-j2\pi/N}X_n^w(k+1)}{2}\right]. \tag{6.46}$$

In this expression the complex exponential factors shift the window back one sample.[19] Scaling by the exponential sinusoid then rotates the whole frame backward also by one sample. From this we can compute the one-sample phase difference at time n, $\Delta_n[\phi(k)]$, as

$$\begin{aligned}\Delta_n[\phi(k)] &= \arg\left\{\frac{X_n^w(k)}{X_{n-1}^w(k)}\right\} \\ &= 2\pi k/N + \arg\left\{\frac{2X_n^w(k) - X_n^w(k-1) - X_n^w(k+1)}{2X_n^w(k) - e^{j2\pi/N}X_n^w(k-1) - e^{-j2\pi/N}X_n^w(k+1)}\right\},\end{aligned} \tag{6.47}$$

and then convert the phase differences to frequencies to Hz,

$$f_n(k) = f_s\left[\frac{k}{N} + \frac{\Delta_n[\phi(k)]}{2\pi}\right]. \tag{6.48}$$

We should be able to note the similarities between Eqs. 6.48, 6.42, and 6.34. All methods compute a frequency deviation term and then add it to a center frequency to obtain the bin frequency value. There is, however, a fundamental distinction between the two IF-based approaches and the phase difference method. Since the former depend on input data from a single time point only, there is no reference to a decimation factor or hopsize D at the point of analysis. An analysis frame may be self-standing, describing the amplitudes and frequencies at an isolated point in time. At synthesis stage, however, we may need to reintroduce the notion of a hopsize so that interpolation can be performed. Thus, if a system is designed around periodically spaced analysis points, then the usual STFT-based approach with inter-frame phase differences may lead to a more straightforward implementation.

6.3.4 Sliding Transform

Another single-sample step approach is given by the sliding DFT (sDFT) (Bradford et al., 2005). This is an alternative method that allows us to get the next frame using an update equation. Starting with the a non-windowed DFT as per Eq. 6.45, which we will call $X_n(k)$, we find the spectrum for time $n+1$,

$$\begin{aligned}
X_{n+1}(k) &= \sum_{m=0}^{N-1} x(n+m+1)e^{-j2\pi km/N} \\
&= \sum_{m=1}^{N} x(m+n)e^{-j2\pi k(m-1)/N} \\
&= e^{j2\pi k/N}\left[\sum_{m=0}^{N-1} x(n+m)e^{-j2\pi km/N} - x(n) + x(n+N)\right] \\
&= e^{j2\pi k/N}\left[X_n(k) - x(n) + x(n+N)\right].
\end{aligned} \tag{6.49}$$

With this result, we may apply windowing in the spectral domain to $X_n(k)$ and $X_{n+1}(k)$ using Eq. 6.44 to get $X_n^w(k)$ and $X_{n+1}^w(k)$, respectively. The frequencies at each bin can then be obtained using

$$f_n(k) = \frac{f_s}{2\pi} \arg\left\{\frac{X_{n+1}^w(k)}{X_n^w(k)}\right\} \tag{6.50}$$

or through taking the inter-frame phase differences. Note that since a shifting complex factor is included in the update equation, the quotient of the transforms gives the bin frequencies directly.

Similarly to the IF methods, the sDFT algorithm relies on the fact that moving the window in time by a very small amount makes the adjacent transforms share most of their data. While the one-sample hopsize method does not actually look for this small difference, and relies on circular shifts to compute the derivatives, the sDFT is actually designed to produce successive transforms, sliding forward by a single sample each time. In fact, it does not even necessarily to be initialized to a start-time spectrum, because it can shift the time-domain data in one sample at time. Therefore it can be employed directly in streaming applications.

The sliding DFT has been used to create a one-sample hopsize implementation of the phase vocoder, the sliding PV (Bradford et al., 2007). While still costly, the algorithm can be implemented in parallel for efficient computation.[20] The advantages of the sliding approach have to do with smoother outputs, fewer restrictions to processes, especially when manipulating the frequencies. The method also has a straightforward inverse transform,

$$x(n) = \sum_{k=0}^{N-1} X(k)e^{j2\pi km/N} = \sum_{k=0}^{N-1} X(k) \tag{6.51}$$

since the sample-by-sample time-domain frame contains only a single sample, and the complex sinusoid terms are all unity for $m = 0$.

6.3.5 Phase Integration

All the methods described here produce analysis data in the amplitude-frequency format. The analysis data can be then used to resynthesize a waveform using an additive synthesis approach. Alternatively, it is possible to use an inverse STFT.[21] In this case, it is necessary to recover the phase spectrum and, from that, put the frequency-domain data back into rectangular form. For this we will take account of the hopsize D, bin center frequency, and sampling rate to compute the inter-frame phase difference,

$$\Delta_i[\phi(k)] = 2\pi D \left[\frac{f_i(k)}{f_s} - \frac{k}{N}\right], \tag{6.52}$$

then use a recursive relation to get the integrated phase spectra[22] for the successive output frames,

$$\phi_{i+1}(k) = \phi_i(k) + \Delta_i[\phi(k)]. \tag{6.53}$$

As long as the hopsize is not more than the maximum for a given synthesis window bandwidth, this will recover the phases correctly for resynthesis. The rectangular output is then formed by $A(k)e^{j2\pi\phi(k)}$. With this we can apply the transform, rotation, window, and overlap-add. This completes the fundamental principles for an analysis-synthesis system and we can move on to implement it.

6.4 Streaming Spectral Processing

The principles developed in the first sections of this chapter form the basis for an implementation of a spectral analysis-synthesis system. The fundamental idea is to develop an architecture that is sufficiently flexible to allow the direct manipulation of amplitudes and frequencies in a streaming fashion. This means taking an input from a time-domain source, analyzing it into a frequency-domain signal, applying one or more processes, then resynthesizing it. While these operational principles are not tied to realtime implementations, they are nevertheless very suitable for such applications. A small amount of latency is inevitable, since we have to accumulate enough samples to have a full frame for analysis. However, in most uses, this does not pose too many difficulties.

6.4.1 The Spectral Analysis–Synthesis Class

As we have already discussed in Chapter 5, designing a class to encapsulate complex processes that hold state is a good idea, and so we will continue to employ it. In fact, we may want to re-factor the fast Fourier transform code given as separate functions as full Python class. There are advantages to this, as there were many fixed operations that were repeated at each function call. We can instead execute them once and then save their results in memory for later application. For example, the reordering code employs a bit-reverse array that is fixed for a given N; we can compute this at construction time, making the actual operation a simple matter of copying data from one array to another. Likewise, the scaling complex exponentials used in the FFT and real to complex/complex to real functions only need to be calculated once.

An outline of the `rdft` class, implementing the real FFT algorithm is shown in Listing 6.1. It implements both forward and inverse transforms, producing/consuming $N/2+1$ bins as before. The conversion and iterative FFT code is similar to the examples already discussed in Chapter 5, with normalization applied to the inverse transform (thus it is omitted from this example). Such an object can then be used to provide DFT transforms for various audio processing applications.

Listing 6.1 Real FFT class in Python.

```
# RDFT class
```

```
class rdft:
  # initialises a rdft object with
  # transform size N
  def __init__(self,N):
    M = N//2
    n = np.arange(0,M//2)
    self.wf = np.exp(-2j*np.pi*n/M)
    self.wi = np.exp(2j*np.pi*n/M)
    n = np.arange(0,M)
    self.wwf = np.exp(-1j*np.pi*n/M)
    self.wwi = np.exp(1j*np.pi*n/M)
    self.bp = np.arange(0,M)
    i, n = 1, M//2
    while i < M:
     for j in range(0,i): self.bp[i+j] = self.bp[j] + n
     i, n = i*2, n//2
    self.M = M

  # bit-reversal
  def reord(self, s):
   N = self.M
   out = np.zeros(N).astype(complex)
   for i in range(0,N): out[i] = s[self.bp[i]]
   return out

  # FFT stage
  def stage(self,s,w,n,k):
   ...
  # radix-2 FFT, iterative
  def fft(self,s,w):
    ...
  # forward transform
  # takes an N-length real input
  # produces an N/2+1 complex output
  def fwd(self,s):
    ...
  # inverse transform
  # takes an N/2+1 complex input
  # produces an N-length real output
  def inv(self,s):
    ...
```

Now we want to design a class that implements all analysis methods explored in Sec. 6.3. To support this, the constructor needs to hold various bits of state:

- An array with band center frequencies: `self.fc = k*fs/N`

- The complex exponentials used in rotation:

```
for n in range(0,OL):
    self.w[n] = 2*np.pi*k*n*D/N
```

- The radians to Hz conversion factor, `self.fac = fs/(2*np.pi)`, and a version that takes account of the hopsize `self.facd = self.fac/D`.
- Windows with built-in normalization, and the time derivative window (IFD method),

```
self.awin = w*N/sum(w)
self.swin = 4*w/(3*OL)
self.dwin = fod(self.awin) # first-order differential
```


- Arrays to hold previous phase spectra for difference method and for integration,

```
self.pphi  = np.zeros(M+1)
self.phi = np.zeros(M+1)
```

- Arrays to hold spectral domain windowing and backward shift (one-sample hopsize method),

```
self.S = np.zeros(M+1).astype(complex)
self.Sm1 = np.zeros(M+1).astype(complex)
```

- Complex exponential factors for backward shift

```
self.shf = np.exp(2j*np.pi/N)
self.shb = np.exp(-2j*np.pi/N)
```

- Complex exponential rotator for sliding DFT, `self.ww = np.exp(2j*np.pi*k/N)`
- The DFT transform object, `self.transform = fft.rdft(N)`

A phase unwrapping operation is required both to get the correct frequency in Hz from the phase differences and to avoid the current phase getting into the stratosphere when integrating. This is implemented as a simple modulo π function,

```
def modpi(x):
    while x >= np.pi: x -= 2*np.pi
    while x < -np.pi: x += 2*np.pi
    return x
```

then vectorized, `self.unwrap = np.vectorize(modpi)`, so it can be applied directly with a phase spectrum array. A first-order difference function is used to provide the derivative of the window function,

```
# first-order diff
def fod(x):
    y = np.zeros(x.size)
    for i in range(0,y.size-1):
        y[i] = x[i] - x[i+1]
    return y
```

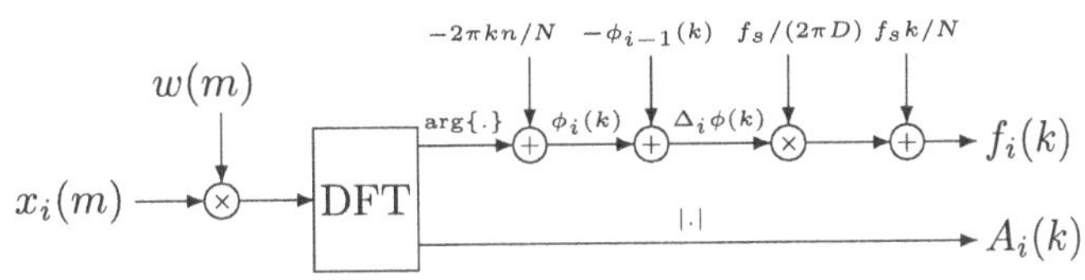

Figure 6.9 STFT spectral analysis method.

from which we may notice that it is computed backwards in time. This is done so we can use the window vector in the forward direction as usual.

6.4.1.1 Short-Time Fourier Transform Method

The STFT-based method can be implemented in two methods, for the analysis and synthesis stages. The former takes in an audio signal frame with N samples, producing two separate real-valued arrays with $N/2+1$ spectral-domain samples each, one containing the bin amplitudes and the other the frequencies in Hz. As discussed earlier, this employs the phase differences method (Fig. 6.9),

1. A windowed DFT is taken.
2. The phase spectrum is obtained using an arctangent call.
3. The complex exponential offset is applied.
4. The phase derivative is computed and unwrapped, and current phases are stored for next time.
5. Frequencies in Hz are produced using Eq. 6.34.
6. Amplitudes are taken from the spectrum and output along with the frequencies.

```
# computes one frame of STFT analysis
# producing two arrays of N/2+1 amplitudes
# and frequencies, respectively
def stft(self,x):
   N = self.N
   X = self.transform.fwd(x[:N]*self.awin)
   phi = np.angle(X) - self.w[self.i]
   d = self.unwrap(phi - self.pphi)
   self.pphi = phi
   freqs = self.fc + d*self.facd
   if self.i is self.ol - 1: self.i = 0
   else: self.i += 1
   return abs(X),freqs
```

Note that we are careful to slice N samples off the input, in case the function is passed a longer array. The code also keeps a count of frames up to the number of overlaps so that the correct phase offset vector can be applied to the phase spectrum. The synthesis method more or less reverses the steps above (Fig. 6.10):

- The inter-frame phase differences are computed from the frequencies in Hz.
- The phase spectrum is obtained through integration and then it is brought down to the principal value range.
- The complex spectrum is recomposed using $a(k)e^{j[\phi(k)+2\pi kn/N]}$, so that the phase rotation is taken into account.

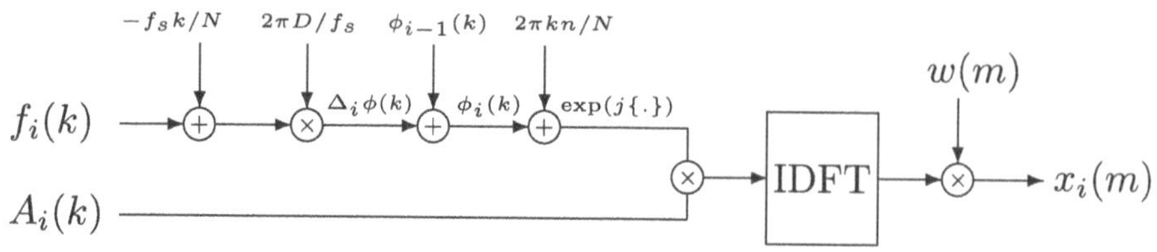

Figure 6.10 STFT spectral synthesis method.

- The inverse transform is taken and windowed to produce the output frame.

```
# Takes two arrays of N/2+1 amplitudes
# and frequencies, respectively
# and produces a time-domain frame
# using the inverse STFT
def istft(self,a,f):
    delta = (f - self.fc)/self.facd
    self.phi += delta
    self.phi = self.unwrap(self.phi)
    X = a*np.exp(1j*(self.phi+self.w[self.p]))
    if self.p is self.ol - 1: self.p = 0
    else: self.p += 1
    return self.transform.inv(X)*self.swin
```

Both the analysis and synthesis methods work on the assumption that they will be given N-size and $N/2+1$-size arrays, respectively, at regular time intervals of D samples. Therefore, it is up to the calling code to the provide the correct input and to implement the overlap-add of the output. In Python, this is a simple matter of employing slices to implement these operations, as in shown in this code excerpt

```
output = np.zeros(L)
for n in range(0,L,D):
    if(L-n < N): break
    amps,freqs = ans.stft(input[n:])
    output[n:n+N] += ans.istft(amps,freqs)
```

where `L` represents the total input buffer size. The loop steps ahead by D samples at a time and an N length of the input is sliced off by the `stft` method. The overlap-add operation is performed in a similar way, starting at time n for N samples, as the frame produced by `istft()` is mixed into the output buffer. Spectral processing code can then operate on the amplitude-frequency data between the analysis and synthesis stages.

6.4.1.2 Instantaneous Frequency Distribution

An IFD method may replace the STFT analysis shown above, producing an output in the same format. The code follows similar lines, but now the frequency estimation is performed using Eq. 6.42 instead. An outline of the steps used in this code is as follows (Fig. 6.11):

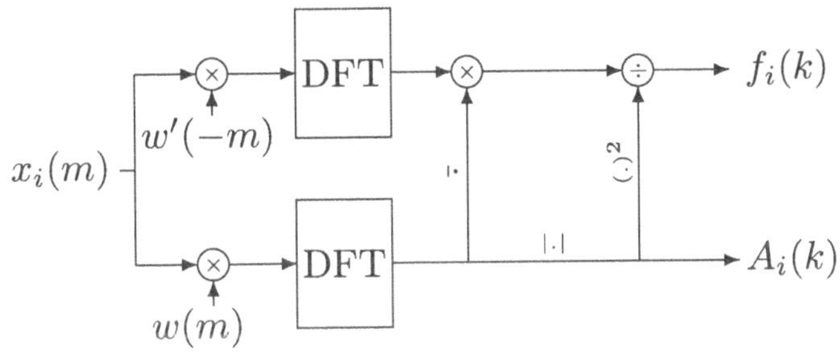

Figure 6.11
IFD spectral analysis method.

- Take two transforms of N input samples, one using the ordinary window, and another using its time derivative.
- Compute the amplitude spectrum.
- Take the quotient of the power spectrum ($|A|^2$) and the product of the derivative transform and the conjugate of the ordinary transform to produce the derivative of the phase spectrum (per-bin frequency deviation values).
- Apply the conversion factor $\frac{f_s}{2\pi}$ and add this to the bin center frequencies.

```
# computes one frame of IFD analysis
# producing two arrays of N/2+1 amplitudes
# and frequencies, respectively
def ifd(self,x):
   N = self.N
   X = self.transform.fwd(x[:N]*self.awin)
   Xd = self.transform.fwd(x[:N]*self.dwin)
   a = abs(X)
   d = np.imag(Xd*X.conjugate())/(a*a)
   freqs = self.fc + d*self.fac
   return a,freqs
```

There are fewer steps in this code, and there is no need for phase unwrapping, as phase differences are not taken. However, given that we have to take two DFTs per time point, this code is less advantageous in a straight streaming processes, where the phase data from adjacent DFTs is reused. However, it may find other applications in cases where single-point transforms are useful.

6.4.1.3 One-Sample Hopsize Method

The one-sample hopsize variation on the IFD has a few differences from the above methods, starting from the fact that it employs a rectangular window DFT. It requires two operations that may be isolated into separate functions for clarity. The first one is frequency-domain convolution by the spectrum of a hanning window,

```
def hann(self,X):
   N = X.size
   for i in range(1,N-1):
    self.S[i] = X[i] - 0.5*(X[i-1]+X[i+1])
   self.S[0] = X[0] - 0.5*(X[1].conjugate()+X[1])
```

```
        self.S[N-1] = X[N-1] - 0.5*(X[N-2]+X[N-2].conjugate())
        return self.S
```

as per Eq. 6.44. We have to take the edges of the frame (DC and Nyquist points) as special cases, since the spectrum is positive-only. Using the conjugate of the nearest bins takes care of this. The next special operation is the backwards window shift, using Eq. 6.46, but missing the complex exponential factor, since it is included in the frequency conversion formula (Fig. 6.12),

```
    def shift(self,X):
        N = X.size
        for i in range(1,N-1):
          self.Sm1[i] = 2*X[i] - (self.shf*X[i-1]
                                  + self.shb*X[i+1])
        self.Sm1[0] = X[0] - 0.5*(self.shb*X[1].conjugate()
                                  + self.shf*X[1])
        self.Sm1[N-1] = X[N-1] - 0.5*(self.shb*X[N-2]
                                  + self.shf*X[N-2].conjugate())
        return self.Sm1
```


With these two operations, the one-sample hopsize analysis method can be implemented using the following steps:

- Take one transform of N input samples, no windowing.
- Compute the convolution with the spectrum of the hanning window.
- Produce the amplitudes.
- Take the quotient of the windowed transform and the backward-shifted transform and compute its arctangent.
- Apply the conversion factor $\frac{f_s}{2\pi}$ and add this to the bin center frequencies.

```
    # computes one frame of one-sample hop analysis
    # producing two arrays of N/2+1 amplitudes
    # and frequencies, respectively
    def onehop(self,x):
        N = self.N
        X = self.transform.fwd(x[:N])
        Xw = self.hann(X)
        a = abs(Xw)
        d = np.angle(Xw/self.shift(X))
```

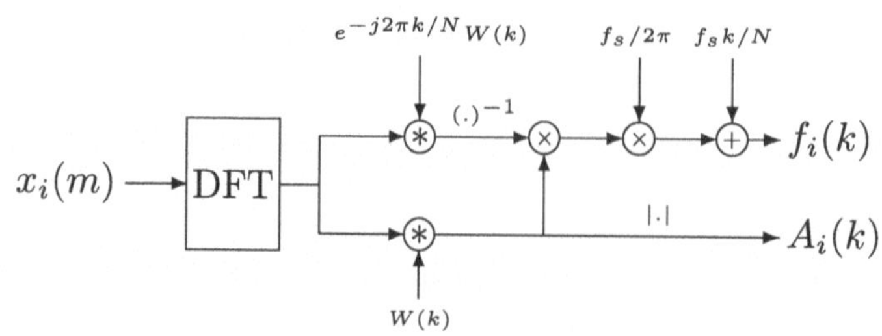

Figure 6.12
One-sample hopsize spectral analysis method.

```
        freqs = self.fc + d*self.fac
        return a,freqs
```

Although this method does not require an extra transform, it has the complication of the windowing in the spectral domain, and so for applications where there will be a sequence of frames, the STFT approach is still simpler. However, similarly to the IFD, for single-point analysis applications the one-sample hopsize method is a good option. It has been employed, for instance, in pitch detection algorithms.

6.4.1.4 Sliding DFT

The sliding algorithm can also be implemented to provide amplitude-frequency analysis, even though its original application is to give a sample-by-sample sequence of transforms as an alternative to employing the FFT. To implement it here, we can reuse the `hann()` function for the spectral-domain windowing, after applying the update equation to get the next spectral frame (Fig. 6.13). Thus, the steps involved are

- Take one transform of N input samples, no windowing.
- Compute the next spectrum, one sample ahead using Eq. 6.49. Note that we need one extra sample from the time-domain input for that.
- Apply the convolution with the hanning window spectrum to each frame.
- Take the quotient of the two spectra.
- Apply the conversion factor $\frac{f_s}{2\pi}$.

```
    # computes one frame of sliding dft analysis
    # producing two arrays of N/2+1 amplitudes
    # and frequencies, respectively
    def sdft(self,x):
        N = self.N
        X = self.transform.fwd(x[:N])
        delta = x[N] - x[0]
        Xn = np.array(self.hann((X - delta)*self.ww))
        X = self.hann(X)
        a = abs(X)
        freqs = np.angle(Xn/X)*self.fac
        return a,freqs
```

The sDFT method is comparable to the one-sample hopsize approach, where instead of taking a step back, we move forward. The differences in computation are small. The same window function is applied to both frames here, but there is an extra complex multiplication per bin, which in the previous method is turned into a real addition. Similar comments apply here concerning comparisons to the STFT-based process, as the technique is more suited to isolated analyses rather than consecutive ones.

6.4.2 Spectral Signals in Csound

Besides the usual time-domain control-rate (`k-`) and audio-rate (`a-`) signal types we have been using, Csound has a specialized data type to hold streaming spectral signals, the *fsig*

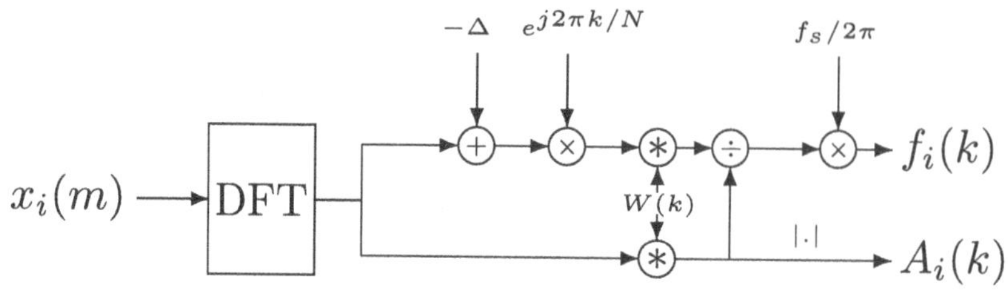

Figure 6.13
Sliding DFT spectral analysis method.

(or `f`- type).[23] This type can hold spectral data of different characteristics, the most common of which is the amplitude-frequency format we are using here. With fsigs, Csound can implement a complete spectral analysis-synthesis system as described in this chapter, in a computational efficient way that permits frequency-domain processing programs to be executed in realtime.

While an fsig can be initialized and used like any other variable, it has a number of special aspects that are worth knowing about:

- It is self-describing, that is, it holds information about the type of data it carries.
- Its main parameters are transform size (N), hopsize (D), window size (normally the same as N), and format (complex (rectangular), amplitude-phase, amplitude-frequency, or tracks).
- It holds its own internal time reference (frame index), which is not exposed but it is key to the spectral analysis-synthesis system operation.

The formats that it can carry are mostly self-explanatory, holding $N/2+1$ bins of real-imaginary, amplitude-phase, or amplitude-frequency data. The exception to these, is the *tracks* format, which holds partial track data that is extracted from amplitude-frequency and amplitude-phase data. In this format, each track models a single sinusoid with a given amplitude, frequency, and phase. This is fairly different from the bin-frame form of analysis data we are using here. To produce it we need to apply further processing to pick peaks and form tracks. We will explore this in detail in Chapter 9.

The fact that fsigs hold their own time index is very important. This means that any opcode operating on them will actually work at an independent rate that is different from the audio and control sampling frequencies used for `a`- and `k`- signals. Furthermore, each fsig stream can have a different update rate, which is determined by its hopsize D, although streams of different update rates cannot be mixed together. Spectral signals are normally produced by one of the various analysis opcodes. These ensure that the internal time index is updated every time a new frame is generated. Audio signals can be recovered by one of the synthesis opcodes.

6.4.2.1 Analysis

A standard streaming stft-based amplitude-frequency analysis method is implemented by the `pvsanal` opcode. This takes in an audio signal, and the parameters desired for a specific fsig,

```
fsig = pvsanal(asig,isize,ihop,iwsize,iwin)
```

where `isize` is the transform size N, `ihop` is the hopsize D, `iwsize` is the window size, and `iwin` is the window type (0 for Hamming, 1 for Hanning, 2 for Kaiser, and a negative value takes the window as defined in a function table).

Note that while it is possible to have window sizes that are larger than the transform size, in practice this is seldom used as it also increases the input/output latency and it does not provide significant improvements in processing in most applications. As discussed before, the minimum hopsize is set according to the window bandwidth, and with the hanning/Hamming windows this is $N/4$. If the hopsize is shorter than one audio block (ksmps) or 10 samples (with any ksmps size), then the sliding DFT algorithm is used, producing sample-by-sample analyses. This of course is generally not appropriate for realtime processing due to the computation requirements.

The transform size and hopsize are the most commonly adjusted parameters in this process. While Csound implements mixed-radix FFTs, they are not as efficient as the usual radix-2 algorithm, and therefore it is advisable to use power-of-two transform/window sizes. As we know already, the chosen transform size is a trade-off between time and frequency resolution. To capture transient changes, smaller windows are better, and to be able to get narrower bands we will need longer analyses. In practice, this translates to a choice in the range of 512 at the shortest to 4,196 at the longest for most applications. Increasing the analysis rate (by decreasing the hopsize) is also possible, and it will improve the quality of some types of manipulations, especially the ones in which the frequency data is modified. The trade-off here is between higher computational requirements and audio processing quality.

In addition to the standard STFT method, an IFD analysis is also provided through the `pvsifd` opcode. This provides two fsig outputs, in amplitude-frequency and amplitude-phase formats, with the other parameters as per `pvsanal`,

```
ffr,fphs pvsifd ain,ifftsize,ihop,iwin
```

This opcode is particularly designed for applications where the phase data is also required (such as sinusoidal track extraction). However, it also works as a drop-in replacement for the STFT-based analysis.

6.4.2.2 Synthesis

Csound provides two basic methods of synthesis of spectral data, inverse STFT and additive synthesis. The former is the most commonly used of the two, as it is also much more efficient for power-of-two transform sizes, while the latter is somewhat more flexible. The opcode

```
asig = pvsynth(fsig)
```

takes the `fsig` and produces an audio signal `asig`, using the inverse STFT, with no need for extra parameters as the spectral type is self describing. In contrast,

```
ares = pvsadsyn(fsrc,inoscs,kfmod[,ioff][,iskp])
```

requires the user to determine the number of oscillators `inoscs` to be employed and includes a frequency modulation parameter `kfmod` that can be used to shift the pitch of the output. It can also take as optional parameters the lowest bin number to use (`ioff`) and a skip value (iskp) that can be used to jump in between bins. The additive synthesis process therefore also includes spectral modification parameters that can be employed for amplitude and frequency modifications.

6.5 Spectral Manipulation

Once we have designed a system to produce spectral signals as described in the earlier parts of this chapter, it is then possible to apply many different types of processing to them. In this section, we will look at different categories of these, which will stretch from standard, *correct* from a signal-processing point of view, to non-standard, but musically interesting manipulations.

6.5.1 Filters

Perhaps the simplest type of processing we may apply to spectral data is to modify the amplitudes solely by scaling them in different ways. This has a filtering effect on the sound. To demonstrate this, let's develop an example using the Python analysis-synthesis class we have designed earlier on. A simple way to design a filter is create a spectral shape and then multiply the amplitude spectrum by this. We have done this before using convolution, and the approach here is going to be similar.

To create a band-pass filter, we can select a shape based on a window; for example, a Hanning or a triangle would be good candidates for this. Note that we are not *windowing* the time domain signal, but using these as the filter amplitude spectrum, directly in the frequency domain. A Python function can be designed to do such filtering,

```
# Takes an amplitude spectrum a
# a bin number k and a window w
# multiplies the spectrum by the
# window
def filter(a,k,w):
 bw = w.size
 k = int(k)
 if k < 0: k = 0
 if k+bw > N//2: bw = N//2 - k
 o = np.zeros(N//2+1)
 o[k:k+bw] = w[:bw]*a[k:k+bw]
 return o
```

where a window of a given size is applied to the amplitude spectrum from bin k upward. The size of the window sets the filter bandwidth (in bins).

We can then insert this function in between the analysis and synthesis steps to implement the filter. By moving the position of the filter, we can make the output spectrum change over time. In the following example, a 50-point Bartlett window is used and the value of k is incremented by one every two frames:

```
k = 0.
```

```
bw = 50
bp = np.bartlett(bw)
for n in range(0,L,D):
     if(L-n < N): break
     amps,freqs = anasyn.stft(input[n:n+N])
     if k >= N//4: k = 0
     else: k += 0.5
     amps = filter(amps,k,bp)
     output[n:n+N] += anasyn.istft(amps,freqs)
```

In Csound, there are various ways to implement this type of process. The most basic of them is use a function table to hold a spectral shape and employ

```
fsig = pvsmaska(fsrc,ifn,kdep)
```


to do the filtering, where a function table `ifn` of size $N/2+1$ is used to modify the amplitudes with depth kdep ($0 \leq depth \leq 1$).

Time-varying filters such as the one in the Python example could be implemented by rewriting the table regularly. We may design our own shapes for that purpose, or derive the amplitude spectrum from another fsig. As should be expected, there is an opcode that can write fsig data to a function table,

```
kflag = pvsftw(fsrc,ifna[,ifnf])
```

where `kflag` can be used to flag if a new data frame has been written to the table. Separate tables, `ifna` and `ifnf`, are employed for amplitude and frequency data, respectively. Alternatively, we could employ two fsigs as inputs, and take the filter amplitudes from one of them,

```
fsig = pvsfilter(fsrc,ffil,kdep),
```

where `ffil` provides the amplitudes directly. Filtering effects of this kind, mixing two arbitrary inputs can create a kind of cross-synthesis that is different from the one we have seen in convolution applications. Here, the frequencies are only taken from the source fsig, and so more of the characteristics of this input are kept by the process.

6.5.2 Blurring

One of the non-standard, but musically interesting effects that can be applied to spectral data is *blurring*.[24] To do this, we average out the spectral data over a certain number of windows. The longer the average, the more blurred is the effect.

A Python function implementing the process can be designed as

```
# blur process
# takes amplitudes a and frequencies f
# buffers ba and bf
# a buffer index n and blur time B in frames
def blur(a,f,ba,bf,n,B):
 ba[n],bf[n] = a,f
```

```
    if n == B-1: n = 0
    else: n += 1
    return sum(ba)/B,sum(bf)/B,n
```

which is a type of a moving average filter applied to the amplitude and frequency signals.

This process gives us an opportunity to show a complete spectral processing Python program developed using the spectral analysis-synthesis class and a couple of standard modules to get audio in and out. The complete code is shown in Listing 6.2. It can be used as a template for other processes (such as, for instance, the filter example discussed earlier). The Python module wave provides access to RIFF-wave audio files, and we use it to get the data into the input buffer, and write out the data from the output into a new file.[25] The code also prints a series of dots to give some feedback showing it is actively processing the data.

Listing 6.2 Spectral blur program in Python.

```
import numpy as np
import specsys
import sys
import wave

# check program parameters, quit if not given
if len(sys.argv) is not 3:
  print('usage: python3 %s <in.wav> <out.wav>' % (sys.argv[0]))
  quit()

# analysis params
N,OL = 1024,4
D = N//OL
w = np.hanning(N)

# input file
wf = wave.open(sys.argv[1], 'rb')
fs = wf.getframerate()
L = wf.getnframes()

# io buffers
input = np.zeros(L)
output = np.zeros(L)

# blur time and buffers
B = 50
ba = np.zeros((B,N//2+1))
bf = np.zeros((B,N//2+1))

# blur process
def blur(a,f,ba,bf,n,B):
```

```
    ba[n],bf[n] = a,f
    if n == B-1: n = 0
    else: n += 1
    return sum(ba)/B,sum(bf)/B,n

# read one channel of an audio file (assuming 16bit data)
norm = 2**15
chn = wf.getnchannels()
s = np.frombuffer(wf.readframes(L)[::chn],dtype=np.int16)
input[:s.size] = s/norm
wf.close()

# analysis-synthesis
anasyn = specsys.sas(w,OL,fs)
print('computing audio')
i = 0
for n in range(0,L,D):
    if(L-n < N): break
    amps,freqs = anasyn.stft(input[n:n+N])
    amps,freqs,i = blur(amps,freqs,ba,bf,i,B)
    output[n:n+N] += anasyn.istft(amps,freqs)
    print('.', end='',flush=True)

# write audio file
wf = wave.open(sys.argv[2], 'wb')
wf.setnchannels(1)
wf.setsampwidth(2)
wf.setframerate(fs)
s = np.ndarray.tobytes((output*(norm)).astype(np.int16))
wf.writeframes(s)
wf.close()
print('\ndone: output written to %s' % (sys.argv[2]))
```


Blurring is implemented in Csound in a similar fashion, by

```
fsig = pvsblur(fsigin,kblurtime,imaxdel)
```

with the difference that the blur time is given in samples. A maximum time `imaxdel` also needs to be supplied so that memory buffers can be properly allocated.

The principle of blurring is effectively a smoothing of the amplitude and frequency functions, a kind of lowpass filtering. Therefore it can also be implemented with other types of filters. The opcode

```
fsig = pvsmooth(fsigin,kacf,kfcf)
```

uses a simple 6dB/octave lowpass filter to implement a similar effect. Two separate filters are used, in fact, acting on amplitudes and frequencies separately. The cutoff frequencies for these are set to kacf and kfcf, and they are given in the range of 0 to 1, as fractions of the fsig frame rate. So the closer these are to 0, the more pronounced the smoothing is.

6.5.3 Tracing

Another non-standard process is provided by *tracing*. This is an extreme type of filtering where bins are zeroed if below a certain threshold, which is set on a frame-by-frame basis by the number of bins we want to keep. Given K as the number of bins we want to preserve, the process is then to search for the K bins with the highest amplitudes, and zero all bins whose amplitude falls below the value of the Kth bin amplitude. The key to this is to have an efficient sorting algorithm to put the bins in descending amplitude order and find the threshold.

In Csound, this is implemented by

```
fsig = pvstrace(fsrc,kn)
```

which only takes how many bins we want to preserve. This is implemented using the C++ standard library nth_element sorting algorithm.

6.5.4 Stenciling

A similar threshold-based process is provided by *stenciling*. Here, we take in an amplitude spectrum as a reference and apply scaling to the bin amplitudes based on this. The comparison is made on a band-by-band basis and if the amplitude is above the reference spectrum, the data is not touched, otherwise we scale it by the requested amount. This can be used as a way to punch holes in the spectrum, or, for more straight applications, to clean signals by using a background noise mask as a reference.

The opcode implementing this process is

```
fsig = pvstencil(fsrc,kgain,klevel,ifn)
```

which takes the masking amplitude spectrum from a function table. It can scale the level of the function (klevel) prior to using it. The amplitudes are modified by the kgain parameter; if this is below 1, the amplitudes of bins falling below the thresholds are reduced, and above 1, they are boosted. So we can possibly use this as a reverse-mask effect, where the background noise can be brought to the fore instead of being reduced.

The mask function table can be constructed in various ways. For creative processing applications, it can be designed to create specific filtering effects, similar to the ones discussed earlier. For noise reduction/masking applications, GEN43 can be used,

```
ifn = ftgen(0,0,isize,43,"mask.pvx",ichn)
```

which takes channel ichn of a PVX analysis file,[26] averages out its amplitudes, and creates the table with $N//2+1$ values (isize needs to be greater or equal to this). This analysis file can be created using the pvanal utility program, which comes with Csound. Or alternatively it can be written from an fsig using the opcode

```
pvsfwrite fsig,ifile
```

6.5.5 Mixing and Demixing

With so many different ways of processing spectra, it may be necessary at times to mix two or more signals. Unless we resynthesize these, add them together, and analyze the result, there is no one-to-one equivalent of mixing time domain signals to mixing spectra. However, it is possible to approximate the effect, and in fact, depending on the conditions, create an ultra-seamless mix.

The principle of spectral mixing is to combine bins from two inputs by selecting the loudest of the two on a bin-by-bin basis. Again, this is somewhat non-standard and may cause horror to some who follow the traditions of signal processing, but it can work very well for music and sound design applications. Csound implements this in

```
fsig = pvsmix(fsig1,fsig2)
```

A somewhat inverse process can be achieved if we have a stereo input, with a mix of inputs panned in different positions. This works by removing the bin data that is common to both channels at a given pan position, leaving behind the spectrum of the input that is panned to that particular direction (Barry et al., 2004). This is implemented by

```
fsig = pvsdemix(fleft,fright,kpos,kwidth,ipoints)
```

where the left and right fsigs are searched for a mixed input at pan position `kpos`, within a certain range of positions `kwidth`. The resolution of the search is given by `ipoints`, which divides each side of the stereo image. The success of the demix process depends on how well separated in terms of panning the different inputs are in the mix, as well as the amount of reflections, reverberation, etc., which may confuse matters.

6.5.6 Frequency Scaling and Shifting

Pitch effects are straightforward to implement in the spectral domain. For example, we may want to shift the pitch of a recording without actually changing the playback speed. To do this, we can scale the frequency data across the spectrum, which changes the pitch of the sound by a given interval. However, we need to be careful about how this is implemented, or otherwise it will not work. The first thing to note is that the bin bandwidths B_k have to observed, and that its frequencies range from $f_k - B_k/2$ to $f_k + B_k/2$, where f_k is the bin center. The width B_k is actually dependent on the frame rate. We have seen this before, when it was noted that the minimum analysis rate is determined by the bandwidth of the particular window employed.

Therefore, depending on the scaling applied, a given component may need to be placed in a different bin after its frequency is modified. Since the bandwidths are equal, this is more likely to happen in the higher bands, but it is simpler not to assume anything and move all frequencies according to the scaling applied. This is not difficult to implement; to get the shifted spectral data $A_{scl}(k)$ and $f_{scl}(k)$ from $A(k)$ and $f(k)$, we can use the following equations, for a pitch shift interval p,

$$\begin{aligned} A_{scl}(pk) &= A(k) \\ f_{scl}(pk) &= pf(k), \end{aligned} \tag{6.54}$$

and since the spectrum is discrete, pk is rounded to the nearest bin.

This can be implemented by a Python function, which as before operates on the frame data produced by the spectral analysis method of the sas class.

```
# freq scaling
# takes amplitudes a
# freqs f and a scaling factor
def fscal(a,f,scl):
  N = a.size
  af = np.zeros(N)
  ff = np.zeros(N)
  for k in range(0,N):
    pk = round(scl*k)
    if pk > 0 and pk < N-1:
     af[pk] = a[k]
     ff[pk] = f[k]*scl
  return af,ff
```


Note that the bins at 0 Hz and Nyquist frequency are untouched. Since the process is executed on a frame-by-frame basis, it is possible to make it vary over time, producing glissandos and vibrato. Csound implements this process using

```
fsig = pvscale(fsrc,kscal)
```

When applying upward pitch shifting, depending on the amount of transposition, it is necessary to increase the minimum analysis frame rate. The application of the scaling factor has the side effect of widening the analysis window bandwidth, and so the criterium employed to set the hopsize needs to be adjusted accordingly. As a rule of thumb, to minimize artifacts[27] and improve the quality of analysis, doubling the frame rate as the transposition ratio exceeds a fifth (3:2) is a good idea.

Another related pitch effect is frequency shifting, which we have seen before implemented using single-sideband modulation. We may recall that it involves a linear offset of all frequency components in the spectrum, which when applied in large amounts renders a harmonic spectrum inharmonic. Only small modification to the frequency scaling equations is required,

$$\begin{aligned} A_{shf}\left(k+s\frac{N}{f_s}\right) &= A(k) \\ f_{shf}\left(k+s\frac{N}{f_s}\right) &= f(k)+s, \end{aligned} \tag{6.55}$$

where we see that a frequency shift by s Hz requires the data to be moved to a bin indexed by $k+sN/f_s$. The corresponding Python code is

```
# freq shifting
# takes amplitudes a
# freqs f, a shift offset
# and the sampling rate fs
def fshift(a,f,shift,fs):
```

```
  N = a.size
  offs = shift*(N-1)*2/fs
  af = np.zeros(N)
  ff = np.zeros(N)
  for k in range(0,N):
    pk = round(k+offs)
    if pk > 0 and pk < N-1:
     af[pk] = a[k]
     ff[pk] = f[k] + shift
  return af,ff
```


Both in the case of frequency shifting and pitch shifting, aliasing is prevented as any transposed data beyond the $N/2+1$ bin will be excluded. The frequency shifting process is implemented in Csound by

```
fsig = pvshift(fsrc,kshift,klowest)
```

where `klowest` can be used to set the lowest frequency to which the shifting will apply. The two processes can be combined to form complete spectral frequency manipulation process.

6.5.7 Spectral Envelope

One of the characteristics of the frequency scaling process outlined above is that it moves both the amplitudes and frequencies. For sounds that depend on a fixed amplitude spectrum, such as the timbre of vowels, this is clearly going to be a problem. Depending on the transposition amount, the result may be similar to a cartoon voice sound, which has been described as the *Donald Duck* effect. In order to counteract this issue, we may want to extract the amplitude spectrum of the original sound, and then re-apply this to the transposed output.

In fact what we want to get is not exactly the amplitude spectrum, but the overall shape of it without the individual peaks, a smoothed-out curve that connects the amplitudes of individual partials. This is called a *spectral envelope*, and can be considered a smoothed version of the amplitude spectrum (Fig. 6.14). We should know what this means: some sort of lowpass filtering needs to be applied to the function. However, in this case, we are looking at a function of frequency, so what we are calling high frequencies in this case (the undulations of the amplitude envelope) are not to do with different periods of time, but of frequency.

At this stage, it is useful then to introduce the principles of the cesptrum transform.[28] The idea here is that a transform of the magnitude spectrum may be taken, revealing its components at different *quefrencies*, which is a term invented to describe the domain of the transformation. In this case, higher quefrencies correspond to the fast fluctuations, the peaks, and troughs defined by the partials. If we remove these, we are left with a smoother curve, which may match closely the spectral envelope we are seeking to obtain.

The cepstrum transform is therefore a DFT of the amplitude spectrum. To get the peaks and troughs closer together, we first take the logarithm of the amplitudes before

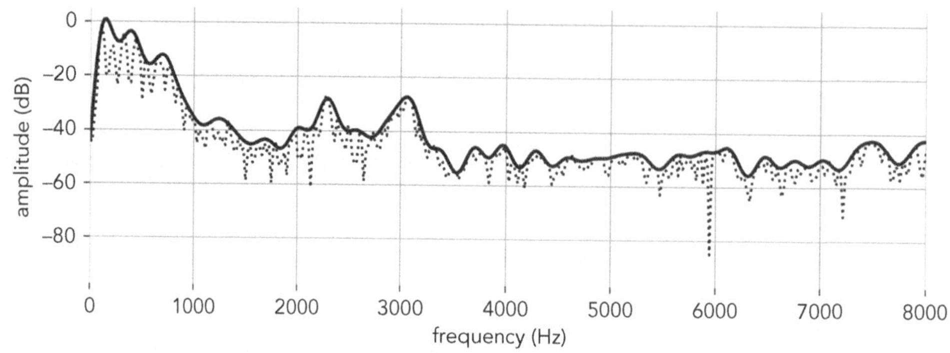

Figure 6.14 Spectral envelope (*solid line*) and the amplitude spectrum (*dots*).

applying the transform. The cepstrum is thus defined,

$$C(q) = \sum_{k=0}^{N-1} \log(|X(k)|)e^{-2\pi qk/N} \qquad 0 \leq q \leq N-1 \qquad (6.56)$$

in which we set the transform size N as the number of magnitude points in the non-negative spectrum. To keep it a power-of-two number, we may exclude the Nyquist amplitude.

The lowpass filter is then implemented by zeroing the higher-order bins. In keeping with the terminology, this is called *liftering*. For the non-negative spectrum, as provided by a real-to-complex transform, we may define it as

$$C_l(q) = \begin{cases} C(q), & 0 \leq q \leq P \\ 0, & P < q \leq N//2 \end{cases} \qquad (6.57)$$

The value of P is determined by how much smoothing is required. Following that, the spectral envelope $E(k)$ is recovered by applying an exponential function to the inverse transform output,

$$E(k) = \exp\left(\frac{1}{N}\sum_{q=0}^{N-1} C_l(q)e^{2\pi qk/N}\right) \qquad 0 \leq k \leq N-1 \qquad (6.58)$$

The amount of liftering determines the kind of spectral envelope we will get. If most bins are zeroed, an overall shape is obtained. As we remove fewer components, more individual peaks may start to become apparent, depending on their separation. A spectral envelope extraction function can be defined by

```
def specenv(a,rdft,P):
    N = a.size - 1
    sm = 10e-15
    env = np.array(a)
    C = rdft.fwd(np.log(a[:N]+sm))
    C[P:] = 0
    env[:N] = np.exp(rdft.inv(C))
    return env+sm
```

which takes an amplitude spectrum, a transform object, the number of cepstrum bins to keep, and returns the spectral envelope

To get a closer approximation, if computation requirements are not an issue, we can use an iterative method called the *true envelope* (Roebel, 2005), which applies the transform, filtering, inverse transform process repeatedly. Results are compared to the amplitudes and the liftering is done again until the spectral envelope matches the amplitude peaks according to a given threshold TH,

```
def true_specenv(a,rdft,P,TH):
   N = len(a) - 1
   sm = 10e-15
   env = np.array(a)
   loga = np.log(a[:N]+sm)
   amp = loga
   check = True
   while(check):
     C = rdft.fwd(loga+sm)
     C[P:] = 0
     env[:N] = rdft.inv(C)
     for i in range(0,N):
      if loga[i] < env[i]: loga[i] = env[i]
      diff = amp[i] - env[i]
      if diff > TH: check = True
      else: check = False
   return np.exp(env)+sm
```


However, for many applications, a single step process may be sufficient to provide a good approximation. With one of these methods, we can attempt to keep the original spectral envelope while transposing the sound. Some spectra, in particular that of vocal sounds, feature specific regions of resonance in the spectrum called *formants*, which are perceptually significant. When transposing signals, these regions are shifted, making the sound change significantly in character. To avoid it, it is important to preserve formants in place. This involves extracting the spectral envelope from the input and applying it to the frequency-scaled spectrum. However, in order to do this, we also need to obtain the spectral envelope of the transposed signal. This can also be seen as a filtering process, defined by

$$A(k) = A_{scl}(k)\frac{E(k)}{E_{scl}(k)} \tag{6.59}$$

where the filter is defined as the ratio of the original spectral envelope and its transposed version. The following code excerpt implements formant preservation using the standard cepstrum method:

```
dft = fft.rdft(N//2)
for n in range(0,L,D):
    if(L-n < N): break
    amps,freqs = anasyn.stft(input[n:n+N])
    envo = specenv(amps,dft,40)
    amps, freqs = fscal(amps,freqs,1.25)
```

```
    envs = specenv(amps,dft,40)
    output[n:n+N] += anasyn.istft(amps*envo/envs,freqs)

output *= max(input)/max(output)
```

Note that depending on the levels of the input audio, it may be necessary to scale the signal to avoid clipping distortion. This is implemented here as a balancing operation, restoring the peak input amplitude to the output signal.

In Csound, the opcodes pvscale and pvshift have special operation modes designed for formant preservation, which are set by optional parameters


```
fsig = pvscale(fsigin,kscal,kmode,kgain,kp)
fsig = pvshift(fsigin,kshift,klowest,kmode,igain,kp)
```

where we have

- kmode – formant extraction method: 0 (the default) means no formant preservation; 1 uses standard cepstrum; 2 employs the true envelope algorithm.
- kgain – gain scaling (1 is the default).
- kp – number of components kept in the liftering process (80 is the default).

With this approach, we are able to scale the frequencies while preserving the general shape of the spectral envelope. As noted earlier, this is particularly important for vocal sounds. The reason for this is that the sound we associate with the spectrum of a particular vowel has well-defined formants. These are more or less fixed: changes in vocal pitch will only affect them slightly in most situations;[29] therefore, we generally want to preserve the original spectral envelope.

Note that we can use a similar technique to shift formants without actually scaling the pitch. By passing a scaling factor and offset factor to the function,

```
def warp(a,dft,scl,shift,fs):
  N = a.size
  envo = specenv(a,dft,40)
  offs = shift*(N-1)*2/fs
  for k in range(1,N-1):
    pk = round(k/scl - offs)
    if pk > 0 and pk < N-1:
      a[k] *= envo[pk]/envo[k]
  return a
```

it is possible to modify the spectral envelope, reapplying it to the input. Depending on the parameters employed, that may have an effect of changing the characteristics associated a vocal sound, in particular as we tend to link age and gender to the position of formants in the spectrum.

This type of spectral warping is implemented in Csound through

```
fsig = pvswarp(fsrc,kscal,kshift[,klowest,kmode,kgain,kp])
```

where a certain amount of scaling (multiplication) and shifting (offset) can be applied to the shape of the spectral envelope. The optional parameters allow a lower bound to be set for the warping, as well as the method of spectral envelope extraction, and gain adjustment.

In general, we may apply formant extraction to general-purpose filtering processes (of which spectral warping is a special case). It is particularly effective to use formant filtering to impose the spectra vocal sounds over any broadband input (with harmonic, inharmonic, or noisy spectra), which is yet another type of cross-synthesis. This is associated with *vocoder* processing, which was originally done employing a bank of bandpass filters. It is possible to realize this very effectively using spectral methods.

In Csound, the `pvsvoc` opcode implements this type of cross-synthesis, taking the spectral envelope (amplitudes) from one input and the frequency data (excitation) from another,


```
fsig = pvsvoc(famp,fexc,kdepth,kgain[,kp])
```

where `famp` and `fexc` are the amplitude and excitation inputs, followed by depth and gain parameters. This uses the standard cepstrum mode and the optional parameter can set the number of cepstral components to be kept.

6.5.8 Morphing

Another interesting effect is given through the interpolation of two spectra, which we may call *morphing*. This allows a composite of two sounds to be created by combining different amounts of amplitudes and frequencies from each. The Csound opcode

```
fsig = pvsmorph(fsig1,fsig2,kaint,kfint)
```

allows the amplitudes and frequencies inputs to be interpolated independently. The parameters `kaint` and `kfint` in the interval $[0, 1]$ control how much of each input is used in the output fsig. Using an envelope to control these, it is possible to create a time-varying morph of the two sounds, which can start at one spectrum and finish at the other.

6.5.9 Spectral Delays

An advantage of the amplitude-frequency representation of the spectrum is that we can flexibly treat a sequence of frames in various ways. If we place the frequency domain signal in a delay line, we can navigate it back and forth, skip frames, stop at a certain point (which causes a spectral *freeze*), play it back at a slower or faster rate, throw away some bins, and so on. Csound provides an environment for doing this with a pair of opcodes. The first one of these creates a memory buffer and writes frames to it,

```
ihandle,ktime pvsbuffer fsig,ilen
```

given a certain delay line length `ilen` in seconds. It provides a time reference, which denotes the current write pointer position, and a handle that can be used to identify the object.

The second opcode provides the means to read the delay line,

```
fsig = pvsbufread(ktime,khandle[,ilo,ihi,iclear])
```

where the delay line is identified by khandle and the position to be read is given by ktime. More than one reader can be used for each delay line, and they can have independent read rates. The readers can be moved to different delay lines at the control rate. Each reader can also read only the bins below and/or above a certain frequency. A clearing flag can be used to clear each output frame before it is used (particularly important if not all bins are being read from).

There are many interesting effects that can be created using a spectral delay line:

- Time-stretch or compress a spectral signal, with no change in pitch. We set a buffer with enough time to hold a good portion of the input signal, then read it at a different rate:

  ```
  ih,kt pvsbuffer fsig, p3
  fdel = pvsbufread(kt*0.75,ih)
  ```

  where the time reference kt is being used, at a fraction of its value, to find the reading position. Note that when the end of the buffer is reached, the time reference wraps around to the beginning.
- Freeze the spectrum at regular times. Using a low frequency oscillator as a gate, we can sample and hold a buffer read position,

  ```
  kgate = lfo(1,2,3)
  kfz = samphold(kt,kgate)
  fdel = pvsbufread(kfz,ih)
  ```

  regularly freezing and unfreezing the spectral signal twice every second (the lfo opcode alternates between 0 and 1 at 2 Hz, closing and opening the sample-and-hold gate).
- Switch between two delay lines, combining two input signals in quick sequence,

  ```
  ih1,kt1 pvsbuffer fsig1, 3
  ih2,kt2 pvsbuffer fsig2, 3
  kgate  = lfo(1,5,3)
  khr = kgate = 0 ? ih1 : ih2
  ftps = pvsbufread(kt1,khr)
  ```

  where the reading is switched back and forth at 5 Hz (as the value of kgate changes from 0 to 1).

Besides these, a number of other approaches can be taken to read the delay lines. For instance, using the optional parameters we can access portions of the spectrum at different times, which can then be mixed together to create complex textures from a single or multiple inputs.

6.6 Timescale Modifications

As we have seen, the spectral analysis-synthesis system we have developed allow us to treat each frame separately and move from one to another at different rates, including stopping (freezing) at one of the time points. This makes the spectral signals model employed here effectively describe an input sound as a collection of short windows of audio kept in spectral form. We can apply this process to develop a flexible system for timescale modification.

For these specific purposes, it is actually possible to simplify the analysis-synthesis system somewhat. We can process data directly in rectangular format, and without the need to convert phase derivatives into frequencies in Hz, we may ignore the periodic rotation of the phase spectra in successive frames. Therefore, the analysis boils down to taking windowed DFTs at each time point, and the synthesis to doing the inverse operation. We will still use the inter-frame phase differences, and integrate them to produce the output, but now the complete process is going to be cast in a slightly different format.[30]

Let's consider an output spectrum $Y(k)$ at time $i+r$, for an arbitrary r. This can be written as

$$Y_{i+r}(k) = A_{i+D}(k)e^{j[\phi_{i+D}(k)-\phi_i(k)+\phi_{i+r-D}(k)]} \tag{6.60}$$

where D is the hop size and $X_i(k) = A_i(k)e^{j\phi_i(k)}$. This is an output consisting of the amplitude spectrum A_{i+D} and the phase spectrum as the sum of the previous output phases ϕ_{i+r-D} and the input phase differences $\phi_{i+D}(k) - \phi_i(k)$.

A complex exponential expression defining the input amplitudes and phase differences (frequencies) can be re-written as

$$\begin{aligned} &A_{i+D}(k)e^{j[\phi_{i+D}(k)-\phi_i(k)]} = \\ &= |X_{i+D}(k)|\frac{X_{i+D}(k)}{X_i(k)}\frac{|X_i(k)|}{|X_{i+D}(k)|} \\ &= X_{i+D}(k)\frac{|X_i(k)|}{X_i(k)}, \end{aligned} \tag{6.61}$$

and we can isolate the phases, kept in exponential form, by dividing the spectrum by its absolute value,

$$e^{j\phi_{i+r-D}(k)} = \frac{Y_{i+r-D}(k)}{|Y_{i+r-D}(k)|} \tag{6.62}$$

Eqs. 6.61, 6.62, and 6.60 are all we need to recast the synthesis expression in rectangular complex form. The output at time $i+r$ becomes

$$Y_{i+r}(k) = X_{i+D}(k)\frac{Y_{i+r-D}(k)}{X_i(k)}\frac{|X_i(k)|}{|Y_{i+r-D}(k)|} \tag{6.63}$$

and we do not need to convert the analysis results into amplitudes and phases, take phase differences, integrate them, and convert them back into real and imaginary pairs. For the particular application we are developing, this saves a lot of operations. However, this is never going to be as flexible and intuitive as the methods we have developed earlier on. They are much more appropriate for general spectral processing applications, as the complex form is much more cumbersome to use.

Nevertheless, Eq. 6.63 allows us to write a timescale modification program based on the following Python code:

```
dft = fft.rdft(N)
win = np.hanning(N)
T = int(timescale*D)
Y = np.zeros(N//2+1).astype(complex)+10e-20
i = 0
for n in range(0,L-(N+D),T):
    X1 = dft.fwd(input[n:n+N]*win)
    X2 = dft.fwd(input[n+D:n+D+N]*win)
    Y = X2*(Y/X1)*abs(X1/Y)
    output[i:i+N] += dft.inv(Y)*win
    i += D
```


where `timescale` is a time dilation factor. If this is greater than 1, the input will jump by a longer step than D, compressing the time. Values below 1 slow the rate of the output. All of these modifications, as before, happen with no changes to the pitch of the sound. Note that the output always progresses using hopsize D, and the taking of two input spectra spaced by D samples guarantees that the phase differences are correct for the integration that happens at the synthesis stage.

The expression in Eq. 6.63 is recursive, as it depends on previous values of $Y(k)$. Since the current value is a product involving the previous output, we need to initialize the first spectrum to a small value so that we do not get zeros as a result. This also prevents a division by zeros.

6.6.1 Phase Locking

Often, and especially with large time-stretching factors, an undesirable artifact is produced in this process, due to the fact that the reconstructed phases of adjacent bins get out of sync, somewhat misaligned in relation to the original spectra. This can introduce beating effects, due to the fact that, as we have seen, a partial may be detected by more than one bin at the same time, as it falls within one analysis bandwidth. A solution for this is to try to keep a certain synchrony by locking the resynthesis phases to the value at the center of peak. A practical solution has been proposed (Puckette, 1995b), in which a three-point average is applied across the output spectrum,

$$Y(k) = X(k) - (X(k-1) + X(k+1)) \tag{6.64}$$

which can be programmed as

```
def lock(X):
    N = X.size
    Y = np.zeros(N).astype(complex)
    Y[0], Y[N-1] = X[0], X[N-1]
    Y[1:N-1] = X[1:N-1] - (X[0:N-2] + X[2:N])
    return Y
```

This can then be applied then to the recursive expression above, in the feedback path, as the current output spectrum is saved for the next iteration.

```
for n in range(0,L-(N+D),T):
    X1 = dft.fwd(input[n:n+N]*win)
    X2 = dft.fwd(input[n+D:n+D+N]*win)
    S = X2*(Y/X1)*abs(X1/Y)
    Y = lock(S)
    output[i:i+N] += dft.inv(S)*win
    i += D
```

6.6.2 Pitch and Timescale

It is also possible to add pitch modifications to this timescale processing algorithm. However, since we do not have the flexibility of modifying individual bins, the frequency-domain method developed earlier is not available to us. We can, nevertheless, apply pitch shifting in the time domain, prior to the analysis step, by resampling the input data at a given scaling factor. For better quality effects, some sort of interpolation is needed to do this; the simplest of these is linear interpolation, based on

$$y = p_i + (p_{i+1} - p_i)x \tag{6.65}$$

where x is the interpolation factor (between 0 and 1), p_i and p_{i+1} are the two samples being interpolated.

A pitch scaling function, which takes data from the input and returns a frame of data for analysis, can be designed as follows:

```
def pscale(input, si, N):
   output = np.zeros(N) + 10e-20
   L, p = input.size, 0.
   if N < L: E = N
   else: E = L
   for n in range(0,E):
      if p < L - 1:
        i = int(p)
        frac = p - i
        output[n] = input[i] + frac*(input[i+1] - input[i])
        p += si
   return output
```

With this in hand, we can modify the timescale program to scale the input pitch by a given factor

```
for n in range(0,L-(N+D),T):
    X1 = dft.fwd(pscale(input[n:],pitchscale,N)*win)
    X2 = dft.fwd(pscale(input[n+D:],pitchscale,N)*win)
    S = X2*(Y/X1)*abs(X1/Y)
    Y = lock(S)
```

```
    output[i:i+N] += dft.inv(S)*win
    i += D
```

The timescale of the output remains independent as we progress through the input array as before, at a rate determined by the time dilation factor. Each frame may take input data resampled at a higher or lower rate depending on the value of `pitchscale`. This works as in the spectral frequency scaling method, in that it is a transposition ratio; values above one give a higher pitch, and below one, a lower pitch, if compared to the original. To avoid possible aliasing in upward shifts, some type of lowpass filtering may be applied prior to this operation.

6.6.3 Csound Opcodes

These ideas are implemented in various opcodes. A complete analysis-synthesis process as in the above examples is provided by two of these, which take their inputs from function tables,

```
asig = mincer(atimpt,kamp,kpitch,ktab,klock[,isize,iol])
asig = temposcal(ktscal,kamp,kpitch,ktab,klock[,isize,iol,
                 ith])
```

whose only difference is the user interface (since they implement the sample principles). The first one of these has an audio rate time position input, so this can be used to jump about the input data in various ways, including freezing time at a given point. The second is designed for typical timescaling applications, as it takes a control-rate time dilation factor. It can also detect onsets (the starting portions of sounds) so not to smear these if time stretching a sound. The `ith` parameter controls this detection process. A version of this opcode that reads directly from an audio file is given by

```
asig filescal(ktscal,kamp,kpitch,Sfile,klock[,isize,iol,ith]
```

There is also an opcode that can produce an fsig from function table audio, more or less in the same way as `temposcal`, with similar parameters,

```
fsig = pvstanal(ktscal,kamp,kpitch,ktab,
                [kdetect,kwrap,ioffs,isize,ihop,ith])
```

with some extra parameters such as an onset detection switch `kdetect`, as wrap-around reading (`kwrap`) and start position offset (`koffs`) controls. As it only takes care of the analysis process, creating an fsig, there is no implementation of phase locking in this opcode.

6.7 The Hilbert Transform

In Chapter 3, we introduced the concept of analytic signals, which form the basis for SSB modulation effects. While it is very straightforward to generate a complex sinusoid, which is the simplest form of such a signal, we need a method that can be applicable to any arbitrary audio signal. The operation to create a quadrature phase-shifted version of a time-domain function, which will provide the imaginary part of an analytic signal, is called a *Hilbert transform*, denoted by $\mathscr{H}$. With this we can then define an arbitrary analytic signal as

$$x_a(t) = x(t) + j\mathscr{H}\{x(t)\} \tag{6.66}$$

We may implement the Hilbert transform as a spectral operation. As we have seen, the spectrum of a real signal contains both negative and positive sides, in complex conjugation. By removing the negative components and doubling the magnitudes, an analytic signal can be obtained from any real waveform. The value at 0 KHz is kept but not scaled. The tools for this job are then a real-to-complex forward transform to produce the non-negative spectrum, and a complex-to-complex inverse transform to output the analytic signal. This can actually be implemented in two ways: using a standard fast convolution algorithm, which we have described in Chapter 5; or through the overlapped analysis-synthesis model developed in this chapter. For this, as before we need to employ windowing to avoid artifacts at the edges of the DFT frames. Since this is a model we are very familiar with, and it is fairly straightforward, we can go ahead and employ it.

So far we have not required the use of complex-to-complex transforms, but now we need it for the resynthesis part. This is fairly simple: a `cdft` class with a similar design to `rdft`, can be developed, but now only applying the DFT, so its forward and inverse methods become fairly minimal,

```
def fwd(self,s):
  return self.fft(s,self.wf)

def inv(self,s):
  return self.fft(s,self.wi)/self.N
```

and the main difference being that the transform size is not halved here and there is no need for any other state beyond the two complex exponential and the bit-reversal arrays. The implementation of the Hilbert transform filter is then simply based on taking a forward transform, copying the data into the first half of an array, and keeping the rest of it as zeros (the negative spectrum). As we resynthesize the data, we have to output both the real and imaginary time-domain outputs representing the analytic signal.

With the Hilbert transform we can now implement a frequency shifter effect based on SSB modulation. For this we can employ a sinusoid to apply a frequency shift to all components of an arbitrary spectrum. This employs a quadrature sinusoid, which can easily be constructed from a cosine and a sine waveform, multiplying an analytic signal produced with the Hilbert transform implementation shown before. Taking the product of an analytical signal $x_a(t) = x_{re}(t) + jx_{im}(t)$ and a complex sinusoid $s(t) = \cos(\omega t) + j\sin(\omega t)$, we have

$$x_{ssb}(t) = \Re\{x_a(t)s(t)\} = x_{re}(t)\cos(\omega t) + x_{im}(t)\sin(\omega t) \tag{6.67}$$

where the result will be the upward frequency shift of all components by $\omega/2\pi$ Hz. Note that to apply downward shift we only need to take the difference of the two terms, instead of the sum. The following code implements frequency shifting using the Hilbert transform,

```
rdft = fft.rdft(N)
```

```
cdft = fft.cdft(N)
win = np.hanning(N)
Xc = np.zeros(N).astype(complex)
shift = 100
ph = 2*np.pi*np.arange(0,L)*shift/fs
cos = np.cos(ph)
sin = np.sin(ph)
for n in range(0,L,D):
    if(L-n < N): break
    X = rdft.fwd(input[n:n+N]*win)
    Xc[0:N//2+1] = X
    an[n:n+N] += cdft.inv(Xc)*win
    re, im = np.real(an[n:n+D]), np.imag(an[n:n+D])
    output[n:n+D] = re*cos[n:n+D] - im*sin[n:n+D]
```


where a 100 Hz complex sinusoid is used to shift all frequencies of the spectrum.

Csound has an FT-based Hilbert transform opcode based on this design, which produces an analytic signal from an arbitrary input `asig`,

```
are,aim hilbert2 asig,isize,ihop
```

The streaming algorithm takes an `isize` analysis frame hopped along `ihop` samples. It can be used to implement a whole variety of effects, including time-varying frequency shifting and AM/FM decomposition.[31]

6.8 The Dynamic Spectrum

This chapter focused on a fundamental aspect of audio design: dynamic spectra. Starting from the platform given by the Fourier analysis theory developed in the previous chapter, we were able to develop a series of methods to capture and modify the spectra of sounds as time passes. This is essential for musical applications: the variations and nuances in timbral quality of sounds plays a vital role in the craft of composition, both in electronic and instrumental music, in fact. Static-spectra sounds are often regarded as unnatural and lifeless (but of course there are always musical applications for these too), and so the ability of treating frequency and time in a balanced way is essential.

Thus we have learned in this chapter that time plays a significant part in the manipulation of the spectra of sounds. We can add some more details into our provisional account of the spectrum that has been built in each chapter:

> Time-varying aspects are very important in the spectra of musical sounds. We can capture these by applying windowed analyses, based on the principles of the short-time Fourier transform. As we do so, we have to balance the time and frequency precision requirements, since the process is always bound by the uncertainty principle. This approach opens the door for a variety of time-varying modifications to the amplitude and frequency parameters of spectra. It allows us also to manipulate the time and frequency dimensions of a sound, independently of each other.

We have seen that all the processes explored in this chapter have a dual quality, as we may interpret them as in-time or in-frequency, depending on the perspective taken. So we chose to introduce the idea of frequency-domain signals to express this duality: a series of spectral analyses grounded on a timeline. It was possible to make a lot of ground by employing this approach, and we were able to base a complete spectral analysis-synthesis system on it. As we did this, we tended to emphasize the frequency dimension of each analysis frame. However, we could equally think of each sub-band, in time, as a separate stream, taking a filtering perspective. This can also be quite productive, and there is a lot that can be said about the subject. We will move toward that direction in the next chapter, to explore the different ways of designing the spectrum with the use of filters.

Notes

1 We should also note note that the sub-band signal is centered at f Hz, and so that this needs to be taken into account. Considering that the center exponential scaling in Eq. 6.3 can be seen as a sinusoidal carrier oscillator at the center frequency, we could exclude it from the phase calculation. Then the sub-band frequency signal becomes a deviation from the band frequency f, that is the modulator. That is the method employed in Flanagan and Golden (1966).

2 Therefore, *phase* vocoder is a bit of a misnomer, since phases are normally discarded in favor of their derivatives. On reconstruction, the original phases are lost, and so if anything is preserved under this vocoder, the phases are not it.

3 The use of the FFT in the implementation of the phase vocoder was first discussed in Portnoff (1976), ten years after the original paper.

4 The first practical use of the phase vocoder in the computer music literature was discussed in Moorer (1978), followed by Dolson (1986) and Moore (1990), whose software implementation was distributed as part of University of San Diego CARL package, ported to Csound and other systems very soon after.

5 When we are at the crossroads of time and frequency, it is often the case that we can change the observing position from one domain to the other.

6 The STFT is more commonly given in continuous-time form, where the sum becomes an integral, but for our purposes here the discrete-time version is more suitable.

7 Moore (1990, 291) derives a similar result in slightly a different way. The approach here follows more directly from the semantics of the STFT.

8 In many cases, we can ignore this rotation. As noted earlier, it only affects the phases and if the process we apply does not modify these (or their derivatives), then we do not need to perform the rotation.

9 Note that this equation essentially describes a cosine carrier whose phase is slowly modulated by $\phi_k()$, as well as linearly offset by $2\pi kiD/N$. This now takes account of the phase rotation, built into the STFT, which adds a positive shift.

10 Also, if additive synthesis is chosen, other interpolation methods can be used. For instance, instead of employing convolution as described here, it is possible to interpolate amplitude and frequency separately, avoiding the overlap-add operation

altogether. The interpolated signals then drive variable frequency oscillators directly. We will return to this method in Chapter 9.

11 We could perhaps have deduced this directly as the proper inverse operation, since it reverses the steps in the analysis stage. However, this would miss the point of exploring the interpolation process, which is the main justification for the windowing at the synthesis phase. Also, the convolution interpretation is essential to support the overlap-add mechanism employed here.

12 That is, not sanctioned by the usual signal processing rules, but producing interesting sounding results.

13 Here we are referring to the spacing of bins, or frequency points. The actual analysis bandwidth will normally be wider than that, since a non-rectangular window is commonly used.

14 Note that we have swapped k and i (turned into continuous time t), since we are now looking at this from the perspective of time. This is actually a partial derivative, but for constant k, we can actually take it as an ordinary derivative function.

15 The reason for using this may be related to the difficulties in computing the arc tangent functions in early computers, which was quite an expensive operation if compared to differences, squares, and division.

16 Back at the start, when we first determined that the filterbank analysis could be interpreted as a STFT, we took the step to remove the complex exponential factor that centred the output at each frequency band. Now, we need to reintroduce it to get the correct frequencies in Hz.

17 Except for the fact that that the window function parameter has an inverse sign, but since the windows considered here are symmetric, there is no difference.

18 However, we should also remember that the window is used in reverse form as per the STFT formulation we obtained before. This is a case where the reversal does matter, since the window is now an odd function. If the time is not reversed, the frequency deviation sign is inverted. We need to pay attention to this when implementing the algorithm.

19 Remembering that a multiplication by $e^{j2\pi k/N}$ rotates the DFT frame by 1 sample. A negative shift rotates it backward. So the shifted window spectrum is now $[-e^{-j2(-1)/N}/2, e^{-j2(0)/N}, -e^{-j2(1)/N}/2]$.

20 See Bradford et al. (2011) for a parallel implementation running in graphic processing units.

21 Strictly, this is an inverse of the phase differences analysis method, although it should also recover a time-domain signal produced by the other approaches.

22 This expression represents a discrete integrator.

23 The fsig system was designed by Richard Dobson and introduced in one of the final revisions of Csound 4. It was based on the existing non-standard spectral format *wsig* (by Barry Vercoe) that existed in Csound since its early versions. This was not suitable for spectral manipulations since it did not include a means of resynthesis. Based on the fsig type, a fairly large set of processing opcodes was then developed in Csound versions 5 and 6.

24 This is one of many effects pioneered by Trevor Wishart, which has contributed a significant number of approaches to audio processing that may not be sanctioned

by signal processing engineers but have nevertheless been great musical tools. See Wishart (1996) for some of the motivation behind these and Wishart (1994) for practical recipes to implement these.

25 The `wave` module is very simple and does not fully support all data formats in RIFF-wave files (no compressed formats, for example). It takes audio data as a Bytes object, so this requires some conversion to/from a Numpy floating-point array. However, despite this, it can be very useful as it is present by default in Python, and so it does not require any extra installation. There are other modules, such as the PySndfile bindings to the libsndfile library, which are more flexible and complete, but these need to be separately downloaded and installed. The code used here assumes soundfiles to be encoded as 16bit integers.

26 These are analysis files used with the Csound streaming code. They can be produced by file-writing opcodes, or using the analysis utility program `pvanal`, which is included in the Csound distribution.

27 Yes, artifacts are produced by this process, but they are generally very low in amplitude and masked in most applications, particularly if the frame rate is carefully considered. This technique may fall short of the exacting standards of classic signal processing, but it is very useful in musical and sound design applications.

28 This term was introduced, along with the other associated concepts, in Bogert et al. (1963). In some texts, the cepstrum is defined using the inverse transform, although the forward transform was used in that original paper. Given that the magnitudes are real-valued, it also makes sense to use the real DFT. In any case, such differences will not impact on the application in question, since the signal is converted back and forth and the phases are not touched in the process.

29 Although the special case of the *bel-canto* technique is an exception to this rule. In this case, singers are trained to make sounds whose lower-order formants are shifted according to the fundamental frequency. The original reason behind this technique was to achieve better projection in a large theater.

30 This method is proposed in Puckette (1995b).

31 For some examples of this, see Goulart et al. (2017). The case of AM/FM decomposition of sub-band analytic signals is in fact the basis for the phase vocoder, as discussed in Sec.6.1.

7

The Spectra of Filters

Filters have been introduced in general terms and used in various applications. We have seen how the notion of filtering can be applied to exclude parts of the spectrum. We have also noted the various configurations in which they can be applied to modify inputs: lowpass, highpass, and bandpass. Therefore it should be expected that filters may be used for spectral manipulation more widely, adding more techniques to the collection we have built so far. In fact, they have been an essential element of sound synthesis and processing since the very early days of electronic music. Different types of fixed and time-varying filters can be very useful to shape the timbre of sounds for various spectral processing applications. So far, and also along very general lines, we have been able to design and apply filters to suit some of our needs. At the start of Chapter 6, we were able to create a filter for sub-band signals, coming up with a lowpass shape (based on the spectrum of a window) and shifting it in frequency to create a bandpass filter. Approaches such as this one are very useful, thus it is a good idea to explore them further, more thoroughly.

As we do so, the concept of spectrum we have been developing will be expanded somewhat. Up until now we have been talking about the frequency-domain representation of waveforms, audio signals, which we have put in the form of complex-valued functions of a real-valued variable, frequency. The step we will take in this chapter is to look at the spectra of filters. This is a more abstract concept of spectrum than what we have grown used to in the preceding chapters, as we move to apply it to processes and operations that act on signals, rather than to signals themselves. Doing this will give us a very powerful set of tools, which allows us to manipulate filters in ways that have not yet been possible.

In the design and implementation of new filters, we will switch back and forth between the time and frequency domains. Filters may be implemented either as spectral operations, as we have seen before, using fast convolution, or as time-domain processes. This will mostly depend on the length and type of filter. For instance, shorter filters can be easily implemented directly in terms of their time-domain equations. Some types have infinite impulse responses, and therefore do not render themselves to a natural implementation as frequency-domain processes. On the other hand, the methods developed in previous chapters will be useful for the cases where we might want to use a more or less arbitrary spectral specification as the basis for a filter.

Spectral Music Design. Victor Lazzarini, Oxford University Press. © Oxford University Press 2021.
DOI: 10.1093/oso/9780197524015.003.0007

7.1 Filters and Delays

Looking back at Sec. 5.3 of Chapter 5, when we introduced the operation of discrete-time convolution, we saw that its direct implementation in the time domain was realized through delay lines. Based on these principles, we can say that an arbitrary filter of the kind we implemented through convolution is generally described as

$$y(n) = \sum_{m=0}^{N} a_m x(n-m), \tag{7.1}$$

which is now N-order filter, and the impulse response length is $N+1$. The filter *order* can also be equated with the longest delay present in the filter.

Note that we have changed the notation a little, instead of using $h(m)$ to denote the impulse response, a_m is used. This is to help introduce new terminology. The sequence a_m is called the set of filter coefficients. This changes the perspective of Eq. 7.1 from a convolution to a *filter equation*, which is defined by a sum of several delayed terms, each one scaled by a given gain a_m. This viewpoint will also pave the way to further conceptual shifts. However, it is always good to remember that $h(m) = a_m$, that is, the set of filter coefficients is an impulse response.

7.1.1 Pure Delays

We can start by looking at some simple cases of Eq. 7.1, to build a picture of how a more general analysis of filters can be developed. Much of what we will see at first recapitulates ideas that have already been introduced. Let's look at the case of a pure delay,

$$y(n) = a_d x(n-d), \tag{7.2}$$

which is the input kept in memory for d samples and then released, scaled by a_d.

This is an *allpass* case; the amplitude spectrum of the output is exactly the same as the input. It is very easy to check this, as we did in other cases, by examining the spectrum of the impulse response. This also allows us to look at the phase spectrum, in which we expect to be where the effect resides. Before we proceed, we need to choose a transform. So far, a number of tools have been developed to perform spectral analysis, but none is completely suited to the job. We could use the DFT, but it is limited to a time window, which brings its own issues. The Fourier transform is also problematic, since it expects continuous time and we have a discrete-time delay.

What we need is a transform that is not limited in time, but takes a discrete input. Extending the domain from $-\infty$ to ∞ makes the transform continuous in frequency. Therefore what we have is a discrete-time (continuous-frequency) Fourier transform (DTFT), which may be given as[1]

$$X(f) = \sum_{n=-\infty}^{\infty} x(n) e^{-2j\pi fn/f_s}. \tag{7.3}$$

This transform is not too different from the others we have used, in particular the Fourier transform, since this is also a continuous function of frequency. We can apply it to the impulse response $h(n)$, defined as

$$h(n) = \begin{cases} a_d, & n = d \\ 0, & n \neq d, \end{cases} \tag{7.4}$$

to get its frequency response,

$$H(f) = \sum_{n=-\infty}^{\infty} h(n)e^{-2j\pi fn/f_s} = a_d e^{-2j\pi fd/f_s}. \tag{7.5}$$

The amplitude response of this filter is confirmed as not dependent on the frequency, since $|X(f)| = a_d$, but the phase response is, as we should have expected. This is given by $\arg\{X(f)\} = 2\pi df/f_s$, and is a linear phase shift: each input component is delayed by the same number of samples. We can show the spectrum of a sinusoid put through this filter to confirm this. As we have seen many times before, this is the equivalent of the product of $H(f)$ and the sinusoid spectrum $X(f)$. If $x(n) = \cos(2\pi f_0 n/f_s)$, a cosine of frequency $|f_0| \leq f_s/2$, we have the following baseband spectrum

$$X(f) = \sum_{n=-\infty}^{\infty} x(n)e^{-2j\pi fn/f_s} = \frac{\delta(f+f_0)+\delta(f-f_0)}{2} \qquad -f_s/2 \leq f < f_s/2, \tag{7.6}$$

and then we can get the spectrum of the output as

$$Y(f) = H(f)X(f) = a_d e^{-2j\pi fd/f_s} \frac{\delta(f+f_0)+\delta(f-f_0)}{2} \qquad -f_s/2 \leq f < f_s/2, \tag{7.7}$$

where the product of the two delta functions centered at $\pm f$ and the complex exponential indicates that the phase of this component is shifted from 0 to something else. Since the meaning of this may not be absolutely clear at this point, we can apply an inverse transform[2] to check the output of the filter,

$$\begin{aligned} y(n) &= \frac{1}{2\pi}\int_{-f_s/2}^{f_s/2} Y(f)e^{2j\pi fn/f_s} df \\ &= a_d \frac{e^{2j\pi fd/f_s}e^{-2j\pi fn/f_s} + e^{-2j\pi fd/f_s}e^{2j\pi fn/f_s}}{2} \\ &= a_d \frac{e^{-2j\pi f[n-d]/f_s} + e^{2j\pi f[n-d]/f_s}}{2} \\ &= a_d \cos(2\pi f[n-d]/f_s), \end{aligned} \tag{7.8}$$

which confirms the phase shift by $-2\pi d/f_s$ and scaling by a_d.

This may look like a roundabout way to conclude that a pure delay of d samples shifts a cosine wave by d samples, and that any amplitude scaling is not dependent on its frequency. While this is true, the journey is more important than the destination. The example above demonstrated that the product of a spectrum by $e^{-j2\pi fd/f_s}$ delays a signal by d samples.[3] This has been seen before in the context of the DFT as a circular shift, but in the context of an open time domain,[4] it becomes simply a time delay. We have therefore developed a method to analyze delays, which are key components of filters, and we can now continue to perfect it to give us a full spectral model for them.

7.1.2 Inverse Comb Filter

We may reapply the method above to other cases, where the amplitude spectrum may not be allpass, and the phase spectrum not necessarily linear. Consider a simple change to the filter,

$$y(n) = x(n) + a_d x(n-d), \tag{7.9}$$

which is minimal and does not appear at first to do much. However, let's apply the same analysis and check it out. The impulse response only differs from the previous case by $h(0) = 1$. Applying the transform, and using the shorthand $\omega = 2\pi f/f_s$ to de-clutter the equations, we get

$$H(f) = 1 + a_d e^{-j\omega d}, \tag{7.10}$$

as the only difference is at $n = 0$, where we have $e^{-j0} = 1$. Now we can look first at $|H(f)|$,

$$\begin{aligned}|H(f)| &= [1 + a_d \cos(-\omega d)]^2 + [1 + a_d \sin(-\omega d)]^2 \\ &= 1 + a_d^2 + 2a_d \cos(\omega d).\end{aligned} \tag{7.11}$$

and note that it is dependent on the frequency f. The absolute amplitude response of this filter is plotted in Fig. 7.1. Note that its shape for $a_d = 1$ resembles that of a comb with its teeth facing down, which gives the filter its name, the *inverse comb filter*. It is interesting to see that a simple modification to the pure delay impulse response can make such a difference. We can explain it intuitively by considering that, given a certain fixed delay d, the addition of the original signal to the delayed copy will cause the signals to combine positively up at certain frequencies (where the two inputs are perfectly aligned in phase), negatively (where they are 1/2-cycle out of phase), and at different levels in between.[5] What is significant about what we have done is that we are able to predict what happens to the amplitude spectrum of any input signal. We can see exactly that the amplitude response maxima are set exactly at kf_s/d and the minima at $[k + 1/2]f_s/d$, k integer, between 0 and $f_s/2$.

However, this is only half the picture, since the phase spectrum can also have an important effect. We can derive it, as before, using

$$\arg\{H(f)\} = \tan^{-1}\left(\frac{-a_d \sin(\omega d)}{1 + a_d \cos(\omega d)}\right), \tag{7.12}$$

which generally makes it a non-linear function, except if we set $a_d = 1$. In this case, the phase shift is linear, albeit with a π flip at the points where the amplitude response is zero, as we can see in 7.2. It is possible to see why this is the case by casting the frequency response in terms of a real-valued amplitude response and a phase response as a complex

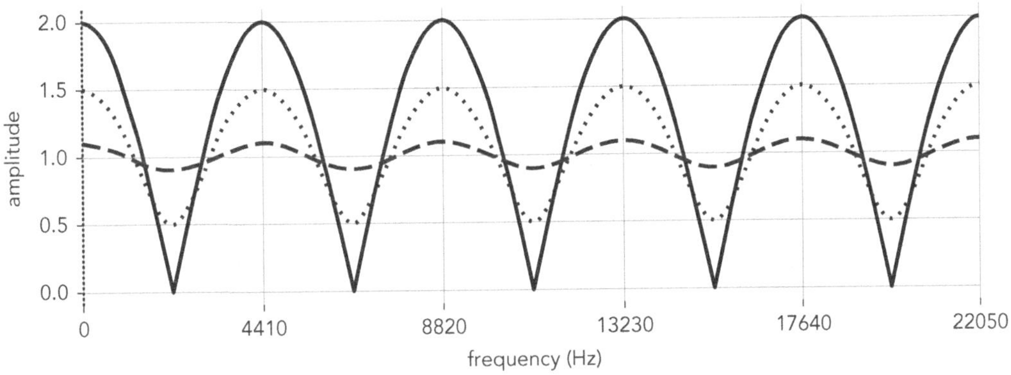

Figure 7.1 Amplitude (magnitude) responses for an inverse comb filter with $a_d = 1$ (*solid line*), $a_d = 0.5$ (*dots*), and $a_d = 0.1$ (*dashes*), using $d = 10$ samples and $f_s = 44{,}100$ Hz.

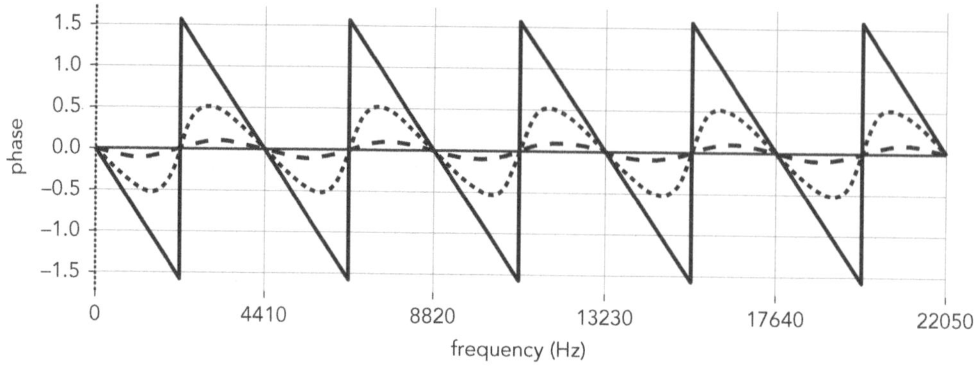

Figure 7.2 Phase responses for an inverse comb filter with $a_d = 1$ (*solid line*), $a_d = 0.5$ (*dots*), and $a_d = 0.1$ (*dashes*), using $d = 10$ samples and $f_s = 44100$ Hz.

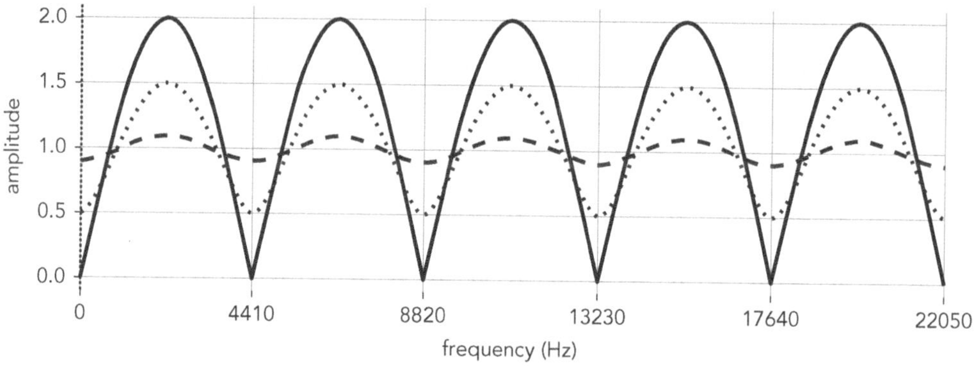

Figure 7.3 Amplitude responses for an inverse comb filter with $a_d = -1$ (*solid line*), $a_d = -0.5$ (*dots*), and $a_d = -0.1$ (*dashes*), using $d = 10$ samples and $f_s = 44,100$ Hz.

exponential,

$$\begin{aligned} H(f) = 1 + e^{-j\omega d} &= e^{-j\omega d/2}\frac{e^{-j\omega d/2} + e^{j\omega d/2}}{2} \\ &= e^{-j\omega d/2}\cos(\omega d/2) \end{aligned} \quad (7.13)$$

where we can see the linear shift by $\omega d/2$. However, in this form, the amplitude spectrum is bipolar; therefore, it will flip the phases periodically by π as it changes sign. If we instead cast the frequency response in terms of magnitudes and angles, this phase reversal would then be reflected in the phase response, since the amplitudes are taken as absolute values.[6]

Another version of the inverse comb filter can be created by reversing the sign of a_d. In this case, the whole frequency response is shifted round by $f_s/2d$, and the minima are now lying at $f_s/2$ (Fig. 7.3). The main characteristic of both filters is how they divide the spectral in equally spaced bands as a result of combining two copies of a signal separated by a certain number of samples. Although we have been able to extract a lot out of the discrete-time Fourier transform method, we need now to expand it a little more so we can understand the underlying structure of these and other filters.

Filters can be programmed in practical applications through the use of delay lines,[7] effectively implementing Eq. 7.1. The particular case of the inverse comb filter provides a good opportunity to demonstrate this. The code in Listing 7.1 shows a Python class implementing it. It holds a delay line, its size, and a read/write position as *state*. The processing method takes an input signal and the coefficient a_d, which may vary on a

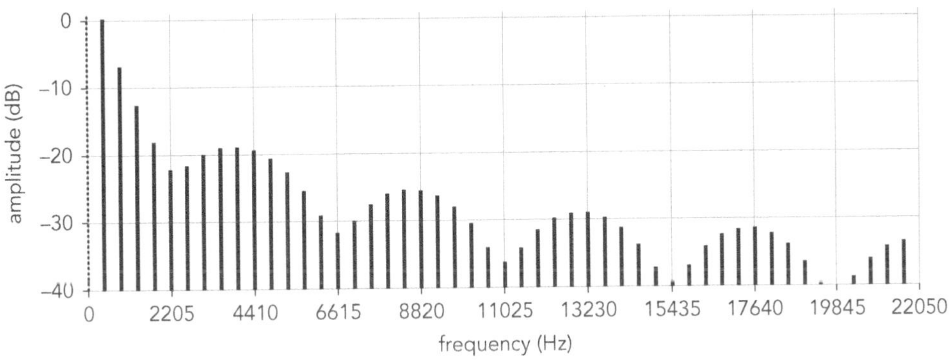

Figure 7.4
The amplitude spectrum (in dB relative to full-scale) of a sawtooth wave filtered with an inverse comb filter with $a_d = 0.5$ and $d = 10$ samples, $fs = 44,100$.

sample-by-sample basis. The result of filtering a sawtooth wave with this code is shown in Fig. 7.4.

Listing 7.1 Inverse comb filter class.

```
class icomb:
  def __init__(self,N):
    self.delay = np.zeros(N)
    self.p = 0
    self.N = N

  def process(self,x,ad):
    xd = self.delay[self.p]
    self.delay[self.p] = x
    if self.p != self.N - 1: self.p += 1
    else: self.p = 0
    return x + xd*ad
```

In Csound, such a filter can also be implemented, for instance, as the UDO

```
opcode IComb,a,aki
 asig,kad,id xin
 aout = delayline(asig,id/sr)
 xout(asig + aout*kad)
endop
```

where the `delayline` opcode takes in a delay in seconds, so we convert the size in samples dividing it by the system sampling rate (`sr`).

7.2 The Z-Transform

The analyses obtained so far were useful to evaluate the amplitude and phase response of filter, which predicts what happens when you multiply an arbitrary spectrum by the filter spectrum. Therefore the frequency response $H(f)$ is effectively the ratio of the output and input spectra in a filtering process, $Y(f)$ and $X(f)$, respectively,

$$H(f) = \frac{Y(f)}{X(f)} \tag{7.14}$$

Therefore, it is possible to call $H(f)$ a *transfer function*, the means to get a given output from a given input. We can now focus on this relationship and try to expand our analysis method. A lot can be achieved if we make f vary in more than one dimension, which will help reveal the structure of $H(f)$ in more detail. Let's begin by recalling its current representation,

$$H(\omega) = \sum_{n=-\infty}^{\infty} h(n)e^{-j\omega n} \tag{7.15}$$

where we use $H(\omega)$ to make the notation clearer.[8] Now we can expand this by turning H() into a function of[9] $e^{j\omega}$,

$$H(e^{j\omega}) = \sum_{n=-\infty}^{\infty} h(n)e^{-j\omega n} \tag{7.16}$$

Finally, we generalise the expression, setting $z = e^{j\omega}$, so that the spectrum now becomes a function of a complex variable.

$$H(z) = \sum_{n=-\infty}^{\infty} h(n)z^{-n}. \tag{7.17}$$

This is called the *z-transform*. It provides the means to derive the transfer function of a discrete-time system and is the essential tool for exploring the spectra of digital filters.

7.2.1 Complex Polynomials

With the z-transform we are actually moving slightly beyond the Fourier theorem. This is a now a complex function of a complex variable and, as we have seen, all the Fourier transform variants are functions of real variables. It is also possible to describe the z-transform as a polynomial.[10] This is a sum of terms involving a variable raised to increasing powers, each one scaled by an associated coefficient, which has the form

$$F(z) = a_0 + a_1 z + a_2 z^2 + ... + a_N z^N \tag{7.18}$$

for an N-order polynomial. In the case of filter transfer functions, we have a complex polynomial, since z is a complex variable[11].

There are some immediate advantages to this representation. The first one is that if we set $z = e^{j\omega}$, then we can interpret multiplication by z^{-n} as a delay by n samples, following our previous conclusions about the complex exponential. This allows us to read Eq. 7.17 as a series of delays and scaling factors that define a filter. Therefore, in the case of an inverse comb filter,[12] we have the following z-transform

$$H(z) = 1 - a_d z^{-d} \tag{7.19}$$

where we made the subtle move $a_n = h(n)$ to denote a sequence of filter coefficients.

7.2.2 Zeros

If we treat the z-transform as a polynomial, we can discover some important aspects about the filter structure. One of them is to find the values where $H(z)$ goes to nothing, or the zeros of the polynomial. This will tell us the frequencies where the product of the filter spectrum and an input spectrum will also be zero. Of course whether this will actually happen depends on the filter. As we saw in the case of the inverse comb filter, zeros in the filter spectrum only occurred when $a_d = |1|$. For other cases, there would be troughs of different amounts, which indicate that the zeros are elsewhere. Their actual position allows us to predict where these dips in the amplitude spectrum are located.

Since the z-transform is a complex polynomial, the location of the zeros can be anywhere in the complex plane. An N-order polynomial has N zeros, so the inverse comb filter has d of these, but where? It's not too difficult to find them, because for $H(z) = 0$, we need $a_d z^{-d} = 1$. Consider first the case where $a_d = 1$, then we need to find a complex number, which raised to $-d$ equals one. We know that $e^{j2\pi} = 0$, and so the d zeros can be found at

$$z_k = e^{j2\pi k/d} \qquad k = 0, \ldots, d-1. \tag{7.20}$$

These are called the n-th *roots of unity*, since they satisfy equation $z^n = 1$. Now for the general case, we have $a_d z^{-d} = 1$, which can then be solved by adjusting the magnitude R of each zero $Re^{j2\pi k/d}$ to $a_d^{1/d}$. Therefore we have

$$z_k = a_d{}^{1/d} e^{j2\pi k/d}, \tag{7.21}$$

which, of course, can be checked by replacing z by z_k for any integer k in Eq. 7.19,

$$H(z_1) = 1 - a_d [a_d{}^{1/d} e^{j2\pi/d}]^{-d} = 1 - e^{-j2\pi} = 0. \tag{7.22}$$

7.2.3 The Z-Transform and the DFT

The z-transform can be seen as an extension of the discrete transforms we have been using up until now. We have actually arrived at it by setting $z = e^{j\omega}$, so it is clear to see that $H(\omega)$ is a special case of $H(z)$, where we are constraining z to be allowed no other values than $e^{j\omega}$. Since this is equivalent to the unit circle, this is the same as *evaluating* (that is, getting the output of) the z-transform *on the unit circle*. We can notate it $H(e^{j\omega})$ or $H(\omega)$, depending on whether we want to emphasize a z-transform perspective (the transfer function) or the frequency response (the filter spectrum). Furthermore, since the DFT $H(k)$ is a special case of $H(\omega)$ for $\omega = 2\pi k/N$, we can go one step further and compute the z-transform on the unit circle for discrete frequency points kf_s/N. For this, we of course need to have a finite impulse response of length N. Such relationships are important to describe the spectra of filters in a complete way.[13]

7.3 Zeros on the Complex Plane

At this point, it may be helpful to introduce a graphical view of the positions of zeros in the complex plane. The value of a z-transform $H(z)$ can be given for any given point on the plane, so we can evaluate it anywhere on it. This is a very similar picture to the one we used to interpret the output of the DFT. In fact, we can reuse Fig. 5.4, and read it as a unit circle in the complex plane, adding the real and imaginary axis. As we noted earlier, evaluating the z-transform at evenly spaced points on the unit circle (that is, setting $z = e^{j2\pi k/N}$) gives the values of the DFT for each bin. Furthermore, since we allow the

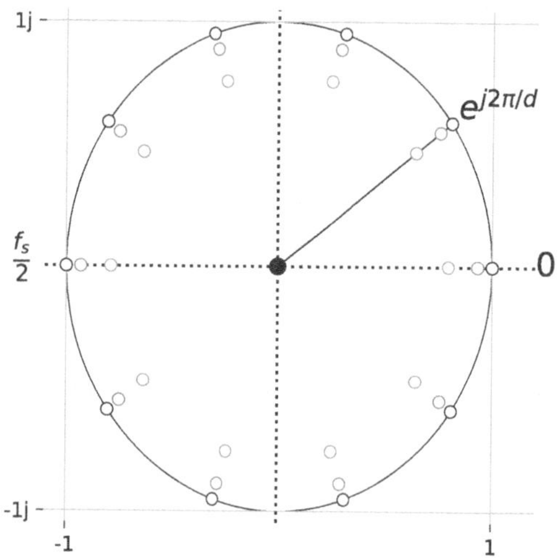

Figure 7.5 Zeros of an inverse comb filter $x[n]-a_d x[n-d]$, with $a_d=1$, $a_d=0.5$, and $a_d=0.1$, using $d=10$ samples.

frequency variable to be move anywhere in the plane, not only circularly on the unit circle, the position of the zeros of a filter spectrum can be located. This is shown in Fig. 7.5, where the zeros of the inverse comb filter are plotted for three values of a_d.

In the complex plane, the positive spectrum occupies the top half of the picture, above the real axis, and the negative spectrum is below it. As in the DFT interpretation, real-valued frequencies lie on the circle, in evenly spaced points if representing discrete (sampled) values, or otherwise continuously. If the frequency is a complex variable z, then it can be represented anywhere on the plane. That is the case of the zeros of the filter transfer function $H(z)$, which are placed at different points, inside of the unit circle in these examples. By virtue of this being a representation of discrete-time spectra, the angles of the unit circle are equivalent to fractions of the sampling rate. For example, 180° or π represents $\pm f_s/2$, 90°, $\pi/2$ is $f_s/4$, and so on.

Note in particular the case where $a_d=1$, where the zeros lie on the unit circle ($e^{j\omega}$). In this filter, the amplitude response does go all the way down to zero, at the exact frequency points $k\omega/d$ (we can check this in Fig 7.3, where ω/d is represented by $f=2205$ Hz). That should tell us that the filter spectrum $H(\omega)$ (or $H(f)$) is zero at those points. If we review the definition of the z-transform, we will see that the DTFT of $h(n)$ is equivalent to $H(z)$ with $z=e^{j\omega}$. This is an important confirmation that the main difference between the two transforms is the use of a complex frequency variable, and that both the DTFT and the DFT (by extension) are particular cases of the z-transform. We can say that the function $H(\omega)$ is only able to identify the zeros of a filter if they lie on the unit circle, which is one of the reasons why we need to make the frequency a complex variable. Another way to put this is to note, as we did before, that the filter spectrum is an evaluation of the transfer fuction given by the z-transform on the unit circle, by setting $z=e^{j\omega}$. Then, we can use the usual spectral analysis tools to compute the frequency response of arbitrary filters, taking the DFT $H(e^{j2\pi k/N})$.

There is more to learn from Fig. 7.5. The location of each zero in relation to the unit circle tells us geometrically how much it affects the spectrum of an input sound at each frequency. Considering that each (infinitesimal) point on the circle represents a given frequency, the amplitude response at that frequency is proportional to the linear distance between a zero and that particular point. So if a zero is placed close to the unit circle at phase angle θ, the amplitude spectrum of an input will have a dip with a minimum at the $\omega = \theta$, as the length of the line connecting a point on the circle and the zero is at its smallest at θ. If we inspect Fig 7.3 and compare it to the positions of the zeros in Fig. 7.5, we will see that this is the case.

We may also wonder where are the zeros in the case of a pure delay. The z-transform of such a filter is just

$$H(z) = z^{-d} \tag{7.23}$$

We can interpret this transfer function as $1/z^d$, in which case the zeros in the denominator make it infinite at $z = 0$. Whichever way this is looked at, it is an allpass amplitude response.

7.3.1 First-Order Filters

To make the geometric picture more complete, let's consider an even simpler case, where there is only one zero, and it is located at $z = -a_1$ (on the real axis). This is a first-order system, since the maximum delay element employed is 1 sample and the polynomial contains only terms up to powers of 1. The z-transform is

$$H(z) = 1 + a_1 z^{-1} = \frac{z + a_1}{z}, \tag{7.24}$$

which allows us to quickly confirm that $H(z)$ is zero when $z = -a_1$. Additionally, $H(z)$ is infinite at $z = 0$, since it has a zero in the denominator at that point.[14] This makes the function infinite at that point and for that this point is also called a *pole*.

We have already seen that the amplitude response is related to the distance from a point in the unity circle to a zero in the transfer function. More precisely in the present case, it is the magnitude of the vector connecting those points. For $a_1 = 1$ (a zero at -1), at 0 Hz, the length of the line connecting the zero and the frequency point is 2, and as the angle ω increases, it gets smaller until at $\omega = \pi$, it is zero. Note that the zero in the denominator does not have any bearing on the amplitude response, since it is equidistant to all points on the unit circle.

What about the phase spectrum? The phase spectrum is the angle of $H(z)$ with $z = e^{j\omega}$. Therefore, from Eq. 7.24, it is equivalent to the differences of two angles $\arg\{e^{j\omega} + a_1\}$ and $\arg\{e^{j\omega}\}$, corresponding to the numerator and denominator of the transfer function. These are the angles that the lines connecting zero and the pole of the filter, respectively, make with the horizontal axis. Therefore, first one is the angle of the vector from $-a_1$ to $e^{j\omega}$, and the other is just ω. So if $a_1 = 1$, the phase spectrum is just

$$\phi(\omega) = \omega/2 - \omega = -\omega/2. \tag{7.25}$$

The amplitude and phase responses for a given frequency point ω are shown geometrically in Fig. 7.6. The length of the solid line, $|e^{j\omega} + a_1|$, is the absolute amplitude response and the angle $\theta - \omega$, the phase response for this filter ($\theta = \arg(e^{j\omega} + a_1)$).

Note that one of the conclusions we may draw from this is that a first-order filter has its zero (and pole) on the real line, or we can also say that it has a real-valued zero. Secondly, moving it to the negative side implies a positive coefficient of the same magnitude,

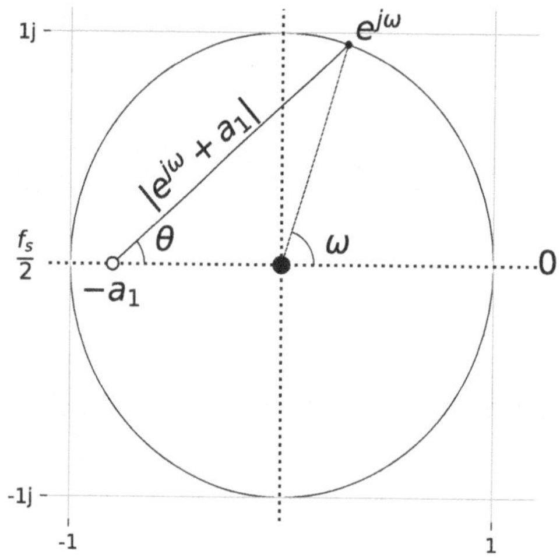

Figure 7.6 A zero at $-a_1$ and the magnitude response as the length of the vector connecting it to a point on the unit circle (*solid line*). The phase response is equivalent the difference $\theta - \omega$.

and a lowpass filter. Conversely, a positive zero makes the filter highpass and the coefficient is negative in that case. These are the only amplitude response curves available to first-order filters.

7.3.2 Second-Order Filters

If we want to create band-reject filters we need to shift the zero circularly off the real axis, making it complex. However, we also need to observe the condition that the impulse response is still real-valued (otherwise we cannot apply it to get a real signal output), so its z-transform needs to have Hermitian symmetry. The only way we can do this is by increasing the order of the filter and having two zeros, which should be in complex conjugation.[15]

Let's try to find a suitable transfer function. We used two methods so far to derive these:

- Starting with an impulse response, we take a z-transform, e.g., for $h(n) = 1, a_1$ we have $H(z) = 1 + a_1 z^{-1}$.
- From the location of the zero(s), we work out the corresponding polynomial. So for a zero at a_1, we have $H(z) = z^{-1} - a_1$.

In the present case, we do not know the IR at the outset, but we know the location of the two zeros. If we consider each separate zero as defining a filter[16] and we feed the output of one into the input of the other, the result is a second-order filter whose transfer function is the product of individual first-order filter transfer functions. So we set the zeros to be in complex conjugation at a given angle, $z_o = Re^{\pm j\omega_o}$. Each individual transfer function has the form $z^{-1} - z_o$, and the z-transform becomes

$$\begin{aligned} H(z) &= (z^{-1} - Re^{-j\omega_o})(z^{-1} - Re^{j\omega_o}) = z - R[e^{-j\omega_o} + e^{j\omega_o}]z^{-1} + R^2 z^{-2} \\ &= 1 - 2R\cos(\omega_o) z^{-1} + R^2 z^{-2}. \end{aligned} \tag{7.26}$$

The positions of these zeros are shown in Fig. 7.7.

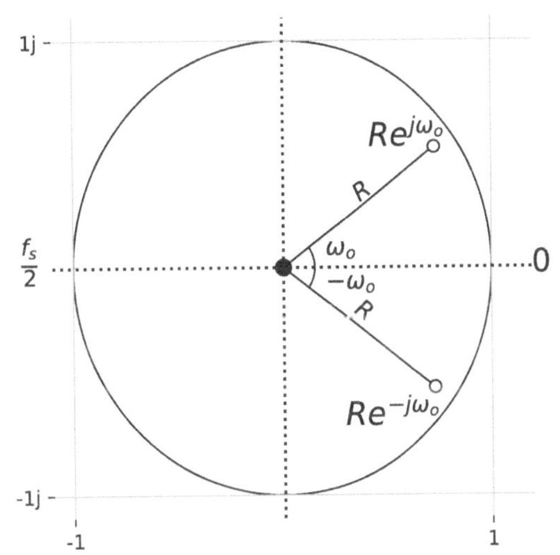

Figure 7.7
The two zeros of a second-order filter transfer function at $z = Re^{\pm j\omega_o}$.

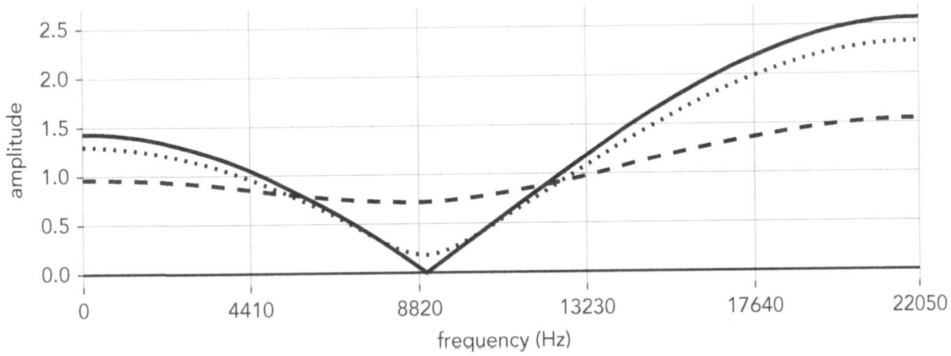

Figure 7.8
Amplitude responses for a second-order filter with $R = 1$ (*solid line*), $R = 0.9$ (*dots*), and $R = 0.5$ (*dashes*), using $f_o = 9{,}000$ Hz and $f_s = 44{,}100$ Hz.

From the transfer function given by Eq. 7.26, we have an impulse response $h(n) = 1, -2R\cos(\omega_o), R^2$, and therefore the filter can be implemented as the convolution

$$y(n) = 1 - 2R\cos(\omega_o)x(n-1) + R^2 a_2 x(n-2) \quad (7.27)$$

We can see that this also follows from the fact that the each term of the polynomial corresponds to a delay of the input signal and its associated coefficient.[17] The frequencies around ω_o will be reduced in amplitude. The radius of the zeros (R) determines how much rejection is achieved, as well as the stopband[18] width. The amplitude responses for this filter, with $f_o = 9$ kHz (denoting zero angles at $\omega_o = \pm 2\pi f_o/f_s$)[19] and various values of R are shown in Fig. 7.8.

Note that the filter only has any significant effect if R is nearly one; in this case the enter of the dip is nearly at ω_o. We can also see that the negative zero has an effect of reducing the left side of the amplitude response curve. This effect becomes reversed if the center frequency exceeds $f_s/4$. Therefore the filter can only have a perfectly balanced band-reject curve if $\omega_o = \pi/2$. In that case, the first-order coefficient is also zero, and the equation becomes a weighted sum of every other sample.[20] In general, filters of this type need to of much higher-order to be have a practical effect.

7.3.3 Minimum Phase

Note that a filter may have zeros anywhere, with any radii. They may be located inside the unit circle, on it, or outside it. If filters have all of its zeros inside the unit circle, as in the case of the inverse comb filter, they are called *minimum phase filters*. These have some particular characteristics that may be helpful in some situations. For example, it is possible to derive their phases from the amplitude spectrum, by taking their Hilbert transform (via the cepstrum transform) and then computing the phase spectrum as the imaginary part of the analytic signal. Starting with a (two-sided) amplitude spectrum $A(k) = |H(k)|$, we first take its cepstrum,

$$C(q) = \sum_{n=0}^{N-1} \log(A(k))e^{-j2\pi qk/N} \qquad 0 \leq q \leq N-1. \tag{7.28}$$

As we have seen, this is a Hermitian spectrum, since the amplitude response is a real-valued function. In order to put it in quadrature, we set $C(q) = 0$ for $N/2 < q \leq N-1$, which removes the negative frequencies. Taking the inverse transform, $\hat{H}(k)$ is produced,

$$\hat{H}(k) = \frac{2}{N}\sum_{q=0}^{N-1} C(q)e^{j2\pi qk/N} \qquad 0 \leq k \leq N-1. \tag{7.29}$$

This is an analytic representation of $\log A(k)$ and the phase spectrum corresponds to the imaginary part of its complex conjugate,[21]

$$\phi(k) = \Im\{\overline{\hat{H}(k)}\}. \tag{7.30}$$

The following Python code excerpt demonstrates these steps, using the NumPy FFT functions:

```
N = H.size
C = np.zeros(N).astype(complex)
C[:N//2+1] = np.fft.rfft(np.log(abs(H)))
Ac = np.fft.ifft(2*C)
phi = -np.imag(Ac)
```

where we take advantage of the real-to-complex DFT for the forward cepstrum, and just take the negative of the imaginary part instead of using the complex conjugate.

We might wonder just how this can work, and for this we may want to elucidate the relationship between amplitude and phase. Putting the $H(\omega)$ as amplitude and phase functions, we have

$$H(\omega) = A(\omega)e^{j\phi(\omega)}. \tag{7.31}$$

Taking its complex logarithm, $\log(H(\omega))$ gives us a result we have seen before,

$$\log(H(\omega)) = \log(A(\omega)) + j\phi(\omega), \tag{7.32}$$

which is the relationship we are looking for. If a filter has a characteristic where the logarithm of its amplitude response and its phase spectrum are in quadrature, it is minimum phase. This is of course not the general case, as it only applies to systems whose zeros are not located outside the unit circle.[22] Therefore to get the phase spectrum we can apply the Hilbert transform to the logarithm of the amplitude spectrum. This needs to be done over the full, two-sided, interval of frequencies.

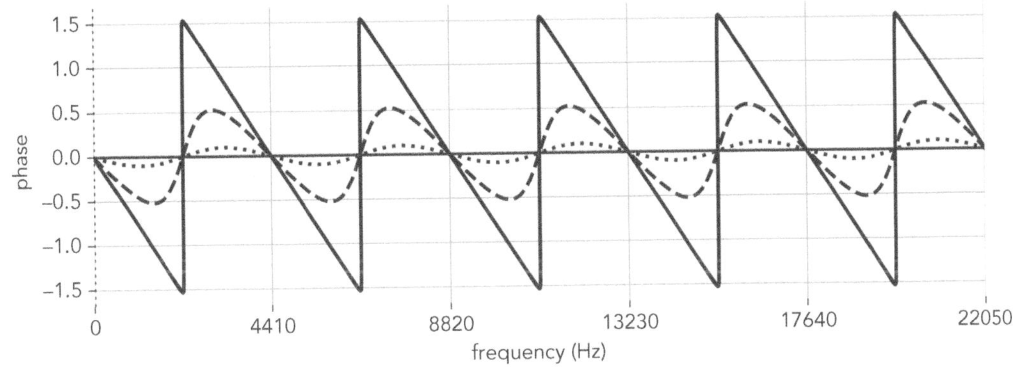

Figure 7.9
Phase responses computed from amplitude spectra, for an inverse comb filter with $a_d = 0.999$ (*solid line*), $a_d = 0.5$ (*dots*), and $a_d = 0.1$ (*dashes*), using $d = 10$ samples and $f_s = 44,100$ Hz.

In Fig. 7.9, the phases of three inverse comb filters, obtained with the method above, are plotted. We can compare these to the original phase spectra from Fig. 7.2. While this looks like a roundabout way to get the phases, which we could have taken directly from $H(k)$, it becomes very useful if we only have the amplitude spectrum to start with. In some cases, we might want to find an intermediate filter between two existing frequency responses, and the best method may be to interpolate the amplitudes and get the phases from these. The method may also be employed in amplitude-only spectral analysis, as yet another method to derive the instantaneous frequencies in addition to the ones discussed in Chapter 6.[23]

Finally, the term *minimum phase* comes from the fact that a filter can be put into a form where the phase offset, across the different frequencies, is at its minimum possible values. We can also say that the range of the phase response is minimal. This is opposed to a filter structure with the same amplitude response, whose phase spectrum has a maximal range, where all zeros are placed outside the unit circle. The two forms of the filter then are said to be minimum phase and maximum phase, respectively.

7.3.4 Linear Phase

Earlier on, in Chapter 5, a criterion for linear phase was put forward with no further justification. It was noted that a filter impulse response of length N (order $N-1$) can be made to have a linear phase delay of $(N-1)/2$ samples. This was shown to be the result of a symmetric impulse response, and a simple example of a first-order filter was used to demonstrate it. We can now look at how we might be able to prove that this is the general case.

To do this, we will concoct a somewhat artificial case: a filter whose impulse response extends into negative time. This is all perfectly possible given that the z-transform is two-sided, but in practice it would mean being able to predict the future. Such a filter would require knowledge of future inputs as well as past inputs, and as such is called *non-causal*. Of course if we have access to all input data beforehand, this can be implemented, but the snag is for the more general case where we do not. In contrast, a *causal* filter depends only on past inputs and does not have any practical restrictions. For this we need an impulse response that is always zero for $n < 0$.

The artificial IR we want to define has one single characteristic (Mulgrew et al., 1999),

$$h(-n) = h(n), \tag{7.33}$$

making it left-right symmetrical about $n = 0$. We can then apply the z-transform,

$$H(z) = \sum_{n=-\infty}^{\infty} h(n)z^{-n} = h(0) + \sum_{n=1}^{\infty} h(n)z^{n} + \sum_{n=1}^{\infty} h(n)z^{-n}. \tag{7.34}$$

Now we set $z = e^{j\omega}$ so we can evaluate it,

$$\begin{aligned} H(e^{j\omega}) &= h(0) + \sum_{n=1}^{\infty} h(n)e^{j\omega n} + \sum_{n=1}^{\infty} h(n)e^{-j\omega n} \\ &= h(0) + \sum_{n=1}^{\infty} h(n)[e^{-j\omega n} + e^{j\omega n}] \\ &= h(0) + \sum_{n=1}^{\infty} 2h(n)\cos(\omega n). \end{aligned} \tag{7.35}$$

The result tells us that we have a real-valued frequency response, and therefore the phases can only take one of two values, 0 or π. To make the filter causal, however, we need to shift it toward the right by $(N-1)/2$ samples so that the IR starts at 0. We know how to do this already: a multiplication by a complex exponential. Therefore for an N-length filter with left-right symmetrical coefficients, we have the resulting frequency response,

$$H(\omega) = e^{-j\omega(N-1)/2} \sum_{n=-\infty}^{\infty} h(n)\cos(\omega n), \tag{7.36}$$

which has a linear phase shift of $[N-1]/2$ samples.[24] The example we used to prove this property implicitly denoted an odd-length impulse response, and for this case the shift is always an integer number of samples. For even-length filters, similar principles apply, except that the phase offset will be fractional instead.

7.4 Filter Design

As we noted earlier, in order to be practical, filters of the kind seen here have lengths of more than a few samples. Very low order equations do make much of an impact, but we would like to keep the filter IR as short as possible to reduce computation. The task of filter design becomes one of optimizing the filter length for a desired frequency response. In particular, steep amplitude responses are only going to be possible with significantly high-order filters. We should know this intuitively, as we have already seen that short transitions or narrow bandwidths in the frequency-domain can only be defined by long time windows. In order to understand these issues, we start by reviewing the two methods we have already used in earlier chapters to design filters.

7.4.1 Time-Domain Method

In Chapter 6, we set off by creating a bandpass filter directly in the time domain with two components:

- A window of some description, providing lowpass amplitude response.
- A sinusoid to shift the center of the passband to any frequency other than DC.

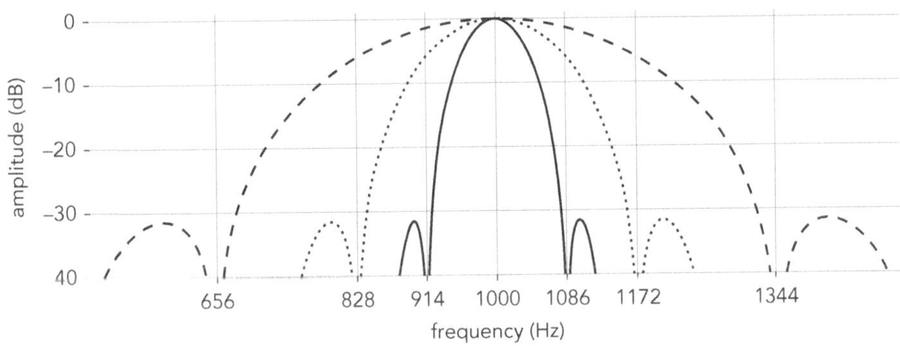

Figure 7.10
Amplitude spectrum (in dB, with reference to full scale) of a hanning-based bandpass filter centred at $f_c = 1{,}000$ Hz, with impulse response lengths $N = 1024$ (*solid line*), $N = 512$ (*dots*), and $N = 256$ (*dashes*) .

Such an impulse response has the form

$$h(n) = w(n)\cos(\omega_c n) \tag{7.37}$$

where, as before, $\omega_c = 2\pi f_c/f_s$.

This works very well for certain applications such as the spectral analysis processes we were looking at in that instance. Generally speaking, the bandwidth and shape of the lowpass filter then defines the characteristics of the amplitude response of the resulting filter. For example, we have seen that the hanning window has a $4f_s/N$ bandwidth (as shown in Fig 6.4). Therefore the length of the impulse response (or the order of the filter) will determine the selectivity of a bandpass filter built using this window. This is demonstrated by the plots of three filter amplitude responses in Fig 6.4, with different IR lengths. Note that besides the main lobe, which is what the defines the filter bandwidth, there are side lobes that are more than 30 dB below the amplitude response maximum.

Of course, this applies to any window shape, as the length of the impulse response determines how narrow the filter is, following from the duality of time and frequency (and the uncertainty principle). An extremely long filter made of a sinusoid and a rectangular may be very selective and pick a very narrow range of frequencies. At infinite time, this is of course the Fourier transform itself. With this approach, the filter design is dependent on the choice of window, and the slope of the transition is dependent on that. The hanning window behaves generally well, but it does have a gentle slope at each side. The same applies to the Bartlett case.

7.4.2 Frequency-Domain Method

We may, instead of employing these fixed-shape filters, decide to work directly in the frequency domain. At the outset, we may even try to implement a brickwall amplitude response, but we would run into trouble immediately. It should first be remembered that such a spectrum is defined by a sinc() window, which has an infinite length. Therefore any frequency response that implies this function as an impulse response will end up with a finite, truncated, form of it. This is where the difficulties start. To illustrate the problem, let's review this frequency-domain approach:

- Design an arbitrary frequency response as the product of an amplitude response and the complex exponential of a linear phase response.
- Take the inverse DFT to obtain the corresponding impulse response.

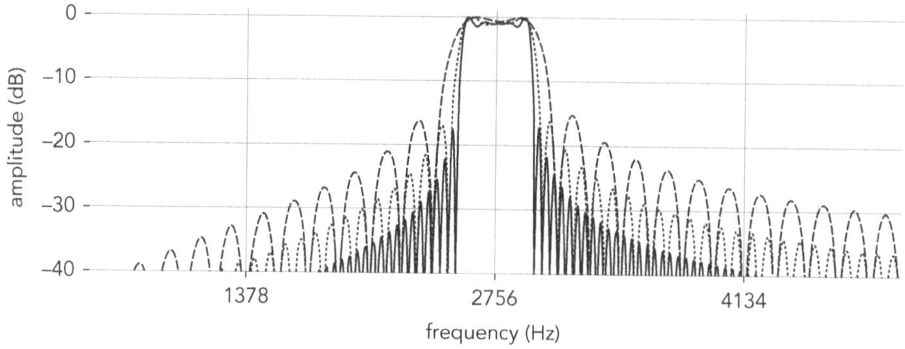

Figure 7.11 Amplitude spectrum (in dB, with reference to full scale) of an attempt to make a brickwall filter centered at $f_c = fs/16$ Hz, using impulse response lengths $N = 1{,}024$ (*solid line*), $N = 512$ (*dots*), and $N = 256$ (*dashes*), with $f_s = 44{,}100$.

So we start with an amplitude response $A(k)$ designed to suit our conditions of passband and stopband, and using Eq. 7.36, create the filter frequency response $H(k)$,

$$H(k) = A(k)e^{-j\omega k(N-1)/2} \tag{7.38}$$

where $\omega = 2\pi/N$.

We will now attempt to design a bandpass filter with a steep transition from passband to stopband, by setting the amplitude response to have value 1 for the passband and zero elsewhere,

$$A(k) = \begin{cases} 1, & \lfloor f_1 N/f_s \rfloor \leq k \leq \lfloor f_2 N/f_s \rfloor \\ 0, & \text{elsewhere} \end{cases} \tag{7.39}$$

where f_1 and f_2 are the frequencies at the edges of the passband. Note that we need the condition $f_2 - f_1 \leq f_s/N$ as the minimum spacing of one DFT bin.

We follow the method above, and get the impulse response. Now we can check the spectrum of these for different lengths and see what we got with this attempt (Fig. 7.11). It is clear to see that the impulse responses are much different from what we have seen before, in particular the shorter length filters. There is a substantial amount of fluctuation, called *ripple* both in the passband and stopband, and none of the filters has the envisage brickwall amplitude response. The larger filter length ($N = 1{,}024$) fares better, but the shorter filters are hopeless.

To diagnose this condition, we may want to have a look at the resulting truncated sinc() impulse responses. They might provide a clue to the issues seen here. These are shown in Fig. 7.12, with respect to the amplitude responses in Fig. 7.11. The first characteristic we notice is that the shorter IRs are more severely truncated. Secondly, and this may be the important aspect, they do not decay to zero at the edges. If this is what is causing the ripples, we may be able to attempt to suppress it by applying a window that is tapered at the ends, such as the ones we use in spectral analysis.

Let's choose the hanning window again and apply it to the impulse responses,

$$h_w(n) = h(n)w(n), \tag{7.40}$$

then we can take a DFT[25] of $h_w(n)$ and check the results again. These are displayed in Fig. 7.13, and show an excellent improvement. Therefore employing windows is usually a good idea when designing filters with steep transitions.

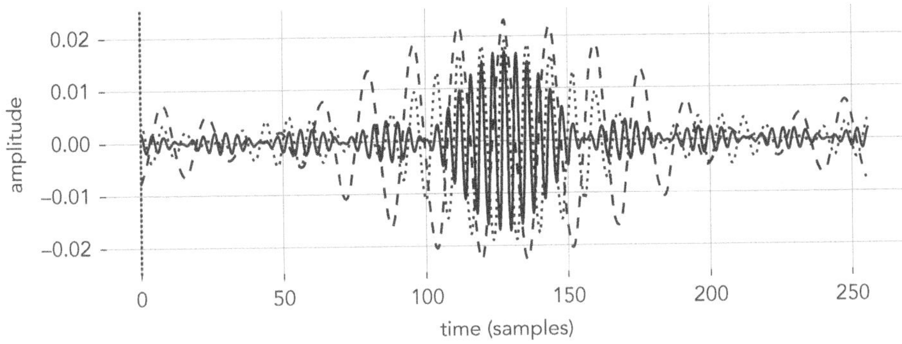

Figure 7.12 Filter impulse responses corresponding to the amplitude responses in Fig. 7.11, $N = 1,024$ (*solid line*), $N = 512$ (*dots*), and $N = 256$ (*dashes*).

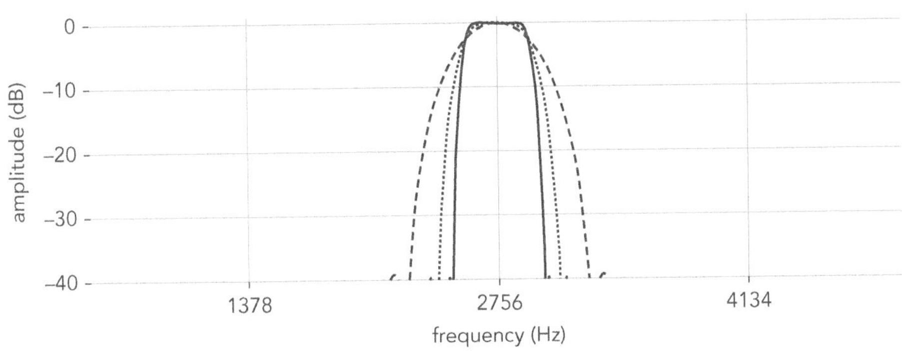

Figure 7.13 Amplitude spectrum of three (hanning-)windowed filters centered at $f_c = fs/16$ Hz, using impulse response lengths $N = 1,024$ (*solid line*), c$N = 512$ (*dots*), and $N = 256$ (*dashes*), with $f_s = 44,100$.

7.4.3 Design Example

With these ideas in hand we can now develop a program to design a filter and apply it to some input audio. We will approach the problem by making a few decisions to start with:

- Lowpass, highpass, and bandpass amplitude response shapes are created with piece-wise linear functions.
- Two frequencies are taken, determining the transition between stop and passband (lowpass or highpass type) or the filter bandwidth (bandpass).
- The IR length is determined by the difference between these two frequencies. Steeper slopes or narrow bandwidths will increase the order of the filter.

The filter IR can be computed using Python, and based on this data we apply a filter in Csound (using the direct convolution opcode `dconv`). First we take the parameters from the user, f_1, f_2 and filter type:

```
import numpy as np
import ctcsound as cs
import sys

if len(sys.argv) != 4:
  print("usage: %d  f1 f2 type\n'' % (argv[0]))
  quit()

# input arguments
f1 = float(sys.argv[1])
```

```
f2 = float(sys.argv[2])
type = int(sys.argv[3])
# swap values if necessary
if f1 > f2: f1, f2 = f2, f1
```

Now with these values in, we can determine the filter length. We will base it on a minimum of twice the passband/transition width (in terms of DFT bins),[26] using the relationship

$$N_{min} = 2\left\lfloor \frac{f_s}{f_2 - f_1} \right\rfloor, \tag{7.41}$$

which gives us the next lines of code:


```
# get filter length
fs = 44100
Nmin = 2*np.floor(fs/(f2 - f1))
N = int(Nmin)
print("Filter length=%d\n"%(N))
```

The next step is to create the amplitude response function from segments of lines. This basically concatenates three line pieces, in the case of lowpass or highpass types, and five segments, for a bandpass filter. The `np.concatenate()` function does this if from a sequence of segments:

```
# lowpass filter
if type == 0:
 k1 = int(np.ceil(f1*N/fs))
 k2 = int(np.ceil(f2*N/fs))
 fr = np.concatenate((np.ones(k1),
                      np.linspace(1,0,k2-k1),
                      np.zeros(N//2-k2)))
# hipass filter
elif type == 1:
 k1 = int(np.floor(f1*N/fs))
 k2 = int(np.floor(f2*N/fs))
 fr = np.concatenate((np.zeros(k1),
                      np.linspace(0,1,k2-k1),
                      np.ones(N//2-k2)))
# bandpass filter
else:
 slp = 2
 k1 = int(np.floor(f1*N/fs))
 k2 = int(np.ceil(f2*N/fs))
 fr = np.concatenate((np.zeros(k1),np.linspace(0,1,slp),
                      np.ones(k2-k1-slp),np.linspace(1,0,slp),
                      np.zeros(N//2-slp-k2)))
```

We need now to compute the frequency response and from this, the impulse response via the inverse DFT, multiplied by a hanning window:

```
H = np.zeros(N//2+1).astype(complex)
ph = np.exp(-1j*np.pi*(N-1) *np.arange(0,N//2))
H[:N//2] = fr*ph
h = np.fft.irfft(H)*np.hanning(N)
```

Now that we have an impulse response, let's fire up Csound to do the filtering. Note that we set the value of its sampling rate to what we defined before (`fs`). We also need to pass the IR length so that an empty table can be created to receive it. It is important not to forget to copy the data into the table, which can be done with a call to `csound.tableCopyIn()`, passing the table number and the array holding the IR.


```
# audio processing
csound = cs.Csound()
csound.setOption('-odac')
csound.setOption('-r=%d'%(fs))
duration = 4
csound.compileOrc('''
// empty table
imp = ftgen(1,0,%d,7,0,%d,0)
instr 1
 asig = diskin:a("fox.wav")
 aout = dconv(asig,ftlen(p4),p4)
   out(aout)
endin
schedule(1,0,%d,imp)
''' % (N,N,duration))
sr=csound.sr()
ksmps=csound.ksmps()
t=np.arange(0,int(sr*duration))
csound.start()
# copy IR into table 1
csound.tableCopyIn(1,h)
for i in t[::ksmps]:
    csound.performKsmps()
```

This software can create very precise filtering to remove particular frequencies, with good pass and stopband behavior. It is quite a practical program, while the sound processing runs in realtime, it can also be adapted fairly easily to work offline, as Csound is very flexible in the way it can be run. Two plots showing amplitude and impulse responses for lowpass and bandpass types are given in Figs 7.14 and 7.15. As can be seen, the amplitude response is smooth and ripple-free to −40 dB in both cases. We should also note that the cutoff frequencies, defined by the half-power point (−3 dB)[27] are at the expected points for the values of f_1 and f_2.

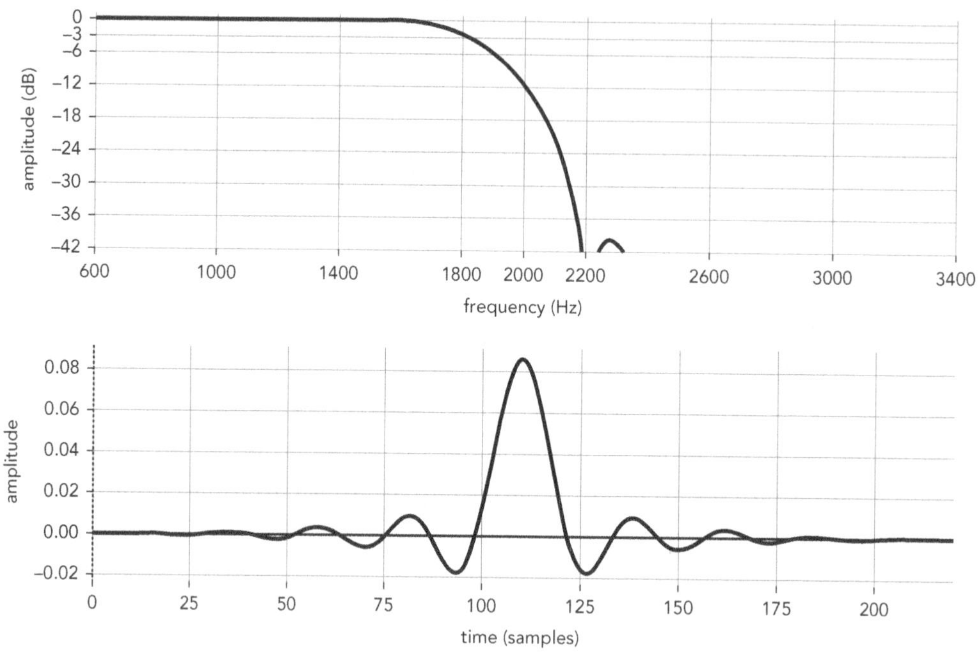

Figure 7.14 Amplitude spectrum of a lowpass filter designed by the Python program, with $f_1 = 1{,}800$ Hz and $f_2 = 2{,}000$ Hz.

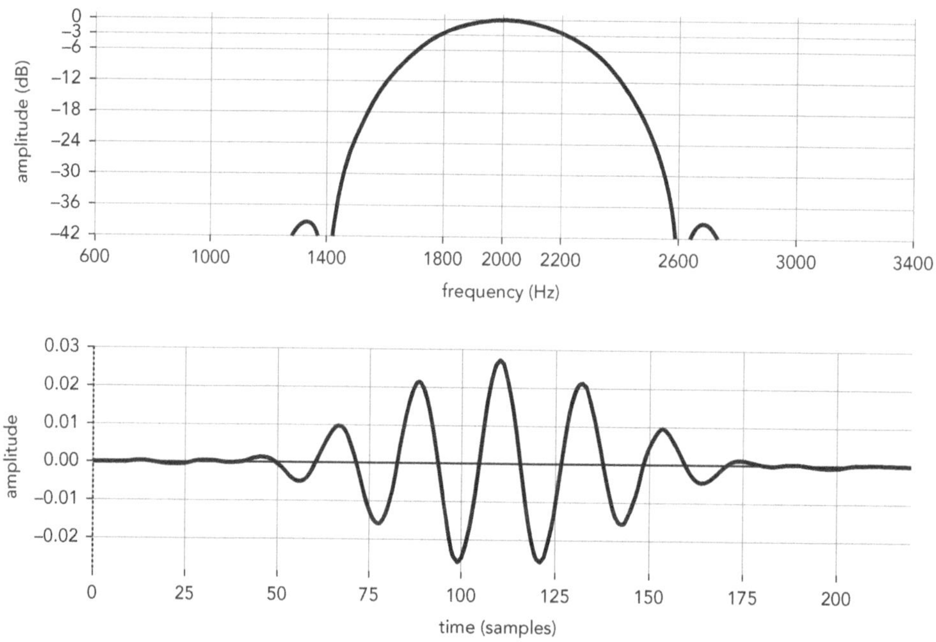

Figure 7.15 Amplitude spectrum of a bandpass filter designed by the Python program, with $f_1 = 1{,}800$ Hz and $f_2 = 2{,}000$ Hz.

7.5 Feedback

The filters we have designed and used so far are somehow lacking in punch. We can remove parts of the spectrum quite well, without messing with the phase relationships, but while this may be very good for spectral analyses, it does not seem to be enough for a variety of musical applications. In particular, it would be good if we could boost some frequencies, creating regions of resonance, without having to make an enormously long filter. It is also essential to be able to have dynamic spectra, so we need time-varying filters. The types of long filters we have designed can be made to change their spectra in the context of the live convolution techniques we explored in Chapter 5. However, for this we need to come up with various ready-made impulse responses, which seems awkward in practice.

7.5.1 Poles

A solution for this is to have a shorter filter, into which we could plug in parameters such as center or cutoff frequency, bandwidth, and so on. We have seen that this is possible within the context of a first- or second-order filter, but the effects of these filters are very limited for any practical use. However, if instead of having a polynomial in the numerator, we have one in the denominator, then as it was shown before, we have *poles* instead of zeros, which are points where the transfer function goes to infinity. If these are located close to the unit circle, then they may have a significant effect on the amplitude response of the filter, boosting the signal at the nearest frequency point.

That looks like a way forward, but so far we only managed to have poles at the origin, from where they cannot be moved. We need to review the fundamental relationships in a transfer function to see how we can change this. Early on, we introduce the notion of the filter spectrum as a ratio between the output and input frequency responses. Rewriting this as a complex transfer function, we have

$$H(z) = \frac{Y(z)}{X(z)}, \tag{7.42}$$

and, magically it seems, we have a ratio of polynomials, and poles may now be available to us. Let's develop this idea a little further by expanding the left-hand side,

$$H(z) = \sum_{n=-\infty}^{\infty} \frac{y(n)z^{-n}}{x(n)z^{-n}}. \tag{7.43}$$

This is the transfer function $H(z)$ as a ratio of z-transforms of the output and input. We know already how to interpret the denominator: it is a sum of copies of the input signal, each one delayed by n samples. Therefore the numerator should similar semantics, but instead of delaying the input, we delay the output. The resulting filter then combines delayed inputs and outputs, and the only way we can achieve this is by employing *feedback*, that is, taking the outputs of the filter and recirculating them together with its inputs. This is also an instance of *recursion*, and so these are called *recursive filters*.

Thus we now need two sets of delays, for the feedback and feedforward signal paths. From what we have discovered in Sec. 7.1, the form of such a filter then is going to be

$$\sum_{m=0}^{N_y} b_m y(n-m) = \sum_{m=0}^{N_x} a_m x(n-m), \tag{7.44}$$

which is an extension of the filter equation (Eq. 7.1) that includes feedback delay terms. So we have two convolutions: one of the input and the other of the output, on left-hand and right-hand side, respectively. We can put this in terms of the current output $y(n)$, a more sensible form of the equation, which allows us to compute it,

$$y(n) = \sum_{m=0}^{N_x} a_m x(n-m) - \sum_{m=1}^{N_y} b_m y(n-m) \tag{7.45}$$

This of course assumes that $b_0 = 1$, which means that there is no extra scaling included in the current output sample. The orders of the feedback and feedforward parts are N_y and N_y, respectively.

Now we can turn to the question of finding a practical transfer function for this filter, which we gave earlier in an outline form as a ratio of two z-transforms. First we use the useful principle that $z^{-m}X(z)$ is equivalent to $x(n-m)$, then we apply this to Eq. 7.44,

$$\sum_{m=0}^{N_y} b_m z^{-m} Y(z) = \sum_{m=0}^{N_x} a_m z^{-m} X(z) \tag{7.46}$$

and define $Y(z)$ as

$$Y(z) = \frac{\sum_{m=0}^{N_x} a_m z^{-m}}{1 - \sum_{m=1}^{N_y} b_m z^{-m}} X(z), \tag{7.47}$$

and since $Y(z) = H(z)X(z)$, the transfer function is

$$H(z) = \frac{\sum_{m=0}^{N_x} a_m z^{-m}}{1 - \sum_{m=1}^{N_y} b_m z^{-m}}. \tag{7.48}$$

This completes the journey, for we now have N_y poles that can be moved around the complex plane quite arbitrarily as far as the equation is concerned. We will see, however, that there are limits to the location of poles. Since they define what we call a *singularity*, following from the fact that $H(z)$ is infinite for certain values of z, the question of stability needs to be considered. Now that we have generic equations for time and frequency domain of filters, we can start to explore the applications and limits of various designs.

7.5.2 Resonators

As a first example, we can have a variation of the band-reject second-order feedforward filter developed earlier,

$$\begin{aligned} H(z) &= \frac{1}{(z - Re^{-j\omega_o})(z - Re^{j\omega_o})} \\ &= \frac{1}{1 - 2R\cos(\omega_o)z^{-1} + R^2 z^{-2}}, \end{aligned} \tag{7.49}$$

using the same analysis principles. However, the zeros are in the denominator now, so they are poles. By analogy, we should expect that this is a bandpass filter, and as R gets closer to 1, there will be a significant boost at the around the frequency ω_o. We have achieved what we hoped for: a short filter that can create regions of resonance in the spectrum. The filter equation we need to compute to implement this is minimal:

$$y(n) = x(n) - b_1 y(n-1) - b_2 y(n-2). \tag{7.50}$$

The advantage of having a short filter that can apply a significant amount of spectral shaping is that we may parameterize it through the use of center frequency and bandwidth controls. These can be applied to modify the coefficients (b_1 and b_2) directly, therefore making the filter time-varying is a cinch. Feedback filters can therefore be employed as powerful tools for dynamically shaping spectra.

To make the second-order filter of Eq. 7.50 a bandpass resonator based on the transfer function of Eq. 7.49, we can first set the pole radius based on a given bandwidth B in

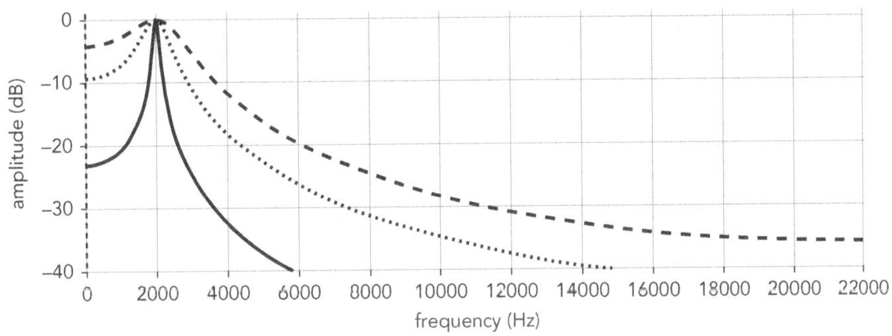

Figure 7.16 Normalized amplitude responses for a second-order resonator filter with $R = 0.99$ (*solid line*), $R = 0.95$ (*dots*), and $R = 0.9$ (*dashes*), using $f_c = 2{,}000$ Hz and $f_s = 44{,}100$ Hz.

Hz[28]

$$R = e^{-\pi B/f_s}. \tag{7.51}$$

We now need to determine the pole angle ω_o, for which the amplitude response peaks at a frequency ω_c. Clearly, if $R = 1$, then $\omega_c = \omega_o$, but otherwise, the filter is not exactly centered at a frequency that corresponds to the the pole angle. We should be able to rely on the amplitude response to find it. However, to make it simpler, we use the power spectrum[29] instead, which avoids taking an unnecessary square root. If this has a maximum at ω_c, its reciprocal will have a minimum at that same position. We are interested at infinitesimal value of ω where this dip takes place. Therefore we work this out as the point where the derivative of the inverse power spectrum goes to zero (Steiglitz, 1996, 90),

$$\frac{d}{d\cos\omega}|H(\omega)|^{-2} = 0, \tag{7.52}$$

which yields the following expression for ω_o[30]

$$\cos(\omega_o) = \frac{2R\cos(\omega_c)}{1+R^2} \tag{7.53}$$

From these relationships and Eq 7.49, we have the filter coefficients as

$$\begin{aligned} b_1 &= -\frac{4R^2\cos(\omega_c)}{1+R^2} \\ b_2 &= R^2 \end{aligned} \tag{7.54}$$

with $\omega_c = 2\pi f_c/f_s$. In Fig. 7.16, we have the plots of amplitude responses of resonator, centered at 2,000 Hz, with different bandwidths. As we can see from the picture, the pole radius R is always very close to 1 for reasonable bandwidth sizes. We should also note the lowpass bias in each amplitude response: due to the close location of the poles on each side of 0 Hz, the curve is lopsided; if the center frequency were above $f_s/4$, the bias would be highpass. Note also that the plots are all normalized to 0 dB; in practice, without any scaling, higher values of R boost input signals inside the passband way beyond 0 dB.

Due to this amplification effect, this filter will also require some amount of input scaling. Depending on how sharp the resonance is (or how close R is to 1), it can significantly boost an input signal and may cause clipping distortion. The amount of resonance is of course signal-dependent; if the input signal has little energy in the passband, the effect is less evident. Some scaling schemes have been proposed. For periodic signals, the following gain adjustment may be applied to the input,

$$g = (1 - R^2)\sqrt{1 - \frac{b_1^2}{4R^2}}, \tag{7.55}$$

which has the effect of normalizing the center of the passband to one. Another scheme normalises the output according to the overall RMS,

$$g = \sqrt{[(R^2+1)^2 - b_1^2]\frac{1-R^2}{R^2+1}} \tag{7.56}$$

Given that scaling is signal dependent, another solution is to apply an envelope follower to its output,

$$s(n) = \frac{\mathrm{RMS}[x(n)]}{\mathrm{RMS}[y(n)]}y(n) \tag{7.57}$$

where RMS[] is a root-mean-square amplitude estimator.[31] With this method, it is possible to balance the output and input signal amplitudes.

This filter is present in Csound as the opcode *reson*, with the three parameters of center frequency, bandwidth, and filter scaling mode as discussed here. In Python, we could design a resonator filter class as

```
class reson:
  def __init__(self):
    self.y1 = 0.0
    self.y2 = 0.0

  def process(self,x,fc,b):
     R = np.exp(-np.pi*b/fs)
     b2 = R*R
     o = 2*np.pi*fc/fs
     b1 = -4*b2*np.cos(o)/(1+b2)
     g = (1-R**2)*(1 - (b1**2)/(4*R**2))**(0.5)
     y = g*x - b1*self.y1 - b2*self.y2
     self.y2 = self.y1
     self.y1 = y
     return y
```

which employs the first one of the two scaling schemes, as well as the equations for setting coefficients $b1$ and $b2$.

The resonator is not only an interesting bandpass filter, but it can also be used to model physical phenomena such as vibration modes in acoustic objects. It can also behave similarly to a spring-mass system, where an object is set into motion and decays as it loses energy due to friction. This is the case for $R < 1$; with $R = 1$ we can have a sinusoidal oscillator, as long as we only feed a single impulse into the filter, since the system is not guaranteed to be stable in that case. The amplitude response of such a filter is a pair of Dirac δ functions at $\pm f_c$, as should be expected from a sinusoid. So we see that R actually controls the decay of the impulse response of this filter, which can be equated to the amount of dampening in a physical system.

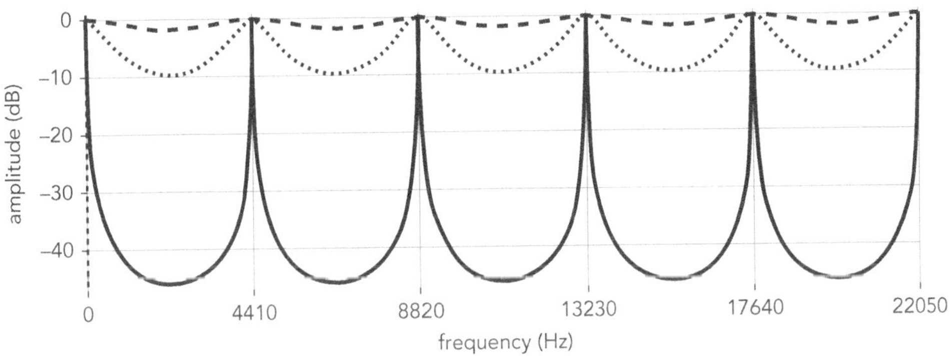

Figure 7.17 Normalized amplitude responses for a comb filter with $b_d = 0.99$ (*solid line*), $b_d = 0.5$ (*dots*), and $b_d = 0.1$ (*dashes*), using $d = 10$ samples and $f_s = 44100$ Hz.

7.5.3 Stability

Since we noted that feedback filters can be unstable, it is worth looking now at the conditions necessary to preserve stability. For this, let's look at a simple case, which is the reciprocal of an inverse comb filter, or simply, the original *comb* filter. We should guess intuitively that its transfer function is similar to the inverse case, but now with a polynomial in the denominator,

$$H(z) = \frac{1}{1 - b_d z^{-d}} \tag{7.58}$$

denoting the filter equation[32]

$$y(n) = x(n) + d_n y(n-d). \tag{7.59}$$

Amplitude responses for this filter are shown in Fig. 7.17 for three values of b_d. As we should have expected, we have a series of upward-facing teeth, resonances, spaced regularly at f_s/d intervals. Depending on the amount of resonance and the delay d, the output of this filter can ring at the filter natural frequency f_s/d as the peaks are all harmonics of this fundamental.

We can determine the stability conditions quite simply by first making the filter unstable, then correcting it. The value of d_n is key: let's set it to 2 and observe what happens to the output if we feed it a single-sample impulse,

$$y(n) = \{1, \ldots, 2, \ldots, 4, \ldots, 8, \ldots, 16, \ldots\}. \tag{7.60}$$

It grows exponentially, and it never stops increasing. It is an infinite impulse response. Filters that have any sort of feedback path are therefore *infinite impulse response* (IIR) filters. If they only have delays of inputs, then they are *finite impulse response* (FIR) filters.

Thus, setting $b_d = 2$ makes the filter severely unstable, due to the fact that gain doubling as the signal is fed back into the delay line. At the edge of stability, we then have $b_d = 1$, and the IR is an impulse train with period d/f_s, which never decays. Therefore we can only guarantee stability if $b_d < 1$. This is illustrated in Fig. 7.18 for three types of filters: stable, unstable, and non-decaying. From this conclusion, we may draw a general rule for feedback filters: none of the feedback paths can have gains that are greater or equal to 1 for the filter to have stability guaranteed. To be meaningful as far as transfer functions are concerned, this condition can be translated as

$$|p_i| < 1 \tag{7.61}$$

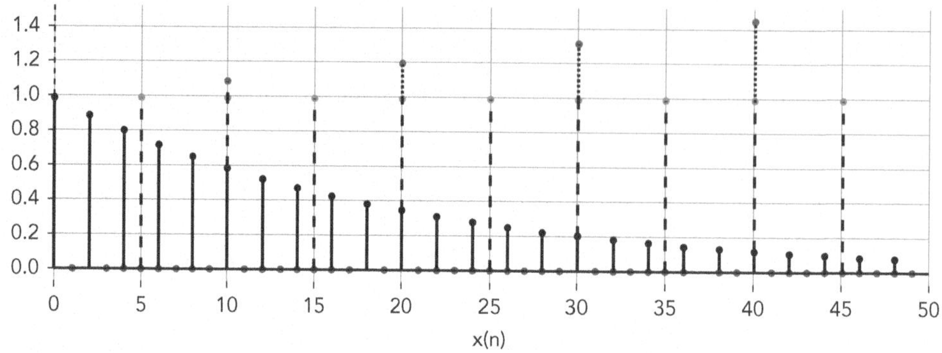

Figure 7.18 Impulse responses for stable (*solid stem, black marker*), unstable (*dots, red marker*), and non-decaying (*dashes, blue marker*) comb filters.

for all poles p_i of a filter. The magnitude of a pole can never exceed 1, or otherwise the system is unstable for all inputs. To guarantee stability for all inputs, it needs to be less than 1.

The comb filter can be implemented using more or less the same components as its inverse, but now instead of putting the input through the delay line, we feed the output into it. This is then mixed with the new samples coming into the filter. A Python class implementing it is shown in Listing 7.2. Given the fact that it can create a series of repeated echoes, when the delay is long, the comb filter is often used as a component in the design of reverberation. For these purposes, some implementations take in a delay time and a reverberation decay time (both in seconds), from which the filter coefficient is computed. This is the case of the opcode `comb` in Csound.

Listing 7.2 Comb filter class.

```
class comb:
  def __init__(self,N):
    self.delay = np.zeros(N)
    self.p = 0
    self.N = N

  def process(self,x,ad):
    yd = self.delay[self.p]
    y = x + yd*ad
    self.delay[self.p] = y
    if self.p != self.N - 1: self.p += 1
    else: self.p = 0
    return y
```

7.5.4 Phase Response

Unlike FIR filters, there is no form of coefficient arrangement that may guarantee a linear phase response for a recursive filter. Therefore, we just have to accept and be aware that the output signal components will have varying timing delays in relation to each other. For example, the phases of a resonator filter demonstrate the typical non-linear phase characteristic of these systems (Fig. 7.19).

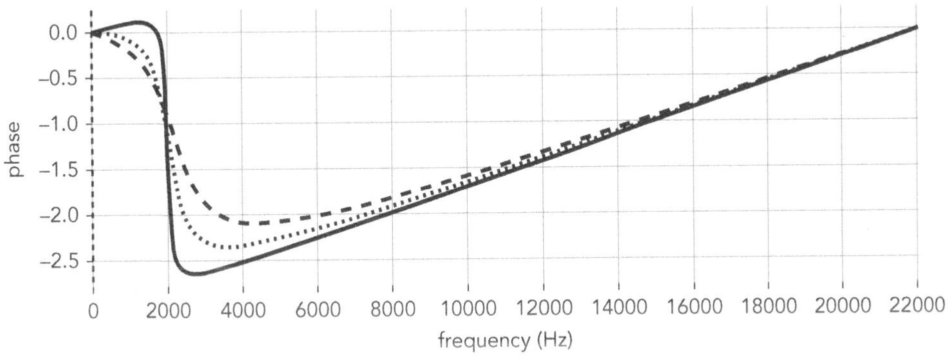

Figure 7.19 Phase responses for a second-order resonator filter with $R = 0.99$ (*solid line*), $R = 0.95$ (*dots*), and $R = 0.9$ (*dashes*), using $f_c = 2{,}000$ Hz and $f_s = 44{,}100$ Hz.

In particular, we should note the phase jump that takes place at the center of the passband. The sharper the resonance, the more accentuated is this flip in the filter phase spectrum. From this, we should expect that components either side of the center frequency will have fairly different phase offsets. Of course, depending on the bandwidth, much of the spectrum would be significantly attenuated; therefore the phase spectrum at the edges of the passband may not have much importance.

7.6 Recursive Filter Design

The approach to design feedback filters is significantly different from what we are used to. We do not attempt to design an impulse response, since that is meaningless as far as the filter structure is concerned. With feedforward filters, this corresponds neatly to a sequence of coefficients, but here the IR is only what the filter spits out when fed with an impulse, and on top of that, it is (theoretically) infinite. We can, of course, have an unknown recursive filter and treat it like a black box, get an IR, truncate it, and construct an FIR filter out of that. This, however, does not go anywhere, as far as achieving a compact, low-order, system.

On one hand, the design task is very hard, but on the other, it is not. If we set up trying to realize an arbitrary amplitude response time from scratch, it is difficult to find a straight method that would be equivalent to design program we developed in Sec. 7.4. However, if we start from structures that are already known and have a given amplitude response, it is much simpler. For example, we already have a basic design for a bandpass filter, which works fairly well. If we want to build a more complex amplitude response with, for instance, two or more resonance regions, we can construct a filter based on the product or sum of two resonator amplitude responses. Since the resonator is parameterized, defining this new filter is a matter of choosing the right parameters, and deciding how we will connect the filters together, in parallel (a sum of frequency responses) or in series (a product). This gives us a head start in terms of design methods for recursive filters.

7.6.1 Parallel and Series Connections

We take a second-order filter as our starting point. This may be the resonator, or another similar structure, since the principles we will develop should apply to other second-order

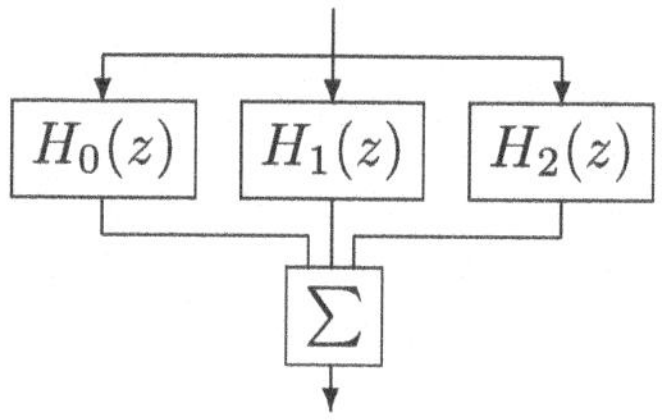

Figure 7.20 Parallel filter connection.

feedback filters. This is the lowest order we can use that includes complex poles, which can be moved inside the unit circle, and therefore it serves as building block for more complex systems.[33] We do not necessarily need to use the same design for all filters connected together, so this approach can be quite flexible.

Let's propose a practical case as a design example, creating a filter to model three vocal tract formants. These can be defined as fixed regions of resonance characterized by three parameters: center frequency, bandwidth, and gain (relative amplitude). These can be mapped straight into the typical bandpass filter controls, as defined above.

7.6.1.1 Parallel Filters

We start with the simplest of the two approaches, three filters set in parallel. This is similar to additive synthesis, where the output is the sum of individual components. For a filter such as the resonator

$$y_k(n) = g_k x(n) - b_{1_k} y(n-1) - b_{2_k} y(n-2), \tag{7.62}$$

parallel connection of two or more filters can be defined as

$$y(n) = \sum_{k=0}^{K-1} a_k y_k(n) \tag{7.63}$$

where a_k is the gain associated to each filter output. The coefficients b_{1_k} and b_{2_k}, as well as the filter scaling factors g_k, can be defined from the required center frequency and bandwidth. A diagram showing the parallel connection of three filters is presented in Fig. 7.20.

The following Python function returns the coefficients for each formant, computed from these parameters:

```
def formant(fc,b,g):
 R = np.exp(-np.pi*b/fs)
 b2 = R*R
 o = 2*np.pi*fc/fs
 b1 = 4*b2*np.cos(o)/(1+b2)
 a = (10**(-g/20))*(1-R**2)*(1 - (b1**2)/(4*R**2))**(0.5)
 return a,b1,b2
```

The frequency response for each formant can be found by applying these coefficients to a second-order transfer function and evaluating it for $z = e^{j\omega}$. We can create another function to do this,

```
def formantFR(fc,b,g,fs):
```

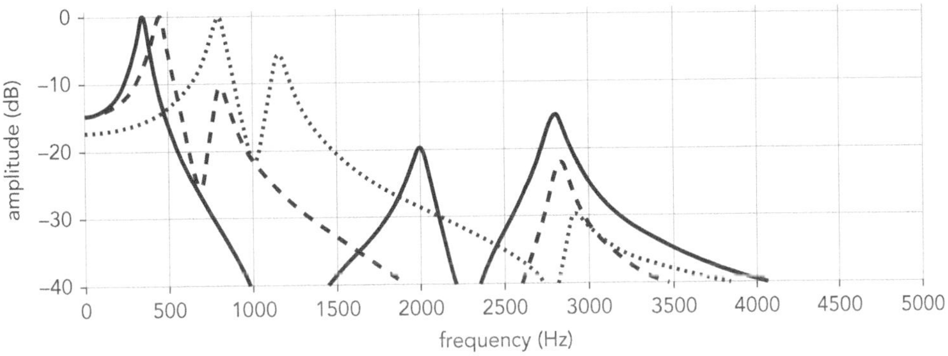

Figure 7.21 Normalized amplitude responses for three parallel-design formant filters, modeling a soprano voice "e" (*solid line*), "o" (*dashes*), and "a" (*dots*)


```
    f = np.arange(0,fs/2)
    w = 2*np.pi*f/fs
    a,b1,b2 = formant(fc,b,g)
    return a/(1 - b1*np.exp(-1j*w) + b2*np.exp(-2j*w))
```

We can look up a table of formants to get parameters for this filter. For example, a soprano "a" can be modeled by center frequencies at 800, 1,150, and 2,900 Hz, with bandwidths of 80, 90, and 120 Hz, and gain of 0, −6, and −32 dB, respectively. The following code generates the frequency response for such a filter:

```
H1 = formantFR(800,80,0,fs)
H2 = formantFR(1150,90,6,fs)
H3 = formantFR(2900,120,32,fs)
H =  H1 + H2 + H3
```

whose amplitude response, along with two other vowel models is plotted in Fig. 7.21.

This works fairly well, but the dips that happen in between formants may be somewhat unnatural, and there is considerable amount of attenuation if the spacing is large. This is caused by unwanted phase interactions when the three signals are mixed, due to the non-linear behavior of the individual phase responses. As the output frequency response is a sum of frequency responses, this may potentially have unanticipated consequences. To illustrate this, we can contrast the amplitude response of the parallel-design filter with the sum of the amplitude responses of individual formant filters (Fig. 7.22). As we can see, the latter approximates a more desirable result, with a smoother, less peaky spectrum, compared to the actual filter amplitude spectrum. Unfortunately, that is not achievable, as the individual filters are not of a linear-phase type.[34]

In fact, we can improve the result somewhat by reversing the polarity of the middle filter, using

$$H(z) = H_0(z) - H_1(z) + H_2(z) \tag{7.64}$$

since we know the phase spectrum of a resonator gets flipped at the center frequency. The amplitude response for this case is shown in Fig. 7.22.

Note that this is the phase behavior of a resonator, and other designs may fare better. We can look at an example of this by modifying the resonator slightly. As we have seen, for center frequencies below $f_s/4$, the resonator has a bias toward a lowpass amplitude response (reversed to a highpass if it exceeds that). We can improve the selectivity on

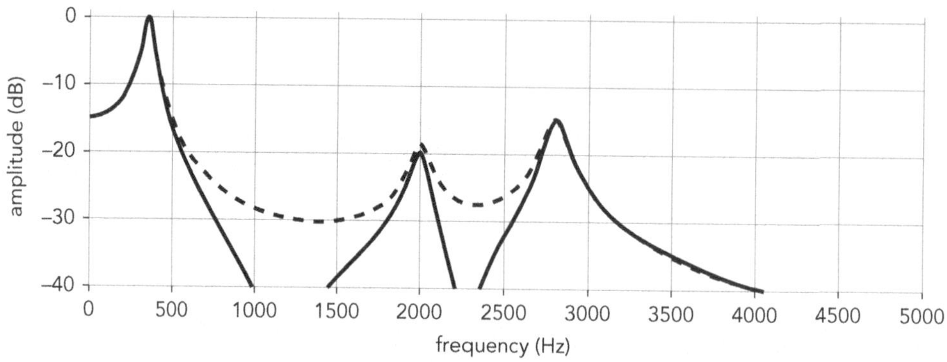

Figure 7.22 Comparison between the amplitude response of a formant filter (*solid line*), and the sum of three individual filter amplitude responses modeling each formant (*dashes*).

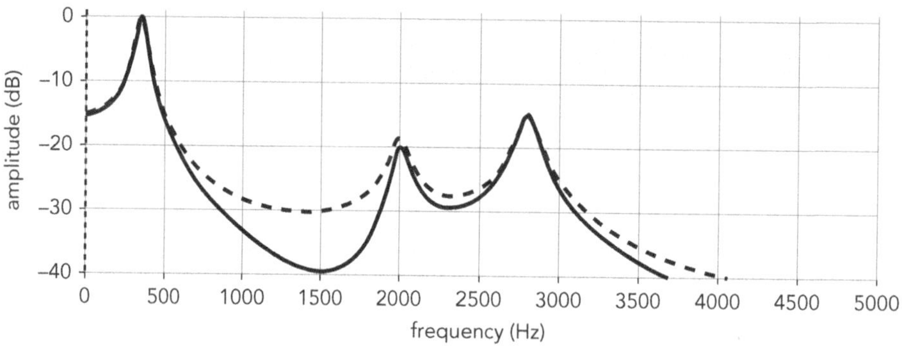

Figure 7.23 Comparison between the amplitude response of a formant filter (*solid line*) whose middle filter has a reversed polarity, and the sum of three individual filter amplitude responses modeling each formant (*dashes*).

the lower side of the center frequency by adding a feedforward term (Smith and Angell, 1982). In the FIR band-reject design of Eq. 7.27, we can see that if we set $a_1 = 0$ and $a_2 = -1$, we have a filter whose zeros are at the DC and Nyquist frequencies. This should have the desired effect of attenuating the lowpass/highpass bias since the zero will make the amplitude response have a dip at both ends of the spectrum.

The transfer function of the modified resonator becomes[35]

$$H(z) = \frac{1 - z^{-2}}{1 - 2R\cos(\omega_c) + R^2} \tag{7.65}$$

excluding the scaling factor. This can be set to $(1 - R^2)/2$ to have the output normalized to unity at the center of the passband. Using the trick of inverting the sign of the middle filter signal, we have a somewhat better result, plotted in Fig. 7.24, when compared to our first attempt.

It is not possible to provide a general rule that will describe all cases of second-order parallel connections. Nevertheless, we should know that once phases are not linearly offset, the amplitude spectrum we get from mixing the outputs of several filters is not the same as the sum of their individual amplitude responses. The phase interactions play an important part in the final result.

7.6.1.2 Series Connections

Alternatively, we may connect filters in series. This means we are feeding the input of one filter with the output of the other and collecting the result after the signal has passed all filters. This is equivalent of having a higher-order filter whose frequency response is the

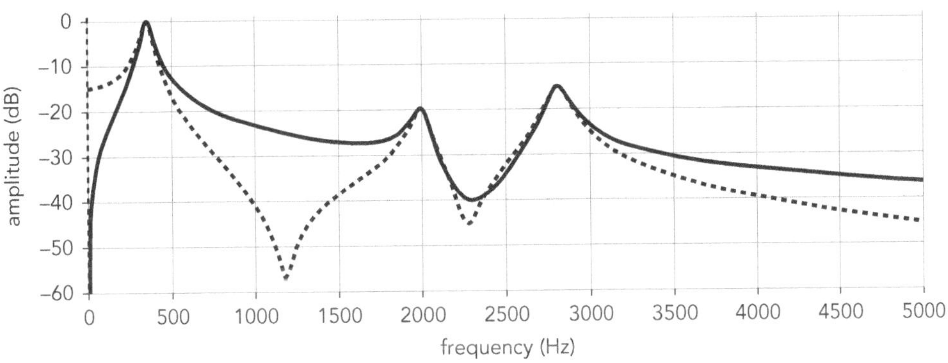

Figure 7.24 Comparison between the amplitude response of an improved formant filter (*solid line*), based on the modified resonator, and the original sum of resonator filter outputs (*dots*) for a soprano "e".

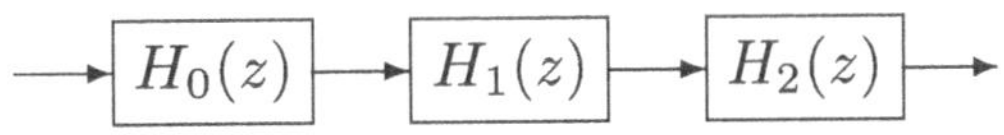

Figure 7.25 Series filter connection.

product of the individual spectra. Therefore, while in the parallel filterbank scheme the total filter is of the same order as the individual components, here it is equivalent to a sum of their orders.

This is not too different from taking two first-order filters and making a second-order filter that is equivalent to the product of their transfer functions. This is how we arrived at the resonator and the band-reject forms. Thus, we can think of a series connection as a form of factorisation of a longer polynomial,

$$\begin{aligned} H(z) &= \left(\frac{1}{1+b_{1_1}z^{-1}+b_{2_1}z^{-2}}\right)\left(\frac{1}{1+b_{1_2}z^{-1}+b_{2_2}z^{-2}}\right)\cdots\left(\frac{1}{1+b_{1_K}z^{-1}+b_{2_K}z^{-2}}\right) \\ &= \prod_{k=1}^{K} H_k(z) \end{aligned} \tag{7.66}$$

where each factor is called a second-order *section*. A diagram of series connection is shown in Fig. 7.25.

One of the advantages of using this approach is that the problems with phase interactions of multiple filters are not there anymore. So the resulting amplitude response is bound to be smoother and without any undue dips. On the other hand, the downside is that since all individual filters are lumped together into a single higher-order system, it is not possible to scale the individual resonance peaks. Also if there is not much overlap between the passbands, the output of the filter might be significantly attenuated. Therefore some sort of envelope following with the input signal used as reference might be needed here.

If, on other hand, the center frequencies are the same or close to each other, steeper transition, or rolloff, may be achieved. This is shown in Fig. 7.26, where second, fourth, and sixth-order filters, constructed from resonator sections, are contrasted. We conclude from this that series connections may be very useful to create highly selective filters.

For the present design case, the creation of formants, the fact that we lost the control of individual filter amplitudes may be problematic. The separate center frequencies and bandwidths can be fed to a bank of resonators in series, resulting in the amplitude responses of Fig 7.27, which may be compared to the parallel design examples.

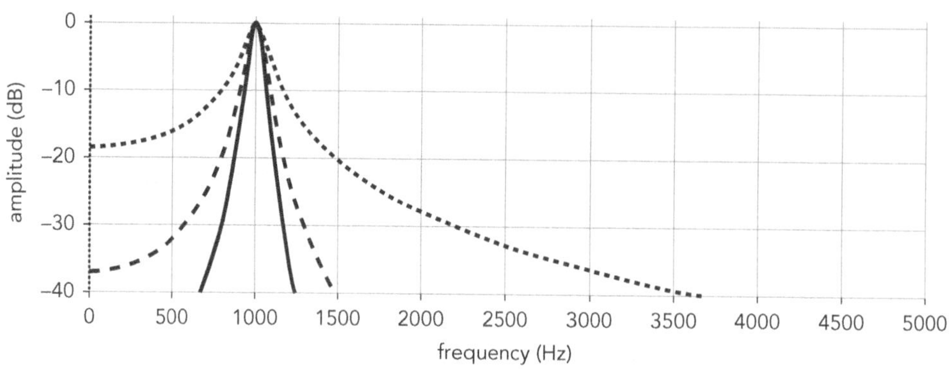

Figure 7.26 Normalized amplitude responses for three bandpass filters using the resonator design, centred at 1000Hz with a 120Hz bandwidth, in sixth-order (solid line), fourth-order (dashes), and second-order (dots) sizes.

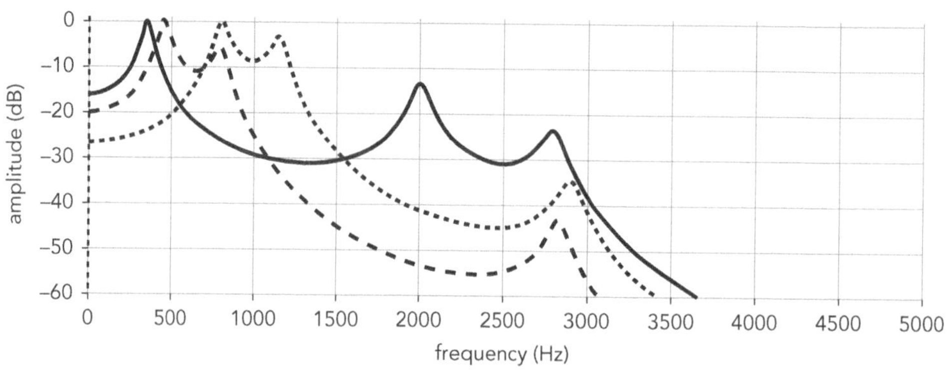

Figure 7.27 Normalized amplitude responses for three series-design formant filters, using center frequencies and bandwidths derived from a soprano voice "e" (*solid line*), "o" (*dashes*), and "a" (*dots*).

As we can see, while we have a much better behavior in the frequency areas in between the formants, the overall rolloff is lowpass, which reflects the nature of the lopsided amplitude response of the resonator design. The higher formants are always lower in amplitude, sometimes substantially so. This may change the perceptual characteristics of these spectra and we may not perceived them as modeling vowels as originally intended. Note that other second-order designs may be more successful to approximate these spectral shapes.

7.6.2 Modeling Physical Systems

We have already noted the fact that recursive filters may be used to model physical systems. The resonator is a simple example of this. For instance, if we bend a metal rod and let it go, there is a restoring that pulls it backward to the original straight position. This, however, overshoots and causes it to continue in the opposite direction, until it reverses the movement, and so on. This back-and-forth movement continues, but decays until the initial energy put into the system gets completely transferred to the environment and it stops. This is the behavior of a spring-mass model.

The resonator impulse response $h(n)$ can be used to demonstrate the similarities between the filter output and such model. We can write it as the recursive expression

$$h(n) = \delta_{0n} + 2R\cos(\omega)h(n-1) - R^2h(n-2), \qquad (7.67)$$

where, as defined before, $\delta_{0n} = 1, 0, \ldots$ is the Kronecker delta sequence. The values of the impulse response $h(n)$ for the first few values are

$$\begin{aligned} h(0) &= 1 \\ h(1) &= R[2\cos(\omega)] \\ h(2) &= R^2[4\cos^2(\omega) - 1] \\ h(3) &= R^3[8\cos^3(\omega) - 4\cos(\omega)] \\ &\ldots \\ h(n) &= R^n w(n). \end{aligned} \tag{7.68}$$

To find $w(n)$, we can set $R = 1$, and observe that the expressions inside the square brackets, which define that function, are recursive. Each term of that sequence is a polynomial in $\cos(\omega)$ of order n. These have the form $U_n(\cos(\omega))$ where $U_n(x)$ is given as (Hochstrasser, 1972)

$$\begin{aligned} U_0(x) &= 1 \\ U_1(x) &= 2x \\ U_{n+1}(x) &= 2xU_n(x) - U_{n-1}(x), \end{aligned} \tag{7.69}$$

which is a well-known recursive expression defining a *Chebyshev polynomial* of the second kind. We can then derive the signal $w(n)$ from the relationship

$$U_n(\cos(\omega)) = \frac{\sin(\omega[n+1])}{\sin(\omega)}, \tag{7.70}$$

which is satisfied by the definition of $U_n(x)$.[36] Therefore, impulse response for the resonator is

$$h(n) = R^n U_n(\cos(\omega)) = R^n \frac{\sin(\omega[n+1])}{\sin(\omega)} \tag{7.71}$$

So in one sweep, we can show that the resonator output with $R = 1$ is a steady sinusoid[37] (which we had noted but not proved) and that the impulse response for $R < 1$ is a decaying sinusoid. This scheme produces a similar audio signal as the discretized spring-mass model discussed in Chapter 3, with losses ($R < 1$) or without ($R = 1$). To demonstrate this, three plots of impulses responses for a resonator tuned to 500 Hz, for various values of R are shown in Fig. 7.28. We can see that a slight change in the loss factor R increases the damping quite substantially.

This is a useful finding from the perspective of sound design. For instance, this tells us that we can use a bank of resonators, tuned to different frequencies as a kind of musical box, modeling the plucking of each bar with an impulse sent into a different filter. We could create more complex timbres by additive synthesis. This is called *modal* synthesis, as each resonator is generating a separate vibration mode.

7.6.3 String Resonators

These ideas can be expanded to model other vibrating objects, such as strings and pipes. As an example, we can look into the former as it has a simple simpler excitation methods. The model for this has to be a filter whose output has multiple peaks, and since the

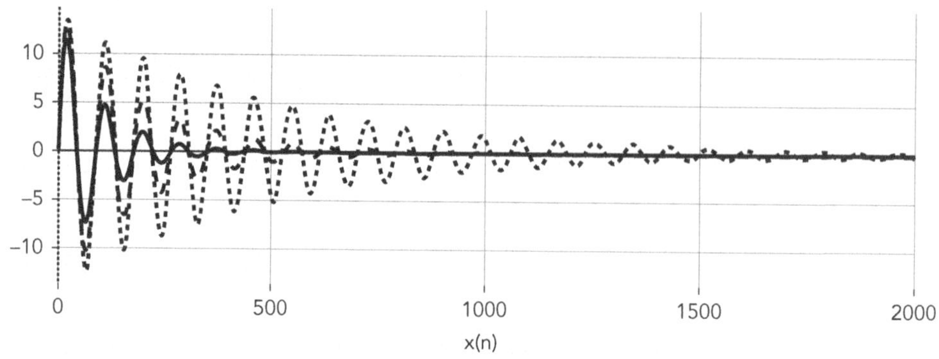

Figure 7.28 Impulse responses for three resonator filters with $R = 0.998$ (dots), $R = 0.995$ (dashes), and $R = 0.99$ (solid line), using $f_c = 500$ Hz and $f_s = 44,100$ Hz.

modes of vibration of a string are harmonics of a fundamental, the peaks should be evenly spaced in the spectrum. That gives us a good starting point: the comb filter. It meets all these initial conditions, and it has the advantage of having a very simple structure.

We already know the impulse response of this filter: it is a series of exponentially decaying impulses, so we do not need to spend time deriving it. We can focus on the magnitude spectrum, which has the characteristic peaks at kf_s/d, and note that while it may represent each longitudinal vibration of a string, it does not model the fact that the higher modes are generally of much lower intensity. The main reason for this is that at each periodic reflection a certain amount of energy is lost, and this is frequency dependent. So in order to reflect this, we can try modifying the comb filter to include this loss factor.

This can be done by inserting a very gentle lowpass filter inside the feedback path. A good candidate is the very basic first-order FIR filter,

$$y(n) = a_0 x(n) + a_1 x(n-1) \tag{7.72}$$

which as we've seen is lowpass in shape for $a_1 > 0$. Its amplitude response is

$$|H(\omega)| = |a_0 + a_1 e^{-j\omega}| = [a_0^2 + 2a_0 a_1 \cos(\omega) + a_1^2]^{1/2} \tag{7.73}$$

We can use $a_0 = 1 - a_1$ to scale the signal so we have a normalized output at 0 Hz. The setting $a_0 = a_1 = 1/2$ gives a good starting point. Taking the comb filter

$$y(n) = x(n) + b_d y(n-d), \tag{7.74}$$

we replace the term $y(n-d)$ by the FIR filter inserted in the feedback path. This creates the following filter

$$y(n) = x(n) + b_d[a_0 y(n-d) + a_1 y(n-d-1)], \tag{7.75}$$

which is our basic model for a *string resonator*. Its z-transform is

$$H(z) = \left[1 - b_d(a_0 z^{-d} + a_1 z^{-d-1})\right]^{-1} \tag{7.76}$$

For sharp resonances b_d should be close to 1, as this coefficient determines the pole magnitudes ($R^d = b_d$). From another perspective, we can also interpret this parameter as a loss factor. Note that even for $b_d = 1$ the filter is guaranteed to be stable for all frequencies except at 0 Hz, since the lowpass feedback gain is < 1 in these cases, but since we do

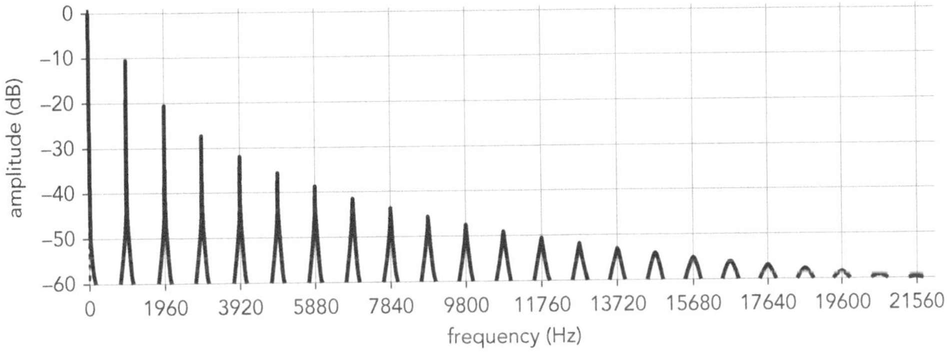

Figure 7.29 Nomalized amplitude response for a string filter tuned to 980 Hz, with $b_d = 0.999$, $a_0 = a_1 = 0.5$, and $f_s = 44,100$ Hz.

not know whether the input spectra has any significant energy at DC, we should keep it below 1. This also tells us that this filter may build quite a bit of DC bias, so it may need to be used in combination with a highpass (lo-cut) filter.

Another key aspect of this filter is the decay time of its impulse response, or how long it will ring for given a short burst of sound as input. For the original comb filter, we can easily estimate this for a given threshold, given that each time it is feedback, the impulse is multiplied by b_d. Over n repetitions, the IR will have decayed by b_d^n. Since n in this case is dependent on the filter delay $1/f_0$ (in seconds), the T_{60}[38] of the impulse response is given as

$$T_{60} = \frac{-60}{f_0 20 \log_{10}(b_d)} \qquad (7.77)$$

In the case of the string resonator, the filter feedback gain is frequency dependent; therefore, we need to replace b_d by its product with the amplitude response of the lowpass filter, $|H_{lp}(f)|$. The T_{60} measurement now is a function of frequency,

$$T_{60}(f) = \frac{-60}{f_0 20 \log_{10}(b_d |H_{lp}(f)|)}, \qquad (7.78)$$

from which we can see that if $|H_{lp}(f)|$ is lowpass, then the decay time is going be shorter at higher frequencies. Moreover, this also affects how quickly the impulse response of different-size filters decays. For $a_0 = a_1 = 0.5$, $|H_{lp}(f)| = \cos(\pi f / f_s)$,[39] which has a significant impact in terms of shortening the decay times as the frequency increases. This is illustrated in Fig. 7.30, where the T_{60} for a string resonator at different frequencies is compared to the fixed decay time of a comb filter. Note that this behavior actually models well a physical object such as a string, whose higher modes of vibration decay faster than the lower ones.

From this we can surmise that b_d controls the overall decay times, in a non-frequency dependent way. To do this in a more sophisticated way, we should also look at adjusting the lowpass filter coefficients a_0 and a_1 (Moore, 1990, 284). We can first rewrite its amplitude response using $a_0 = 1 - a_1$,

$$|H(\omega)| = [(1 - a_1)^2 + 2(1 - a_1)a_1 \cos(\omega) + a_1^2]^{1/2}. \qquad (7.79)$$

Now we can look for a value a_1 that would give us a frequency-based control over the decay time. We start with Eq. 7.78, set $f_0 = f$, as we are looking at the fundamental frequency here specifically, and define a decay rate in dB as $D_{dB} = 60/T_{60}$. Now we cast in

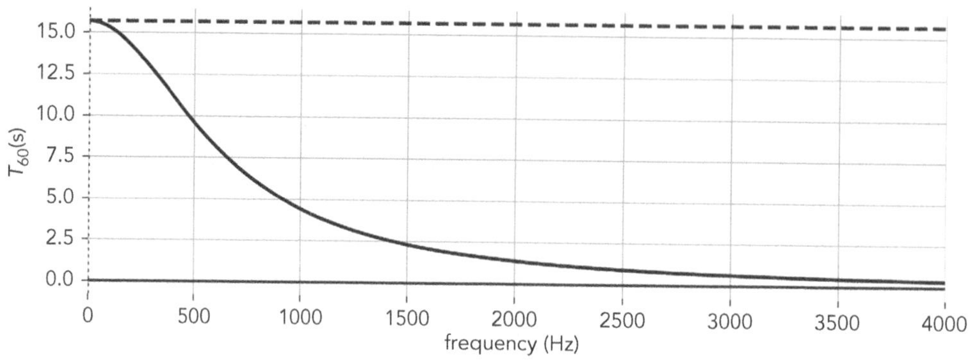

Figure 7.30 String resonator −60 dB decay times for a $f_0 = 440$ Hz (*solid lines*), compared to flat comb filter decay time (*dashes*), with $b_d = 0.999$ and $f_s = 44,100$ Hz.

terms of the lowpass amplitude response,

$$|H_{lp}(f)| = \frac{10^{\frac{-D_{dB}}{20f}}}{b_d} \tag{7.80}$$

which connects a desired D_{dB} to a frequency f at which we want to achieve this decay. Clearly, the issue at hand is that higher frequencies may have a decay that is too short for our applications. This is due to the action of the lowpass filter. We can try to lengthen the decay by making the filter less selective at high frequencies. For this case, we need a value of a_1 that will flatten the amplitude response. Setting $b_d = 1$, we can solve Eq. 7.80 for a_1:

$$\begin{aligned} |H_{lp}(f)|^2 &= [10^{-D_{dB}/(20f)}]^2 \\ (1-a_1)^2 + 2(1-a_1)a_1\cos(\omega) - a_1^2 &= 10^{-D_{dB}/(10f)} \\ [2-2\cos(\omega)]a_1^2 + [2\cos(\omega)-2]a_1 + 1 - 10^{-D_{dB}/(10f)} &= 0, \end{aligned} \tag{7.81}$$

using the quadratic formula for $x = a_1$,

$$x = \frac{-b \pm [b^2 - 4ac]^{1/2}}{2a}, \tag{7.82}$$

where

$$\begin{aligned} a &= 2 - 2\cos(2\pi f/f_s) \\ b &= 2\cos(2\pi f/f_s) - 2 \\ c &= 1 - 10^{-D_{dB}/(10f)} \end{aligned} \tag{7.83}$$

The value we are seeking for coefficient a_1 should be between 1/2 (standard lowpass response) and 1 (allpass). Since the second-order equation has two roots, we choose the one within that range. To achieve an arbitrary decay time for a given frequency, we start with $a_1 = 0.5$ and $b_d = 1$. If we need to shorten the decay, the lowpass filter can be left untouched and b_d is adjusted; Eq. 7.78 can be used to set it. Otherwise, we can apply Eq. 7.81 to lengthen it. The following Python function implements these ideas:

```
# streson coef setting
# takes a decay rate in dB/s
# the filter fundamental f
# and the sampling rate
```

```
# returns coefficients a1,bd
def streson_coef(dec,f,fs):
 g = 10**(-dec/(20*f))
 ar = np.cos(np.pi*f/fs)
 if g > ar:
  cosf = np.cos(2*np.pi*f/fs)
  a = 2 - 2*cosf
  b = 2*cosf - 2
  c = 1 - g**2
  d = (b**2 - 4*a*c)**(1/2)
  s1 = (-b + d)/(a*2)
  s2 = (-b - d)/(a*2)
  if s1 >= 0.5: a1 = s1
  else: a1 = s2
  bd = 1
 else:
  a1 = 0.5
  bd = g/ar
 return a1,bd
```


Of course we have not ignored the fact that $b_d = 1$ may make the filter unstable. However, if problems arise we can always reduce this value slightly so that no matter what we do, the filter will not explode. The overall decay rate may be slightly shortened, but if we choose a value that is close enough to 1, this should be only a minimal reduction. Employing a highpass filter may also help prevent any issues with significant energy at DC.

One question we have not yet discussed, but have been assuming an answer all along, is about the actual fundamental frequency of the string resonator. We know that for the comb filter case, this is f_s/d, but what we have here is different. The addition of the lowpass filter places and additional delay in the feedback path, so we have to assume the fundamental frequency is slightly lower. We can compute this by looking at the lowpass phase response. It is an FIR filter, so if the coefficients are symmetric about their center, we have a linear phase response of $1/2$ sample (as the filter is 2 samples long). Therefore if $a_0 = a_1$, the total delay is $d + 1/2$ and the fundamental frequency is $f_s/(d+1/2)$ samples. If, however, we change a_1 to make the decay longer at a given frequency, the phase response becomes non-linear. We can calculate this for any frequency f, but for low frequencies, the total delay can actually be approximated to

$$d_{total} \approx d + a_1, \tag{7.84}$$

which can be used then to determine the fundamental frequency of the filter, since we are using a_1 in the $[0.5, 1]$ range at all times. The phase spectra for three values of this coefficient is shown in Fig. 7.31, where we can see that the linear approximation assumed above is fairly accurate.

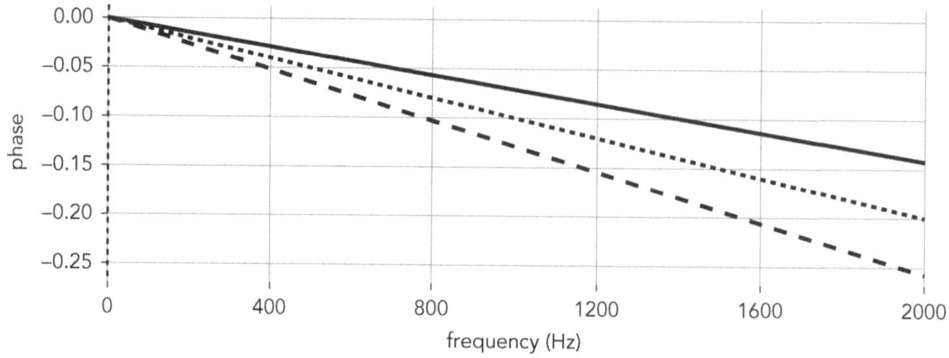

Figure 7.31 First-order FIR lowpass phase spectra for $a0 = 1 - a1$ and $a1 = 0.5$ (*solid line*), $a1 = 0.7$ (*dashes*), and $a9 = 0.9$ (*dots*), showing approximate linear behavior at low frequencies ($f_s = 44,100$ Hz).

7.6.4 Allpass Filters

Allpass filters have a flat amplitude response, and therefore their action is solely on the phase spectrum. We have already seen some cases of allpass filters, but these were linear phase, as in the pure delay example with which we started this chapter. The most interesting types of allpass filters are those that exhibit particularly useful non-linear phase behavior. For this we will can employ a pole-zero combination, since we know that poles allow short filters to be constructed, but on their own they cannot realize a flat amplitude response. Allpass systems can be used in a number of applications involving delays, and interestingly enough in filter combinations where they contribute to an overall amplitude response that is not allpass.[40]

7.6.4.1 Tuning Filter

We will look first at an example where an allpass phase shift can be harnessed to provide fractional delays to tune the string resonator we have just designed. Since that has a delay of $d + a_1$ samples, we can only use fundamental periods that vary in whole number of samples: $(d + a1)/fs$, $(d + 1 + a1)/fs$, $(d + k + a1)/fs$, and so on. To fix this we need to provide the extra fractional delay that can tune the filter exactly. This is the role of the first-order allpass filter: to provide a phase shift of zero to one samples at a given fundamental frequency. Of course, since its phase spectrum is non-linear, higher frequencies will be offset by other amounts, but this should not be problematic in general.

The working principle behind the pole-zero allpass is to counterbalance the action of a zero with a pole (or vice-versa), so that the amplitude response is kept flat across the spectrum. so for a given zero with magnitude R, we want to add a pole with the same angle and magnitude $1/R$ (Fig 7.32). For the filter to be stable, the pole needs to be inside the unit circle, which means that in this case $R > 1$ and the zero is placed somewhere outside the unit circle. In a first-order allpass, poles and zeros are real, their angle is either 0 or π, so the pole location is within the $(-1, 1)$ interval. The general form of the transfer function for an N-order filter is

$$H(z) = \prod_{i=1}^{N} \frac{z^{-1} - R_i e^{-j\omega_i}}{z^{-1} - R_i^{-1} e^{-j\omega_i}} \tag{7.85}$$

which, for any value of ω_i and R_i, $|R_i^{-1}| < 1$, has an allpass frequency response, noting that any pole or zero that is complex needs to have a complex conjugate pair for the filter to have real-valued coefficients.[41]

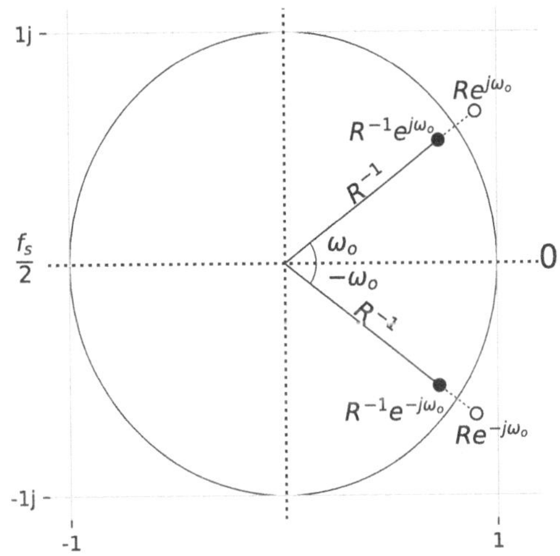

Figure 7.32
The poles and zeros of a second-order allpass filter at $z = R^{-1}e^{\pm j\omega_0}$ and $z = Re^{\pm j\omega_0}$, respectively.

From this, we can now develop a first-order allpass. We start by setting $c = -R$ so that the filter coefficient is in the correct form.[42] The transfer function becomes

$$H(z) = \frac{z^{-1} + c}{z^{-1} + c^{-1}}. \tag{7.86}$$

The next step is to make the filter have unity gain. If we observe the stability condition $|c^{-1}| < 1$, then the coefficient c is above 1, and the input signal gets amplified by a certain amount. We can multiply the transfer function by c^{-1} to scale it so that the amplitude response has no effect whatsoever,[43]

$$H(z) = \frac{c + z^{-1}}{1 + cz^{-1}}. \tag{7.87}$$

This has the added bonus of making the expression even simpler. The allpass filter equation can therefore be implemented as

$$y(n) = cx(n) + x(n-1) - cy(n-1). \tag{7.88}$$

Noting the stability condition of $|c| < 1$, we can go ahead and find a relationship between the phase response $\phi(\omega)$ and the filter coefficient, so that we can start with a fractional delay at ω and derive c. Once we have this, we can create a filter that gives the required phase shift at an arbitrary fundamental frequency ω. The frequency response is

$$\begin{aligned} H(\omega) &= \frac{e^{-j\omega} + c}{1 + ce^{-j\omega}} = \frac{e^{j\omega/2}[e^{-j\omega/2} + ce^{j\omega/2}]}{e^{j\omega/2}[e^{j\omega/2} + ce^{-j\omega/2}]} \\ &= \frac{e^{-j\omega/2} + ce^{j\omega/2}}{e^{j\omega/2} + ce^{-j\omega/2}} = \frac{z}{\bar{z}} \\ &= e^{j2\theta}, \end{aligned} \tag{7.89}$$

which is proved to be allpass with unity gain. Since $z = Ae^{j\theta}$, the phase response can then be defined as

$$\phi(\omega) = \arg(e^{j2\theta}) = 2\arg(e^{-j\omega/2} + ce^{j\omega/2}). \tag{7.90}$$

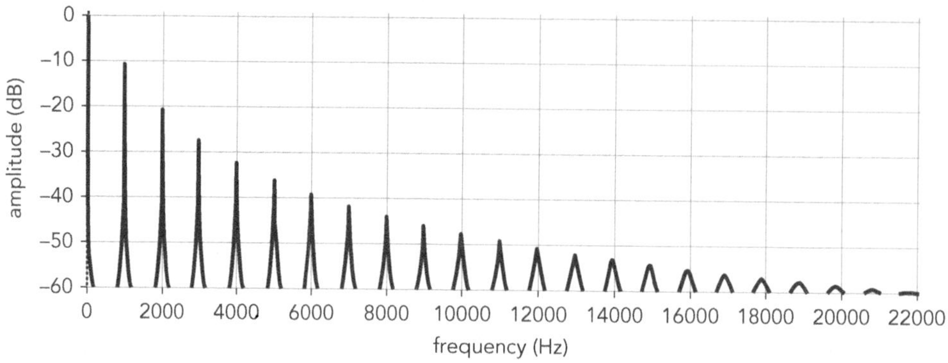

Figure 7.33 Normalized amplitude response for a string filter tuned to 1,000 Hz, with $b_d = 0.999$, $a_0 = a_1 = 0.5$, and $f_s = 44,100$ Hz.

While we can get the exact phase shift by computing this equation, it is possible to find an approximation that works for low frequencies (small values of ω),

$$\phi(\omega) = 2\tan^{-1}\left(\frac{-(1-c)\sin(\omega/2)}{(c+1)\cos(\omega/2)}\right) = 2\tan^{-1}\left(\frac{-(1-c)}{(c+1)}\tan(\omega/2)\right)$$
$$\approx \frac{-(1-c)}{(c+1)}\omega = -K\omega \tag{7.91}$$

where the final steps are given by $\tan(\omega/2) \approx \omega/2$ and by consequence $\tan^{-1}(-K\omega) \approx -K\omega$, both possible because of the very small values of ω and K. To get the delay in samples from the phase response $\phi(\omega)$ (in radians), we only need to divide it by the radian frequency ω. Therefore K is the phase delay we are looking for, and a little rearrangement gives us the simple formula

$$c = \frac{1-K}{1+K}, \tag{7.92}$$

which is what we have been looking for. Since $c < 1$, it should also be set in the range $0 \leq K < 1$, ensuring that the filter stable and the delay is a fractional value.

Now we can apply this filter to tune the string resonator. The approach is to insert it in the feedback signal path, before or after the lowpass (it does not matter since these are linear systems). The corresponding transfer function becomes

$$H(z) = \left[1 - b_d(a_0 z^{-d} + a_1 z^{-d-1})\left(\frac{c + z^{-1}}{1 + cz^{-1}}\right)\right]^{-1} \tag{7.93}$$

which has of course quite a complex frequency response. The amplitude response for this filter with $f_0 = 1,000$ Hz is shown in Fig. 7.33.

As expected, there is no change in the amplitude response curve. However, now it is possible to tune the filter to frequencies that do not correspond to integer delay line lengths, as is the case of an $f_0 = 1,000$ Hz at $f_s = 44,100$ Hz, which corresponds to a period of 44.1 samples. Also, we can see that, due to the non-linear phase response of the allpass filter, the higher partials deviate slightly from the harmonic series. As can be seen from Fig. 7.34, the delays produced by the allpass filter tend to increase with frequency, making higher harmonics flatter. This effect is more pronounced with higher values of c (or lower K). As the pole/zero pair get closer to the unity circle, the non-linearity of the phase spectrum is more apparent. The fractional delay error at low frequency is larger at

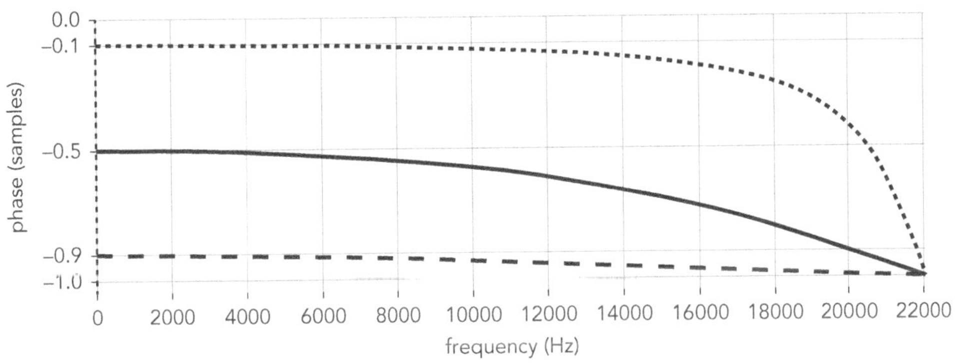

Figure 7.34 Allpass phase response in samples for $K = 0.5$ ($c = 0.33$, *solid line*), $K = 0.1$ ($c = 0.82$, *dots*), and $K = 0.9$ ($c = 0.05$, *dashes*), with $f_s = 44100$ Hz.

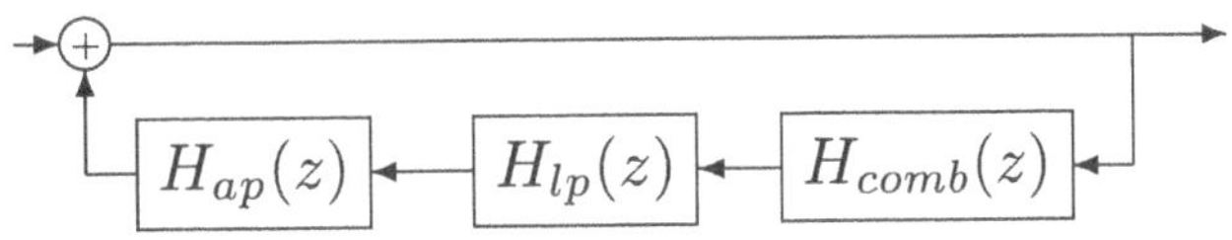

Figure 7.35 The string resonator as a series connection of three filters, pure delay, lowpass, and allpass, in a feedback arrangement.

$K = 0.5$, although it only starts to have any significance above 5 kHz, which is beyond the f_0 limit of most musical instruments.

We now have all components to program a string resonator. We can proceed as before, by creating a self-contained processor as a Python class. In this way, any number of strings can be modeled in parallel. This is shown in Listing 7.3, where we have a longer object initialization step, since we need to compute all the filter coefficients according to the user parameters. Note that we are employing the lowpass coefficient setting function developed earlier. The allpass coefficients are calculated using Eq. 7.92. The string resonator itself of course is built from three components (Fig.7.35):

- lowpass filter:

$$w_{\text{LP}}(n) = y(n-d)$$
$$y_{\text{LP}}(n) = a_0 w_{\text{LP}}(n) + a_1 w_{\text{LP}}(n-1)$$

- allpass filter:

$$w_{\text{AP}}(n) = y_{\text{LP}}(n) - c w_{\text{AP}}(n-1)$$
$$y_{\text{AP}}(n) = c w_{\text{AP}}(n) + w_{\text{AP}}(n-1)$$

- comb filter:

$$y(n) = x(n) + b_d y_{\text{AP}}(n)$$

The allpass filter equation is split into two expressions, which allows the reuse of the single-sample delay.[44] This filter is potentially unstable if fed with signals with significant energy at DC and a slow decay rate is sought. Limiting b_d to less than unity should protect this, but does nothing against potential build up of DC bias.

Listing 7.3 String resonator class in Python.

```
# Class streson
# string resonator
# fundamental freq f (Hz)
```

```
# decay rate dec (dB/sec)
# sampling rate fs
class streson:
  def __init__(self,f,dec,fs):
      a1,bd = streson_coef(dec,f,fs)
      d = fs/f
      di = int(d)
      if di + a1 > d: di -= 1
      K = d - (di + 0.5)
      self.bd = bd
      self.a1 = a1
      self.c = (1 - K)/(1 + K)
      self.N = di
      self.delay = np.zeros(self.N)
      self.p = 0.
      self.lpd = 0.
      self.apd = 0.

  def process(self,x):
    c = self.c
    a0 = 1 - self.a1
    a1 = self.a1
    bd = self.bd
    yd = self.delay[self.p]
    lp = (a0*yd + a1*self.lpd)
    self.lpd = yd
    wa = lp - c*self.apd
    ap = self.apd + c*wa
    self.apd = wa
    y = x + ap*bd
    self.delay[self.p] = y
    if self.p != self.N - 1: self.p += 1
    else: self.p = 0
    return y
```


A similar filter is present in Csound, the opcode `streson`, which employs the tuning filter, but does not offer a fine control over the decay rate (or the decay lengthening code). It is nevertheless a useful tool to construct tuned resonators, which may be used to model sympathetic resonances on strings. A sound synthesis packaging of this filter is given in the extended Karplus-Strong algorithm.

7.6.4.2 Allpass Reverberators

Higher-order allpass filters can also be used in reverberation as an alternative or in combination with comb filters. From the principles outlined earlier, it is easy to see that the

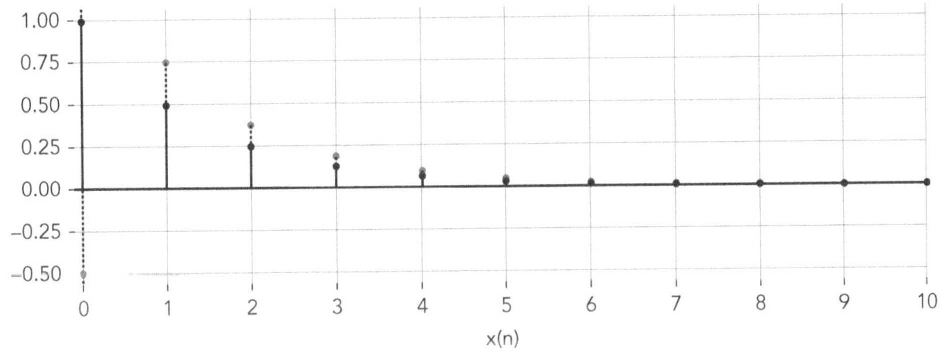

Figure 7.36 Impulse responses for allpass (*red marker, dots*), and comb (*black marker, solid stem*) filters, with $g = 0.5$.

comb filter transfer function can be modified to create an allpass frequency response,

$$H(z) = \frac{z^{-d} - a_0}{1 - b_d z^{-d}} \tag{7.94}$$

with $a_0 = b_d$. This is generally implemented with the following pair of equations:

$$\begin{aligned} w(n) &= x(n) + gw(n-1) \\ y(n) &= w(n-1) - gw(n) \end{aligned} \tag{7.95}$$

which, as before, allows a single delay line to be reused for both feedback and feedforward signal paths. An allpass implementation in Python can be provided as the following class:

```
class apass:
  def __init__(self,N):
    self.delay = np.zeros(N)
    self.p = 0
    self.N = N

  def process(self,x,g):
    wd = self.delay[self.p]
    w = x + wd*g
    y = wd - w*g
    self.delay[self.p] = w
    if self.p != self.N - 1: self.p += 1
    else: self.p = 0
    return y
```

The advantage of having an allpass reverberator is its flat amplitude response, of course, which does not color the input. However, as we have seen in the case of the string filter, there is a decay time associated with its impulse response, which, if long, may cause the filter to ring at its fundamental frequency. This is the *transient* behavior of the filter, which happens in response to abrupt changes in the output. For example, if an input sound has a sharp/quick decay, ringing might occur if the feedback gain is close to 1. In practice, these allpass filters are used with much lower values of g than it is the case with comb filters. This is illustrated by Fig. 7.36, where impulse responses for $g = 0.5$ are shown for both allpass and comb filters.

7.6.4.3 Minimum Phase-Allpass Decomposition

Another application of allpass filters is to modify the phase response of a filter so that it can be put into a minimum-phase form.[45] For example, if we have a filter transfer function $H(z)$ with a zero outside the unit circle, we can decompose it into an allpass filter, with a zero and a pole balanced out and a minimum phase filter, where the zero is situated inside the unit circle. Since the allpass does not have any influence in the filter magnitude, the minimum phase component will preserve the same magnitude response as $H(z)$.

For example, the second-order FIR filter,

$$H(z) = a_0 + a_1 z^{-1} + a_2 z^{-2} = (z^{-1} - Re^{j\omega_z})(z^{-1} - Re^{-j\omega_z}) \tag{7.96}$$

is maximum phase for $R > 1$. The minimum-phase form of this filter then can be obtained by using the following allpass filter,

$$H_a(z) = R^2 \frac{(z^{-1} - Re^{j\omega_z})(z^{-1} - Re^{-j\omega_z})}{(z^{-1} - R^{-1}e^{j\omega_z})(z^{-1} - R^{-1}e^{-j\omega_z})} \tag{7.97}$$

and dividing $H(z)$ by it,

$$H_m(z) = \frac{H(z)}{H_a(z)} = \frac{(z^{-1} - R^{-1}e^{j\omega_z})(z^{-1} - R^{-1}e^{-j\omega_z})}{R^2} \tag{7.98}$$

We have then obtained a minimum phase filter that has the same magnitude response as the original filter, but, of course, different phase response. Note that the process may be applied in reverse form to get a maximum phase filter instead. In this case, this is obtained by multiplying a filter by a suitable allpass arrangement. The main snag with this method is that it depends on the factorization of a transfer function, something that may always not be practical. Alternatively, we can derive a minimum phase filter by the method given in Sec. 7.3.3, which provides a phase response that is derived from an existing amplitude response.

7.6.5 The Channel Vocoder

We complete this section by looking at a classic filterbank design: the channel vocoder. This is effectively an analysis-synthesis system, which employs two sets of filters: one for estimating the amplitude spectrum of an input signal at a given band, and the other to realize the extracted an amplitude response based on this analysis data. The two banks of filters are normally identical, and the synthesis filters are usually fed some sort of spectrum-rich, broadband signal, which may be pitched or unpitched. The vocoder has been implemented in many different ways, with different filter designs, but the principles behind all its various instances is fairly similar. A vocoder filterbank band is depicted in Fig 7.37, where we see that the synthesis filter takes an arbitrary spectrum-rich excitation signal $x(n)$, and an input sound $s(n)$ is fed to the analysis filter. The amplitude detected at this band is then imposed on the synthesis filter output. A parallel set of these bands is then able to apply the spectral envelope taken from the input onto the excitation signal.

From Fig 7.37, we can surmise that there are two main dependencies needed to construct a vocoder. The first of these is a bandpass filter. We should choose a flexible, fairly selective filter. From what we have seen so far, either a second- or a fourth-order pole-zero resonator should be a good starting point. There are other bandpass IIR alternatives,

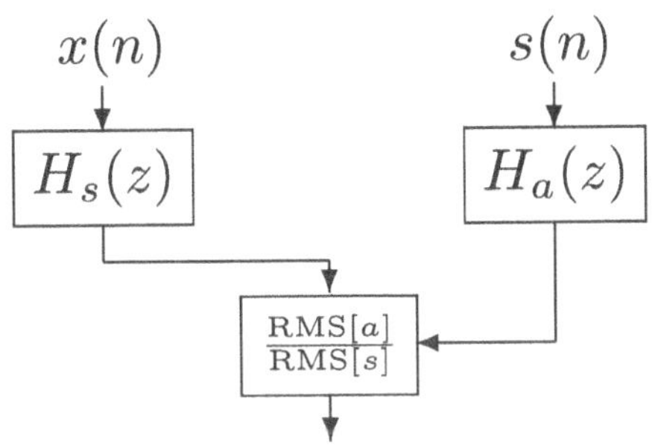

Figure 7.37
A channel vocoder band, composed of two filters and an envelope follower.

but we have not explored them yet. FIR filters are less likely to be good choices, as they are going to be far more demanding in computational terms. Of course, the phase vocoder is an efficient FIR-based system, but it has a number of specific characteristics, such as equal-bandwidth filters, which in some ways make it less flexible than the approach we want to develop here.

The second element needed for the channel vocoder implementation is an envelope follower of some description. This is needed to extract the amplitude signal from an analysis filter and apply it to the signal coming out of the synthesis filter. This detection takes the form of an RMS estimation, and the envelope follower balances two signals, the output of the analysis and the output of the synthesis on each band, by taking streaming measurements of their amplitudes. We need to see how we could construct such a processor. Remembering that the RMS is the defined as the square root of the mean of squared signal samples, it appears that we have all the steps at hand, except for the averaging part.

A clue to this comes from the fact that the continuous-time definition of RMS employs an integral as part of the mean estimation. We know what this means: a lowpass filter. In fact, the digital integrator can be defined as a first-order recursive filter,

$$y(n) = x(n) + y(n-1), \tag{7.99}$$

which has a lowpass amplitude response. In practice we might want some sort of *leaky* integrator, which has a coefficient attached to $y(n-1)$ so that we can control the amount of smoothing applied. Even better if we could determine the cutoff frequency or similar parameter. For this, we may start with a first-order IIR filter of a standard design,

$$y(n) = a_0 x(n) - b_1 y(n-1). \tag{7.100}$$

This filter has a single pole that is either close to the Nyquist frequency (highpass) or DC (lowpass, which is what we want). Its transfer function is

$$H(z) = \frac{a_0}{1 + b_1 z^{-1}}, \tag{7.101}$$

from where we can see that the pole location is at $-b_1$. The coefficient a_0 can be used to scale the transfer function,

$$a_0 = 1 + b_1, \tag{7.102}$$

which makes the amplitude response unity at DC. We can now define the coefficient b_1 by finding the angle ω_c where the amplitude response falls to the half-power, -3 dB point.

Since power is the square of the magnitude, we have

$$|H(\omega_c)|^2 = \frac{1}{2} \tag{7.103}$$

which gives us

$$\frac{(1+b_1)^2}{1+2b_1\cos(\omega_c)+b_1} = \frac{1}{2} \tag{7.104}$$

Solving for b_1, we have the quadratic equation

$$b_1^2 + [4+2\cos(\omega_c)]b_1 + 1 = 0, \tag{7.105}$$

which gives us two results,

$$b_1 = \pm\sqrt{[2-\cos(\omega_c)]^2 - 1} - 2 + \cos(\omega_c). \tag{7.106}$$

Of the two solutions, one puts the pole inside the unit circle and the other outside. We, of course, have to pick the smallest,

$$b_1 = \sqrt{[2-\cos(\omega_c)]^2 - 1} - 2 + \cos(\omega_c), \tag{7.107}$$

which is the value of b_1 for a given cutoff frequency ω.

This is a very gentle lowpass filter, with a 6 dB/octave slope since it is based on a first-order equation. Now we can use it with a low (< 10 Hz) cutoff frequency to provide the averaging need for RMS estimation. This can be computed on a sample-by-sample basis using the operator $\text{RMS}[x(n)] = f_{\text{rms}}(n)$, defined as

$$f_{\text{rms}}(n) = \sqrt{\text{LP}[x(n)^2]}, \tag{7.108}$$

where $\text{LP}[x(n)]$ is the output of a first-order lowpass filter acting on a signal $x(n)$. Note the inherent non-linearity of this operation, as it involves the squaring of an input signal. The function $f_{\text{rms}}(n)$ provides the sample-by-sample envelope of a signal.

The vocoder is constructed from a collection of parallel channels, where the output is the sum of all synthesis filters $s_k(n)$, to which the envelopes extracted from the analysis filters $a_k(n)$ is applied.[46]

$$y(n) = \sum_{k=0}^{N} s_k(n)\frac{\text{RMS}[a_k(n)]}{\text{RMS}[s_k(n)]} \tag{7.109}$$

Each filter is bandpass, which can be designed in different ways. As we have noted, we can use the pole-zero resonator to implement the filter banks,

$$s_k(n) = \frac{1-R_k^2}{2}[x(n)-x(n-2)] + 2R_k\cos(\omega_k) - R_k^2, \tag{7.110}$$

with $a_k(n)$ also similarly defined. The values for ω_k and R_k come from the center frequency and bandwidth of each filter.

Since our perception of frequencies is nearly logarithmic, it is best to space the filters using some sort of interval measure, based on an equal division of the octave. This gives us the number of required bands depending on the number of octaves we want to cover. A typical spacing is 1/3 octave, giving filter frequencies $f_k = 2^{k/3}f$. Typically 4 to 8 octaves are covered (12 to 24 bands) from a base frequency f.

The bandwidths B_k of each band are determined as a ratio of the center frequency and a sharpness factor, Q,

$$B_k = \frac{f_k}{Q}. \tag{7.111}$$

Higher values of Q make the bands narrower. We can set this according to how much overlap we want adjacent filters to have, or how resonant we want each band to be. Depending on the bandpass filter used, a Q in the range of 50–100 may be appropriate for musical applications.

We can now design a vocoder class in Python with the components we have developed so far. From a macro perspective, this is fairly straightforward, just an implementation of Eq. 7.109. The code is shown in Listing 7.4. The object state holds $2N$ filters in two separate banks, for analysis and synthesis. It also has N envelope followers. We hold these into three lists, which can then be iterated on in the process function. Bands are spaced according to a requested number per octave, and this, together with the total number of filters, then determines the width of the complete vocoder filter. A Q factor is supplied independently of this, to narrow or widen the analysis-synthesis bands. Processing takes an excitation signal into which an amplitude spectrum will be imposed.

Listing 7.4 Channel vocoder class in Python.

```
# Channel Vocoder
# with minimum freq minf Hz
# filter spacing S per octave
# filter Q and number of bands N
# sampling rate fs
class vocoder:
 def __init__(self,minf,S,Q,N,fs):
   self.af = []
   self.sf = []
   self.ef = []
   f = minf
   for k in range(0,N):
     f = minf*2**(k/S)
     self.af.append(bandpass(f,Q,fs))
     self.sf.append(bandpass(f,Q,fs))
     self.ef.append(follower(10,fs))

 def process(self,ex,sp):
     k = 0
     y = 0.0
     for f in self.ef:
        ya = self.af[k].process(sp)
        ys = self.sf[k].process(ex)
        y += f.process(ys,ya)
        k += 1
     return y
```

This code depends on two other classes, an envelope follower and a bandpass filter. The first of these can be implemented as

```
class follower:
```

```
  def __init__(self,fc,fs):
    c = 2 - np.cos(2*np.pi*fc/fs)
    self.b1 = (c*c - 1)**(0.5) - c
    self.y1 = 0.
    self.y2 = 0.

  def process(self,x1,x2):
    a0 = self.b1 + 1
    b1 = self.b1
    self.y1 = (x1**2)*a0 - self.y1*b1
    self.y2 = (x2**2)*a0 - self.y2*b1
    return x1*self.y2**(1/2)/(self.y1**(1/2)+1e-10)
```


which uses Eq. 7.108. We take care to add a minimal quantity to the input signal RMS so that a division by zero is avoided.

The bandpass filter class implements a second-order pole-zero resonator (with zeros at DC and Nyquist frequencies), with fixed frequency and bandwidth set at initialization time:

```
class bandpass:
  def __init__(self,fc,Q,fs):
    b = fc/Q
    R = np.exp(-np.pi*b/fs)
    o = 2*np.pi*fc/fs
    self.b2 = R*R
    self.b1 = -2*R*np.cos(o)
    self.g = (1-R**2)/2
    self.d1 = 0.
    self.d2 = 0.

  def process(self,x):
    w = self.g*x - self.b1*self.d1 - self.b2*self.d2
    y = w - self.d1
    self.d2 = self.d1
    self.d1 = w
    return w
```

If we prefer to use a fourth-order filter instead, we can create it by simply composing a new class out of two bandpass2 objects,

```
class bandpass4:
  def __init__(self,fc,Q,fs):
    self.f1 = bandpass(fc,Q,fs)
    self.f2 = bandpass(fc,Q,fs)

  def process(self,x):
    return self.f2.process(self.f1.process(x))
```

which proves again how useful the concept of object-oriented programming can be for sound and music computing applications.

The vocoder can of course also be implemented in Csound, which would make it very efficient and suitable for realtime processing of audio. For this, Csound has a good range of second-order bandpass filters that can be employed. The design we used here is implemented by the `resonz` opcode. The `balance` opcode uses the same approach developed earlier. Therefore, a channel vocoder band can be programmed as

```
aa = resonz(asig,kf,kf/iQ,1)
as = resonz(aexc,kf,kf/iQ,1)
ab = balance(as,aa)
```

The main difference is that in Csound, a filterbank cannot be implemented in a control-rate loop. The alternatives are to make each band an instrument, and use an i-time loop to schedule them, or use recursion instead (as we did in the additive synthesis example from Chapter 3). We can also create a recursive UDO, which then calls itself to instantiate the parallel filters (Lazzarini et al., 2016). We will return to the channel vocoder in the context of sound design and Csound in Chapter 11.

7.7 Time-Varying Filters

Creative applications of fixed filters are somewhat limited, since they generally lack the capacity to produce dynamic spectra. In the previous section, we saw some examples that went against this statement, in the form of filters that can model physical systems, but those represent exceptions that confirm the rule. The channel vocoder is of course a special case of a time-varying filter: from the perspective of each band, it is composed of a set of fixed bandpass filters; however, as a whole, it is a filter whose amplitude response is dynamically changing. Similarly, we have seen that long FIR convolution can produce time-varying spectra as a property of its cross-synthesis dimension, even though it is based on a fixed filter.

Time-varying filters are essential for a form of sound synthesis called *source-modifier* synthesis, commonly known by the rather inaccurate name of *subtractive* synthesis.[47] In this model, we have a rich sound source, which may be one of the classic waveforms studied in Chapter 4, a sound derived from a recording (a *sample*, not to be confused with the samples of digital audio), a form of broadband noise, or actually anything. Filters play the part of the modifier, which shapes the spectrum. However, for any modification to have some audible interest, it really needs to change over time (otherwise we could have just started with the static spectral description we wanted and thrown away the modifier). While we have already seen time-varying convolution as an example of dynamic filters, the application here demands a parametric approach to changes in the spectrum, which are awkward to achieve with FIR filters. However, they are not hard at all for the shorter recursive systems we have explored earlier.

Parametric variation of filters tends to be concentrated on changes to the center or cutoff frequencies of filters, although bandwidths can also be modified. The types of signals used to control these parameters are usually separated into two larger classes: modulation and envelopes. The former is normally represented by continuously varying signals, which may be periodic or aperiodic; for example, a sinusoid at a given frequency (which may vary over time) is a modulation signal, and so is a noise source of some description. The second form is given by single-period functions of time, which are usually defined in a piecewise manner. The typical example is the trapezoid envelope made of an onset, steady-state, and decay segments. There is of course a gray area between these two types of signals, since a function of time with many segments can be considered both an envelope and a modulation source.

In source-modifier synthesis, one typical example is the creation of brass-like tones by coupling the cutoff frequency of a lowpass filter with the amplitude envelope. This makes the sound get brighter, and have more energy at higher frequencies, as it gets louder. This kind of dynamic spectrum is typical of brass instruments, which exhibit that particularly in their onset phase (the start of a note). Many other characteristics can be developed by designing envelopes that make the spectrum change over time. The use of multiple filters can also be harnessed to create a multi-parametric approach to spectral composition. For example, we can design formant transitions between different vowel configurations with a filterbank-based instrument such as the ones seen in Sec. 7.6.

7.7.1 Allpass Phasers

Likewise, modulation can be a very useful way to shape spectra. As an example, we could look at the design of allpass phase shifters, which use an input control to move the poles to different positions in the complex plane. When their outputs are combined with the original signal, the phase shifts create peaks and troughs in the output spectrum. This effect can be used in various ways: multiple peaks can be moved in coordination, or independently if more than one modulation signal is used. We start with a second-order allpass, which is handy because we know we can make higher-order filters with more poles by connection more of these in series. From the general definition of an allpass, Eq. 7.85, we have, after scaling by $1/R^2$, the unity-gain second-order allpass,

$$H_a(z) = \frac{(R - e^{j\omega_o}z^{-1})(R - e^{-j\omega_o}z^{-1})}{(1 - Re^{j\omega_o}z^{-1})(1 - Re^{-j\omega_o}z^{-1})} = \frac{R^2 - 2R\cos(\omega_o)z^{-1} + z^{-2}}{1 - 2R\cos(\omega_o)z^{-1} + R^2z^{-2}} \tag{7.112}$$

To get higher-order filters, we can connect these sections in series. The allpass plays the part of the phase shifter, and when combined with the input signal, we have the following transfer function,

$$H(z) = a + bH_a(z) \tag{7.113}$$

which can be normalized to unity gain if we set $a = 1 - b$. Maximum effect can then be achieved with $b = 0.5$. We need to be able to tune this filter to various center frequencies and bandwidths. The following formulae apply,

$$\begin{gathered} R = e^{-\pi B/f_s} \\ \cos(\omega_o) = \frac{1 + R^2}{2R}\cos(\omega_c) \end{gathered} \tag{7.114}$$

for bandwidth B and center frequency[48] $\omega_c = 2\pi f_c/f_s$. A plot of the magnitude response of a sixth-order phase shifter with $Q = 10$ and center frequencies at 2,000, 4,000, and 8,000 Hz is shown in Fig. 7.38.

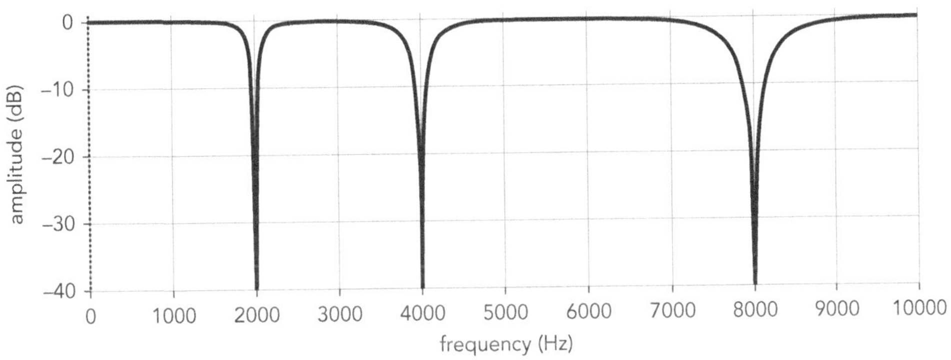

Figure 7.38
Amplitude response of a sixth-order phase shifter with $Q = 10$ and center frequencies at 2,000, 4,000, and 8,000 Hz, with $fs = 44,100$ Hz.

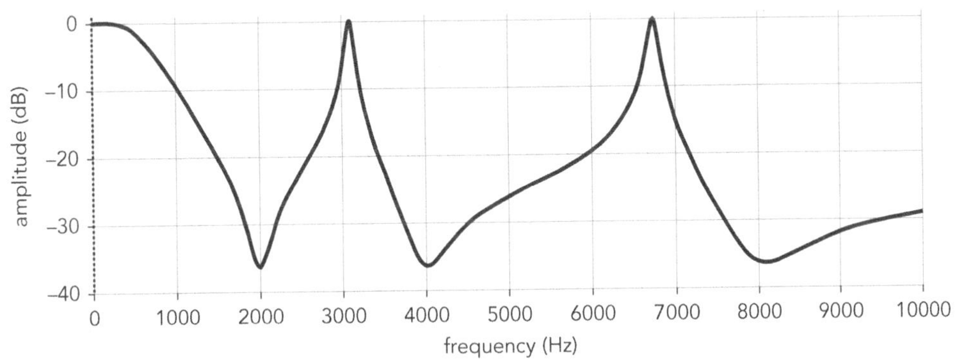

Figure 7.39
Amplitude response of a sixth-order feedback phase shifter with $Q = 10$ and $g = 0.97$ and center frequencies at 2,000, 4,000, and 8,000 Hz, with $fs = 44,100$ Hz.

Another implementation of the phaser shifter employs feedback of the output back into the input, which is described by the transfer function

$$H(z) = \frac{a}{1 - gH_a(z)} \tag{7.115}$$

where we can set $a = 1 - g$ to normalize the input to unity, and $g < 1$ to guarantee stability. In this case we will get a shallower and wider bandwidth for a given Q, as shown in Fig. 7.39. In many ways, this is a more desirable effect, since the audible result of very narrow and deep troughs is not as strong.

Other combinations may also yield interesting possibilities. For instance, reversing the sign of the feedback gain g turns the filter into a bandpass filter with a sharp peak at the center frequency. Keeping a set to $1 - g$ normalizes the filter to unity at the center frequency. Another very nice result is given by the following transfer function (Regalia and Mitra, 1987):

$$H(z) = 1 + H_a(z) + g(1 - H_a(z)), \tag{7.116}$$

which is a filter which has unity gain at the passband, but is boosted by g at the center frequency. This is extremely useful: it allows us to construct higher-order filters in series connections with individual gain control for each one of them. A typical application is to build a graphic equalizer with a bank of these, but there are many other possible creative applications. The amplitude response of this filter is shown on Fig. 7.40 for three different Q settings.

Returning to the phaser implementation, what we need is to provide now is a low-frequency periodic signal. This modulates the center frequency of the phaser to provide the time-varying spectral changes characteristic of this effect. Such a signal can be simply

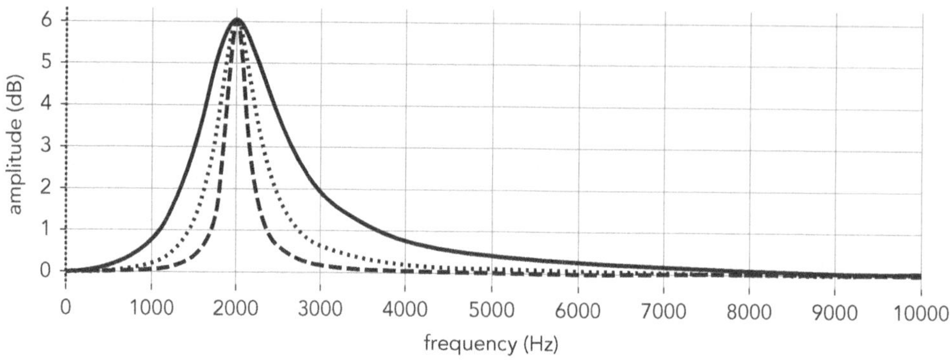

Figure 7.40 Amplitude response of a second-order allpass-based equalizer filter, centered at 2,000 Hz, with $Q = 2.5$ (*solid line*), $Q = 5$ (*dots*), and $Q = 10$ (*dashes*), $f_s = 44,100$ Hz.

implemented by a sinusoidal oscillator, such as the one defined by the equations,

$$x(n) = \frac{M}{2}[1 + \sin(\theta(n))]$$
$$\theta(n+1) = \theta(n) + 2\pi\frac{f_m(n)}{f_s} \tag{7.117}$$

where M is the modulation width and $f_m(n)$ is the modulation frequency in Hz. The following Python code implements a second-order phaser class,

```
# phase shifter
# sampling rate fs
class phaser:
  def __init__(self,fs):
    self.d1 = 0.
    self.d2 = 0.
    self.fdb = 0.
    self.fs = fs
    self.ph =  0.0

  def coeff(self,fc,B):
   o = 2*np.pi*fc/self.fs
   c2 = np.exp(-2*np.pi*B/self.fs)
   c1 = -(1 + c2)*np.cos(o)
   return c1, c2

  def filter(self,x,fc,Q,g):
   out = 0.0
   s = self.fdb*g + x*(1-g)
   c1, c2 = self.coeff(fc,fc/Q)
   w = s - c1*self.d1 - c2*self.d2
   y = w*c2 + c1*self.d1 + self.d2
   self.d2 = self.d1
   self.d1 = w
   self.fdb = y
```

```
    return y

  # process method
  # takes signal x
  # mod rate r
  # mod width m [0,1]
  # Q and feedback gain g
  def process(self,x,r,m,Q,g):
    s = (np.sin(self.ph) + 1)*m/2
    self.ph += 2*np.pi*r/self.fs
    if self.ph >= 2*np.pi: self.ph -= 2*np.pi
    if self.ph < 0: self.ph += 2*np.pi
    fmax = self.fs/4
    return self.filter(x,s*fmax,Q,g)
```

The design is based on the feedback type of Eq. 7.115, and it offers control over the modulation rate, width (over the interval [0,1]), Q, and feedback gain. This code can be expanded to use higher-order phase shifters by adding other second-order sections. The output of the last filter then gets placed into the feedback path. The filter center frequencies need to be carefully organized so that they are always well spaced apart for optimal effect.

7.7.2 Audio-Rate Coefficient Modulation

The typical time-varying filters such as the phase shifter studied above normally employ modulation rates that are at sub-audio frequencies (< 20 Hz). However, it is possible to extend these into the audio range, as long as we keep an eye on the question of filter stability. This opens some interesting possibilities for sound synthesis and processing. Let's consider a simple case, which can be described by the following equation (Kleimola et al., 2011):

$$s(n) = (m(n) + a)\frac{c(n)}{b}, \tag{7.118}$$

involving a carrier signal $c(n)$ and a modulator $m(n)$. This describes different types of amplitude modulation. If $a = 0$ and $b = 1$, we have ring modulation. If $a = b \neq 0$, then the equation describes a standard type of AM. We can re-write this equation using $a = b = 1$, and a sinusoidal carrier $c(n) = \cos(\omega_0 n)$, and a delayed output as $m(n)$

$$y(n) = \cos(\omega_0 n)[1 + y(n-1)] \tag{7.119}$$

with a fundamental frequency $\omega_0 = 2\pi f_0$. This can be both described as feedback AM (FBAM) and as filter with an audio-rate modulation applied to one of its coefficients. Analyzing this as an FBAM expression, we can expand it as

$$\begin{aligned} y(n) &= \cos(\omega_0 n) + \cos(\omega_0 n)\cos(\omega_0[n-1]) + \\ &\quad \cos(\omega_0 n)\cos(\omega_0[n-1])\cos(\omega_0[n-2]) + \ldots \\ &= \sum_{k=0}^{\infty}\prod_{m=1}^{k}\cos(\omega_0[n-m+1]), \end{aligned} \tag{7.120}$$

from which we can glimpse that, if we consider this as a filter, some magic has happened. A sinusoidal input was turned into an output with several harmonics. The filter has become a sound generator itself.

From another perspective, we can put Eq. 7.119 in more general terms as an IIR filter,

$$y(n) = x(n) + a(n)y(n-1) \tag{7.121}$$

where we for the specific case of FBAM, $x(n) = a(n) = \cos(\omega_0 n)$. This is, similar to the phase shifter case, a periodic linear time-varying (PLTV) filter. However, the audio-rate modulation requires us now to take more care in the analysis. While in the former case we could consider the static transfer functions as fixed moments in time without much consequence, now we need to develop a means to analyze the frequency response as a fully time-varying function, that is, as a function of both frequency and time.

To do so, we start with a simpler case, a non-recursive PLTV of length N,

$$y(n) = \sum_{k=0}^{N} a_k(n)x(n-k), \tag{7.122}$$

where we have the signals $b_k(n)$ as time-varying coefficients. The impulse response of this filter, measured at time n, is

$$h_n(m) = \sum_{k=0}^{N} a_k(n)\delta(n-k-m). \tag{7.123}$$

From this, we can obtain the time-varying transfer function for the filter as

$$H_n(z) = \sum_{k=0}^{N} a_k(n)z^{-k}. \tag{7.124}$$

If we freeze time at n, then the impulse and frequency responses of a non-recursive PLTV can be equated to that of a time-invariant system. However, for an IIR filter, the situation is more complicated. Starting again with a time-varying impulse response, we have the following expression for the first-order recursive filter of Eq. 7.121,

$$h_n(m) = \begin{cases} \prod_{i=m+1}^{n} b(i) = \frac{g(n)}{g(m)}, & m < n \\ 1, & m = n \\ 0, & m > n \\ 0, & m < 0, \end{cases} \tag{7.125}$$

where

$$g(n) = \begin{cases} \prod_{i=1}^{n} b(i), & n > 0 \\ 1, & n = 0. \end{cases} \tag{7.126}$$

The transfer function for this filter is now

$$H_n(z) = \frac{1 + \sum_{k=1}^{N-1} h_n(n-k)z^{-k}}{1 - g(N)z^{-N}} \tag{7.127}$$

where N is the period (rounded to the nearest sample) of the coefficient-modulating signal ($a(n)$). For a modulator whose period is $N_0 = \lfloor f_s/f_0 + 1/2 \rfloor$, we have

$$H_n(z) = \frac{1 + \sum_{k=1}^{N_0-1} a_k(n)z^{-k}}{1 - b_{N_0} z^{-N_0}} \tag{7.128}$$

where $a_k(n)$ and b_{N_0} are given by

$$a_k(n) = \prod_{m=1}^{k} \cos(\omega_0[n-m+1]) \tag{7.129}$$

and

$$b_{N_0} = \prod_{m=1}^{N_0} \cos(\omega_0 m) \tag{7.130}$$

This IIR system is equivalent to a filter of length N_0, consisting of a series connection of a PLTV FIR or order $N_0 - 1$ and a comb filter with a fixed coefficient b_{N_0},

$$y(n) = x(n) + \sum_{n=1}^{N_0-1} a_k(n)x(n-k) + b_{N_0}y(n-N_0). \tag{7.131}$$

As we can see, the stability condition is that $|a_{N_0}| < 1$. The FIR section and its time-varying coefficients are responsible for the generation of the partials. We can also define a modulation width factor A so that we can control the amount of spectral richness introduced by the modulation,

$$y(n) = \cos(\omega_0 n)[1 + Ay(n-1)]. \tag{7.132}$$

With A, it is possible to generate dynamic spectra, from a sinusoid to a fully modulated signal with many partials. The value of A is limited by two factors: aliasing, as the signal bandwidth increases with it; and stability, since it can push b_{N_0} beyond the limit. This is an important condition, which we can redefine as

$$|Ab_{N_0}| < 1. \tag{7.133}$$

In general, for usual values of f_0 the output will become aliased before the system becomes unstable. At $f_s = 44,100$, maximum values for A range from 1.8 at 100 Hz to 0.3 at 4 kHz. Also, the gain if an FBAM system can vary significantly with A. To control this, we can apply the envelope following technique shown before, where we balance the RMS of input and output signals. The ideas developed above allow us to define a general first-order PLTV filter for any arbitrary signal as

$$y(n) = x(n)[1 + Ay(n-1)]. \tag{7.134}$$

In this case, the points made earlier about stability and aliasing will vary according to the input signal, so experimental may be needed. Note that theory developed here applies for periodic inputs (as well as quasi-periodic), as otherwise the filter is not a PLTV system. Similar ideas can be followed for higher-order filters. However, with these, the question of stability becomes more complex and requires further analysis (Cherniakov, 2003).

7.7.3 Delay Time Modulation

Another class of time-varying filters are the ones where we apply a modulation to the delay time. Of course, this is something that takes place within a variable allpass context as well, since there is fractional delay shift applied, for example, in the case of the phaser effect. However, we can actually make much larger changes to the delays in a filter, by making them longer or shorter over time. A typical example is that of a variable-delay comb filter, which implements the *flanger* effect. In this arrangement, the delay time is

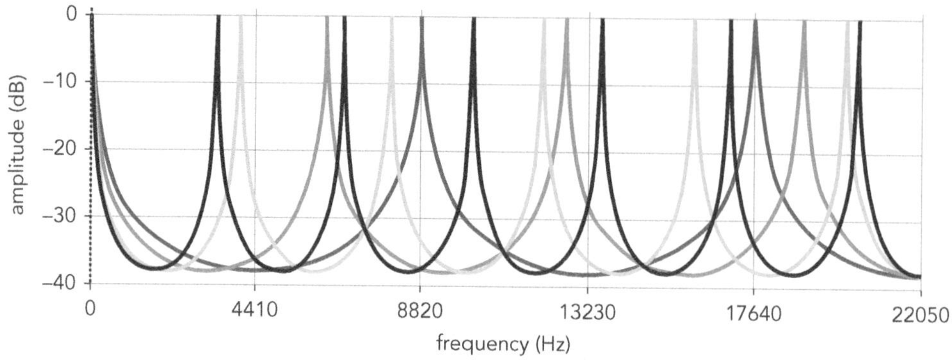

Figure 7.41 Amplitude response curves for a variable comb filter at four moments in time.

moved from near zero to around 2–5 milliseconds, and the effect is a changing in the spacing of the comb peaks and troughs. Some have likened this effect to a barber pole: as the order of the filter decreases, peaks move downward and disappear off the DC end of the spectrum (Fig. 7.41). The time-varying transfer function describing this effect can be cast as

$$H_n(z) = \frac{1}{1 - b_d z^{-f(n)}} \tag{7.135}$$

where $f(n)$ is an arbitrary function with a non-negative range. Typically it is a raised sinusoid, but it can also have other shapes such as a unipolar triangle, or it may be an envelope of some description.

In order to have a smooth output, we need to be able to treat the non-fractional delay times that will arise from this. We have seen already a method to obtain precise delays with allpass filters, but there is a snag: in the change from maximum to minimum delays as the delay crosses the sample boundary will cause artifacts in the output. So we need to find another method. The typical approach uses a polynomial of a given order; the simplest is given by linear interpolation, but fourth-order (cubic) interpolation is also employed in some applications. For the flanger effect, we can possibly get away with using a first-order polynomial. This finds the intermediate value between two adjacent samples in the delay line as

$$y = x_p + (p \bmod 1)(x_{p+1} - x_p), \tag{7.136}$$

where p is the integer part of the reading position index, and $p \bmod 1$ is its fractional part. This is the equivalent of placing a first-order lowpass filter in the delay line loop, but its effect on the spectral magnitudes is very gentle.

A flanger class can be implemented as follows:

```
# Flanger class
# takes max delay maxd in seconds
# and sampling rate fs
class flanger:
  def __init__(self,maxd,fs):
     self.N = int(maxd)
     self.delay = np.zeros(self.N)
     self.p = 0
```

```
        self.ph = 0
        self.fs = fs

    def filter(self,x,d,g):
      rd = self.p - d
      if rd < 0: rd += self.N
      if rd >= self.N: rd -= self.N
      rpf = rd
      rp = int(rpf)
      frac = rpf - rp
      if rp != self.N - 1: yd1 = self.delay[rp+1]
      else: yd1 = self.delay[0]
      yd = self.delay[rp]
      yd = yd + frac*(yd1 - yd)
      y = x + yd*g
      self.delay[self.p] = y
      if self.p != self.N - 1: self.p += 1
      else: self.p = 0
      return y

    # process method
    # takes input signal x
    # mod rate x
    # mod amount m [0,1]
    # and feedback gain g
    def process(self,x,r,m,g):
      s = (np.sin(self.ph) + 1)*m/2
      self.ph += 2*np.pi*r/self.fs
      if self.ph >= 2*np.pi: self.ph -= 2*np.pi
      if self.ph < 0: self.ph += 2*np.pi
      d = self.N*(1 - s)
      if d <= 1: d = 1
      if d >= self.N: d = self.N
      return self.filter(x,d,g)
```

It uses a similar modulation signal as in the phaser example, but now it applies to the delay time. We take care to make sure the minimum delay is a single sample, and that the maximum delay does not exceed the allocated memory size. Delay line modulation is not only very useful for implementing time-varying filters (as demonstrated here), but is also employed in various pitch shifting and modulation applications. For example, an FIR version of the flanger effect described here can be used to create the *chorus* effect. In this case, we do not employ a feedback path, but mix the input signal with the variable delay line output, employing in a typical application slightly longer minimum delay times (5–30 ms) and a slightly wider modulation range, so that the compound effect is both of a delay and a pitch nature. An example of this is shown in Chapter 11.

7.8 A Generalized Concept of Spectrum

We started this chapter by introducing the final transforms of the Fourier family, the discrete-time Fourier transform and its inverse. This proved to be too limited for the development of a theory of filter spectra, and so we had to push the boundaries by making frequency complex-valued, and with this, introduce the z-transform. This was presented as a discrete-time, complex-frequency transform that allowed us to peek at the internal structure of filters. It was also shown that the discrete-time Fourier transform actually represents a special case where the complex frequency variable is set to $z = e^{j\omega}$. A completely new theoretical toolset was developed: complex polynomials, zeros and poles, the z-plane. We were able to design new filters from more or less arbitrary specifications, both in fixed and time-varying forms.

As we have done before, we can improve on our continuously evolving definition of spectrum. The ideas developed in this chapter allow us to add that

> By generalizing the concept of frequency, it is possible to describe the structure of sound-processing units called filters. This allows us to analyze in detail the effect of these on input audio and predict their output based on a set of parameters that control them. As a result, we have expanded the original concept of spectrum so that it not only refers to audio signals but can also be used to describe processes that act on them, producing some sort of modification.

This chapter concludes the discussion of Fourier-based methods. At this stage, we have not only reached the edges of this theory, but started stepping out of it. The next chapter of this book will move away from the linear world of the Fourier transform and relations, to look at other alternative ways to describe and manipulate the spectrum. In any case, all the principles and mechanisms developed so far provide a solid foundation for what is to come.

Notes

1 We have seen this equation briefly when introducing the STFT in Chapter 6.

2 The inverse discrete-time Fourier transform is an integral sum in the interval $[-f_s/2, f_s/2]$, which as we know determines the digital baseband. This completes the set of Fourier transforms:

(1) Fourier transform: $x(t) \Rightarrow X(f)$
(2) Inverse Fourier transform: $X(f) \Rightarrow x(t)$
(3) Discrete-frequency Fourier transform: $x(t) \Rightarrow X(k)$
(4) Fourier series: $X(k) \Rightarrow x(t)$
(5) Discrete Fourier transform: $x(n) \Rightarrow X(k)$
(6) Inverse discrete Fourier transform: $X(k) \Rightarrow x(n)$
(7) Discrete-time Fourier transform: $x(n) \Rightarrow X(f)$
(8) Inverse discrete-time Fourier transform: $X(f) \Rightarrow x(n)$

Note that the discrete-frequency Fourier transform is similar to the original method of obtaining the trigonometric series coefficients, and that is the Fourier transform spectrum sampled in frequency.

3 In fact, this can also be verified in the time domain, if an *analytic* signal is employed. Delaying such a signal is equivalent to

$$e^{j2\pi fn/f_s} \times e^{-j2\pi fd/f_s} = e^{j2\pi f[n-d]/f_s}$$

and to make it an audio signal we just use $\Re\{e^{j2\pi f[n-d]/f_s}\} = \cos(2\pi f[n-d]/f_s)$.

4 Since this is a continuous-frequency transform, it is not trying to model a periodic waveform that repeats forever; therefore the domain is open. In the DFT, where this is the case, the domain is closed to [0, N−1], and circular, x(n+N) = x(n).

5 This effect also happens in some environments where there are delayed sound reflections combined with a direct signal, something also commonly called the *comb filter effect*.

6 Here is a case where we are not equating magnitude and amplitude, using the latter term in a more general sense. We could also put it in another way, by calling the amplitude spectrum a purely real frequency response, where the phase can only be one of two values (0 or π), and the spectrum is a sum of cosines.

7 However, depending on the number of non-zero coefficients, we may need to use fast convolution, as seen previously.

8 From now on, we will mostly use $\omega = 2\pi f/f_s$, as we have done before to simplify the notation. Since this makes $\omega = \pi$ represent the Nyquist frequency $f_s/2$, any angle between 0 and π is within the digital baseband. Fractions of 2π then can represent fractions of the sampling rate. We can recover the frequency f in Hz by dividing ω by 2π and multiplying the result by the sampling rate.

9 Note that this is only just a choice of notation, as the expression still represents the DTFT.

10 It represents a particular kind of what is known as a *Laurent series* (Penrose, 2005, 157), although this terminology is more commonly used in mathematical physics. There are many connections between the spectral theory we are developing here for audio signals and that used in modern physics.

11 We might have noticed that in the z-transform we have the variable raised to negative powers (depending on the sign of n). This does not change the description, all it does is place the polynomial in the denominator of a fraction (since $z^{-n} = 1/z^n$).

12 The second type we discussed, $x(n) - a_d x(n-d)$.

13 Remembering that the difference between the DTFT and DFT is the fact that the latter is time-limited to N samples, and this implies sampling the digital baseband in N frequency points. So the z-transform is two steps more general than the DFT, as it is continuous in frequency as well as having a complex domain. Note also that many authors use the $H(e^{j\omega})$ to represent the Fourier transform as a special case of the z-transform without specifying what form it is exactly. To simplify things, we will use $H(\omega)$ representing the DTFT, which is a much cleaner notation. The DTFT can readily be computed by the DFT for a given filter size, if needed.

14 These denominator zeros at the origin are also present in the inverse comb filter case, but we did not need to consider them since they did not affect the amplitude response.

15 Note that is already the case in the inverse comb filter, where each complex zero has a conjugate pair.

16 With a complex-valued impulse response if the zero is not real-valued.

17 The general form of the second-order filter is $y(n) = a_0 + a_1x(n-1) + a_2x(n-2)$, and so in this case we have $a_0 = 1$, $a_1 = -2R\cos(\omega_o)$ and $a_2 = R^2$.

18 The stopband is the bandwidth of a band-reject filter, the inverse of the passband in a bandpass filter.

19 The frequency f_o is at the center of the stopband when $R = 1$.

20 From this we have

(1) First-order sum, $x(n) + x(n+1)$, is a lowpass filter.
(2) First-order difference, $x(n) - x(n+1)$, is a highpass filter.
(3) Second-order sum, $x(n) + x(n+2)$, is a band-reject filter.
(4) Second-order difference, $x(n) - x(n+2)$, is a bandpass filter.

All of these filters will have a pass band of $f_s/4$, so we can say that their cutoff or center frequency is also fixed at $f_s/4$.

21 There are actually three other different ways to get the correct results:

(1) Change the sign of the *arg*{} function to −*arg*{}.
(3) Zero the positive spectrum and keep the negative frequencies (plus DC).
(3) Swap the forward by the inverse transform and vice-versa, when taking the cepstrum transform.

Of these, the second and third are suboptimal because they would require the use of the complex-to-complex FFT in the forward direction (or in the second case, use the conjugate of the positive spectrum). The first one is probably the simplest approach overall.

22 This is given by the requirement that $\log H(z)$ must be analytic outside the unit circle. Having points at which this is zero breaks this condition (Oppenheim et al., 1999, 346).

23 It may be a surprise to see that the Fourier transform of a real signal has this characteristic, but if we review, for instance, the method used to derive the real-to-complex FFT, where we broke it down into two sequences shifted by π, we will see that this indeed makes sense.

24 Note the similarities of this and how we worked out the phase response for the inverse comb filter in Eq. 7.13. Here we also have a bipolar amplitude, or a purely real frequency response, shifted by a complex exponential.

25 To make the results clear and comparable, we are using a 16,384-point DFT all along, padding the IRs with zeros.

26 This number was not picked at random; it stems from the hanning window bandwidth.

27 The power spectrum is defined as the square of the amplitude spectrum $|H(\omega)|^2$, so the half-power point is where the amplitude drops by 3 dB.

28 This is a very good approximation of the real value of R for the typical use case where this is close to unity. The actual value of R for bandwidth B is found by finding the frequency where $|H(\omega)|^2$ falls to 1/2 of the maximum power. This value is equivalent to half the cutoff frequency of a lowpass filter whose pole is on the real line.

29 See note 27.

30 The steps to get this result are as follows. From the filter transfer function,

$$H(z) = \frac{1}{(z - Re^{-j\omega_o})(z - Re^{j\omega_o})},$$

we first work out the reciprocal of the power spectrum,

$$|H(\omega)|^{-2} = |e^{j\omega} - Re^{-j\omega_o}|^2 \times |e^{j\omega} - Re^{j\omega_o}|^2$$

$$= [1 - 2R\cos(\omega_o - \omega) + R^2][1 - 2R\cos(\omega_o + \omega) + R^2]$$

$$= 1 + 2R^2 + R^4 - 2R[\cos(\omega_o - \omega) + \cos(\omega_o + \omega)] - 2R^3[\cos(\omega_o - \omega) + \cos(\omega_o + \omega)]$$

$$+ 4R^2 \cos(\omega_o - \omega))\cos(\omega_o + \omega)$$

$$= 1 + 2R^2 + R^4 - 2R[1 + R^2][\cos(\omega_o - \omega) + \cos(\omega_o + \omega)] + 4R^2 \left[\frac{\cos(2\omega_o) + \cos(2\omega)}{2}\right]$$

$$= 1 + 2R^2 + R^4 - 4R[1 + R^2]\cos(\omega_o)\cos(\omega) + 4R^2[\cos^2(\omega_o) + \cos^2(\omega) - 1].$$

$$= 1 - 2R^2 + R^4 - 4R[1 + R^2]\cos(\omega_o)\cos(\omega) + 4R^2[\cos^2(\omega_o) + \cos^2(\omega)]$$

$$= 4R^2\cos^2(\omega) - 4R[1 + R^2]\cos(\omega_o)\cos(\omega) + 4R^2\cos^2(\omega_o) + (1 - R^2)^2$$

The final line of this equation can be read as a second-order polynomial in $\cos(\omega)$, or alternatively in $\cos(\omega_o)$. To find an expression connecting the two, we can hold one fixed and differentiate with respect to the other, which means finding an expression for the infinitesimal increment of that cosine (remembering that we are working in continuous frequency). A quadratic function differentiated once should give us a simple linear relationship. All terms except for those involving $\cos(\omega)$ are constant, so the derivative is

$$\frac{d}{d\cos(\omega)}|H(\omega)|^{-2} = \frac{d}{d\cos(\omega)}\left[4R^2\cos^2(\omega) - 4R[1 + R^2]\cos(\omega_o)\cos(\omega)\right]$$

$$= 8R^2\cos(\omega) - 4R[1 + R^2]\cos(\omega_o).$$

Now we can make this a linear equation to find $\cos(\omega_o)$ for a given frequency $\omega = \omega_c$ where the dip takes place,

$$8R^2\cos(\omega_c) - 4R[1 + R^2]\cos(\omega_o) = 0$$

$$\cos(\omega_o) = \frac{2R\cos(\omega_c)}{1 + R^2}$$

This appears a little complicated at first, but if we review the steps, it is fairly straightforward and it follows the path of least mathematical resistance. Here's the summary:

- We knew that the amplitude response had a peak at ω_c, but we did not have the exact pole angle corresponding to that. A peak is hard to quantify, so we look at a dip instead, which can be equated to 0. The reciprocal of the amplitude gives us that.
- The amplitude response involves a square root, which is not needed since we are not looking for a particular value, just a 0. The power spectrum eliminates the square root and is 0 at the same point as the amplitude response.
- We need to know the infinitesimal value of ω where the dip takes place. We then work the power spectrum out to a point where we have the simplest expression to differentiate.

- This turns out to be a second-order polynomial of the cosine of the angle, so we go with that. We are using a cosine in the filter equation, but if we ever needed ω_c, we could have just extracted it anyway using the inverse cosine.

Finally, we may wonder why not just use ω_o directly. It is true that for $R = 1$, the two angles are the same, but that also can make the filter unstable, so R is typically used in the range $[0, 1)$. What we have just derived shows that there is a small, but maybe significant, tuning difference. This is due to the influence of the complex conjugate pole, particularly if ω is close to 0 or close to π (0 Hz or Nyquist frequencies) at the edges of the non-negative digital baseband. We can also see this influence in the amplitude response curve, which shows less attenuation below the center frequency, for frequencies below $f_s/4$.

31 This operator is defined later in Sec. 7.6.5.

32 In typical applications, comb filters are actually implemented by a pair of equations,

$$w(n) = x(n) + d_n w(n-d)$$
$$y(n) = w(n-d),$$

which put a pure delay of d samples after the actual feedback filter. The practical difference is that in this case we are tapping the output of the delay line, whereas with the feedback equation we are getting the output signal just before it re-enters the delay. In the latter case, there is no initial lag, and the input goes straight to the output mixed with the feedback signal. In the former, we must wait until it goes through the delay line. The pure delay of course has no bearing on the amplitude response, but the transfer function must include it to be correct.

33 Incidentally, this has many similarities with how we use sinusoids as the fundamental component with which to build a spectrum. We have already seen that a resonator, in particular, can serve as a sinusoidal oscillator, so the connections are all there.

34 This demonstrates that the linear phase characteristic of FIR filters can be highly desirable.

35 Note that since we have put two zeros at the edges of the positive spectrum, the filter center frequency behaves much better and can be equated to the pole angle.

36 Chebyshev polynomials of the first and second kinds are interesting mathematical objects with a lot of relevance to musical signal processing. They can be used in non-linear distortion synthesis, as they neatly produce a harmonic of an input sinusoid ($U_n(\cos(\omega t)\sin(\omega t)$, for example, produces $\sin([n+1]\omega t)$).

It is interesting that they also make an appearance here describing the impulse response of a resonator. However, this could have been predicted, since the ordinary generating function for U_n is

$$\sum_{n=0}^{\infty} U_n(x)t^n = \frac{1}{1 - 2tx + t^2}$$

which for $x = \cos(\omega)$ and $t = z^{-1}$ is equivalent to the z-transform of $h(n)R{-}n$, which is the resonator transfer function for poles at the unity circle.

37 Of course, as long as it is only fed a single non-zero sample.

38 This is a common measurement of decay time: the time a sound takes to decay to -60 dB of its original amplitude (or 1/1000).

39 Note that the $T_{60}(f)$ function is also dependent on the sampling rate f_s, since the cutoff frequency of the simple lowpass varies with it.

40 We saw an example of this when mixed the output of different resonators. Non-linear phase combinations can be very useful to realize dips and peaks in the spectrum; after all that is how filters work in a general sense.

41 This is based on the general form of a filter transfer function, where the poles and zeros have been factored out,

$$H(z) = \frac{\prod_{i=1}^{M} z^{-1} - Z_i}{\prod_{i=1}^{N} z^{-1} - P_i}$$

which we have used before to describe second-order filters.

42 Why? What is the correct form? There are two things we want from this filter: a fractional delay ($0 \leq d < 1$ samples), and a small phase deviation (or error) at low frequencies. We know we cannot have linear phase, but we can make it as close as possible for lower frequencies, where, in the case of the string filter, the output signal will have higher amplitudes. When we compare first-order allpass pole locations, we see that this desired behavior takes place when their angle is π rather than 0, or we can say $R < 0$. Therefore, we may as well start with the correct coefficient sign to help us get the correct result at the end. The phase spectra of two filters, with $R = -0.5$ and $R = 0.5$, are shown in Fig. 7.42. Note that in the former case, the phase curve is closer to linear at lower frequencies, which is an important result. The other filter shows a much wider (and less linear) phase spectrum at lower frequencies.

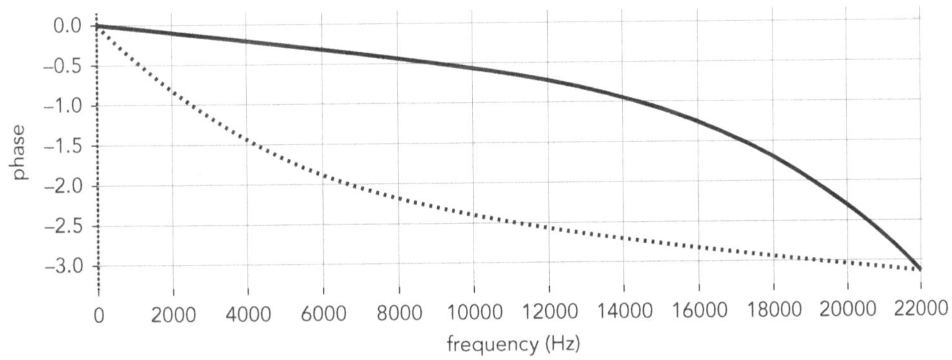

Figure 7.42
Allpass phase response for $R = -0.5$ (*solid line*) and $R = 0.5$ (*dots*), with $f_s = 44,100$ Hz.

43 This looks like a minor detail, but it is actually very important. The plan is to use this filter to provide a fractional delay in the string resonator feedback loop. Without unity gain, it could push the feedback gain over the stability threshold, rendering the filter useless.

44 Note that other filters with zeros and poles can use this arrangement. For example, the second-order resonator with zeros at DC and Nyquist frequencies,

$$y(n) = x(n) - x(n-2) - b_1 y(n-1) - b_2 y(n-2),$$

can be rewritten as

$$w(n) = x(n) - b_1 w(n-1) - b_2 w(n-2)$$
$$y(n) = w(n) - w(n-2).$$

The idea is that we have two filters in series, an all-pole ($w(n)$) followed by an all-zero ($y(n)$). We have already seen that the total transfer function corresponds to the product of the individual filter transfer functions. The different forms of implementing filters are called *filter topologies*.

45 Oppenheim et al. (1999, 249).

46 Some care may be needed to handle the case where the $s_k(n)$ signal has an RMS of zero.

47 No subtraction is involved.

48 Note the similarities between the estimation of the pole angle here and in the resonator case. While the center frequencies get attracted by the complex conjugate pole in an all-pole filter, they are shifted in the opposite direction in this allpass arrangement.

8

Non-Linear Synthesis of Spectra

In this chapter, we will proceed to examine different ways in which we can synthesize spectra through various non-linear methods. Before we begin, we may want to review briefly the definition of non-linearity, which we have introduced in Chapter 3. We noted that if we apply a non-linear transfer function to an input signal, its amplitude will be distorted in some way. It is also possible to think of this as a mapping from an input to an output: if it involves a straight line, then we have a linear transfer function resulting in some sort of gain modification. In any other case, the shape of the input waveform might be modified, depending on its amplitude and the waveshaper. In general, the simplest form of non-linear distortion synthesis is a sinusoid mapped by a function $f()$,

$$s(t) = f(\cos(2\pi ft)) \tag{8.1}$$

where the transfer function is not linear. A demonstration of this process is shown in Fig. 8.1, where a cosine wave is distorted by a non-linear waveshaping function, $f(x) = x + 5x^2 - 3$, producing an output with more than one partial. Transfer functions such as this one can be designed so that the output is precisely determined, given a known input (generally a sinusoid). The focus of this chapter is to explore the various methods to which the basic principle of non-linear distortion can be applied.

At this stage, it is important to flag some slight changes in the way time is treated in this chapter. Most of the mathematics we will explore here works fairly independently of whether discrete or continuous-time signals are employed. So, in order to simplify matters, in many equations we will employ t in seconds, instead of the usual n denoting time in samples. This can be of course easily translated to a discrete representation by the formula $n = tf_s$ if we wish, but it is generally not necessary. Of course, the computer implementation of all synthesis equations eventually translates the expressions into their digital forms. Furthermore, we are not going to ignore the fact that the synthesized signals need to be bandlimited to avoid aliasing. However, it is far simpler to present the equations in a form that is more general. This means that we do not need to continually include a f_s factor in all expressions, and that frequency can be directly translated into the usual unit of cycles per second or Hz. In the occasions where it matters to show that we are in fact working in discrete time, we can revert to the usual time notation.

Spectral Music Design. Victor Lazzarini, Oxford University Press. © Oxford University Press 2021.
DOI: 10.1093/oso/9780197524015.003.0008

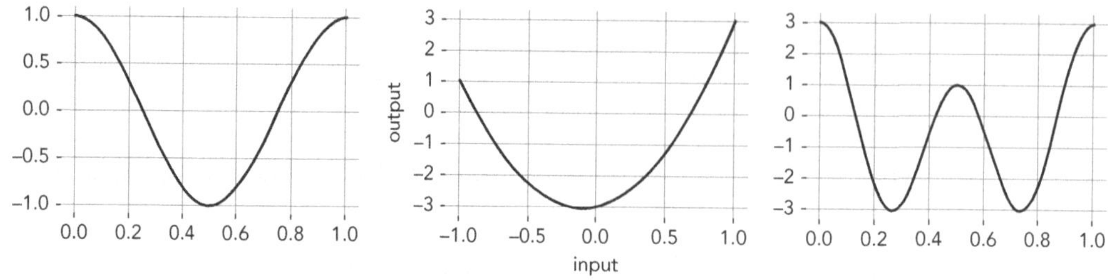

Figure 8.1 Non-linear distortion as the mapping of a sinusoid (*left*) by a non-linear function $f(x) = x + 5x^2 - 3$ (*middle*), producing a new waveform shape.

8.1 Closed-Form Synthesis Formulae

We begin fairly close to the territory we have threaded in the last few chapters, with the well-known Fourier series of one of the classic waveforms. A bandlimited pulse signal with frequency f and all harmonics up to the N partial can be described as the following sum of sinusoids,

$$c(t) = 1 + 2\sum_{k=1}^{N} \cos(k\omega) = \sum_{k=-N}^{N} e^{jk\omega}, \tag{8.2}$$

where $\omega = 2\pi ft$.[1]

As we know from before, this is a purely real spectrum, and so when the complex sinusoid terms of negative and positive frequencies are added up, the imaginary part of each component vanishes. Another way to look at this sum is as a geometric series: the complex sinusoid is raised to increasing powers up to the power of N. This is a useful way to put things, since we can use well-known relationships to obtain a closed form for the expression.

We know that a two-sided finite sum involving powers of a can be described by the following formula,

$$\sum_{k=-N}^{N} a^k = a^{-N}\left(\frac{1-a^{2N+1}}{1-a}\right), \tag{8.3}$$

which can neatly be applied to our case. We first substitute $a = e^{j\omega}$, then simplify the expression to give us a synthesis formula for the bandlimited pulse waveform,

$$\begin{aligned} c(t) &= \sum_{k=-N}^{N} e^{jk\omega} = e^{-j\omega N}\left(\frac{1-e^{j\omega[2N+1]}}{1-e^{j\omega}}\right) \\ &= \frac{e^{-j\omega N} - e^{j\omega[N+1]}}{1-e^{j\omega}} = \frac{e^{-j\omega N - j\omega/2} - e^{j\omega[N+1]-j\omega/2}}{e^{-j\omega/2} - e^{j\omega/2}} \\ &= \frac{e^{-j\omega[N+1/2]} - e^{j\omega[N+1/2]}}{e^{-j\omega/2} - e^{j\omega/2}} = \frac{\sin\left(\omega\left[N+\frac{1}{2}\right]\right)}{\sin\left(\frac{\omega}{2}\right)}. \end{aligned} \tag{8.4}$$

This is a great result in computational terms. It means that for the price of two sine wave oscillators we get a complete spectrum of harmonics. There is even some small element of tonal control, in that we can change N to determine how many components we will produce. While there may be some small audible plucking artifacts, it is also possible to vary N over time (the formula certainly does not prohibit it). This is also a well known result, called the *Dirichlet kernel*.[2] If we object to having the DC built into the formula,

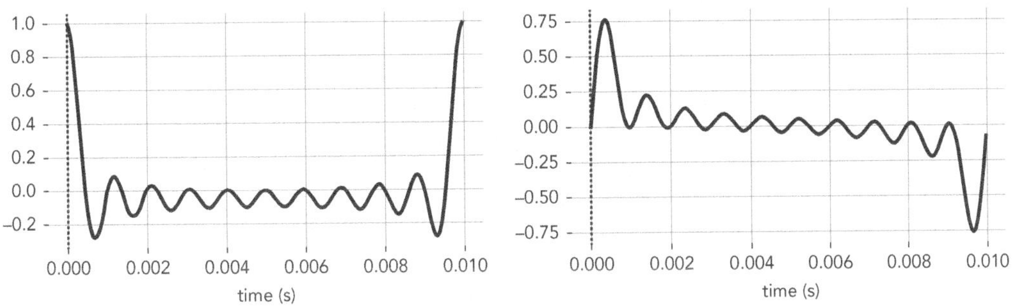

Figure 8.2 Cosine (*left*) and sine components (*right*) of a quadrature bandlimited pulse waveform generated using Eqs. 8.5 and 8.6, with $f = 100$ Hz and $N = 10$.

we can remove it by subtracting 1 from the formula. Additionally, the formula can be normalized through the scaling factor $1/2N$,

$$c(t) = \frac{1}{2N}\sum_{k=1}^{N}\cos(k\omega) = \frac{1}{2N}\left[\frac{\sin(\omega[2N+1]/2)}{\sin(\omega/2)} - 1\right]. \tag{8.5}$$

Note that we have based the sines in both the numerator and denominator on the same phase angle $\omega/2$, which denotes half the fundamental frequency, to bring out the relationship between these two signals (which we will explore later). This is now in the form of one of Lagrange's trigonometric identities.[3] The other one, to complete the set, is

$$s(t) = \frac{1}{2N}\sum_{k=1}^{N}\sin(k\omega) = \frac{1}{2N}\left[\frac{\cos(\omega/2) - \cos(\omega[2N+1]/2)}{\sin(\omega/2)}\right]. \tag{8.6}$$

With these two expressions, we can create a bandlimited pulse of arbitrary phase, as well as the analytic signal $c(t)+js(t)$ (Fig. 8.2). We can implement these expressions fairly easily using sinusoidal oscillators. Since they all depend on the same phase angle, we can synthesize ω first, and then feed it to the sines and cosines, scaling it as required. This keeps all signals phase-locked, which is important for the formulae to work properly. That way, we can modulate the frequency as much as like without any concern. Another key aspect of these synthesis formulae is that N needs to be an integer, since the relationships are not valid if N is real. As we noted, this can be varied in time as well, providing a plucking effect as each new harmonic is added to the mix.

When implementing these formulae, we should note that they have a singularity for $\sin(0)$ and $\sin(\pi)$, so we should take measures to protect against that. A clean-sounding solution[4] for this is to decide on a threshold x and then apply the expression

$$y(t) = \begin{cases} c(t), & \sin(\omega/2) > |x| \\ 1, & \sin(\omega/2) \leq |x|, \end{cases} \tag{8.7}$$

since the singularity occurs at the points where the waveform is at its maximum value. In the case of $s(t)$, the singularity occurs as the waveform is zero, so we replace the output with zero instead.

This is the method employed in the C code for the `buzz` opcode, which implements the cosine formula. A similar approach can be taken when programming these formulae in Python or Csound. In the latter case, we will need to process sample-by-sample in order to apply the check, instead of proceeding in blocks of ksmps samples. To do this we need to set the local ksmps to 1 in a UDO or an instrument, using the `setksmps` opcode. As an example, we can develop a quadrature pulse synthesiser. This employs two UDOs, one to compute the cosine and sine formulae, and another one to wrap it so. This approach isolates the sample-by-sample code, which is less efficient, from the rest, keeping it to the minimum necessary number of operations. In order to produce the phase signal ω, we employ the phasor generator, which produces a normalized phase by integrating a frequency input (as discussed in Sec. 3.1.2). It needs to be scaled to the correct range in radians, for which we can use the M_PI constant provided by Csound.

```
opcode Lagrange,aa,ak
 setksmps(1)
 aph,kn xin
 kdn = sin(aph)
 if abs(kdn) > 0.0002 then
  aphn = aph*(2*kn + 1)
  ac = sin(aphn)/kdn - 1
  as = (cos(aph) - cos(aphn))/kdn
 else
  ac = 2*kn
  as = 0
 endif
 xout(ac,as)
endop
```



```
opcode QuadPulse,aa,kkk
 ka,kf,kn xin
 aph = $M_PI*phasor(kf)
 ac,as Lagrange aph,int(kn)
 ksc = ka/(2*kn)
 xout(ac*ksc,as*ksc)
endop
```

The QuadPulse UDO take as parameters the amplitude, frequency, and number of harmonics, respectively. The first output contains all the cosine phase partials and the second the sines. If only the cosine pulse is required, the more efficient option is to use the buzz opcode.

With this UDO, it is also possible to create inharmonic spectra if we employ a single-sideband modulation method. For this we need a sinusoid in quadrature so we can implement the expression

$$y(t) = \cos(\omega_m)c(t) - \sin(\omega_m)s(t), \quad (8.8)$$

where $\omega_m = 2\pi f_m t$, and f_m is the modulation signal (or shift) frequency. For example, the following Csound code uses this approach to create a stretched pulse wave spectrum, with a sine wave oscillator modulating the number of partials:

```
instr 1
 ia = p4 // amp
 ifr = p5 // lowest freq
```

```
  imax = p6 // max freq
  kn = 2 + (oscili(0.5,0.2) + 0.5)*imax/ifr
  kfm = ifr/sqrt(2);
  ac,as QuadPulse ia,ifr,kn
  acos = oscili(1,kfm,-1,.25) // cos
  asin = oscili(1,kfm) // sine
  ash = acos*ac - asin*as // ssb
  aout = linen(ash,0.01,p3,0.1)
    out(aout)
endin
schedule(1,0,30,0.5,300,4000)
```

The effect is to create an arpeggio of inharmonic partials (note the use of $1/\sqrt{2}$ as the stretch ratio), as they appear and disappear with the changes in kn.

8.1.1 Generalized Summation Methods

In order to understand the mechanism involved in the production of so many partials from two sinusoids in Eqs. 8.5 and 8.6, it is worth attempting to modify these and observe the results. Using the cosine pulse formula, we can insert a means of scaling the sine wave in the denominator, so we can change the waveform from a pulse into something else,

$$\frac{\sin(\omega[2N+1]/2)}{a+(1-a)\sin(\omega/2)}. \tag{8.9}$$

We removed the scaling factor and DC offset to simplify things. Since this formula is only used as an experiment, and not as a final synthesis expression, we can ignore these.

The a parameter therefore allows us to control how much the output is going to look like a pulse waveform. For $a = 0$, we have a pulse with N components, and with $a = 1$, the output is a sine wave. In between, we have different stages of waveshaping, as the sine wave is coerced into a pulse shape. However, we will need to proceed carefully, as some values of a will cause problems. In particular, values in the range $0 < a \leq 0.5$ will cause the output to grown significantly at some point, and the denominator will also be zero at one or more locations in the complete waveform cycle. To avoid this, we can use values of a above 0.5. This is shown in Fig. 8.3. We can see that as a gets closer to 0.5, the waveform starts to resemble a pulse and as a is increased, the troughs become shallower and we approximate a sinusoid. The conclusion we reach is that the working mechanism of this expression involves two components. The first one is a sinusoid, and the second is another function that creates an envelope, giving the outer shape of the waveform. In Fig. 8.4, we set the numerator of Eq. 8.9 to 1 and plot a single period of the resulting function. This shows the effect of the shaping signal.

As we noted, we need to be careful with the values of a, or otherwise the formula will stop working. With $a > 0.5$, however, it is possible to use it as synthesis method. The resulting spectrum for $a = 0.51$ is shown in Fig. 8.5 (*right*), where we can see several sideband partials around a center frequency. To understand this result, we follow the reasoning introduced above, breaking the expression into a sinusoid and a shaping function. The numerator is a sine wave tuned to $[2N+1]\omega/2$. The other part can be cast as a

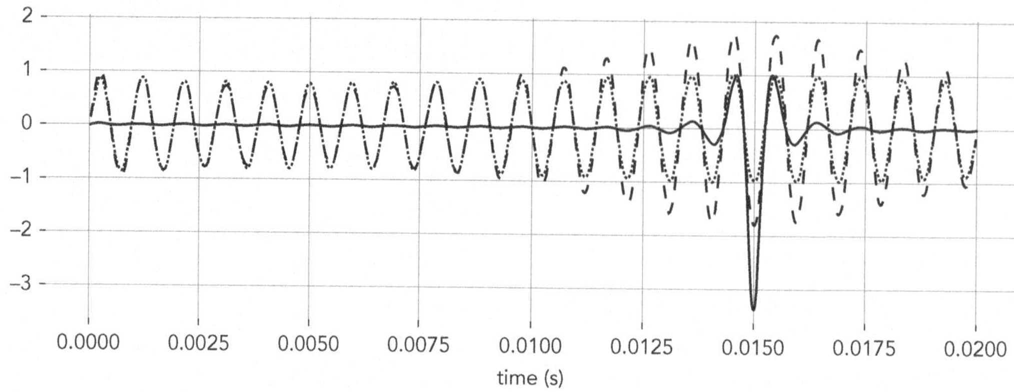

Figure 8.3 Plots of Eq. 8.9 with $a = 0.501$ (*solid line*), $a = 0.7$ (*dashes*) and $a = 0.95$ (*dots*), normalized to unity peak amplitude.

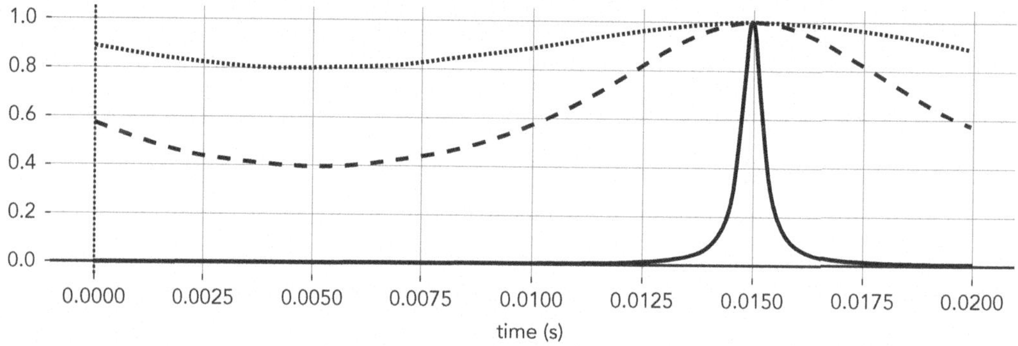

Figure 8.4 Plots of Eq. 8.9 with the numerator set to 1 and $a = 0.501$ (*solid line*), $a = 0.7$ (*dashes*) and $a = 0.95$ (*dots*), normalized to unity peak amplitude.

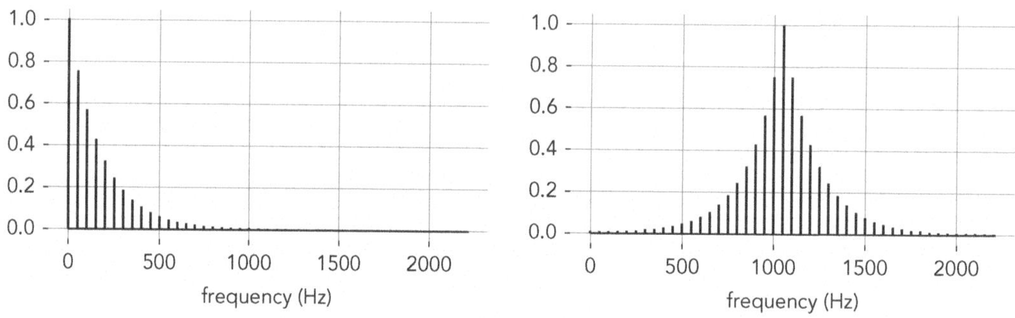

Figure 8.5 Normalized spectra of Eq. 8.10 (with $x = \sin(\omega/2)$) (*left*), and Eq. 8.9 (*right*), with $\omega = 2\pi 100t$, $N = 10$, and $a = 0.51$.

mapping of a sine wave by the function

$$f(x) = \frac{1}{a + (1 - a)x}. \tag{8.10}$$

This function distorts its pure sinusoidal input to produce a spectrum with several partials. This is shown in Fig. 8.5 (*left*). When we multiply this by another sine wave, the result as we know is that the spectrum of $1/[a + (1 - a)\sin(\omega/2)]$ is shifted to be centred at a new frequency. If we tune the distorted sinusoid to 50 Hz and set $N = 10$, then the spectrum will be composed of harmonics of 50 Hz with a peak at 1,050 Hz. This demonstrates how a closed-form formula can use a few sinusoids to generate a waveform whose spectrum has several partials. The principle at work in the bandlimited pulse is similar, although with different parts contributing to the final result.

However, while we can use this expression to generate different waveforms, this is not yet in a well-defined form to provide a useful synthesis method. Firstly, we have the stability difficulties created by setting a to certain values. Secondly, we do not know exactly the amplitudes and phases of the partials in the resulting spectrum. As with the bandlimited pulse, we would like to have closed-form formulae that are equivalent to an exact known mix of partials. Thankfully, there are a number of such expressions, which can be readily applied to sound synthesis.[5] The first one of these produces a decaying spectrum, whose width is controlled by the a and N parameters,[6]

$$\sum_{k=0}^{N} a^k \sin(\omega + k\theta) \\ = \frac{\sin(\omega) - a\sin(\omega - \theta) - a^{N+1}[\sin(\omega + (N+1)\theta) - a\sin(\omega + N\theta)]}{1 - 2a\cos(\theta) + a^2}. \tag{8.11}$$

Note that also that we also have two frequency controls in this formula, represented by $\omega = 2\pi f_1 t$ and $\theta = 2\pi f_2 t$. This allows us to create various harmonic and inharmonic spectra, since these two parameters are independent. As a measure of the type of spectrum that is generated, we can put forward the ratio $n_1{:}n_2$ with $n_2 f_1 = n_1 f_2$. If both n_1 and n_2 are small integers, we have a harmonic spectrum; if one (or both) of them is irrational, we have an inharmonic spectrum. This formula works well to provide a bandlimited spectrum with N partials, whose rolloff is determined by a. Higher values of this parameter will make the spectral envelope slope less accentuated, and for $a > 1$, a highpass shape is created (depending of course on the relationship between ω and θ). Note that for the case $a = 1$, we need to check for zeros in the denominator.[7]

One snag is the question of normalization: the peak amplitude of the waveforms generated will vary depending on the parameters selected. A scaling factor can be applied (Moorer 1976),

$$\sqrt{\frac{1-a^2}{1-a^{2N+2}}}, \tag{8.12}$$

but it does not fully work to normalize the output for all parameter combinations, although it helps to keep the output away from a major overflow. However, some attention needs to be paid to the ranges of parameters employed. In particular as a gets close 1 and N is large, some further control may be needed. The complete Csound code for this synthesis expression is shown in the opcode[8]

```
opcode SineSum,a,kkkk
 kf1,kf2,ka,kn xin
 aph1 = 2*$M_PI*phasor(kf1)
 aph2 = 2*$M_PI*phasor(kf2)
 kn = int(kn)
 aw = sin(aph1)
 awt = ka*sin(aph1 - aph2)
 awtn = ka*sin(aph1 + kn*aph2)
 awtn1 = sin(aph1 + (kn+1)*aph2)
 adn = 1 - 2*ka*cos(aph2) + ka*ka
 as = (aw - awt - ka^(kn+1)*(awtn1 - awtn))/adn
 ksc = sqrt((1 - ka*ka)/(1 - ka^(2*kn + 2)))
```

```
    xout(as*ksc)
endop
```

This formula produces what can be termed a *single-sided* spectrum. An alternative version can be used to produce sideband partials around a center frequency,

$$\sin(\omega) + \sum_{k=1}^{N} a^k[\sin(\omega + k\theta) + \sin(\omega - k\theta)] = \sin(\omega)\left[\frac{1 - a^2 - 2a^{N+1}[\cos([N+1]\theta) - a\cos(N\theta)]}{1 - 2a\cos(\theta) + a^2}\right] \quad (8.13)$$

for which a scaling factor can be computed as

$$\sqrt{\frac{1 - a^2}{1 + a^2 - 2a^{2N+2}}} \quad (8.14)$$

We should also note the multiplication by $\sin(\omega)$ which shifts a 0 Hz-centered spectrum to be placed around f_1 Hz, as we have observed in the earlier modifications of the bandlimited pulse formula. Therefore we can again conceptually break this formula into two components of a ring-modulation expression, $C(t)M(t)$ (Puckette, 1995a). The latter is a *modulator* signal, with a spectrum possibly composed of several partials (depending on a and N), whose frequency is determined by θ. The former is a simple sine wave *carrier*.

An interesting aspect of this double-sided formula is that it can produce formants whose slope and bandwidth are controlled by a and N. For this we decide on a formant center frequency f_c and fundamental frequency f_0. The two frequencies f_1 and f_2 (as before, relative to the phase angles ω and θ, respectively) are then set as

$$f_2 = f_0$$
$$f_1 = \left\lfloor \frac{f_c}{f_0} + \frac{1}{2} \right\rfloor f_2. \quad (8.15)$$

The fundamental frequency f_0 determines f_2, and the f_1 is set to a harmonic n near the formant center frequency (denotes the floor function). Since the spectra produced is in the sine phase, we can mix two or more of these to produce a vocal emulation without the phase interaction difficulties we had when using recursive filters.

These formulae are both bandlimited, so we can rely on the parameter N to avoid aliasing. However, this also makes the expressions more complicated than they should be. We can simplify them considerably by removing the terms that are dependent on N, making them non-bandlimited. With careful settings of a, we may be able to avoid issues with aliasing distortion. For the one-sided formula we have then

$$\sum_{k=0}^{\infty} a^k \sin(\omega + k\theta) = \frac{\sin(\omega) - a\sin(\omega - \theta)}{1 - 2a\cos(\theta) + a^2}, \quad (8.16)$$

with a simpler scaling factor

$$\sqrt{1 - a^2}. \quad (8.17)$$

To complement this, we have the non-bandlimited formula for the double-sided spectrum,

$$\sin(\omega) + \sum_{k=1}^{\infty} a^k[\sin(\omega + k\theta) + \sin(\omega - k\theta)] = \sin(\omega)\left[\frac{1 - a^2}{1 - 2a\cos(\theta) + a^2}\right], \quad (8.18)$$

and its scaling factor,

$$\sqrt{\frac{1-a^2}{1+a^2}}. \tag{8.19}$$

In general, these non-bandlimited formulae are more efficient (as they have fewer operations) and, since the spectral shape is only dependent on a, simpler to control. In particular, the double-sided formula is a very flexible synthesis expression, which depends only on a single sinusoidal carrier and a relatively simple modulator. Moreover, replacing $\sin(\omega)$ by $\cos(\omega)$ puts the whole spectrum in the cosine phase, which may also be useful if we need a quadrature expression. As we noted before, the double-sided equation can be used to synthesize formants; with only a to consider, we can define this parameter in terms of the ratio of the fundamental frequency f_0 and bandwidth B, (Lazzarini, 2017, 122)

$$a = e^{-f_0/B} \tag{8.20}$$

where, as before, we set $f_2 = f_0$ and $f_1 = nf_2$.

Finally, it is useful at this point to stress again the connection between these formulae and non-linear distortion, which we have touched on earlier. We can recast the modulation signal in Eq. 8.18 as a waveshaper acting on a sinusoidal wave (Puckette, 1995a). This gives the synthesis formula a more compact definition as[9]

$$s(t) = A\sin(\omega)f\left(B\sin\left(\frac{\theta}{2}\right)\right) \tag{8.21}$$

where we have

$$f(x) = \frac{1}{1+x^2}, \tag{8.22}$$

$$A = \frac{1+a}{1-a}, \tag{8.23}$$

and

$$B = \frac{2\sqrt{a}}{1-a}. \tag{8.24}$$

Note that the amount of distortion (and the number of spectral components generated) varies according to B; if this is low, very little distortion occurs and fewer partials are present in the output spectrum. This is an important feature of non-linear distortion methods, in that the richness of the spectrum is dependent on the input signal level. Of course, we know this intuitively, since B is dependent on a, and we can see in the series expansion that as it approaches 1, we have an increased number of partials with significant amplitude.

This expression gives us an efficient way to compute the double-sided formula. We can create a function table in memory with enough resolution to hold $f(x)$ in a given interval, then access it through interpolated table lookup. This process is implemented in the following Csound code, which is composed of three UDOs. The first one runs only at initialization time to produce a function table containing $f(x)$ at a given resolution,

```
opcode FTab,i,ii
 iL,imx xin
 ifn = ftgen(0,0,iL,7,0,iL,0)
 ii = 0
```

```
  inc = imx/iL
  ix = -imx/2
  while ii < ftlen(ifn) do
   tablew(1/(1+ix*ix),ii,ifn)
   ix += inc
   ii += 1
  od
  xout(ifn)
endop
```

The second opcode produces a signal corresponding to the modulation signal in the synthesis formula. It calls FTab to initialize the table with $f(x)$ in the $[-2000, 2000)$ interval. This is set according to how close to 1 we want a to get, determining the maximum amplitude for the sine wave signal that is shaped by that function. If this exceeds that range, the output of $f(x)$ is limited to $x = |2000|$.[10] The sine wave signal is then used as an index to lookup the function values in the table using tablei. Note that indexing is normalized (0,1), and offset by 0.5, so that $x = 0$ is set to the middle of the table. The input signal needs to be scaled by the total bipolar range of $f(x)$ as stored the table (defined by ires).


```
opcode Mod,a,kk
 kf,ka xin
 ires = 4000
 ifn = FTab(ires*10,ires)
 aph = $M_PI*phasor(kf)
 aw = sin(aph)
 k1ma = (1-ka)
 kA = (1+ka)/k1ma
 kB = 2*sqrt(ka)/k1ma
 amod = tablei:a(kB*aw/ires,ifn,1,.5)
 xout(kA*amod)
endop
```

Finally, the summation formulae for cosine and sine-phase signals is given by two opcodes that depend on Mod to produce the modulator signal. They apply this in ring modulation with a cosine or sine carrier, and scale the signal,[11]

```
opcode DCosSum,a,kkk
 kf1,kf2,ka xin
 aph = 2*$M_PI*phasor(kf1)
 amod = Mod(kf2,ka)
 asig = amod*cos(aph)
 kasq = ka*ka
 ksc = (1 - 2*ka + kasq)/(1+kasq)
 xout(asig*ksc)
endop
```

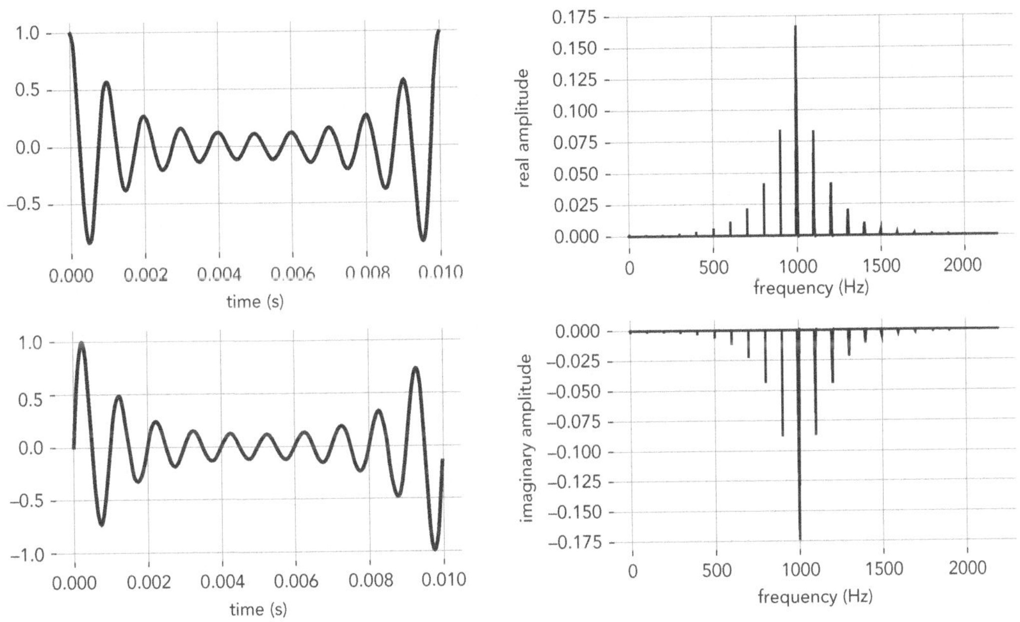

Figure 8.6 Waveform and spectra of cosine and sine quadrature double-sided summation formulae with $f_1 = 1000$ Hz, $f_2 = 100$ Hz, and $a = 0.5$.


```
opcode DSinSum,a,kkk
 kf1,kf2,ka xin
 aph = 2*$M_PI*phasor(kf1)
 amod = Mod(kf2,ka)
 asig = amod*sin(aph)
 kasq = ka*ka
 ksc = (1 - 2*ka + kasq)/(1+kasq)
 xout(asig*ksc)
endop
```

The waveform and spectra for these quadrature double-sided synthesis formulae are shown in Fig. 8.6. We should note that there could be a significant difference between the sine and cosine spectra if the lower sideband partials fall below 0 Hz. In this case, given the Hermitian symmetry of a real signal spectrum, they are reflected back into the positive side. Since cosines are even functions, the negative frequency waveforms will have the same polarity (sign) as the positive frequencies. If the reflected partials are aligned (as would be the case with the frequency settings used in Fig 8.6), they will add constructively to the positive spectrum. On the other hand, sine functions are odd, and thus the negative frequencies will be in reverse polarity and add destructively to any positive spectrum partials they are reflected over. This of course only happens with higher values of a, which create wider sidebands, and ratios of f_1 to f_2 in which partials may overlap.

We have now shown how complex waveforms[12] with many partials can be created out of a single sinusoid by employing non-linear distortion methods. We have also demonstrated that it is possible to describe the spectra produced by these synthesis techniques via closed-form formulae. This is particularly helpful for us to determine what type of distortion function may be appropriate. The following sections of this chapter will delve deeper into the various known synthesis methods that can be grouped under the category of non-linear spectral synthesis.

8.2 Frequency and Phase Modulation Synthesis

The choice of waveshaping functions is quite wide, since the basic requirement is to be non-linear, which is shared by many. Of course, as we have seen, it is important that we obtain a well-described spectrum at the output; otherwise it becomes hard to control the process in terms of sound design. One advantage we have is that the series expansion of many fundamental functions has already been worked out, as a product of centuries of work in mathematical physics. A classic example is given by the waveshaping expression

$$s(t) = \cos\left(z\sin(\theta)\right), \tag{8.25}$$

where the output is the mapping of a sinusoid by another sinusoid, which is scaled by the variable z. From what we know, we should expect this to be a signal whose fundamental frequency is determined by θ, featuring, depending on the value of z, a number of harmonics.

We can define precisely the resulting spectrum, since this is an expression that has been very well studied, as it arises in many other fields of science.[13] Its expansion into a series, in its most general form,[14] is

$$\begin{aligned} e^{\pm jz\sin(\theta)} &= J_0(z) + \sum_{n=1}^{\infty}(\pm 1)^n J_n(z)\left[e^{jn\theta} + (-1)^n e^{-jn\theta}\right] \\ &= J_0(z) + 2\sum_{n=1}^{\infty} J_{2n}(z)\cos(2n\theta) \pm 2j\sum_{n=0}^{\infty} J_{2n+1}(z)\sin([2n+1]\theta). \end{aligned} \tag{8.26}$$

From this, we derive the following sums (Watson, 1944, 22)

$$\cos\left(z\sin(\theta)\right) = J_0(z) + 2\sum_{n=1}^{\infty} J_{2n}(z)\cos(2n\theta) \tag{8.27}$$

and

$$\sin\left(z\sin(\theta)\right) = 2\sum_{n=0}^{\infty} J_{2n+1}(z)\sin([2n+1]\theta), \tag{8.28}$$

as well as, with $\eta = \pi/2 - \theta$,

$$\cos\left(z\cos(\eta)\right) = J_0(z) + 2\sum_{n=1}^{\infty}(-1)^n J_{2n}(z)\cos(2n\eta) \tag{8.29}$$

and

$$\sin\left(z\sin(\eta)\right) = 2\sum_{n=0}^{\infty}(-1)^n J_{2n+1}(z)\cos([2n+1]\eta). \tag{8.30}$$

Therefore we have a set of trigonometric series, resembling a Fourier series, except for the fact that it employs a set of coefficients $J_n(z)$, which are functions of the input signal amplitude z, and are associated with the different terms of the expansion. These are known as *Bessel coefficients*, and are defined by the following equation[15]

$$J_n(z) = \frac{1}{2\pi}\int_0^{2\pi} \cos\left(n\theta - z\sin(\theta)\right)d\theta. \tag{8.31}$$

which has an alternative solution that is more appropriate for computing its values,[16]

$$J_n(z) = \sum_{m=0}^{\infty} \frac{(-1)^m \left(\frac{z}{2}\right)^{n+2m}}{m!(n+m)!} \tag{8.32}$$

where $m!$ denotes the factorial of m, $m(m-1)(m-2)\ldots(1)$.

8.2.1 Phase Modulation

Thus, the expression we started with, which uses a cosine waveshaper, gives us the even harmonics of θ. Its complement is the sine waveshaper, which provides the odd harmonics. If we mix the two, we have the signal

$$\begin{aligned} &\cos(z\sin(\theta)) + \sin(z\sin(\theta)) = \\ &J_0(z) + 2\left[\sum_{n=1}^{\infty} J_{2n}(z)\cos(2n\theta) + \sum_{n=1}^{\infty} J_{2n+1}(z)\sin([2n+1]\theta)\right], \end{aligned} \tag{8.33}$$

giving us a full spectrum. In this expression, we should note that the harmonic phases alternate between 0 and $-\pi/2$, which is not of great concern. In any case, we are not completely finished with it. Going one step further, it is possible to transform this one-sided formula into fully flexible double-sided expression. We already have a method: multiplying it by a sinusoid centered at another frequency. Before we do that, it is good to choose the phase of the sinusoid so we can produce a uniform phase response. Moreover, we should be mindful of the following relationship,[17]

$$J_{-n}(z) = (-1)^n J_n(z), \tag{8.34}$$

which tells us that the odd-order negative Bessel coefficients are equivalent to the positive ones with reversed sign, and that the even-numbered coefficients are the same, regardless of the sign of n. We know where this comes from, since we have seen this type of symmetry before in spectral analysis, where the coefficients for cosine components form an even function of frequency, and the ones for sine components make up an odd function. Similar mechanisms are at play here.

We do not need to use the same modulator for the separate cosine and sine terms. This allows us to chose their phase so that we end up with a cosine or a sine-only spectrum. For example, using a $\cos(\omega)$ for the cosine waveshaper and a $\sin(\omega)$ for the sine, we have

$$\cos(\omega)\cos(z\sin(\theta)) = \frac{\cos(\omega - z\sin(\theta)) + \cos(\omega + z\sin(\theta))}{2} \tag{8.35}$$

and

$$\sin(\omega)\cos(z\sin(\theta)) = \frac{\cos(\omega - z\sin(\theta)) - \cos(\omega + z\sin(\theta))}{2}. \tag{8.36}$$

Now we can mix these, but changing the sign of the sine term so that we end with the sums of angles rather than the differences,

$$\cos(\omega)\cos(z\sin(\theta)) - \sin(\omega)\cos(z\sin(\theta)) = \cos(\omega + z\sin(\theta)), \tag{8.37}$$

which we can expand as

$$\cos(\omega + z\sin(\theta)) = J_0(z)\cos(\omega) + \sum_{n=1}^{\infty} J_{2n}(z)\left[\cos(\omega - 2n\theta)) + \cos(\omega + 2n\theta)\right]$$
$$+ \sum_{n=1}^{\infty} J_{2n+1}(z)\left[\cos(\omega + [2n+1]\theta) - \cos(\omega - [2n+1]\theta)\right] \tag{8.38}$$
$$= \sum_{n=-\infty}^{\infty} J_n(z)\cos(\omega + n\theta).$$

This uses the relationship given by Eq. 8.34 to give a very compact spectral representation, as a double-sided formula with components at $\omega \pm n\theta$, where $\omega = 2\pi f_c t$ and $\theta = 2\pi f_m t$ (called the carrier and modulator frequencies, respectively). The choice of modulators has paid off, as we have ended up with a uniform cosine-phase spectrum. We have also discovered a very compact synthesis equation, which does not use any ring modulation. Instead, this non-linear distortion formula implements a synthesis technique known as *phase modulation* (PM). This is a very important method, first introduced to computer music by John Chowning (1973), and subsequently implemented in many variations, both in software and hardware forms.[18]

The key to understanding the spectra of phase modulation is found in the Bessel coefficients, $J_n(z)$, and their shape. The first ten of these are plotted in Fig. 8.7. For each sideband coefficient n, we have an associated Bessel coefficient of order n (upper side) or $-n$ (lower side), according to Eq. 8.38. We can get the negative order coefficients using the relationship in Eq. 8.34. As we can see from the plots, these functions are bipolar and vary with z, so we can use that as a single spectral control, determining the bandwidth of the signal. Since z controls the amount of modulation applied, it is also called the *index of modulation*. With low values of z most coefficients are zero or close to zero, they only start appearing as that parameter grows. Note also that, with the exception of $J_0(0)$, all absolute values are always under 1. This is a great advantage, which indicates that the signal amplitude should be more or less under control. In fact, if we examine the synthesis formula, we will come to the conclusion that it is normalized with a peak amplitude of 1, since there are no scaling factors involved. This of course makes it easier to handle than the summation formulae we have seen earlier.

Note also that an important effect imparted by the Bessel coefficients is a fluctuation of the amplitude of each sideband as z is varied. This gives the typical phase-modulation spectral evolution, which is both an asset and a liability. If we are trying to emulate the natural changes in partial amplitudes found in acoustic instruments, this spectral wobbling is not a desirable feature. On the other hand, if we are seeking synthetic timbres of arbitrary designs, this can be used to create interesting complex spectral changes. In any case, the fact that the frequency-domain characteristics of PM can be controlled with few parameters is very useful.

8.2.2 Signal Bandwidth and Aliasing

Note that the PM equation produces a signal that is not guaranteed to be bandlimited; therefore aliasing is a concern. In order to protect against this, we should try to identify the highest component with any significant amplitude. Again, the key to this is found in

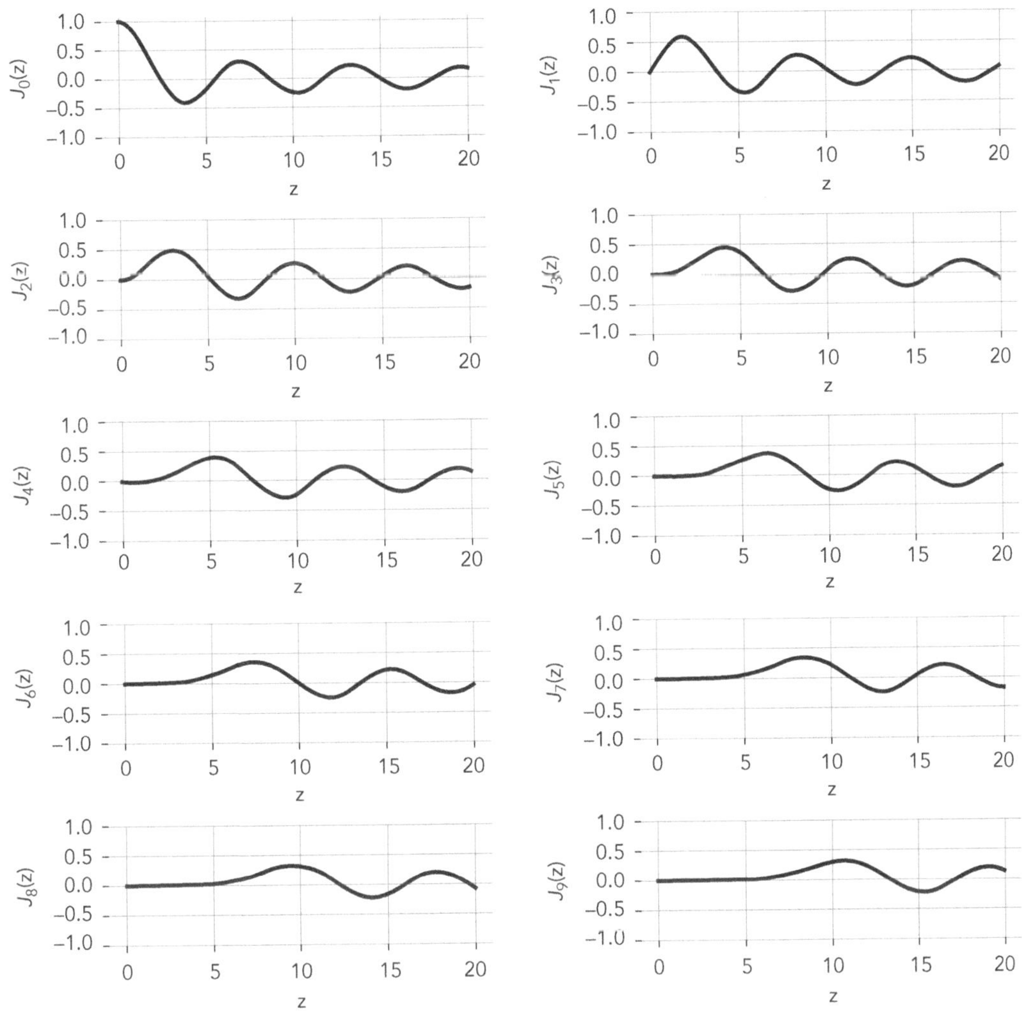

Figure 8.7
Bessel coefficients $J_n(z)$, of orders 0 to 9.

the Bessel coefficients scaling each sideband. The upper sideband partials are the ones we should focus on, scaled by the positive-order coefficients. A cursory analysis of Fig. 8.7 tells us that higher-order coefficients are very close to zero until z reaches a certain magnitude. A method to find the highest significant component is to check at what value of z $J_n(z)$ is of a sufficient absolute value to make sideband n start to feature with some prominence in the spectrum.

A practical way to do this is consider the extension of the Bessel coefficients as functions of two variables $J_v(z)$, where v is allowed to take any real value. These are known as Bessel functions. We can now look at their values when we set $v = z + r$, for various values of r. This allows us to estimate the value of z at which the sideband n will be of a certain amplitude. This is shown in Fig. 8.8, where we plot the 0 dB full-scale magnitude for $J_{z+r}(z)$, with r set to various values. For reference, we also plot the $J_z(z)$ case, which is the value of the Bessel function of order z for a z input. This is unity (0 dB) for J_0, but it falls as the total waveform energy gets distributed to the various sidebands.

We should read the results as follows. Each plot line relates to sideband $z + r$ and its amplitudes for various values of z. For example, sideband $z + 3$ (dots) with $z = 10$ has an amplitude of -30 dB. This refers to the frequency $f_c + 13f_m$. This analysis gives us some

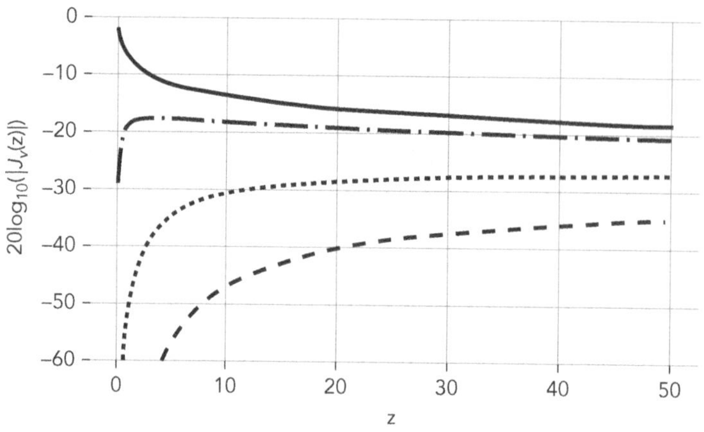

Figure 8.8 Plots of $20\log_{10}(|J_v(z)|)$ for $v = z$ (*solid*), $v = z+1$ (*dash-dots*), $v = z+3$ (*dots*), and $v = z+5$ (*dashes*).

useful results. The plot of $v = z+1$ shows that sideband $z+1$ is about -20 dB down from full scale, which in many cases can still be considered significant. The case of $v = z+3$, suppression of sideband $z+3$ is about 30 dB, whereas $z+5$ gives us better than 40 dB for $z < 20$. Therefore we should consider that for a given value of z, the highest significant component is somewhere at sidebands $z+3$ to $z+5$. This should give us a good measure to avoid objectionable aliasing distortion. Setting a limit to z that is dependent on the frequency of the highest significant bandwidth and the sampling rate will prevent any issues in this regard.

8.2.3 Carrier to Modulator Ratio

In addition to z, which controls the spectral bandwidth, the only other two parameters of phase modulation are the carrier and modulator frequencies. These determine, similarly to summation formula synthesis, the partial frequencies and the overall characteristic of the spectrum as harmonic or inharmonic. For example, simple ratios of f_c:f_m will lead to a set of harmonic components. Chowning provided a basic recipe to determine the resulting spectrum (Chowning, 1973, 528–529). Setting the carrier to modulator ratio as

$$\frac{f_c}{f_m} = \frac{N}{M} \tag{8.39}$$

where N and M are integers with no common factors, we have

$$f_o = \frac{f_c}{N} = \frac{f_m}{M} \tag{8.40}$$

From this he also observed that

- f_c is always the Nth harmonic in the series.
- If $M > 1$, every Mth harmonic will be missing.
- For $N \neq 1$, with small indices the partial containing f_0 may have little or no energy.
- If N or M or both are not integers, but irrational numbers, the resulting spectrum will be inharmonic.

Another consideration when setting the values of f_c and f_m is the reflected negative-frequency sidebands. In a harmonic spectrum, they overlap with partials in the positive side of the spectrum, and can either add constructively or destructively, depending on the value of $J_n(z)$. Since the spectrum is solely composed of cosines, there is no difference in polarity between the negative and positive spectrum, and thus the result is solely dependent on the coefficients of the overlapping partials.

8.2.4 Implementation

The Csound implementation of phase modulation can take different forms. We can first reproduce the synthesis formula as is, to provide a reference for any further modifications,

```
opcode PMC,a,kkk
 kfc,kfm,kz xin
 aphc = 2*$M_PI*phasor(kfc)
 aphm = 2*$M_PI*phasor(kfm)
 apm = cos(aphc + kz*sin(aphm))
   xout(apm)
endop
```


This is the cosine carrier formula, which we have developed in this text. A simple variant is Chowning's original expression, which uses a sine wave carrier,

```
opcode PMS,a,kkk
 kfc,kfm,kz xin
 aphc = 2*$M_PI*phasor(kfc)
 aphm = 2*$M_PI*phasor(kfm)
 apm = sin(aphc + kz*sin(aphm))
   xout(apm)
endop
```

These two formulae can form a quadrature pair, as they produce cosine-and sine-only spectra. This implementation follows the form we have used before when programming the other closed-form synthesis expressions earlier in the chapter. However, as we have also demonstrated, it is possible to use table lookup to provide an efficient implementation of such formulae. In this case, we can use a sine wave oscillator as the modulation source, and a table reader to produce the phase modulation signal,

```
opcode PM,a,kkki
 kfc,kfm,kz,ifn xin
 aphc = phasor(kfc)
 aphm = oscili(kz/(2*$M_PI),kfm)
 apm = tablei:a(aphc + aphm,ifn,1,0,1)
   xout(apm)
endop
```

Note that since the table lookup index is normalized to $[-1,1]$, we need to scale z by $1/2\pi$ to get it in the equivalent range of values as given in the original expression. This is a more generic implementation, since the carrier waveform is given in a function table, and so we can swap it from a cosine to a sine without needing to change the implementation. In fact, we can also use any other waveform as a carrier, resulting in a more complex spectrum. In Chapter 11, we will develop yet another form of implementation, based on the concept of a PM operator, which is very modular and can be used to create complex modulation topologies.

8.2.5 Frequency Modulation

The terms *phase* and *frequency modulation* (FM) are often used interchangeably. In Chowning's paper, the latter is used, even though the actual formula describes the former. In most software and hardware implementations, it is far more common to see the word FM as a descriptor, probably due to the original paper and subsequent deployment of the technology in digital *synthesizers*. However, it is important to point out the differences, and to demonstrate how the same formula can be programmed both as PM and as FM. Remembering from Sec. 3.1.2 that the phase and frequency are related by integration and differentiation. So if we start with the phase of carrier signal that is being modulated,

$$\phi(t) = 2\pi f_c t + z\sin(2\pi f_m t) \tag{8.41}$$

the continuous-time instantaneous frequency of that signal is (Moore, 1990)

$$\frac{d}{dt}\phi(t) = \frac{d}{dt}[2\pi f_c t] + \frac{d}{dt}[z\sin(2\pi f_m t)] = 2\pi\left[f_c + z f_m \cos(2\pi f_m t)\right] \tag{8.42}$$

This tells us that if we want to produce a signal that corresponds to the output of the PM synthesis expressions developed earlier, now through the use of FM, we need to

- Use a cosine wave modulator, $m(t) = \cos(\theta)$.
- Set the carrier frequency to $f_c + z f_m m(t)$.

The quantity zf_m, the product of the modulation frequency and the index of modulation, is also known as the frequency deviation Δ_f. This is the absolute amount of frequency variation that offsets the carrier frequency f_c. In sub-audio frequency modulation, this is equivalent to the vibrato width; it is possible to equate FM synthesis to a wide and fast vibrato applied to a carrier signal.

With these relationships in hand, we can now proceed to implement FM synthesis. For this, we can use two oscillators, one in the cosine phase and the other employing an arbitrary carrier waveform. The following opcode produces a signal that is equivalent to the generic PM example, with the addition of an amplitude control,

```
opcode FM,a,kkkki
 ka,kfc,kfm,kz,ifn xin
 afm = oscili(kz*kfm, kfm,-1,0.25)
 acar = oscili(ka,kfc+afm,ifn)
   xout(acar)
endop
```

Simple FM synthesis implementations have the same characteristics and attributes discussed in relation to PM. However, the former method is in general not as flexible as the latter, as it poses problems for some interesting cases such as the stacking of modulators (using the output of a FM expression to modulate another), or the use of feedback modulation, both of which are used in many synthesis applications. The reason for this is that any such connections may disrupt the integration performed by the oscillators resulting in unwanted side effects (Pinkston, 2000). Therefore it is far better to avoid any difficulties by always implementing phase modulation.

8.2.6 Splitting Sidebands

It is possible to split the PM formula into components to gain control over the individual sidebands. As we have seen earlier in Eq. 8.26, the cosine and sine waveshapers can produce gapped spectra, only the odd or the even sidebands. Therefore the two formulae,

$$s_e(t) = \sin(\omega)\cos(z\sin(\theta)) \tag{8.43}$$

and

$$s_o(t) = \sin(\omega)\sin(z\sin(\theta)), \tag{8.44}$$

can produce even and odd sidebands separately. Note that the first produces sine-phase, and the second, cosine-phase partials. We can now combine these with the principle of single-sideband modulation to produce upper, lower, even, and odd split sideband signals. As we have seen, the Hilbert transform can be applied to produce negative or positive-only frequencies,

$$s_-(t) = s(t) - j\mathscr{H}\{s(t)\} \tag{8.45}$$

and

$$s_+(t) = s(t) + j\mathscr{H}\{s(t)\}. \tag{8.46}$$

Therefore we can replace single-sideband modulation instead of the ring modulation used in Eqs. 8.43 and 8.43. This allows the upper and lower sidebands to be split and we have the four separate signals,

$$\begin{aligned}
s_{u,e}(t) &= \sin(\omega)\cos(z\sin(\theta)) + \cos(\omega)\mathscr{H}\{\cos(z\sin(\theta)\} \\
&= \sum_{n=0}^{\infty} \varepsilon_{2n} J_{2n}(z)\sin(\omega + 2n\theta) \\
s_{u,o}(t) &= \sin(\omega)\sin(z\sin(\theta)) + \cos(\omega)\mathscr{H}\{\sin(z\sin(\theta)\} \\
&= \sum_{n=0}^{\infty} \varepsilon_{2n+1} J_{2n+1}(z)\cos(\omega + [2n+1]\theta) \\
s_{l,e}(t) &= \sin(\omega)\cos(z\sin(\theta)) - \cos(\omega)\mathscr{H}\{\cos(z\sin(\theta)\} \\
&= \sum_{n=0}^{\infty} \varepsilon_{2n} J_{2n}(z)\sin(\omega - 2n\theta) \\
s_{l,o}(t) &= \sin(\omega)\sin(z\sin(\theta)) - \cos(\omega)\mathscr{H}\{\sin(z\sin(\theta)\} \\
&= \sum_{n=0}^{\infty} \varepsilon_{2n+1} J_{2n-1}(z)\cos(\omega - [2n+1]\theta)
\end{aligned} \tag{8.47}$$

where we introduced a factor ε_n,[19] which is 1 for $n = 0$ and 2 for $n \neq 0$.

We can program these formulae in Csound using the Hilbert transform operator developed in earlier chapters. The resulting opcode produces four outputs simultaneously, which can be then be used, mixed, and further processed in a variety of ways.

```
opcode SpSB,aaaa,kkkk
 ka,kfc,kfm,kz xin
 am = oscili(kz/(2*$M_PI),kfm)
 acm = tablei:a(am,-1,1,0.25,1)
 asm = tablei:a(am,-1,1,0,1)
 are,aie hilbert2 acm,1024,256
 aro,aio hilbert2 asm,1024,256
 ac = oscili(ka/2,kfc)
 ad = oscili(ka/2,kfc,-1,0.25)
 aeu = are*ac + aie*ad
 aou = aro*ac + aio*ad
 ael = are*ac - aie*ad
 aol = aro*ac - aio*ad
 xout(aeu,aou,ael,aol)
endop
```


8.2.7 Feedback

The principle of feedback modulation, which we have employed in Chapter 7 in the guise of amplitude modulation leading to audio-rate time-varying filters, can also be applied to phase modulation synthesis. In this case, we have a formula where the output is fed back to offset the phase of a carrier oscillator, expressed by (Lazzarini, 2017, 196)

$$y(n) = \sin(\omega(n) + zy(n-1)) \tag{8.48}$$

where we have[20] $\omega(n) = 2\pi fn/f_s$. We can describe the phase of this signal by

$$\phi(n) = \omega(n) + \sin(z\phi(n-1)), \tag{8.49}$$

in which case the expansion of $\sin(\phi(n))$ becomes

$$2\sum_{k=1}^{\infty} \frac{J_k(kz)}{k} \sin(k\omega(n)) \tag{8.50}$$

The feedback PM expression is yet another closed form of a sine summation. If set $z = 1$, the Bessel coefficients become

$$2\frac{J_k(k)}{k} \approx \frac{1}{k} \tag{8.51}$$

which approximates the $1/f$ sawtooth wave spectrum. The output of this equation is a harmonic spectrum, whose bandwidth is dependent on z, as in the other cases of phase modulation. Likewise, it is not a bandlimited formula, and therefore we have to control the modulation index carefully to avoid aliasing.

This formula can be implemented as a UDO in Csound. Since we require a 1-sample delay of the output as the modulator signal, it is necessary again to process it on a sample-by-sample basis, so we isolate the feedback expression from the rest of the code into a separate function. This is then called by feedback PM opcode. Note also that we have to scale the modulation index by $1/2\pi$ as before, since the table lookup is normalized.

```
opcode FB,a,aki
 setksmps(1)
 ay init 0
 aph,kz,ifn xin
 ay = tablei:a(aph+ay*kz,ifn,1,0,1)
 xout(ay)
endop

opcode FPM,a,kki
 kf,kz,ifn xin
 aph = phasor(kf)
 apm = FB(aph,kz/(2*$M_PI),ifn)
 xout(apm)
endop
```


8.2.8 Complex PM

It might be useful now to examine the case where either the modulator or the carrier is a complex signal, that is, not a single-component, pure sinusoid. We have seen in the expansion of the closed-form formula that many components are generated from two sinusoids. We should expect that as we make the signals more complex, the number of partials in the spectrum can grow significantly. We can start by examining the case of a multiple-component carrier signal. Consider a periodic function, which has the Fourier series

$$f(x) = \sum_{n=0}^{N} a_n \cos(nx), \tag{8.52}$$

to which we apply a phase signal $\phi(t)$, producing the waveform $s(t) = f(\phi(t))$. The phase modulation synthesis case is where we have

$$\phi(t) = \omega(t) + z\sin(\theta(t)). \tag{8.53}$$

Substituting this expression back into the series of Eq. 8.52 we have

$$s(t) = \sum_{n=0}^{N} a_n \cos(n\omega(t) + nz\sin(\theta(t))) \tag{8.54}$$

which describes the modulation of each term in the series, or each partial. What we have is a multiple-carrier setup but with increased modulation injected in the higher components. The series expansion becomes

$$s(t) = \sum_{n=0}^{N} a_n \sum_{k=-\infty}^{\infty} J_k(nk)\cos(n[\omega + k\theta]). \tag{8.55}$$

This tells us that with only a few partials in the carrier waveform, we can end up with a very rich spectrum. We should also be careful with aliasing, considering that the modulation index acting on the higher components is a multiple of the z used in the modulation signal. This type of complex PM can be easily implemented by employing a complex wavetable as an input to the phase modulation opcode developed earlier. We can derive this wavetable from a recording of an audio signal, if we wish, but we should note that the fundamental frequency driving the phase signal $\phi(t)$ is the basis for computing the frequencies in Eq. 8.55, and not the original fundamental frequency of the original sound. Once in a wavetable, the phase signal determines the playback rate. A more flexible approach for this case, however, is to use an adaptive method, which we will explore later.

The case of multiple-component modulators can be considered by generalizing the PM formula and then adding each component at a time.[21] Starting with a PM expression,

$$\cos(\omega + z_1 \sin(\theta_1)) = \sum_{n_1=-\infty}^{\infty} J_{n_1}(z_1) \cos(\omega + n_1\theta_1), \tag{8.56}$$

we make the replacement $\omega = \omega + z_2 \sin(\theta_2)$, which gives us

$$\sum_{n_1=-\infty}^{\infty} J_{n_1}(z_1) \cos(\omega + z_2 \sin(\theta_2) + n_1\theta_1). \tag{8.57}$$

To get the series expansion for this closed-form formula, we first transform it by setting $\Omega = \omega + n_1\theta_1$,

$$\sum_{n_1=-\infty}^{\infty} J_{n_1}(z_1) \cos(\Omega + z_2 \sin(\theta_2)), \tag{8.58}$$

and then expand the inner modulation expression,

$$\begin{aligned} &\sum_{n_1=-\infty}^{\infty} J_{n_1}(z_1) \sum_{n_2=-\infty}^{\infty} J_{n_2}(z_2) \cos(\Omega + n_2\theta_2) = \\ &\sum_{n_1=-\infty}^{\infty} \sum_{n_2=-\infty}^{\infty} J_{n_1}(z_1) J_{n_2}(z_2) \cos(\omega + n_1\theta_1 + n_2\theta_2) \end{aligned} \tag{8.59}$$

This basic principle can be applied to further modulating signals, giving the general expression

$$\sum_{n_1=-\infty}^{\infty} \cdots \sum_{n_K=-\infty}^{\infty} \left[\prod_{k=1}^{K} J_{n_k}(z_k)\right] \cos\left(\omega + \sum_{k=1}^{K} n_k\theta_k\right). \tag{8.60}$$

From the perspective of this equation, we can see that the spectrum can very quickly become saturated with components if more than a few modulators are used. Sidebands will appear as the sum and difference of the carrier and all the modulating frequencies scaled by a product of Bessel coefficients. Therefore, it is essential that we limit the amount of modulation that is applied by each modulating source.

The main motivation behind complex modulation is to use very small indices of modulation, which produce spectra depending on lower-order Bessel coefficients whose values rise in the positive range. This allows smoother timbral evolutions, which are better suited to emulate acoustic instrumental sounds such as brass or strings. To implement this synthesis technique, we only need to add one or more modulation sources to the generic PM or FM expression. A very good example of this is discussed in Chapter 10, where complex FM is applied to produce tightly controlled spectra in the piece *Stria* by John Chowning. Alternatively, if we stack one or more PM signal generators, we can also obtain a similar effect, where the index of modulation in the topmost opcodes will determine the number of modulation terms in the PM signals down the chain.

8.2.9 Exponential FM

The type of phase and frequency modulation that we have been developing so far offsets the carrier frequency in a linear way. For example, if the deviation Δ_f is x Hz, an oscillator f_c will vary from $f_c - x$ to $f_c + x$. Due to this fact, these methods are often described as *linear* FM, although they of course implement a form of non-linear distortion. This is mostly used to distinguish them from a different form of FM, which arises from analogue synthesis technologies, where voltage-based controls are often to respond to volt per octave signals. This technique is called *exponential* FM (Hutchins 1975). In such an arrangement, a frequency f doubles proportionally to a unit change of voltage v, $f \propto 2^v$. For an exponential FM signal with phase $\phi(t)$, the instantaneous frequency can therefore be characterized as (Timoney and Lazzarini, 2009)

$$\frac{d}{dt}\phi(t) = \omega 2^{v(t)} \tag{8.61}$$

where $\omega = 2\pi f_c t$ and $v(t)$ is a modulation signal defined by a constant offset v_0 and a sinusoidal modulator $\cos(\theta)$ with $\theta = 2\pi f_m t$, of amplitude v_m,

$$v(t) = v_0 + v_m \cos(\theta). \tag{8.62}$$

This allows us to the put the instantaneous frequency as

$$\frac{d}{dt}\phi(t) = \omega 2^{v_0 + v_m \cos(\theta)} = \omega_{ce} e^{\log(2) v_m \cos(\theta)} \tag{8.63}$$

where $\omega_{ce} = e^{V_0 \log(2)}$. This gives an expression containing an exponential modulation term of the form $e^{k\cos(\theta)}$. We can expand this using Eq. 8.26, making the replacement $z = -jk$. This leads to Bessel coefficients of a purely imaginary argument, also called *modified* Bessel coefficients $I_n(z)$, defined as (Watson, 1944, 77) as

$$I_n(z) = j^{-n} J_n(jz). \tag{8.64}$$

With this, we can now derive the expression

$$e^{k\cos(\theta)} = \sum_{n=-\infty}^{\infty} I_n(k)\cos(n\theta) \tag{8.65}$$

with which we can derive the modulation factor as

$$e^{\log(2)v_m \cos(\theta)} = I_0(\log(2)v_m) + 2\sum_{n=1}^{\infty} I_n(\log(2)v_m)\cos(n\theta), \tag{8.66}$$

where we used the relationship $I_n(k) = I_{-n}(k)$. This gives us the following result for the instantaneous frequency

$$\frac{d}{dt}\phi(t) = \omega_e[I_0(\log(2)v_m) + 2\sum_{n=1}^{\infty} I_n(\log(2)v_m)\cos(n\theta)]. \tag{8.67}$$

We can now integrate the instantaneous frequency to obtain a phase signal that drives a sinusoid,

$$\sin(\phi(t)) = \sin\left(I_0(\log(2)V_m)\omega_{ce} + 2\sum_{n=1}^{\infty} \frac{\omega_{ce} I_n(\log(2)v_m)}{n\theta}\sin(n\theta)\right). \tag{8.68}$$

This provides us with an equivalent phase modulation expression, which is helpful for considering the differences between linear and exponential FM. We can analyze this result as follows:

- A sinusoidal modulator in exponential FM has a similar effect to a multi-component modulator in linear FM.
- The carrier frequency is not fixed; it changes with the modulation amplitude v_m, unlike in linear FM, where it is independent of any modulation applied.

Another useful aspect of Eq. 8.68 is that it can be used to derive the spectrum of exponential FM using the theory we have developed for linear FM. For this, the carrier phase can be given as

$$\omega_e = I_0(\log(2)V_m)\omega_{ce}. \tag{8.69}$$

The complex modulation signal has a different index for each partial. We can define these as

$$z_n = \frac{2\omega_{ce} I_n(\log(2)v_m)}{n\theta} \tag{8.70}$$

With these we can make Eq. 8.68 somewhat more compact as

$$\sin\left(\omega_e + \sum_{n=1}^{\infty} z_n \sin(n\theta)\right), \tag{8.71}$$

for which a PM spectrum expansion can be given using the complex modulation formula

$$\sum_{n_1=-\infty}^{\infty} \cdots \sum_{n_\infty=-\infty}^{\infty} \left[\prod_{k=1}^{\infty} J_{n_k}(z_k)\right] \sin\left(\omega_e + \sum_{k=1}^{\infty} kn_k\theta\right) \tag{8.72}$$

where the number of significant modulator terms depends on the amount of modulation v_m. From these equations, we can see that the spectrum can become quite complex with only a small amount of modulation. This is generally not very problematic in analogue implementations, but any digital emulation of this effect will have to put stringent controls in place to avoid aliasing.

8.3 Phase Distortion Synthesis

Phase Distortion (PD) is variant of PM where we apply a non-linear mapping to the phase signal that is used to read a sinusoidal wavetable (Timoney et al., 2011). It was developed as a response to the success of the early digital synthesizers employing FM synthesis. In opposition to the perceived complexities of programming a FM synthesizer, PD implementations laid out the controls for this synthesis method emulating a typical source-modifier arrangement, where an oscillator block was followed by a wave modifier unit (controlling timbre) and an amplifier (controlling the amplitude envelope). This of course was just fantasy, since the whole method can be described by a single equation (as in PM synthesis). We can lay out the fundamentals of PD synthesis as described in its original formulation as follows. Starting with a sinusoid controlled by a modulo-1 phase $\phi(t)$,

$$s(t) = -\cos(2\pi\phi(t)), \tag{8.73}$$

we apply a non-linear distortion function $f(x)$ to the phase signal,

$$s(t) = -\cos\left(2\pi f(\phi(t))\right). \tag{8.74}$$

Depending on the shape of the $f(x)$ function, we may have different results. A typical distortion function employed in this method is a piecewise linear function composed of

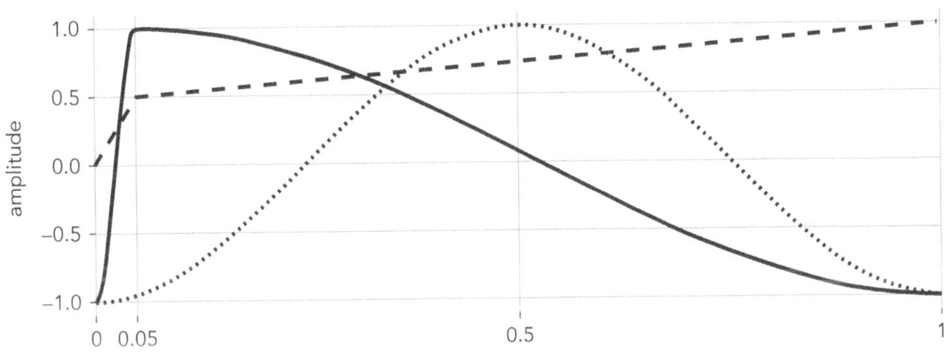

Figure 8.9 Phase distortion waveform (*solid line*), created from an inverted cosine (*dots*), by a non-linear phase lookup function (*dashes*) made of two linear segments inflected at $d = 0.05$.

two segments, such as the one defined by

$$f(x) = \begin{cases} \frac{x}{2d}, & x < d \\ \frac{1}{2}\left[1 + \frac{x-d}{x+d}\right], & x \geq d \end{cases} \tag{8.75}$$

in the interval $[0,1)$. The important parameter in phase distortion is the position of the inflection point d, which can make the sound vary from a sinusoid ($d = 0.5$), when $f(x)$ becomes linear, to a complex wave with many components ($d < 0.5$). A PD waveform is shown in Fig. 8.9 as a product of distorting the phase lookup function $\phi(t)$, with a piecewise linear function inflected at $d = 0.05$. We can see that this generates a sawtooth-like shape, for which we should expect to have a (non-bandlimited) decaying spectrum. Aliasing might be an issue with high fundamental frequencies, depending of course on the sampling rate employed.

To show that this is effectively a form of complex PM, we can split $f(x)$ into a linear and a non-linear term, $x + g(x)$. The non-linear term is then defined as

$$g(x) = \left[\frac{1}{2} - d\right] \begin{cases} \frac{x}{d}, & x < d \\ \frac{1-x}{1-d}, & x \geq d \end{cases} \tag{8.76}$$

which is effectively a non-bandlimited sawtooth-shape wave that has an inflection at d and a peak amplitude of $1/2 - d$.

We have now the means to put the PD formula in terms of PM, and employ the usual analysis tools derive its spectrum. The expression describing this becomes, at the limit (d→0),

$$s(t) = \sin\left(\omega + \sum_{k=1}^{\infty} \frac{\sin(k\omega)}{k}\right) \tag{8.77}$$

and the resulting series is then

$$\sum_{n_1=-\infty}^{\infty} \cdots \sum_{n_\infty=-\infty}^{\infty} \left[\prod_{k=1}^{\infty} J_{n_k}\left(\frac{1}{k}\right)\right] \sin\left(\omega\left[1 + \sum_{k=1}^{\infty} kn_k\right]\right) \tag{8.78}$$

which is form of complex PM with the modulator frequencies set to multiples of the carrier frequency.

8.3.1 Vector Phase Shaping

Other types of distortion functions may be employed, with more than one inflection point, for instance. In fact we can expand the principle of phase distortion into a more

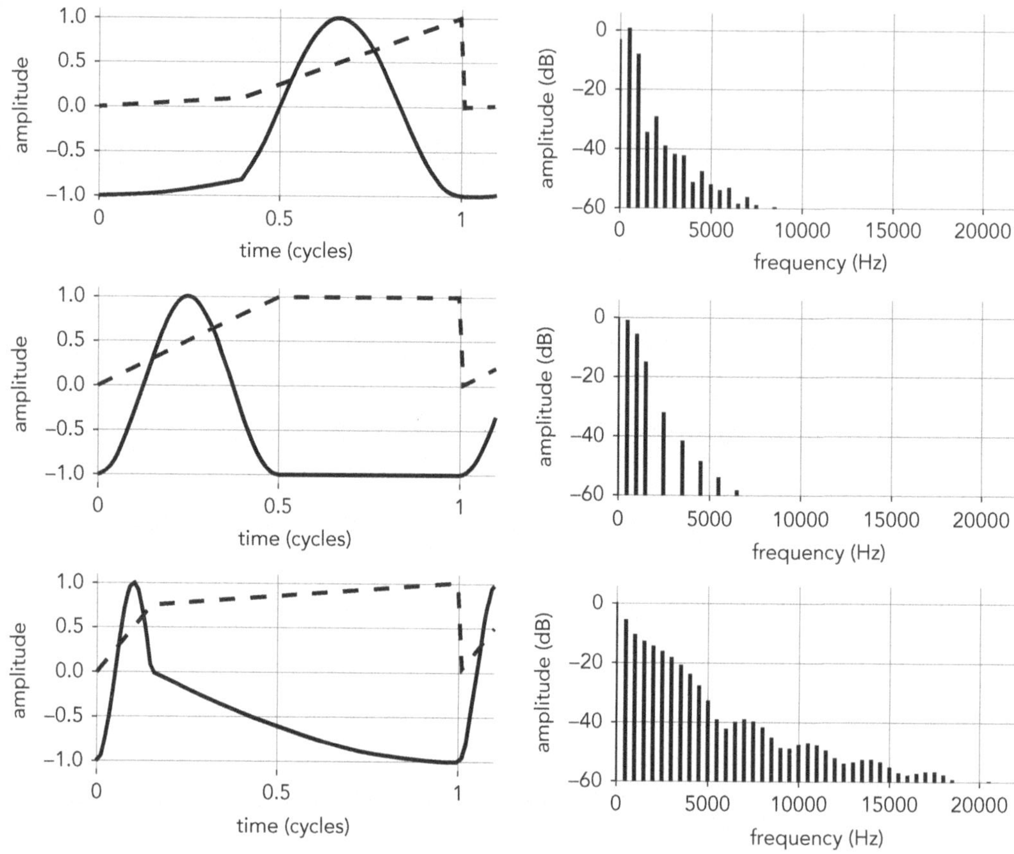

Figure 8.10 VPS distortion functions (*dots*), waveforms (*solid*), and spectra using $p = (0.4, 0.1)$ (*top*), $p = (0.5, 1)$ (*middle*), and $p = (0.15, 0.75)$ (*bottom*).

general form called *vector phase shaping* (VPS) synthesis (Kleimola et al., 2009). In PD, the inflection point d controls the horizontal position of the piecewise linear function joint (see Fig. 8.9). In this case, the vertical position of the bend is fixed at 0.5. We can remove this constraint, expressing the inflection point as a vector

$$p = (d, v), \tag{8.79}$$

where d is the horizontal, and v the vertical, component defining the inflection position. A two-dimensional phase distortion function can then be given as

$$\phi_{d,v}(x) = \begin{cases} \frac{vx}{d}, & x < d \\ (1-v)\frac{(x-d)}{(1-d)} + v, & x \geq d, \end{cases} \tag{8.80}$$

which is equivalent to PD for $v = 0.5$. Three plots showing different shaping functions, their corresponding waveforms, and spectra are given in Fig. 8.10.

As can be seen, various configurations are possible. Different types of spectra can be obtained by moving the inflection point freely in two-dimensional space. Some positions produce particular effects. For example, if we set $v = 1$, as shown in the middle panel of Fig. 8.10, d controls the duty cycle of the waveform. Its pulse width can be modified from a narrow impulse up to a full-cycle cosine at $d = 1$. Pulses of various widths can be generated by varying this parameter, and pulse width modulation is possible by controlling this parameter using time-varying input (Fig. 8.11).

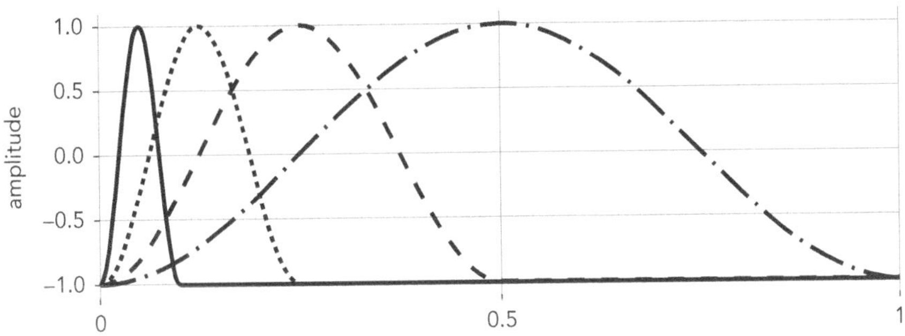

Figure 8.11
Various pulse widths produced by varying the value of d in the vector $p = (d, 1)$.

Some vectors can be arranged to produce sinusoidal output. We know that a simple linear phase function can be obtained with $v = d = 0.5$, and similarly with other values of v and $d = 1$, such as $v = 1, 2, 3 \ldots n$. Other cases are governed by the following relationship:

$$\left|\frac{v}{d}\right| = \left|\frac{1-v}{1-d}\right|, \tag{8.81}$$

for integral values of v/d. The explanation is simple, in that the derivative of the phase on both sides of the inflection point has the same magnitude, which is all we need for a linear phase lookup when using an even function such as the cosine waveform. The resulting frequency is a multiple of the fundamental, $f_0 v/d$. This can provide some very interesting musical effects when dynamically modifying the distortion vector; as with certain positions of the inflection point, the complex spectrum can suddenly become a single sinusoid. This can be equated to some sort of non-linear filtering effect.

A straight implementation of VPS requires sample-by-sample processing, so again, we split the code so that block processing can be used as much as possible. This code expects d to be in the usual range $[0, 1)$, but v can be any real number. The opcode also implements standard phase distortion if we clamp v to 0.5.

```
opcode VPS,a,akk
 setksmps 1
 asig,kd,kv xin
 ksig = asig
 if ksig < kd then
  aout = asig*kv/kd
 else
  aout = kv + (1-kv)*(asig-kd)/(1-kd)
 endif
 xout(aout)
endop

instr 1
 kd = oscili:k(0.5,2) + 0.5
 kv = oscili:k(0.5,1.5) + 0.5
 aph = VPS(phasor(p5),kd,kv)
 asig = p4*tablei:a(aph,-1,1,0.25,1)
 out(linenr(asig,0.1,0.1,0.01))
```

```
endin
schedule(1,0,10,.5,110)
```

The phase shaping opcode VPS is only used to demonstrate the implementation of the technique. For practical applications, Csound provides vps, which implements the same equations, but, as a built-in opcode, it is much more efficient.

8.4 Modified Frequency Modulation Synthesis

We can now develop a variation on the PM/FM synthesis, which has an expansion employing modified Bessel coefficients (Lazzarini 2010). This has a number of advantages over the original PM method, which we will outline below. If we start with the PM expression,

$$\cos(\omega + z\cos(\theta)) = \Re\{e^{j\omega + jz\cos(\theta)}\}, \tag{8.82}$$

and make the replacement $z = -jk$, which makes the modulation index purely imaginary, we have

$$\Re\{e^{j\omega + k\cos(\theta)}\} = e^{k\cos(\theta)}\cos(\omega). \tag{8.83}$$

As we have seen, the modulation factor $e^{k\cos(\theta)}$ can be expanded as a series of partials scaled by the modified Bessel coefficients $I_n(z) = j^{-n}J_n(jz)$. This gives us the double-sided spectrum

$$e^{k\cos(\theta)}\cos(\omega) = \sum_{n=-\infty}^{\infty} I_n(k)\cos(\omega + k\theta). \tag{8.84}$$

Since the synthesis formula is based on ring-modulation, the output is not going to be normalized as in the PM case. Therefore we need to complete this with a scaling factor. It is possible to use e^{-k} for this purpose, completing the Modified FM (ModFM) synthesis expression,

$$ae^{k\cos(\theta)-k}\cos(\omega) = \frac{a}{e^k}\sum_{n=-\infty}^{\infty} I_n(k)\cos(\omega + n\theta). \tag{8.85}$$

When we compare the plots of normalized modified Bessel coefficients (Fig. 8.12) against the standard Bessel coefficients (Fig. 8.7), we can see that the bipolar wobble that characterizes the latter is not present anymore. The coefficients are monotonically decreasing with k, as well as being unipolar. For a given k, their value falls as their order increases, $I_{n+1}(k) < I_n(k)$, which is another welcome characteristic. The spectrum is more predictable, especially also considering that negative and positive-order coefficients have the same value. This means that any folded negative-frequency partials will always sum constructively with any overlapping positive components.

8.4.1 Phase-synchronous ModFM

ModFM is particularly useful in the synthesis of resonance, since it produces a neat cosine-based spectrum with well-behaved dynamic characteristics. By making a small modification to the implementation, we can produce a method that is highly effective for producing formants and other types of resonance effects. Using a strictly integral carrier to modulator ratio f_c:f_m and driving the formula from a single phase signal source,

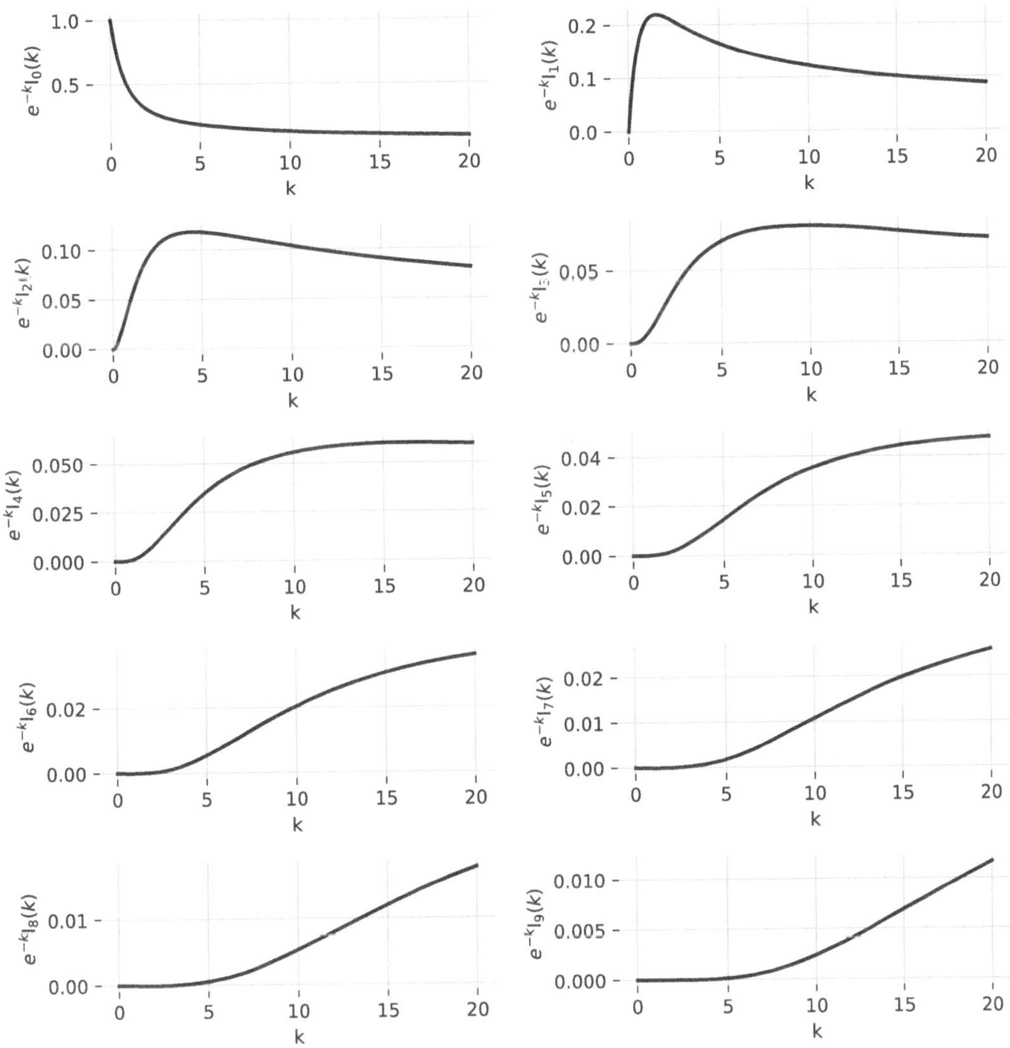

Figure 8.12 Normalized modified Bessel coefficients $e^{-k}J_n(k)$, of orders 0 to 9.

we can create a phase-synchronous version. The relationship between a resonance center frequency f_f, the carrier frequency, and the center frequency can be put as

$$f_c = nf_m = \left\lfloor \frac{f_f}{f_m} + \frac{1}{2} \right\rfloor f_m. \tag{8.86}$$

In this arrangement the spectrum is strictly harmonic, with a peak at f_f, which is harmonic of the fundamental frequency. We can keep the synchronous phase between the carrier and modulator, but generalize the synthesis expression to allow the peak resonance frequency to be set smoothly to any frequency. Additionally we can reintroduce the possibility of inharmonic spectra without destroying the carefully arranged phase-synchronous relationships. This requires the use of two carriers, and a shift frequency, $\omega_s = 2\pi f_s t$,

$$ae^{k\cos(\omega_0)-k}\left[(1-x)\cos(n\omega_0+\omega_s)+x\cos([n+1]\omega_0+\omega_s)\right], \tag{8.87}$$

which has the expansion

$$\frac{a}{e^k}\sum_{n=-\infty}^{\infty} I_m(k)\left[(1-x)\cos((n+m)\omega_0+\omega_s)+x\cos((n+m+1)\omega_0+\omega_s)\right]. \tag{8.88}$$

The composite carrier signal is the linear interpolation of the two carrier outputs, determined by the resonance center frequency. Therefore the value of x in these equations is the fractional part of f_f/f_m,

$$x = \frac{f_f}{f_m} - \left\lfloor \frac{f_f}{f_m} \right\rfloor. \tag{8.89}$$

The value of k can be linked to the bandwidth of the resonance region using a similar expression to Eq. 8.20, which was given for the case of the double-sided summation formula. The value of k is approximated as

$$k \approx \frac{2g}{(1-g)^2} \tag{8.90}$$

with g defined as

$$g = 2^{-f_0/(0.29B)}. \tag{8.91}$$

Phase synchronous ModFM can be implemented as a Csound opcode in the following form:

```
opcode PSModFM,a,kkkk
 kamp,kfo,kfc,kbw xin
 kg = 2^(-kfo/(.29*kbw))
 kk = 2*kg/(1-kg)^2
 kf = kfc/kfo
 kfin = int(kf)
 kx = kf - kfin
 aph = phasor(kfo)
 amod = tablei:a(aph,-1,1,0,1)
 aexp = exp(kk*amod-kk)
 acos1 = tablei:a(aph*kfin,-1,1,0.25,1)
 acos2 = tablei:a(aph*(kfin+1),-1,1,0.25,1)
 acos = (kx*acos2 + (1-kx)*acos1)*aexp
    xout(kamp*acos)
endop
```

where the input parameters are amplitude, fundamental frequency, center frequency, and bandwidth.

If we examine the phase-synchronous ModFM expression, we can infer that resonance can be synthesized as a heterodyning operation of two phase-synchronous signals. One of these, running at a lower frequency, forms the outer envelope of a waveform, and the other, at a multiple of that fundamental, is responsible for the resonance peak. The former is a multi-partial modulator, and the latter, a sinusoid. We have already noted this as a characteristic of distortion synthesis methods at the very start of the chapter. We demonstrate it again in Fig. 8.13, where a phase synchronous ModFM waveform is plotted alongside the modulation signal. We can interpret a single waveform cycle in this arrangement as a resonant filter impulse response. It also works the other way: we can consider a filter impulse response as a single cycle of a resonant waveform made by the product of a modulating signal and a sinusoid. For example, we have seen that the

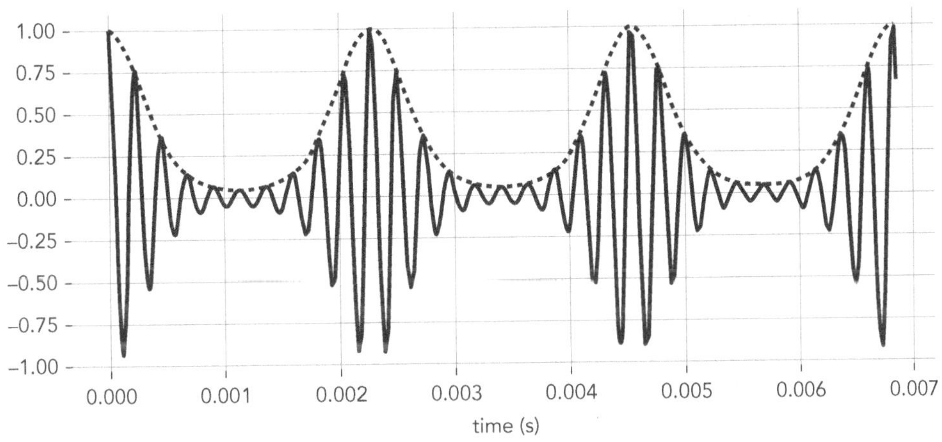

Figure 8.13 A phase-synchronous ModFM waveform, with $f_0 = 440$ Hz and formant frequency set to $10f_0$ (*solid line*), and the modulator waveform (*dots*) as its outer envelope.

resonator impulse response can be described as

$$s(n) = R^n \frac{\sin(\omega_c[n+1])}{\sin(\omega_c)}, \tag{8.92}$$

where, for $0 < R < 1$, we can interpret the exponentially decaying signal R^n as a modulator, and if we truncate and repeat this function, we get a resonant periodic waveform (Lazzarini and Timoney, 2013). We will return to the topic of synthesis of resonance in Chapter 11.

8.4.2 Extended ModFM

We may extend the original ModFM formula so that it can produce a variety of spectra with different configurations. We can add a second phase modulation signal with the same modulation frequency, producing a hybrid of the PM and ModFM formulae,

$$ae^{rk\cos(\theta)-rk}\cos(\omega + sk\sin(\theta)), \tag{8.93}$$

where two new parameters, s (in the range $[-1,1]$) and r ($[0,1]$) are introduced to control the spectral symmetry. The series expansion for this formula is now

$$\begin{aligned} &\frac{1}{e^{rk}} \sum_{n=-\infty}^{\infty} I_n(rk)\cos(n\theta) \sum_{m=-\infty}^{\infty} J_m(sk)\cos(\omega + m\theta) = \\ &\frac{2}{e^{rk}} \sum_{n=-\infty}^{\infty} \sum_{m=-\infty}^{\infty} I_n(rk)J_m(sk)\left[(\omega + [m+n]\theta) + (\omega - [m-n]\theta)\right]. \end{aligned} \tag{8.94}$$

While this appears to be a complicated expression, in the extreme ranges of the parameters r and s, it reduces to well-known expressions. For $r = 0$ and $s \neq 0$, we have the classic PM formula. In the case of $r = s = 1$, we have the curious situation where the Bessel coefficients vanish and the series becomes simply one-sided,

$$e^{k\cos(\theta)}\cos(\omega + k\sin(\theta)) = \sum_{n=0}^{\infty} \frac{k^n}{n!}\cos(\omega + n\theta), \tag{8.95}$$

where we excluded the normalization factor for simplicity. The remaining case $s = -r = -1$ produces a similar one-sided spectrum, but now featuring only the lower sidebands. Plots of four different combinations of r and s are shown in Fig. 8.14. By dynamically

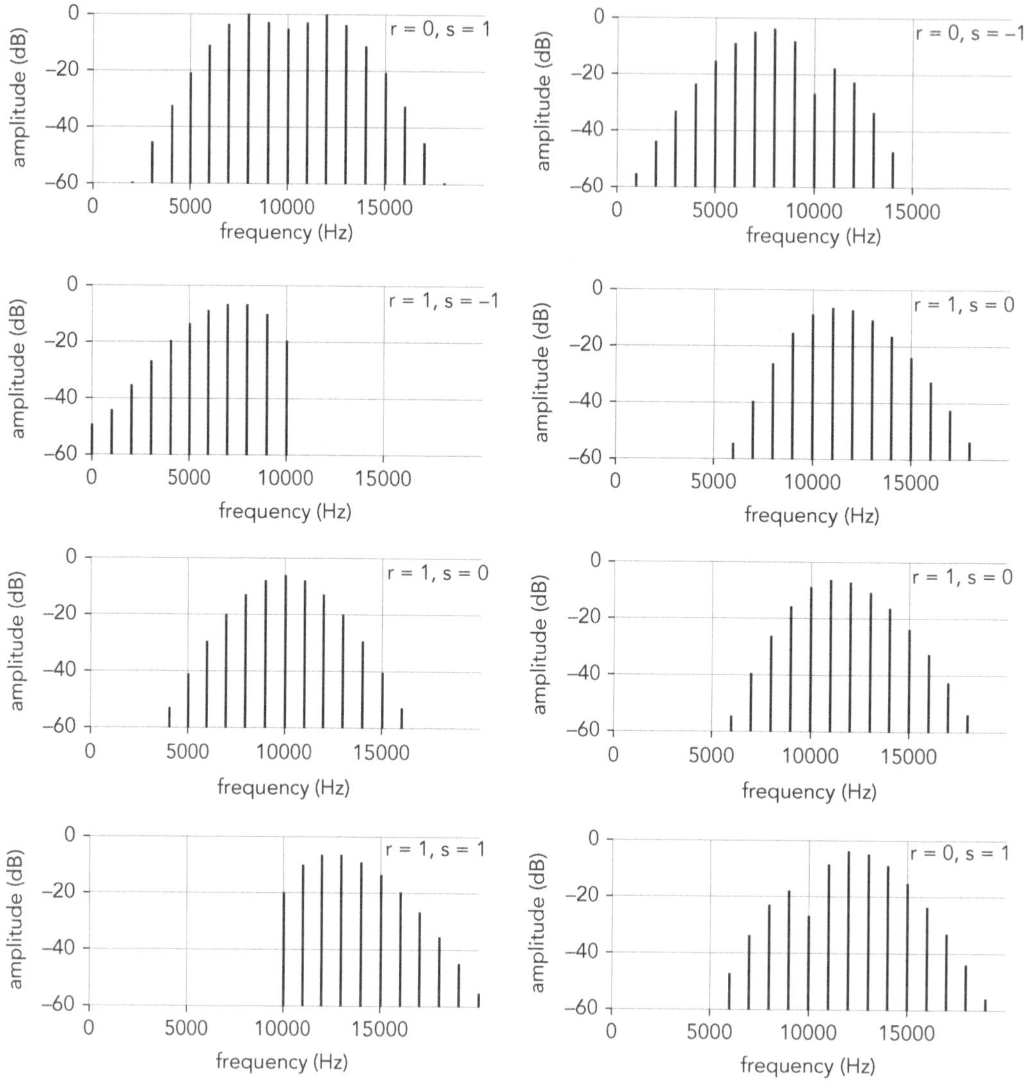

Figure 8.14 Spectra of extended ModFM, for various settings of r and s, with $k = 3$, $f_c = 10,000$ Hz and $f_m = 1000$ Hz.

varying the values of r and s we can produce sonically interesting changes in the symmetry of the spectrum. As in VPS, we can control these in a 2-D space, which allows various timbral transitions to be performed.

An opcode implementing extended ModFM in Csound is shown below. Note that, for $s = 0$, this also implements the original formula, and, as discussed earlier, for $r = 0$, we have PM. So this opcode can be regarded as a generic phase modulation/modified frequency modulation synthesizer.

```
opcode EModFM,a,kkkkkk
 kamp,kfc,kfm,kk,krr,ks xin
 aphw = phasor(kfc)
 apht = phasor(kfm)
 acos = tablei:a(apht,-1,1,0.25,1)
 asin = tablei:a(apht,-1,1,0,1)*kk*ks/(2*$M_PI)
 aexp = exp(krr*kk*(acos-1))
```

```
acar = tablei:a(aphw + asin,-1,1,0.25,1)
   xout(kamp*acar*aexp)
endop
```

8.5 Polynomial Waveshaping

It is possible to define non-linear waveshaping functions in various ways. As we have seen, we can use standard trigonometric functions such as cosines and sines, which give us the different flavors of phase modulation. We can also employ linear piecewise functions, with a different number of segments and inflection points, as we did in phase distortion. However, one of the problems brought about by the hard edges in such functions is that they make the spectrum non-bandlimited. This is not such a problem per se, as we can limit the amount of modulation to avoid aliasing, but it is compounded by the fact that it may not be easy to derive a closed-form solution in such cases. A better approach for a generic design of waveshapers is to employ polynomials to define these functions, as we did in the first example of this chapter (Fig. 8.1).

Let's start with the simplest non-linear case, the second-order polynomial given by

$$f(x) = x^2, \tag{8.96}$$

to which we feed a cosine wave, $f(\cos(\omega))$. This can be expanded using a well-known trigonometric identity,

$$f(\cos(\omega)) = \cos^2(\omega) = \frac{1+\cos(2\omega)}{2}. \tag{8.97}$$

We can see that the resulting signal has two components with the same amplitudes, at DC and twice the original frequency. Therefore if the polynomial was set to $f(x) = x + x^2$, we have a complex signal with three partials (0 Hz, f_0, $2f_0$). Let's increase the polynomial order to 3, $f(x) = x^3$. This produces the following waveform

$$f(\cos(\omega)) = \cos^3(\omega) = \frac{3\cos(\omega)+\cos(3\omega)}{4}, \tag{8.98}$$

a wave with two components, but now composed of the odd harmonics instead. Therefore, for $f(x) = x^3 + x^2$, we will have the first three partials of the harmonic series plus DC in the resulting output. Intuitively, we can derive the following two general rules:

- The highest partial produced by a polynomial of order n is nf_0.
- Even-power terms produce only even harmonics and odd-power terms produce only odd harmonics.

The first one of these can be easily proved. Consider the complex-valued sinusoid $e^{j\omega}$. It follows that $(e^{j\omega})^n$ is $e^{jn\omega}$ if n is an integer. Therefore we have the general expression,

$$(\cos(\omega) + j\sin(\omega))^n = \cos(n\omega) + j\sin(n\omega), \tag{8.99}$$

which is known as the *DeMoivre's law*. The second can be proved by finding a general principle guiding these expansions.

In order to find this rule, we can go a little further and consider higher powers of $\cos(\omega)$. With a fourth-order polynomial $f(x) = x^4$ we have the following series,

$$\cos^4(\omega) = \left(\frac{1+\cos(2\omega)}{2}\right)^2 = \frac{3+4\cos(2\omega)+\cos(4\omega)}{8}, \tag{8.100}$$

for which we used the relationship $(1+x)^2 = 1+2x+x^2$. This is an instance of the more general *binomial theorem*,

$$(1+x)^n = \sum_{k=0}^{n} \binom{n}{k} x^k, \tag{8.101}$$

where the binomial coefficient is defined as

$$\binom{n}{k} = \frac{k!}{(n-k)!k!} \tag{8.102}$$

From this, we can now derive a rule for the cosine powers of even numbers,

$$\cos^n(\omega) = \frac{\binom{n}{\frac{n}{2}} + 2\sum_{k=0}^{n/2-1}\binom{n}{k}\cos([n-2k]\omega)}{2^n} \tag{8.103}$$

If we comparing this equation and the series for $\cos^3(\omega)$, we can arrive at the following formula for odd powers,

$$\cos^n(\omega) = \frac{2}{2^n}\sum_{k=0}^{(n-1)/2}\binom{n}{k}\cos([n-2k]\omega). \tag{8.104}$$

These two general expressions for even and odd powers prove the second rule we intuitively proposed.[22] Incidentally, this second rule has another, more general consequence. Since polynomials containing only even powers produce even transfer functions, we can surmise that any even waveshaper will produce a spectrum with even harmonics when driven by a sinusoid signal. Conversely, odd transfer functions generate only odd components.[23] With these formulae now in hand, we can now determine the spectrum for any polynomial. For example, we can take the following polynomial,

$$f(x) = x^5 - 2x^4 + 3x^3 - x^2, \tag{8.105}$$

and apply a $\cos(\omega)$ as an input. The resulting series expansion is

$$\begin{aligned}&\frac{\cos(5\omega)+5\cos(3\omega)+10\cos(\omega)}{16} - \frac{2\cos(4\omega)+8\cos(2\omega)+6}{8} + \\ &\frac{3\cos(3\omega)+9\cos(\omega)}{4} - \frac{\cos(2\omega)-1}{2} = \\ &\frac{1}{16}[\cos(5\omega) - 4\cos(4\omega) + 17\cos(3\omega) - 24\cos(2\omega) + 46\cos(\omega) - 12].\end{aligned} \tag{8.106}$$

Another interesting aspect of these formula is that for a given power, the highest-frequency term is always set to unity (disregarding the signal scaling), which means that we can predict easily the intensity of the highest harmonic. This might be important when judging the aliasing conditions of a given polynomial.

8.5.1 Dynamic Spectra

We have been able to derive the spectrum for a steady-state sinusoid of unity amplitude. As we keep insisting, any interesting musical application requires the spectrum to change over time. Thankfully, this is something that is practically built into the wave-shaping method, as modifying the amplitude of the input will access different ranges of the distorting function, producing different spectra. Let's examine the case of a 5th-order polynomial function $f(x) = c_5x^5 + c_4x^4 + c_3x^3 + c_2x^2 + c_1x + c_0$, now driven by $z\cos(\omega)$,

$$c_5z^5\frac{\cos(5\omega)+5\cos(3\omega)+10\cos(\omega)}{16} + c_4z^4\frac{\cos(4\omega)+4\cos(2\omega)+3}{8}$$
$$+ c_3z^3\frac{\cos(3\omega)+3\cos(\omega)}{4} + c_2z^2\frac{\cos(2\omega)-1}{2} + c_1z\cos(\omega) + c_0$$
$$= h_5(z)\cos(5\omega) + h_4(z)\cos(4\omega) + h_3(z)\cos(3\omega) + h_2(z)\cos(2\omega) + h_1(z)\cos(\omega) + h_0(z). \quad (8.107)$$

This gives us the function $h_n(z)$ determining the amplitude of harmonic n,

$$\begin{aligned}
h_5(z) &= \frac{c_5z^5}{16} \\
h_4(z) &= \frac{c_4z^4}{8} \\
h_3(z) &= \frac{c_3z^3}{4} + \frac{5c_5z^5}{16} \\
h_2(z) &= \frac{c_2z^2}{2} + \frac{4c_4z^4}{8} \\
h_1(z) &= c_1z + \frac{3c_3z^3}{4} + \frac{10c_5z^5}{16} \\
h_0(z) &= c_0 + \frac{c_2z^2}{2} + \frac{3c_4z^2}{8},
\end{aligned} \quad (8.108)$$

from which we can see that the amplitude of each partial varies non-linearly with z. To design a distortion function, we normally determine a target steady-state spectrum for $z = 1$ and then vary the amplitude of the driving sinusoid to obtain a dynamic spectral evolution.

Using the cosine expansion formulae, we can derive a general expression for the $h_n(z)$ functions, based on the order N of the polynomial employed,

$$h_n(z) = \varepsilon_n \begin{cases} \sum_{k=(n-1)/2}^{(N-1)/2} \binom{2k+1}{k-(n-1)/2} \frac{c_{2k+1}z^{2k+1}}{2^{2k+1}}, & n \text{ odd} \\ \\ \sum_{k=n/2}^{N/2} \binom{2k}{k-n/2} \frac{c_{2k}z^{2k}}{2^{2k}}, & n \text{ even,} \end{cases} \quad (8.109)$$

where ε_n is defined as before, as 1 for $n = 0$, and 2 for $n \neq 0$. This can also be used to define the harmonic weights for a steady-state spectrum by setting $z = 1$.

The example in Fig. 8.15 shows the harmonic amplitudes for an order-5 polynomial designed to have a steady-state spectrum with a decaying slope between −6 and −12 dB/octave, with DC suppressed. This particular polynomial manages to achieve a well-behaved spectral evolution, with all partials, except for DC, rising without any particular fluctuations. It is not guaranteed, however, that this is going to be the case for every polynomial.

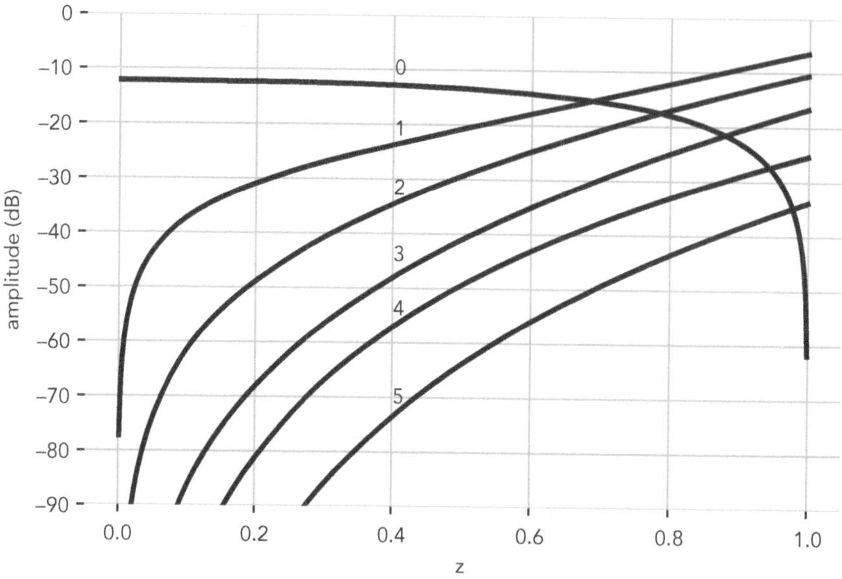

Figure 8.15 Normalized amplitudes for harmonics 0 to 5 produced by the polynomial $f(x) = 36x^5 + 48x^4 + 20x^3 + 18x^2 + 15x - 27$ driven by a sinusoid $z\cos(\omega)$.

8.5.2 Normalization

In order to normalize the output of the waveshaper, we can consider the maximum absolute value of $f(1)$ and $f(-1)$, then use the reciprocal of this value as a scaling factor,

$$a = \begin{cases} |1/f(1)|, & |f(1)| \geq |f(-1)| \\ |1/f(-1)|, & |f(1)| < |f(-1)|, \end{cases} \tag{8.110}$$

for $af(\cos(z\omega))$. Varying the cosine input amplitude will have two effects, on the overall amplitude and on the spectral weights of each harmonic. This is often a desirable feature, since in many instrumental sounds changes in brightness and the presence of harmonics are coupled with the amplitude envelope.

However, it is also possible to separate the amplitude and the timbral envelope by using a scaling function $a(z)$ that is dependent on z. This can be constructed by examining the polynomial and creating a curve that is inversely proportional to the maximum absolute values found at the various values of x in the interval employed. In this case, the output amplitude will be kept steady while z changes.

8.5.3 Implementation

A Csound opcode implementing polynomial waveshaping can be constructed quite simply by laying out the chosen polynomial expression and feeding it a cosine signal,

```
opcode F,a,ai[]
 ax,ia[] xin
 kcnt = 1
 ay = 0
 while kcnt < lenarray(ia) do
  ay += ia[kcnt]*ax^kcnt
  kcnt += 1
```

```
 od
 xout(ay+ia[0])
endop

opcode F,i,ii[]
 ix,ia[] xin
 icnt = 1
 iy = 0
 while icnt < lenarray(ia) do
  iy += ia[icnt]*ix^icnt
  icnt += 1
 od
 xout(iy+ia[0])
endop

instr 1
   icf[] fillarray -27,15,18,20,48,36
   ia = p4
   ifo = p5
   if1 = abs(F(1,icf))
   ifm1 = abs(F(-1,icf))
   iscl = if1 > ifm1 ? if1 : ifm1
   kz linen 1,0.1,p3,0.1
   asig = oscili(kz,ifo,-1,0.25)
   asig1 = F(asig,icf)*ia/iscl
   outs(asig1)
endin
```


The F opcode takes the coefficients as an i-time array, then computes the polynomial result using a control-rate loop. All processing is done on a ksmps-block basis, so despite all the arithmetic involved, it is not too complex. Note that we employ two versions of F to be run at perf- and init-time. The first one is used in the synthesis part, and the other to determine a scaling factor. This programming approach is called opcode *overloading*, where the same name is given to functions with different parameter types. Csound then picks the most appropriate when the code is executed.

We can complement this opcode with a Python function that can be employed to design the polynomial. This implements Eq. 8.109 and determines the amplitude of each harmonic n given a vector a with polynomial coefficients, a cosine amplitude z, scaled to a maximum amplitude[24] x,

```
from scipy import special as sp
# Polynomial computation
def f(a,x):
   s = 0.
   for i in range(0,a.size):
     s += x**(i)*a[i]
```

```
   return s

# Computes the harmonic n
# weight for polynomial
# coeffs c and cosine
# amplitude z
# scaled for amplitude x
def h(c,n,z,x=1):
 N = c.size - 1
 if n > N: return 0
 if n > 0: e = 2
 else: e = 1
 s = 0.
 test = n/2 - n//2
 if test != 0:
  for k in range((n-1)//2, (N-1)//2+1):
   i = 2*k + 1
   s += sp.binom(i, k - (n - 1)/2)*c[i]*(z**i)/2**i
 else:
  s = 0.
  for k in range(n//2, N//2+1):
   i = 2*k
   s += sp.binom(i, k - (n)/2)*c[i]*(z**i)/2**i
 f1 = abs(f(c,x))
 fm1 = abs(f(c,-x))
 if f1 > fm1: scl = f1
 else: scl = fm1
 return e*s/scl
```


This function can be used to determine both the steady-state amplitudes, and their evolution, if z is given as a NumPy array. A Python waveshaper design program can be written using this code; the resulting coefficients can then be plugged back into a Csound array for synthesis.

We can improve the efficiency of the processing code by using table lookup, as we did in the case of the double-sided summation formula. Csound has a handy GEN routine (GEN 3) that computes a polynomial in a given interval, taking its coefficients as its input. This is complemented by GEN 4, which computes a normalizing function for varying amplitude values of the driving sinusoid. With these, we can write an alternative waveshaping opcode, which takes a polynomial as a function table, and uses table lookup to compute the output values.

```
opcode W,a,aki
 as,kz,ifn xin
 ifs = ftgen(0,0,ftlen(ifn)/2,4,ifn,1)
 kscal = table:k(kz,ifs,1)
 asig = table:a(as/2,ifn,1,0.5)
```

```
 xout(asig*kscal)
endop

instr 1
   ia = p4
   ifo = p5
   ifn = ftgen(0,0,1024,3,-1,1,-27,15,18,20,48,36)
   kz linen 1,1,p3,0.1
   asig = oscili(kz,ifo,-1,0.25)
   asig1 = W(asig,kz,ifn)*ia
   outs(asig1)
endin
```

Besides the efficiency improvement of using a table lookup method, another major difference is that normalization is now a function of z, and so there is a decoupling of the amplitude and timbre envelopes. In this implementation, the parameter kz is mostly affecting the distribution of the energy in the spectrum, and we can apply a separate signal to shape its amplitude.

8.5.4 Chebyshev Polynomials

While the Python code we developed is a handy tool to help the design of waveshaping functions, it would be better if there was a way to go directly from a steady-state spectrum to coefficients. Again, classic mathematical physics comes to our rescue in the form of Chebyshev polynomials.[25] We have already seen these make an appearance in the formula for the impulse response of feedback resonators. They also provide us with a very good method to define a waveshaping function on the basis of a set of harmonic weights.

A Chebyshev polynomial of the first kind of order n, $T_n(x)$, has the useful characteristic that when it is fed with a cosine wave, it produces a single harmonic n times the original frequency (Hochstrasser, 1972),

$$T_n(\cos(\omega)) = \cos(n\omega). \tag{8.111}$$

We can assemble a waveshaping function as a sum of weighted $T_n(x)$ to define a steady-state spectrum of our choice,

$$f(x) = \sum_{n=0}^{N} a_n T_n(x), \tag{8.112}$$

where a_n are the harmonic amplitudes of a target spectrum. As before, it is possible to construct a perfectly bandlimited spectrum.

Chebyshev polynomials can be generated recursively. Starting with the first two,

$$T_0(x) = 1 \qquad \text{and} \qquad T_1(x) = x, \tag{8.113}$$

we have the next ones defined as

$$T_{n+1}(x) = 2xT_n(x) - T_{n-1}(x). \tag{8.114}$$

Therefore we have $T_2(x) = 2x^2 - 1$, $T_3(x) = 4x^3 - 3x$, etc. With this method, we can construct a series of polynomials to compute Eq. 8.113.

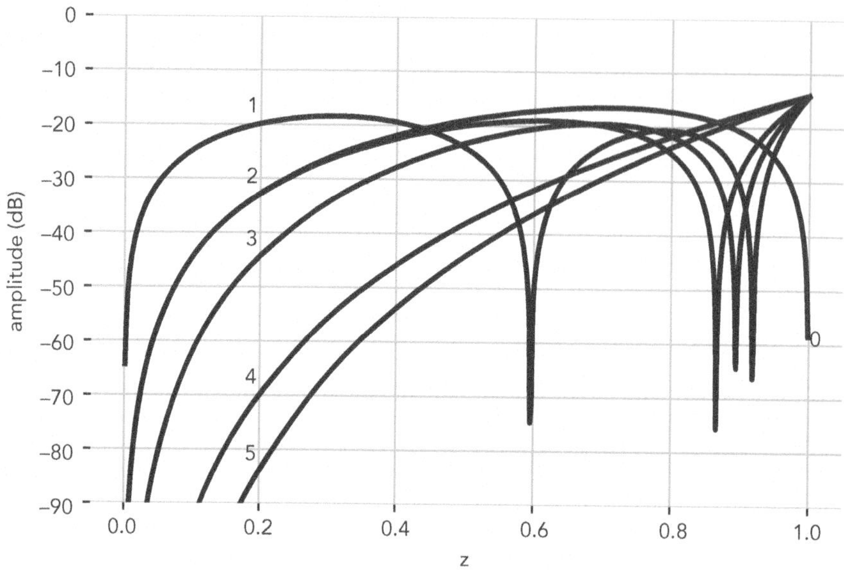

Figure 8.16 Normalized amplitudes for harmonics 0 to 5 of a pulse wave produced by the polynomial $f(x) = 3x - 6x^2 - 16x^3 + 8x^4 + 16x^5$ driven by a sinusoid $z\cos(\omega)$.

As we did before, by varying the amplitude of the driving sinusoid, we can get dynamic spectra. However, the evolution of the amplitudes of each component may not follow a smooth trajectory, due to the interactions of the different terms of the polynomial. While the steady-state amplitudes are guaranteed, the harmonics my evolve in unpredictable ways. For example, consider the spectrum of a pulse wave with five partials (and zero DC). Combining the first five Chebyshev polynomials with equal weight, we arrive at the following function,

$$f(x) = 3x - 6x^2 - 16x^3 + 8x^4 + 16x^5. \tag{8.115}$$

Plotting the evolution of the five harmonics (Fig. 8.16), we can see that indeed, for $z = 1$, we have a bandlimited pulse spectrum. However, we can also see that the evolution of the lowest harmonics is not smooth. The two highest harmonics have a much better trajectory, since they are only affected by a single polynomial term. Therefore for a more complex polynomial, we should expect a certain unsmooth transition from $z = 0$ to $z = 1$. We have a similar situation to PM synthesis, where a wobble is observed in the spectrum. Again, if the intention is to emulate some natural/acoustic timbral evolution, this is not going to be very successful.

However, it is possible to mitigate these effects by playing with the harmonic phases. In this case, we can only flip them (by π), since the resulting spectrum is always in the cosine phase. The pattern $+, +, -, -$ when applied to a Chebyshev-polynomial generated spectrum can mitigate an otherwise wobbly timbral evolution. The result of setting this pattern to the steady-state partials results in a smoother dynamic spectrum, as shown in Fig. 8.17. As we should have expected, we have a different polynomial function now, defined by

$$f(x) = 2 + 9x - 10x^2 - 24x^3 + 8x^4 + 16x^5, \tag{8.116}$$

where we note that the two highest-power terms are unchanged.

To implement this, we can use the generic waveshaper code developed earlier. Csound provides GEN 13 to construct a polynomial table from user-provide partials. This takes

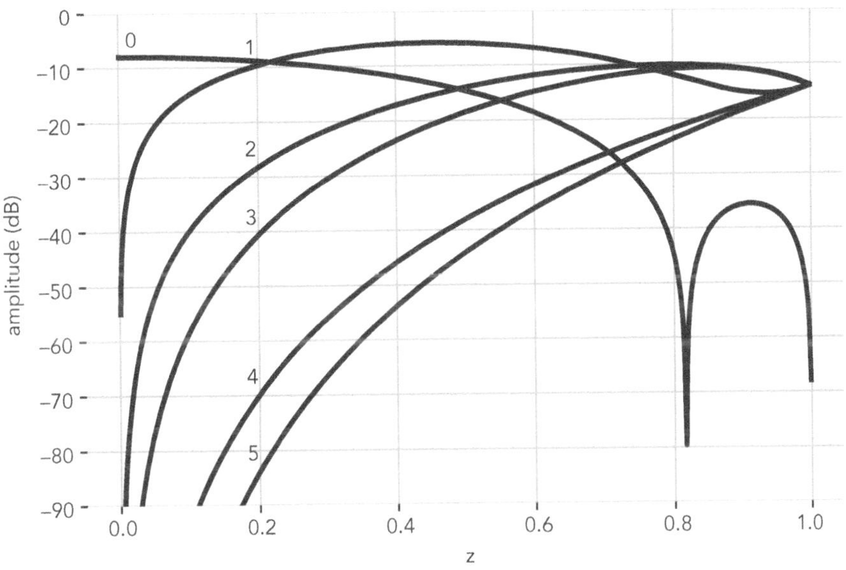

Figure 8.17 Normalized amplitudes for harmonics 0 to 5 of a pulse wave produced by the polynomial $f(x) = 2 + 9x - 10x^2 - 24x^3 + 8x^4 + 16x^5$ driven by a sinusoid $z\cos(\omega)$.

the maximum absolute value m of x and evaluates the polynomial in the interval $[-m, m]$, as well as the amplitude of driving cosine for which the spectrum is being matched. The following code provides an example:

```
instr 1
   ia = p4
   ifo = p5
   ifn = ftgen(0,0,1024,13,1,1,0,1,0,-1/3,0,1/5,0,-1/7,0,1/9)
   kz linen 1,1,p3,0.1
   asig = oscili(kz,ifo,-1,0.25)
   asig1 = W(asig,kz,ifn)*ia
   outs(asig1,asig1)
endin
```

where the spectrum of a square wave with harmonics up to $9f_0$ is created via waveshaping using the `W` opcode and GEN 13. As before, the z parameter controls the timbre more or less independently from the overall signal amplitude.

8.5.5 Quadrature Waveshaping

The Chebyshev polynomials of the first kind provide us with a very useful method to generate a given spectrum. As we noted, however, while we can match an amplitude spectrum, we are limited to cosine-phase partials; therefore, it is not possible to produce an arbitrary phase spectrum. It would be useful if we could produce signals in quadrature, as we did before, by applying a polynomial that generates sine-phase components from the same driving cosine. Thankfully, we do not need to look very far. In Chapter 7, we saw that the Chebyshev polynomials of the second kind, $U_n(x)$, can produce partials in sine phase when driven by a cosine wave. Their closed form is[26]

$$U_{n-1}(\cos(\omega))\sin(\omega) = \sin(n\omega). \tag{8.117}$$

As before, we can combine these to give us a waveshaping function that will produce a certain sine-based spectrum,

$$g(x) = \sum_{n=1}^{N} b_n U_{n-1}(x). \tag{8.118}$$

The Chebyshev polynomials of the second kind also obey a recursive relationship, which allows us to find the higher-order functions. Starting from the bottom cases,

$$U_0(x) = 1 \qquad \text{and} \qquad U_1(x) = 2x, \tag{8.119}$$

we have the same expression to generate the following ones,

$$U_{n+1}(x) = 2xU_n(x) - U_{n-1}(x). \tag{8.120}$$

The significance of these results is that we can now use a combination of waveshapers to synthesize a harmonic spectrum with an arbitrary phase response. Using $f(x)$ as defined in Eq. 8.112, we have

$$s(t) = f(\cos(\omega)) + \sin(\omega)g(\cos(\omega)), \tag{8.121}$$

and the weights a_n and b_n now define the phase of each partial as well as their amplitude. As we have seen before, these are the Fourier coefficients of a given harmonic series.

The spectrum of any signal produced with the waveshaping process we have seen so far in this section is harmonic. Since we are starting with a periodic input, there is no possibility of achieving anything else. However, as we have seen countless times before, we can include a shift sinusoidal factor to recenter the spectrum at a frequency other than 0 Hz. This is of course the heterodyne method of ring modulation. We then have a similar case to PM, where we can determine if the spectrum is harmonic or inharmonic depending on the ratio between the modulator (shift) and the carrier (waveshaper) frequency. Generally, the same principles developed for that case apply here.

Note also that if we set $a_n = b_n$, we have an analytic signal,

$$\hat{s}(t) = f(\cos(\omega)) + j\sin(\omega)g(\cos(\omega)), \tag{8.122}$$

which allows us to apply single sideband modulation instead of ring modulation, producing one-sided spectra. Stretched and compressed spectra can then be created, opening up many possibilities for spectral design. Combining the quadrature waveshaper signal with more than one ssb modulators of different frequencies will allow us to create interesting dynamic spectra with various harmonic and inharmonic characteristics.

The following opcode generates the quadrature waveshaping signals when given two tables generated using sums of Chebyshev polynomials of the first and second kind:

```
opcode QuadShape,aaaa,kkkii
 kf1,kf2,kz,ifc,ifs xin
 aph = phasor(kf1)
 aphs = phasor(kf2)
 ac = table:a(aph,-1,1,0.25,1)
 as = table:a(aph,-1,1,0,1)
 acosw = table:a(kz*ac/2,ifc,1,0.5)
 asinw = as*table:a(kz*ac/2,ifs,1,0.5)
```

```
 acosw = balance(acosw,ac/2)
 asinw = balance(asinw,ac/2)
 are = table:a(aphs,-1,1,0.25,1)
 aim = table:a(aphs,-1,1,0,1)
 aup = are*acosw - aim*asinw
 alo = are*acosw + aim*asinw
 xout(acosw,asinw,aup,alo)
endop
```

The output consists of four signals, the waveshaper cosine and sine outputs, and the upper and lower sidebands of a SSB modulation by a sinusoid tuned to `kf2`. The outputs of this opcode can be mixed in various ways to produce different types of spectra. The function table generator GEN14 can be used to produce a polynomial using weighted sums of $U_n(x)$. Note that we are not employing a scaling function, instead we are using an envelope follower. The reason for this is that scaling the waveshaper signal introduces a gain factor in the sine component, reducing its amplitude. In this case, we are not implementing Eqs. 8.121 and 8.122, since $f(\cos(\omega))$ and $\sin(\omega)g(\cos(\omega))$ are not scaled by the same factor. Additionally, GEN 13 and 14 cannot be normalized either. Since this is done by default, we need to switch it off by passing negative GEN values to `ftgen()`, as shown in the code excerpt


```
ifc = ftgen(0,0,1024,-13,1,1,0,1,0.5,0.25)
ifs = ftgen(0,0,1024,-14,1,1,0,1,0.5,0.25)
ac,as,aup,alo QuadShape kfr,kfsh,kz,ifc,ifs
```

where the outputs `ac` and `as` are in quadrature when `kz=1`.

8.6 Other Distortion Functions

Beyond polynomial functions, it is of course possible to use any other non-linear curve as a waveshaper. If the transfer function has a known polynomial expansion, called a Taylor's series, then we can use the tools developed for polynomial waveshaping to analyze the resulting spectrum. For example, the absolute value function $|x|$, which we saw earlier in Chapter 3, has a known series in the form of

$$|x| = \frac{2}{\pi} + \frac{4}{\pi}\sum_{n=1}^{\infty}(-1)^{n-1}\frac{T_{2n}(x)}{4n^2-1} \tag{8.123}$$

for x in the interval $[-1,1]$. With this we can then derive the output of $|\cos(\omega)|$ as

$$\frac{2}{\pi} + \frac{4}{\pi}\sum_{n=1}^{\infty}(-1)^{n-1}\frac{\cos(2n\omega)}{4n^2-1} \tag{8.124}$$

This is of course another example of a non-bandlimited spectrum, which is prone to aliasing. Another related function is the *signum*, defined as

$$\mathrm{sgn}(x) = \begin{cases} -1, & x < 0 \\ 0, & x = 0 \\ 1, & x > 0, \end{cases} \tag{8.125}$$

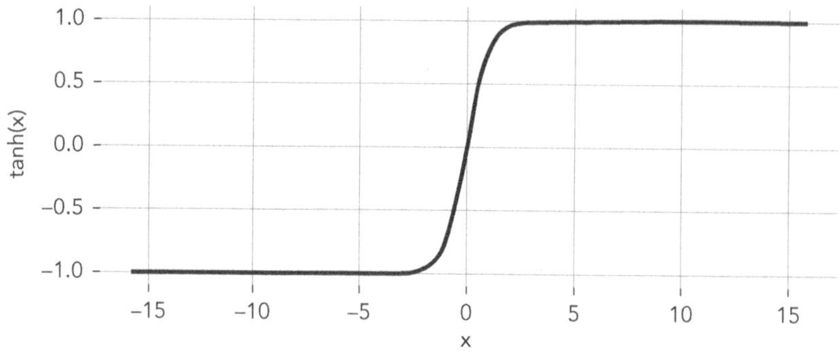

Figure 8.18 The $\tanh(x)$ function in the interval $-5\pi \leq x \leq 5\pi$.

which can also be defined (for non-zero x) as $|x|/x$. If we apply a sinusoid to this function, we have a non-bandlimited square wave output. It is of course a extreme case of *hard clipping* distortion. If we smooth the transition around 0, we can develop more useful functions, which will produce various degrees of soft clipping.

This is where we come to the tanh() function (Lazzarini and Timoney, 2010). This is a sigmoid-type function, that is, it displays an s shape as shown in Fig. 8.18. As can be seen, it smooths out the signum shape, providing a gradual transition around $x = 0$. For low input amplitudes, this transfer function can provide a nearly linear amplification with low levels of distortion. The tanh() function also offers a good method for the synthesis of a square wave via waveshaping, even though it is also theoretically non-bandlimited. If used with care, the higher harmonics are fairly low in intensity. Its Taylor's series, for $x = [-\pi/2, \pi/2]$, is defined as (Zucker, 1972)

$$\begin{aligned}\tanh(x) &= \sum_{n=1}^{\infty} \frac{2^{2n}(2^{2n}-1)B_{2n}x^{2n-1}}{(2n)!} \\ &= x - \frac{x^3 - 0.4x^5}{3} - \frac{17x^7}{315} + \ldots\end{aligned} \tag{8.126}$$

where B_i is the Bernoulli[27] number i,

$$B_{2n} = (-1)^{n+1}\frac{2(2n)!}{(2\pi)^{2n}}\sum_{m=0}^{\infty}\frac{1}{m^{2n}} \tag{8.127}$$

If we drive a tanh() function with a sine wave with amplitude $\pi/2$, we get the following spectrum:

$$\begin{aligned}\tanh\left(\frac{\pi}{2}\sin(\omega)\right) &= \sum_{n=1}^{\infty}\frac{2^{2n}(2^{2n}-1)B_{2n}(\pi/2)^{2n-1}\sin^{2n-1}(\omega)}{(2n)!} \\ &= \sum_{n=1}^{\infty}\frac{2^{2n}(2^{2n}-1)B_{2n}(\pi/2)^{2n-1}}{(2n)!}\left[\frac{2}{2^{2n-1}}\sum_{k=0}^{n-1}(-1)^{n-k-1}\binom{2n-1}{k}\sin([2n-2k-1]\omega)\right] \\ &= \sum_{n=1}^{\infty}\sum_{k=0}^{n-1}(-1)^{n-k-1}\frac{2B_{2n}(2^{2n}-1)(\pi/2)^{2n-1}}{k!(2n-k-1)!n}\sin([2n-2k-1]\omega)\end{aligned} \tag{8.128}$$

While this appears overly complicated, there are two important things that can be observed from this equation and the effect of tanh() as a waveshaper:

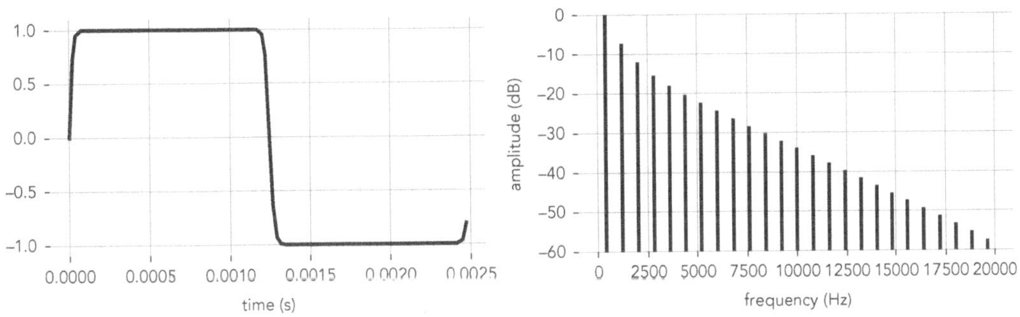

Figure 8.19 Waveform and normalized spectrum for $\tanh(5\pi \sin(\omega))$, with $f_0 = 400$ Hz and $f_s = 44,100$ Hz.

- Only odd harmonics are present (which we already knew from its Taylor's series and the fact that tanh() is an odd function).
- The scaling factors for the ascending harmonic series get small very quickly. We can see this by considering the terms for $k = 0$, and noting that the denominator grows quite steeply with n.

Therefore, for all practical purposes, with a sinusoidal driver of up to $\pi/2$ amplitude, the output spectrum of tanh() waveshaping is fairly bandlimited, or perhaps at least with good aliasing suppression (all of this is of course dependent on f_0 and and f_s). In fact, this shows also that we can increase the amplitude somewhat beyond the $\pi/2$ point, approximating a smooth square wave. While we cannot fully rely on Eq. 8.128, since it does not apply anymore, it nevertheless gives us a hint toward the kind of spectrum we will get. For example, a 400 Hz sine wave with 5π amplitude produces a full spectrum with suppressed aliasing at $f_s = 44,100$ Hz (Fig. 8.19). As we increase the frequency, we should reduce the amount of modulation. For $\tanh(z \sin(2\pi f_0 t))$, the following relationship applies (Lazzarini and Timoney, 2010, 34):

$$z_{max} = \frac{12000}{f_0 \log_{10} f_0} \tag{8.129}$$

where z_{max} is the optimal value to better approximate a square wave with good aliasing suppression at $f_s = 44,100$ Hz. This can be modified according to changes in sampling rate.

Implementing the tanh() waveshaper can be done using a direct application of the function or through the use of table lookup, as we have shown before. Csound provides the tanh() function as part of its opcode collection as well as a named GEN (`"tanh"`) to create a function table within a user-defined interval.

8.7 Adaptive Modulation Methods

The non-linear spectral synthesis techniques we have seen until now were based on distorting sinusoids. It is possible of course to drive a waveshaper with an arbitrary signal, producing a much more rich spectrum if the input is itself a complex waveform. However, in digital implementations we should be mindful of keeping aliasing distortion at a minimum. Even with fairly bandlimited waveshapers, high harmonics of an arbitrary input may cause objectionable aliasing distortion. The principles developed earlier in relation to pure sinusoids can provide a handy guide for determining the limits of waveshaping and modulation for complex input signals.

Additionally, it is possible to develop some of the modulation techniques seen earlier in a different form so that they can be applied to arbitrary inputs, in, for instance, an adaptive processing scenario. The term *adaptive* here means that we may be able to extract parameters from the input signal to control the synthesis/processing operation. For example, we can very easily use an input from an acoustic instrumental source as a modulator in PM synthesis. If we extract the f_0 from this input, we can employ different f_c:f_m ratios to produce various types of spectra. The difficulty, as we have seen when discussing complex phase modulation, is that only a few components may already saturate the spectrum with partials, so deriving a modulation from an input may require a significant amount of filtering.

8.7.1 Adaptive Frequency Modulation

Another possibility is to use an arbitrary input as a carrier, for which an alternative implementation of PM is needed, since we are not synthesizing the carrier signal from scratch. What we require is a method to modulate the phase of an input signal. In Chapter 7, we saw a special type of filter that passes all frequencies with the same amplitude, but modifies the phase of its input, called an *allpass* filter. We can try to apply these for PM synthesis. Since the amounts of modulation employed here might be significant, the best choice of allpass is the pure delay provided by a higher-order FIR filter. As we know, this is implemented by a delay line; to modulate the phase, we can employ a time-varying delay as we did, for instance, to develop the flanger effect. This is the basis of a modulation method called *Adaptive FM* (Lazzarini et al., 2008).

Let's develop this idea, considering a sinusoidal signal $x(t) = \sin(\omega t)$ going into a delay line, and an arbitrary delay time modulation source, such as

$$m(t) = \delta_{max} D(t), \tag{8.130}$$

where δ_{max} is the maximum delay and $D(t)$ is a unipolar signal with amplitude a, $0 \leq a \leq 1$. The modulated output can be described as

$$y(t) = \sin(\omega[t - \delta_{max} D(t)]). \tag{8.131}$$

We can determine the instantaneous frequency $\omega_i(t)$ of $y(t)$ as the derivative of the sine function phase:

$$\omega_i(t) = \omega - \frac{d}{dt} D(t) \delta_{max} \omega, \tag{8.132}$$

from which we obtain the instantaneous frequency $f_i(t)$ of the output signal in Hz as

$$f_i(t) = f_c - \frac{d}{dt} D(t) \delta_{max} f_c. \tag{8.133}$$

If we employ a modulator based on a unipolar cosine wave with peak amplitude 1, $D(t) = 0.5 + 0.5\cos(\omega_m t)$), then $f_i(t)$ can be defined as

$$f_i(t) = f_c + \delta_{max} \pi f_c f_m \sin(2\pi f_m t) \tag{8.134}$$

and from this we have the maximum frequency deviation $\Delta_f = \delta_{max}\pi f_m f_c$. From here onward, we can develop this as a form of PM, using the modulation index z defined as

$$z = \frac{\Delta_f}{f_m} \tag{8.135}$$

In practical implementations, we will need to define a minimum delay time to δ_{min}, therefore determining the delay modulation width as $\Delta_d = \delta_{max} - \delta_{min}$. From this we can finally define delay modulation in terms of z and f_c

$$\Delta_d = \frac{z}{\pi f_c} \tag{8.136}$$

We can now apply the tools developed earlier to determine the output spectrum of delay time modulation. Furthermore, by using a variable delay line to implement this, we can use any arbitrary input as a carrier. Therefore we now have the following expression for a modulated delay line input with N partials of amplitudes a_n, frequencies ω_n and phases ϕ_n:

$$\begin{aligned} x(t) &= \sum_{n=0}^{N-1} a_n \cos(\omega_n + z_n \sin(\theta) + \phi_n) \\ &= \sum_{n=0}^{N-1} \sum_{k=-\infty}^{\infty} a_n J_k(z_n) \cos(\omega_n + k\theta + \phi_n). \end{aligned} \tag{8.137}$$

Since z_n is equal to $\Delta_d \pi f_n$, we have the known result from complex carrier PM, where $z_n = z f_n / f_c$, and higher partials are modulated more intensely relative to lower ones.

As we noted earlier, we can extract the fundamental frequency from the arbitrary input (provided it is reasonably periodic), which allows us to set a f_c:f_m ratio and determine the amount of modulation given a modulation index z. A Csound opcode implementation can be provided as

```
opcode AdFM,a,akk
 as,krt,kz  xin
 kcps,kamp ptrack as,512
 km = kz/(2*$M_PI*kcps)
 adt = km + oscili(km,kcps/krt,-1,0.25) + 1/kr
 adp delayr 1
 adel = deltap3(adt)
 delayw(as)
 xout(adel)
endop
```

where the parameters are the f_c:f_m ratio (`krt`) and the index of modulation (`kz`). The code employs a cubic interpolation variable delay line to ensure that the modulated output is reasonably clean of artifacts. A similar approach can be developed for other types of PM synthesis, including, for instance, the ModFM and split-sideband techniques.[28]

8.8 The Non-Linear Spectrum

We began this chapter looking at some ideas that are closely connected to the Fourier theorem, in the form of the Dirichlet kernel, and the Lagrange's identities. Very soon we were able to cast these and their extensions in terms of non-linear distortion expressions. From there onward, we went on to look at a number of new ways of synthesizing dynamic spectra, which were fairly different from the additive and inverse-transform approaches we had been employing before. The majority of these were very economical in terms of computing resources, combining a few simple generators in various ways. Of course, it is always theoretically possible to reproduce the spectra described by the various series expansions shown here. However, even if this were possible in practice, it would be of no benefit to do so, as the non-linear distortion methods provide a much simpler and more elegant model.

The theory of Bessel coefficients (and functions) played an important part in much of the mathematics discussed here. Equally important was the theory of Chebyshev polynomials, used in the design of transfer functions. This not only plays an important part in distortion synthesis, but it also forms a bridge back to the Fourier theorem. Many connections can be established between the two, the Dirichlet kernel being one of them.

We can now add to our developing set of ideas about the spectrum by saying that

> Non-linear distortion provides a range of methods to design different types of spectra composed of harmonic or inharmonic components which are an alternative to the additive approaches stemming out of the Fourier theorem. They are also not directly accessible through that theory, although a number of connections with it can be established. In that sense, they form a complementary body of ideas about the spectrum, which extends the toolset of spectral music design.

We have now built a substantial collection of methods, techniques, and principles about the spectrum. In doing so, we have been mostly dealing with what is described as discrete spectra, by developing synthesis methods that generate or process distinct partials, particularly in this chapter. It is time we move on to consider in special detail the types of spectra where individual components may not be discernible. This will complement all of the techniques seen so far and provide an even broader idea of what the spectrum of musical sounds may comprehend. We are heading in that direction in the next chapter.

Notes

1 Note that, for simplicity, we will use a notation that employs the phase $\omega(t)$ (etc.) denoting $2\pi ft$, whenever this is possible and univocal. This of course is a function of time and the complete notation should be $\omega(t)$, but when this is implicit, we can make the notation simpler by omitting the time parameter.

2 This function is actually fundamental in the Fourier theory, since it can be used to prove the validity of a Fourier series. If a function $f(x)$ can be decomposed as a Fourier series, then

$$f(x) = \lim_{x \to \infty} \frac{1}{2\pi} \int_{-\pi}^{\pi} \frac{\sin\left([x-t]\left[n+\frac{1}{2}\right]\right)}{\sin\left(\frac{x-t}{2}\right)} f(t)dt$$

also holds (Watson, 1944, 581).

3 Like this one, many of the formulae we will encounter in this chapter are based on classic mathematical physics expressions.

4 Adding a small value to the denominator produces objectionable artifacts, so it is not recommended here.

5 These have been known for a long time as closed-form trigonometric formulae. In Jolley (1961), we find an extensive list of these; for example, the single-sided sine sum is the trigonometric series summation in Eq. 455 (pp.84–85), and a cosine version is shown as Eq. 454. There are also many other expressions in that volume that may be useful in sound synthesis. The idea of using them for synthesis was brought into the computer music literature by Andy Moorer (1977, 1976).

6 This and the other formulae may be derived using a similar method to the bandlimited pulse, through the use of complex exponentials.

7 Note that for $a = 1$ and $\theta = \omega$, this formula reduces to the bandlimited pulse equation.

8 The subsequent formula can be easily implemented by modifying this code.

9 This is how the modulator formula can be derived from the waveshaping expression,

$$\frac{1+a}{1-a}f\left(\frac{2\sqrt{(a)}}{1-a}\sin(\omega/2)\right) = \left(\frac{1+a}{1-a}\right)\frac{1}{1+\frac{2a[1-\cos(\omega)]}{1-2a+a^2}}$$

$$= \left(\frac{1+a}{1-a}\right)\frac{(1-a)^2}{1-2a\cos(\omega)+a^2} = \frac{1-a^2}{1-2a\cos(\omega)+a^2}.$$

10 The maximum value chosen for x in this case allows a to reach 0.9995, which would define a very narrow pulse, very likely to cause aliasing in any case, so it constitutes a safe limit.

11 Note that we are using a slightly modified scaling method, which works better for the ranges of values of a that are more musically useful. This scaling factor is

$$\frac{(1-a)^2}{1+a^2}.$$

12 We use the term *complex* as in multi-partial, in opposition to simple, pure, single-component signals. This is not to be confused with the use of the word in relation to complex numbers. Audio signals are of course always real-valued.

13 The mathematics leading to the solution for this expression in terms of a series expansion can be traced back to Johann Bernoulli (1694), with many mathematicians involved in studying these, including Daniel Bernoulli, Lagrange, Fourier, Euler (who used this to model vibrations on a membrane), and others. This has important applications in the study of planetary motion, which was the focus of attention of Bessel as the basic results relating to the functions that now carry his name (Watson, 1944, 10–13).

14 This result is obtained as follows. The generating function for Bessel coefficients $J_n(z)$ is

$$e^{\frac{z}{2}\left[t-\frac{1}{t}\right]},$$

which can be expanded into a Laurent series on t (we have seen that the z-transform is also an example of such series),

$$e^{\frac{z}{2}\left[t-\frac{1}{t}\right]} = \sum_{n=-\infty}^{\infty} t^n J_n(z).$$

If we set $t = \pm e^{j\theta}$, the left-hand side is $e^{\pm jz\sin(\theta)}$ and the series expansion becomes

$$e^{\pm jz\sin(\theta)} = J_0(z) + \sum_{n=1}^{\infty}(\pm 1)^n J_n(z)\left[e^{jn\theta} + (-1)^n e^{-jn\theta}\right],$$

from where we get the result as a trigonometric series (Watson, 1944, 22). This result uses the relationship $J_{-n}(z) = (-1)^n J_n(z)$, which is proved below.

15 As with the other expressions, this can be derived from the generating function $e^{(z/2)[t-1/t]}$ and its Laurent series expansion. The method given by Watson (1944, 20) is as follows. Multiply both sides of the expansion by t^{-n-1}, and apply contour integration counterclockwise on a loop γ encircling the origin,

$$\frac{1}{2\pi i}\oint_{\gamma} t^{-n-1} e^{\frac{z}{2}\left[t-\frac{1}{t}\right]}dt = \sum_{m=-\infty}^{\infty}\frac{J_m(z)}{2\pi i}\oint_{\gamma} t^{m-n-1}dt$$

$$\frac{1}{2\pi i}\oint_{\gamma} t^{-n-1} e^{\frac{z}{2}\left[t-\frac{1}{t}\right]}dt = J_n(z).$$

Now we take the contour to be the unit circle, $t = e^{-j\theta}$, to obtain

$$J_n(z) = \frac{1}{2\pi}\int_{\alpha}^{2\theta+\alpha} e^{j[n\theta - z\sin(\theta)]}d\theta.$$

Finally, setting $\alpha = -\pi$, we have

$$J_n(z) = \frac{1}{2\pi}\int_0^{\pi}\left[e^{j[n\theta - z\sin(\theta)]} + e^{-j[n\theta - z\sin(\theta)]}\right]d\theta$$
$$= \frac{1}{2\pi}\int_0^{2\pi}\cos\left(n\theta - z\sin(\theta)\right)d\theta.$$

16 This is also derived from the generating function $e^{(z/2)[t-1/t]}$ (Watson, 1944, 15),

$$e^{\frac{zt}{2}}e^{-z2t} = \sum_{r=0}^{\infty}\frac{\left(\frac{z}{2}\right)^r t^r}{r!}\sum_{m=0}^{\infty}\frac{\left(-\frac{z}{2}\right)^m t^{-m}}{m!} = \sum_{n=-\infty}^{\infty} t^n J_n(z)$$

from which, for $n \geq 0$, we have the result,

$$J_n(z) = \sum_{m=0}^{\infty}\frac{(-1)^m\left(\frac{z}{2}\right)^{n+2m}}{m!(n+m)!}$$

.

17 In the generating function, we can replace t by $-1/t$ to get

$$e^{\frac{z}{1}\left[\frac{-1}{t}+t\right]} = \sum_{n=-\infty}^{\infty}(-t)^{-n}J_n(z) = \sum_{n=-\infty}^{\infty}(-t)^n J_{-n}(z),$$

which proves the relationship in Eq. 8.34 (Watson, 1944, 15).

18 As we have seen, there are other possible waveshaping configurations, which will result in slightly different spectra. Chowning's formula actually uses a sine waveshaper, leading to the expansion

$$\sin\left(\omega + z\sin(\theta)\right) = \sum_{n=-\infty}^{\infty} J_n(z)\sin(\omega + n\theta).$$

19 *Neumann's factor* (Watson, 1944, 22).

20 Since we will be employing 1-sample delays, here we revert to using n as time in samples explicitly.

21 This approach follows Lebrun's (1977) direct derivation.

22 Since this is all related to the binomial theorem, we can also lookup the solution in Pascal's triangle,

$$
\begin{array}{ccccccccccccc}
 & & & & & & 1 & & & & & & \\
 & & & & & 1 & & 1 & & & & & \\
 & & & & 1 & & 2 & & 1 & & & & \\
 & & & 1 & & 3 & & 3 & & 1 & & & \\
 & & 1 & & 4 & & 6 & & 4 & & 1 & & \\
 & 1 & & 5 & & 10 & & 10 & & 5 & & 1 & \\
1 & & 6 & & 15 & & 20 & & 15 & & 6 & & 1
\end{array}
$$

where the coefficients for each n-power expansion can be found starting from the middle of row n (counting from 0) and progressing to right. In the case of even n, the first coefficient is also divided by 2. The divisor associated with each row is 2^{n-1}.

23 It is fairly easy to prove that $f(x) = ax^n$ is even if n is even, and odd if n is odd. Therefore, polynomials containing terms of only one kind or the other are even or odd.

24 We are also using the SciPy library `special` which provides a handy function to look up binomial coefficients.

25 The first use of these to construct waveshaping functions should be credited to LeBrun (1979), and the remainder of this section is based on ideas from the cited work.

26 Note the similarity between this formula and the Dirichlet kernel,

$$U_{2n}(\cos(\omega/2)) = \frac{\sin([2n+1]\omega/2)}{\sin(\omega/2)},$$

bringing us back to where we started off.

27 These numbers are due to Jacob Bernoulli, *Ars Conjectandi* (1713).

28 See Lazzarini (2017, Chapter 5) for a comprehensive survey of adaptive non-linear distortion methods.

9

Noise

In this chapter we will try to tackle a very difficult problem: the spectra of noise. Put it simply like this, it appears strange to devote attention to something we usually want to avoid, but it is very important to tease out all the nuances that exist in relation to the subject. Firstly, we are using the word *noise* as a blanket description for everything that does not easily conform to the descriptions and parameters of spectra we have seen up until now. Perhaps a better word to nominate this group of signals and sounds is *non-deterministic*, which defines well their character but has a less immediate meaning. In any case, these are very important to both music and sound design, and therefore we should give them some special attention.

We may remember from Chapter 2 that these were the types of sounds that had been banished from Helmholtz's theories of acoustics, hearing, and music. They were not considered to be useful for any sort of proper music, maybe only for novel, comedic, or lesser types of artistic endeavor. Therefore they did not merit any further attention. While this was motivated by some aesthetical considerations, it nevertheless saved him significant effort, since the signals he went on to study were far more well behaved and predictable; the mathematics and physics he had at hand could cope with those. It would have been another story to examine in detail what happens, for instance, in the immediate onset of a flute sound or when a violin bow brushes against a string without producing a clear pitch. While there is some small mention of turbulence at the start of woodwind tones, these phenomena are left aside as unimportant in relation to the stable harmonic spectra of instrumental tones. As we mentioned before, Helmholtz's ideal was a steady mixture of sinusoids, probably with a gentle decaying spectrum so the fundamental is well emphasized, with which tonal-function harmonies can be delivered. Anything else was not *Musik*, not worth the trouble of further consideration.

Thankfully, this is not the picture we have of music-making these days. The majority of music that is experienced in everyday listening, and much of what is performed in live concerts, has plenty of these pesky non-deterministic sounds as integral elements. The ubiquitous drum kit, whether acoustic or electronically generated, is a source of a variety of these non-deterministic spectra, but that is not, by any means, the only one. Often the unnoticed blend of noisy elements is part of the individual sounds we admire and praise. For the old-timers, the example of Lester Young's breathy saxophone tone is a great example, made up of a mixture of turbulent air sounds with steady partials, articulated with care, and shaped with much expressivity to deliver his very personal sound. For the 21st century, we can think of the various uses of noisy spectra employed in electronic dance music and similar genres, each with specific spectral characteristics

Spectral Music Design. Victor Lazzarini, Oxford University Press. © Oxford University Press 2021.
DOI: 10.1093/oso/9780197524015.003.0009

that mix and contrast with the other more stable sounds. Likewise, there are thousands of other examples, even before we start mentioning contemporary concert music, where of course noise has been a key element for a hundred years.

In general, we have not been concerned with the nature of the signals we have been analyzing so far. The spectral methods developed in earlier chapters were agnostic insofar as they did not attempt to model or predict the input signal. However, by analyzing everything as a composite of complex sinusoids, we should expect that spectra with clearly defined partials would be described more precisely. By lumping noise with sinusoids, we are manipulating everything using the same tools; while this can work well, sometimes we might want to separate the spectrum into deterministic (well-defined partials) and non-deterministic (noise) components. In this chapter, we will first explore the nature of the latter, and then look for ways to model the spectrum as a sum of these two parts.

9.1 Random Processes and Noise Signals

It is useful at this point to find some ways to describe the various characteristics of noisy sounds, that is, of non-deterministic signals. We can begin as usual, with what we know, and then develop some further methods as we go along. Firstly, the main focus of attention is the distribution of energy in the spectrum; since the nature of the signals is random, there is not really much point in trying to track frequencies and phases of individual components. Therefore, to start with, the measurements we need to take stem from the squared amplitude, or power, spectrum of a waveform,

$$W(f) = \left| \int_{-\infty}^{\infty} x(t) e^{-j2\pi f t} dt \right|^2, \tag{9.1}$$

for the continuous-time case, and

$$W(f) = \left| \sum_{-\infty}^{\infty} x(n) e^{-j2\pi f n/f_s} \right|^2, \tag{9.2}$$

for the discrete-time case. Note that, to be as general as possible we have defined these in terms of continuous frequency, as we did when analyzing filters. Of course, when it comes down to computing the spectra we can set a limit to the time window we are considering and evaluate these in terms of equally spaced frequency points as an N-point DFT.

9.1.1 Centroid and Bandwidth

The power spectrum, as we know, provides a general shape that describes the distribution of energy per frequency.[1] We can use that to extract further details about some generic qualities of a sound, which may be of deterministic or non-deterministic nature. The first measure we can look for is the center of gravity, so to speak, of the spectral energy distribution. For a N-point DFT[2] of an audio signal we have

$$C = \frac{f_s}{N} \frac{\sum_{k=0}^{N/2} kW(k)}{\sum_{k=0}^{N/2} W(k)}, \tag{9.3}$$

which is called the spectral *centroid*.

This measure can be employed to give us an indication of spectral brightness, since higher values will indicate that on average, the distribution of energy is biased towards high frequencies, whereas lower values indicate otherwise. We can use this to perform a secondary analysis where we look for how much the energy is spread in the spectrum. This is the spectral bandwidth B, defined as

$$B = \sqrt{\sum_{k=0}^{N/2} W(k) \left(k\frac{f_s}{N} - C\right)^2}, \tag{9.4}$$

computed around the centroid C.

The bandwidth gives us a measure of the spread of the spectrum,[3] which is an important perceptual aspect of a sound. Note that for both the centroid and bandwidth we have employed the bin center frequencies as a reference. For deterministic spectra, we could have employed more finely defined instantaneous frequencies, but for the current case, these are not relevant.

9.1.2 Probability Distribution and Density

From the definitions above, we can say that the centroid is therefore a weighted average, or *mean*, of the amplitude spectrum. The bandwidth is a a measure of how much the values are spread away from the mean, or the *standard deviation*. These are terms that stem from statistical analysis, which is of course highly relevant here since noise is a stochastic phenomenon: something that may have a probability distribution, but cannot be precisely predicted.

A probability (or cumulative) distribution function $P(x)$ is defined as the probability that a random variable X will assume a value $X \leq x$ (Papoulis and Pillai, 2002, 375). It allows us to analyze the trends involved in a random process. From the probability distribution, we can estimate the odds of a random variable assuming a particular value, which is called the probability *density*. This is equivalent to the slope of $P(x)$, or its derivative. It is convenient to treat all these functions as continuous, even though when we come to implement any random process we will obviously be using discrete signals.

$$p(x) = \frac{d}{dx}P(x). \tag{9.5}$$

For stationary signals, we can put the discussion of random processes in terms of expected values $E[x]$ of a variable[4] x, which is also equivalent to the mean μ,

$$\mu = E[x] = \int_{-\infty}^{\infty} xp(x)dx \tag{9.6}$$

The variance σ^2 is a also an important characteristic of a random process; it measures the width of the spread of the values around the mean. It can be defined as

$$\sigma^2 = E[x^2] - \mu^2 = \int_{-\infty}^{\infty} x^2 p(x)dx - \mu^2, \tag{9.7}$$

and the standard deviation σ is the square root of this quantity. Note that the variance reduces to the *mean squared value*, $E[x^2]$, if the mean is zero.

The meaning of these parameters of random processes can be seen quite clearly in one of the most commonly-observed continuous probability distributions, the *Gaussian*

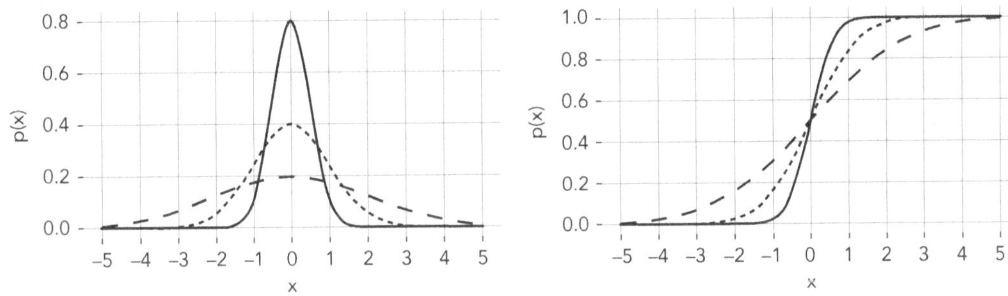

Figure 9.1
Normal distributions: probability density functions (*left*) and probability distribution functions (*right*) for $\sigma = 0.5$ (*solid*), $\sigma = 1$ (*dots*), and $\sigma = 2$ (*dashes*), and $\mu = 0$.

or *normal* distribution. This is generally given by the following density function,

$$p(x) = \frac{1}{\sigma\sqrt{2\pi}} e^{-\frac{(x-\mu)^2}{2\sigma^2}} \tag{9.8}$$

which reduces to $1/\sqrt{2\pi} e^{-x^2/2}$, when the mean is zero and the standard deviation is 1. This is called the *standard* normal deviation.

The probability density of the normal distribution shows that values away from the mean are less likely to occur than closer to it. The probability of an outcome within a given range is equivalent to the area under the curve, which is the integral of $p(x)$, the probability distribution function,

$$P(x) = \frac{1}{\sigma\sqrt{2\pi}} \int_{-\infty}^{x} e^{-\frac{(t-\mu)^2}{2\sigma^2}} dt. \tag{9.9}$$

Plots of the density and cumulative functions are shown in Fig. 9.1, for three values of σ and zero mean.

Note that the probability of a specific, precise, value goes to zero as the area under the curve of the probability density curve gets increasingly small. Here we find a typical characteristic of a stochastic process: it may be possible to model how the values are distributed, and the likelihood of certain outcomes, but it is not possible to predict the exact outputs.

9.1.3 Power Spectrum Density

The mean, standard deviation, and variance are parameters that provide a certain insight into random processes, but they may be of limited use in analyzing signals. We can go one step further and look at spectral aspects of probability theory. In Sec. 9.1.1, we approached this using the tools of spectral analysis directly on an arbitrary source (deterministic or not) to get a measurement of these generic parameters of a time-domain signal. The notion of the spectrum of a random process is, however, more abstract than what we are used to in the case of the spectra of audio signals.

Firstly, we should consider the samples of any given noise waveform are *realizations* of a random variable. If we consider the random process as embodying all the possibilities of an outcome, the waveform is made of up of the one that took place. Therefore we now change the focus from realization to process. For this, we need to we need to extend our analysis tools, since the Fourier transform of a random sequence does not exist.[5] We begin by defining a new signal, called the *autocorrelation*, as

$$R_{xx}(\tau) = \int_{-\infty}^{\infty} x(t)x(t+\tau)dt. \tag{9.10}$$

which may be interpreted as a measure of the similarity of the sequence with itself for different values of the time lag τ. If there is a certain amount of correlation in the signal, it will show up at various values of τ. We can apply this, for instance, to a deterministic input, and we will observe a peak at $R_{xx}(T)$ in the case of steady periodic signals with frequency $1/T$.

In the case of non-deterministic signals, we can use this notion to define an autocorrelation function (Papoulis and Pillai, 2002, 376)

$$r_{xx}(\tau) = E[x(t)x(t+\tau)] = \int_{-\infty}^{\infty}\int_{-\infty}^{\infty} x(t)x(t+\tau)p(x(t),x(t+\tau))dx(t)dx(t+\tau), \quad (9.11)$$

where now we are considering the probability[6] of x at times t and $t-\tau$, and since we have two signals to consider, $p()$ is a *joint* probability density function. This is defined for stationary signals, where we only need to consider τ. We can also put forward an autocovariance function,

$$k_{xx}(\tau) = E[(x(t)-\mu(x(t)))(x(t+\tau)-\mu(x(t+\tau)))] = r_{xx}(\tau)-\mu(x(t))\mu(x(t+\tau)), \quad (9.12)$$

from which we can see again that if the mean is zero, the autocorrelation and autocovariance are the same. The autocorrelation function is an even function of τ, $r_{xx}(\tau) = r_{xx}(-\tau)$, and $r_{xx}(0)$ gives us the mean squared (also the variance for a zero-mean process).

From the autocorrelation function we can now obtain the *power spectrum density* (Papoulis and Pillai, 2002, 408), which is its Fourier transform,[7]

$$S_{xx}(f) = \int_{-\infty}^{\infty} r_{xx}(\tau)e^{-j2\pi f\tau}d\tau, \quad (9.13)$$

and by consequence

$$r_{xx}(\tau) = \int_{-\infty}^{\infty} S_{xx}(f)e^{j2\pi f\tau}df. \quad (9.14)$$

As an aside, it is interesting to note the similarities between the definition of the power spectrum density (Eq. 9.13) and the Wigner-Ville distribution (Cohen, 1989)

$$S_t(f) = \int_{-\infty}^{\infty} x\left(t+\frac{\tau}{2}\right)\bar{x}\left(t-\frac{\tau}{2}\right)e^{-j2\pi f\tau}d\tau, \quad (9.15)$$

which, alongside with the Gabor approach used throughout this book, is one of the original time-frequency analysis methods. For stationary processes, it is easy to show that this is exactly equivalent to the power spectrum density.[8]

Moreover, since we have indicated that the autocorrelation function is even, the power spectrum density $S_{xx}(f)$ is purely real, and we only need to employ a half transform to obtain it,

$$S_{xx}(f) = \int_{-\infty}^{\infty} r_{xx}(\tau)\cos(2\pi f\tau)d\tau. \quad (9.16)$$

Note that this is a different definition of power spectrum from what we discussed earlier, as the squared amplitude (or magnitude) spectrum, even though the two are related. The power spectrum of a noise waveform, as computed, for instance, by the DFT, can be equated to an estimate of the power spectrum density of a random process defined by probability distribution.[9] It is also important to consider that the power spectrum density as the Fourier transform of an autocorrelation function applies strictly to random stationary processes, where the mean and autocovariance do not vary over time.

For example, consider a uniformly distributed random-signal with a zero-mean, whose values lie in the interval $[-1, 1]$. Since the values that this signal assume are supposed to be uncorrelated, the autocorrelation function can be defined by a Dirac δ, whose Fourier transform, as we know, is unity. The power spectrum density of this signal is therefore a straight line, which does not vary with frequency. This may be defined as $1/f^0$ or *white* noise, in reference to light, as in a stochastic sense it contains all frequencies of the audio spectrum. The case of the Gaussian distribution that we have noted above is an example of a random process in the time domain that generates noise with such a power spectrum density.

9.1.4 Fractional Noise

We can characterize different types of noise by how their spectra falls with frequency. The example of uniformly distributed random processes as having a flat spectrum is an extreme case of a large class of signals defined by $S_{xx}(f) \propto f^{-\nu}$, where ν is a non-negative real number.[10] Among these, we have the case of $\nu = 2$, which defines *Brownian* noise (Mandelbrot and van Ness, 1968). A signal of this type can be created by integrating a waveform produced by a uniform distribution. In the case of digital signals, we may remember that an integrator can be implemented by a first-order IIR lowpass filter,

$$y(n) = x(n) + y(n-1). \tag{9.17}$$

If the signal $x(n)$ is derived from a uniform distribution, then the output of this filter is f^{-2} noise. We also may recall that such an integrator has a f^{-1} amplitude response, which confirms the result. The power spectrum slope is therefore 6 dB/octave. Note that the use of such a filter in practical applications may result in an accumulation of DC bias (as we know, this filter is not strictly stable at 0 Hz).

The value of ν determines how constrained the random process, and by consequence, the noise waveform, is. In the extreme end of the range, $\nu = 0$, there is no constraint, and the samples of a digital signal are independent of each other. Another notable case is $\nu = 1$, or f^{-1} noise (Mandelbrot, 2006), whose power spectrum decays by 3 dB/octave.[11] It is a much less steep slope and results in a constant energy distribution per octave. For this reasons, it finds many applications as source for source-modifier synthesis methods in music and sound design. To distinguish it from the flat-spectrum white noise, such signals are often called *pink* noise.

Using this description of a random signal spectrum, we can use other values of ν to change the amount of constraint to generate different types of signals. One way to characterize them is as fractional Brownian motions, with reference to different amounts of interdependence between current and past values of a random signal. For values $0 < \nu < 2$, these may be described as *fractional* noise, as they are a fractional integral of white noise (Wyss, 1991).

9.1.5 Spectral Moments

The mathematical concept of *moment* allows us to characterize the shape of a function (Papoulis and Pillai, 2002, 146–147). This can be applied to the probability density of a random process, as well as to the power spectrum density. The general definition of a moment n of a function $f(x)$ about a value c is defined as

$$\mu_n = \int_{-\infty}^{\infty} (x-c)^n f(x)dx. \tag{9.18}$$

By setting $c = 0$ and $n = 0$, we can see that the degree zero moment of a probability density function is its *mean*. This is also called a *raw* moment. For $n > 0$, setting $c = \mu$ gives us the moments about the mean. These are the *central* moments, of which the second-degree moment is equivalent to the variance σ^2. Also in probability theory, a third type of moment is defined as

$$\tilde{\mu_n} = \frac{\mu_n}{\sigma^n} \tag{9.19}$$

for $n > 0$. This is called the *normalized* or *standard* moment.

The first degree standard moment is always zero, and the second is one, since, as we noted earlier, the second central moment is the variance. The third and fourth moments are called *skewness* and *kurtosis*. When applied to a probability density function, these measure the asymmetry of the distribution, and the propensity to produce outliers (related to the tail of the distribution), respectively. For the Gaussian distribution, the skewness is zero (it is symmetrical about the mean) and the kurtosis is 3, since its central moment is $3\sigma^4$.

We can apply these principles to derive *spectral* moments (Sweitzer et al., 2004) for a power spectrum density $S_{xx}(f)$ (or an estimate of it),

$$\mu_n = \int_{-\infty}^{\infty} (f-f_c)^n S_{xx}(f)df. \tag{9.20}$$

Applying this, as we did before, to the DFT of a waveform, we have the spectral moment m_n defined as

$$m_n = \sum_{k=0}^{N/2} (f_k - f_c)^n W(k), \tag{9.21}$$

where f_k is the frequency of bin k, f_c is the mean, and $W(k)$ is the squared amplitude, or power, spectrum. We can see how this relates to the previously defined centroid and bandwidth parameters. If we use the centroid C as the mean, the variance then is squared bandwidth.[12] From these definitions we can get an estimate of the skewness and kurtosis of the power spectrum by using the bandwidth B in place of σ,

$$\tilde{m_n} = \frac{1}{B^n} \sum_{k=0}^{N/2} (f_k - C)^n W(k). \tag{9.22}$$

We have already seen the meaning of the centroid and bandwidth in relation to the spectrum of a noise signal. The skewness can have a similar interpretation to what we have noted in relation to a probability distribution, that is, a measure of the asymmetry of the spectrum. A negative skew points at a concentration of energy toward higher frequencies, for example, a high-pass shape, with a longer tail on the lower side of spectrum. A positive skew reverses this shape. A skew of zero, as in the normal distribution, indicates a symmetric curve. Kurtosis is related to how much energy is spread away from the mean, providing a measure of the shape of the tail of the spectrum. This is often compared to the kurtosis of the Gaussian distribution, giving rise to the term *excess* kurtosis, $\tilde{\mu_4} - 3$.

We can compute these spectral moments from an N-point power spectrum (e.g. as derived from a DFT estimate of the power spectrum density) and the sampling rate. From this, the code can then return the centroid, bandwidth, skewness, and kurtosis. This is done by the following Python function,

```
# takes a power spectrum
# with N/2 + 1 bins
# and the sampling rate
# returns spectral centroid, bandwidth,
# skewness, and kurtosis
def specmoments(S,fs):
    N = S.size
    fo = fs/(2*(N-1))
    C = 0
    D = 0
    k = 0
    for s in S:
        C += s*k
        D += s
        k += 1
    C = fo*C/D
    B = 0
    k = 0
    for s in S:
        B += s*(fo*k-C)**2
        k += 1
    B = np.sqrt(B)
    Sk = 0
    k = 0
    for s in S:
        Sk += s*(fo*k-C)**3
        k += 1
    Sk = Sk/B**3
    Kt = 0
    k = 0
    for s in S:
        Kt += s*(fo*k-C)**4
        k += 1
    Kt = Kt/B**4
    return C,B,Sk,Kt
```


In Fig. 9.2, we apply this code to analyze three different power spectrum functions. The first two are ideal spectra derived from a Gaussian curve (with $\sigma = 500$ and $\mu = f_s/4$) and a raised inverted cosine (a Hanning window) of width $f_s/4$ centered at $f_s/8$ Hz ($f_s = 44,100$ Hz). The other analysis is taken from the DFT of a white noise source filtered by a resonator tuned to 8,000 Hz. As can be seen, the kurtosis and skew of the ideal curves meet, as expected, the values defined for such functions, zero skewness, and kurtosis of

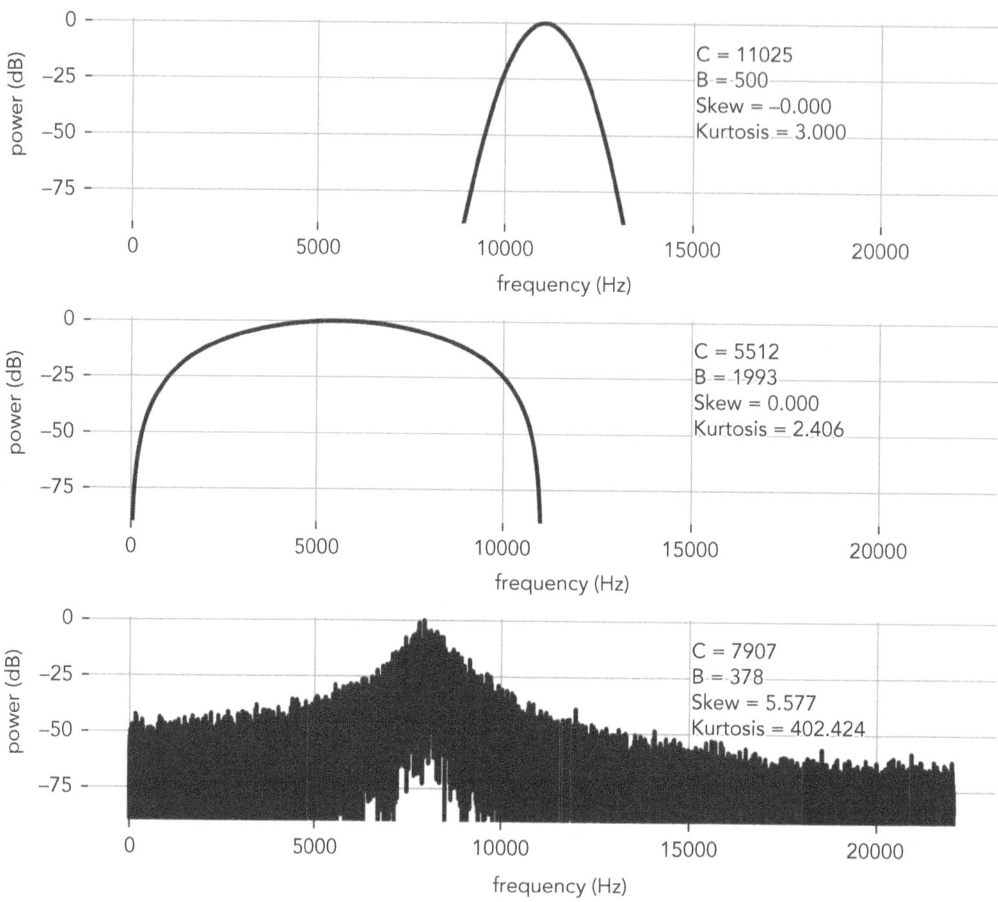

Figure 9.2 Three power spectra (dB, normalized) and their respective spectral moments: Gaussian curve (*top*), raised inverted cosine (*middle*), and filtered noise waveform (*bottom*).

(3 and 2.4, respectively). Note that the bandwidths, also as per definition, are equal to the standard deviation, and the centroids to the mean.

These analyses when applied to an observation of a noise waveform give a certain indication of the shape of the wave. In the present case, the centroid is close but slightly lower than the filter center frequency. This is due to the fact that the resonator passes more signal in its lower half, if it is centered below $f_s/4$ (reversing this effect for higher center frequencies). This is also reflected in the positive value for skewness, as the tail of the spectrum is to the right and the energy is concentrated on the left. The high value for the kurtosis indicates the rounded, tailed, shape of the spectrum.

9.2 Computing Noise

The difficulty with noise in a computational environment is that our tools are designed to be deterministic as much as possible. Of course, at the edges they might fail to be, with precision errors in real number arithmetics as an example where things might go wrong (most of these errors can of course be put in deterministic terms), and so on. However, the spirit is that we want to compute our expressions in ways that are well-defined, producing an expected result. If we want to deviate from that, we need to do in an algorithmic way. We can also attempt to measure some physical parameter that has a non-deterministic behavior, discretize it, and derive a signal out of this. It is, however, most common to approach it from a computational perspective rather than a sampling one.

The issue at hand is *randomness* and how to produce it. Most programming languages provide a function that is purposely defined to yield a random number sequence, that is, a series of results that are not correlated in some obvious way. This is of course fake randomness, because as we have anticipated, computers are not generally designed to produce non-deterministic results. Therefore we have a *pseudo-random* sequence of numbers, generated by some algorithm that will, within a given limit, yield numbers that may be considered to have a random nature. A typical approach in such algorithms is to generate a long set of numbers that are uncorrelated, according to some formula. The sequence will eventually repeat, and so it is not truly random, but it may be long enough to give an impression of randomness.

9.2.1 Random Number Generators

If we are interested in the actual numbers, the repetition may be problematic: we may not want, for instance, to have the same sequence every time we run a program. They may be uncorrelated, but they are always the same. However, if the focus is on the apparent randomness and not in the specific values of a signal, that is really not a problem. This is of course what matters to us when trying to produce a noisy spectrum: if the algorithm produces a fairly uncorrelated sequence, then we are not going to have any concentration of energy at specific frequencies and it would sound just as noisy as, for instance, a sampled non-deterministic source. The C language `rand` function is the classic example of a pseudo-random number generator that could be sufficient for our applications. Alternatively, a better noise source is also provided by some operating systems such as the UNIX-like systems in the from of the `/dev/random` or `/dev/urandom` pseudo-devices, which can be opened and read by programs.

In Python, the NumPy library provides a module dedicated to generating various types of random sequences. For example, the following lines produce a signal based on a uniform distribution, that is, where the likelihood of any particular value within a given range is the same as any other:

```
import numpy.random as rnd

fs = 44100
t = np.linspace(0,1,fs)
noise = 2*(rnd.rand(t.size) - 0.5)
axs.plot(t,noise,'k', linewidth=1)
```

whose result is plotted in Fig. 9.3.

Since our use of random numbers is to extract a noisy signal, and we are interested in a flat and well-distributed spectrum, we can largely ignore the details of pseudo-random algorithms. We can them as black boxes with an output whose behavior we understand. If the sequence is long enough, audible time patterns are unlikely to appear, and we should have a signal that will for all practical purposes be similar to real noise signals

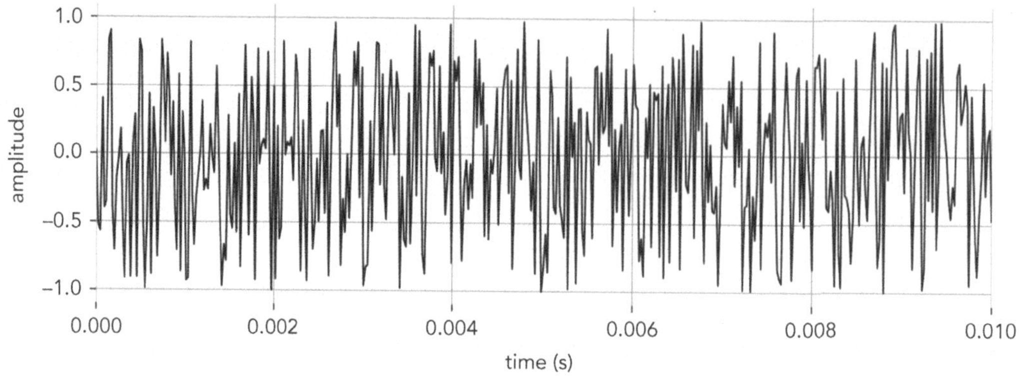

Figure 9.3 Noise signal waveform generated by a Python program.

holding random numbers. However, some pseudo-random number generator algorithms may not pass this test. For example, Csound's original `rand` algorithm is based on a 15-bit sequence, which is too short. When lowpass filtered, it reveals an audible time pattern (similar to a rhythm), which is objectionable. The alternative 31-bit algorithm provides a much better result, with no such patterns emerging. Comparing, for instance, the output of this two instruments, which uses integration as a method of lowpass filtering, generating f^{-2} noise

```
instr 1
 a1 = rand:a(p4/100)
 a2 = integ(a1)
 out(a2)
endin

instr 2
 a1 = rand:a(p4/200,0,1)
 a2 = integ(a1)
 out(a2)
endin
```

It is possible to hear a pattern in the case of the first instrument, which employs the original algorithm. The second instrument does not have any such audible artifacts (besides a steady rise in DC caused by the integration), as it employs a better number generator (selected by the third, optional, parameter).

Therefore, we can assume that we have a reasonably good source of such numbers, in a normalized bipolar form (no DC bias), from any of these generators. This is similar to the assumptions we have been making about, for example, built-in mathematical functions such as sin() or cos() from system libraries, that is, that they produce pure sinusoidal, that are not distorted in any way, within of course the precision limits of floating-point representation. With this out of the way, we can examine the various nuances of noise and non-deterministic signal generation that are of interest to sound design and music.

9.2.2 Sample and Hold

Continuous-time white noise, as an analogue signal, can be produced by circuits that use thermal noise as a random voltage source. We can think of digital noise derived from a pseudo-random number generator as equivalent to sampling such a signal. This is a slightly different notion, although related, to what we developed before in Chapters 3 and 4. A good way to describe this process from a continuous-time perspective is as a *sample-and-hold* operation, where we pick up a random number, hold it for one sampling period, then pick another, etc., to generate a digital white noise signal. The consequence of this is that we are effecting the convolution of a rectangular function with the noise waveform. Therefore the amplitude spectrum of the resulting signal is the product of these two spectra, the ideal $1/f^0$ white noise and a sinc() function,

$$A(f) = \mathrm{sinc}\left(\frac{f}{f_s}\right) \tag{9.23}$$

This is very close to a flat spectrum, there is only a slight reduction of around -4 dB at the Nyquist frequency. White noise as a signal is not very useful on its own, but can serve as a source for modification processes. Furthermore, by re-applying the sampling process we outlined earlier, it is possible to constrain the signal and reduce its bandwidth. A sample-and-hold noise generator operating on a sampled signal has the following amplitude spectrum,

$$A(f) = \mathrm{sinc}\left(\frac{f}{f_h}\right)\mathrm{sinc}\left(\frac{f}{f_s}\right), \tag{9.24}$$

where f_h is the sample-and-hold frequency employed. This has the effect of reducing the energy of the high-frequency components of the noise waveform. It is quite an effective way to produce bandlimited noise, but we can go one step further. If instead of holding the values steadily for the duration of the sample-and-hold period, we interpolate linearly between the successive random values, the randomness is further constrained. This produces an even more bandlimited spectrum.

A linear interpolation filter operating on successive samples of a waveform has a $\mathrm{sinc}^2()$ amplitude spectrum.[13] When we apply this to a sampled white noise waveform, we get

$$A(f) = \frac{\mathrm{sinc}^2\left(\frac{f}{f_h}\right)}{\mathrm{sinc}\left(\frac{f}{f_s}\right)} \tag{9.25}$$

The power spectra of sampled white-noise, and sample-and-hold and linear interpolation bandlimited noise, are shown on Fig. 9.4. As we can see, there is very good attenuation of high frequencies when we interpolate between random values, as the peak of the first side lobe is about -30 dB down from full scale. In both cases of bandlimited noise, we can see that the frequency of the sample process determines the bandwidth of the output. It can be considered as proportional to the cutoff frequency of a lowpass filter applied to the white noise source.

In Csound, the two bandlimited noise generators are implemented by `randh` (sample-and-hold) and `randi` (interpolation). They have similar parameters to the white noise source `rand` with an extra frequency control.

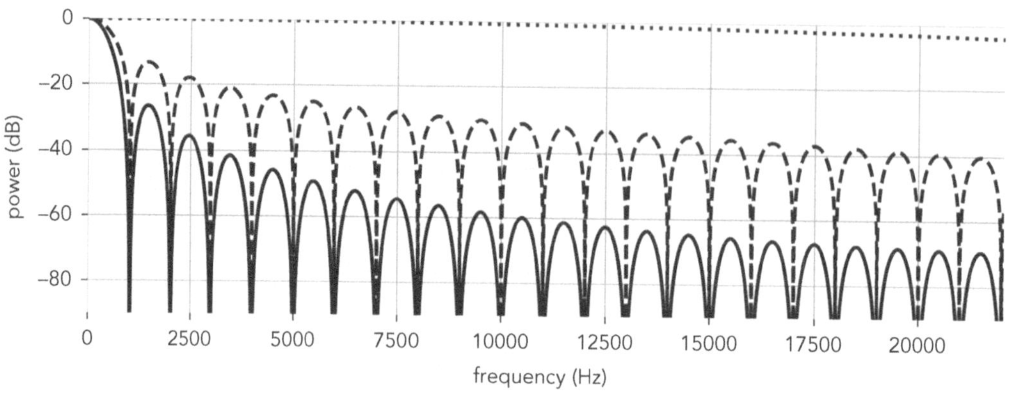

Figure 9.4 Power spectra from three digital noise sources: pseudo-random number generator (*dots*), sample-and-hold (*dashes*) and linear interpolation (*solid line*) bandlimited noise generators, with $f_h = 1{,}000$ Hz and $f_s = 44{,}100$ Hz.

9.2.3 Heterodyning

The bandlimited noise generators are strictly lowpass, which may limit their usefulness in synthesis applications. However, as we have seen before, we can easily make a lowpass into a bandpass spectrum by heterodyning. Using a sinusoid of frequency f_c, we multiply it by the output of a bandlimited noise generator to shift the amplitude spectrum by f_c. This can be used to construct various types of distributed spectra instruments. For example, the following opcode employs this method, together with an exponential decay envelope to produce percussive sounds:

```
opcode Perc,a,kkki
 ka,kcf,kbw,id xin
 ae = expon(1,id,0.001)
 an = randi:a(ka,kbw)
 as = oscili(ae*an,kcf)
 xout(as)
endop
```

Spectra with more than one peak can use a combination of these heterodyne bandlimited noise sources. Since both the center frequency and bandwidth can be made to vary in time, it is also possible to produce dynamic spectra with this method. The various distortion synthesis methods developed in Chapter 8 can also be applied to generate modulation signals with more than one component. Furthermore, we may derive synthesis parameters from frequency-domain measurements of the centroid and bandwidth.

9.2.4 Filtered Noise

It is also possible to apply filtering to a noise source in order to approximate a given spectrum. In this case, we normally start with a wideband source, which may be, for instance, digital white noise. For example, if we need to generate a signal with a $1/f$ power spectrum, that is, pink noise, there are two main approaches:

- Use an algorithm that produces an approximation of the random number distribution.
- Starting with white noise, apply a filter with a $1/\sqrt{f}$ spectral envelope.

We can easily construct such a filter from its amplitude response using the FIR design methods we have developed before. With Csound, we can write the spectrum to a table, then use GEN 53 to compute the filter coefficients. The filter is implemented using direct convolution, to which we apply a white noise source. Note that the output is not normalized and will vary with different filter sizes.

```
opcode Pink,a,ki
 ka,il xin
 if1 = ftgen(0,0,il,7,0,il,0)
 ic = 1
 tablew(1,0,if1)
 while ic < il do
  tablew(1/(sqrt(ic)),ic,if1)
  ic += 1
 od
 if2 = ftgen(0,0,il*2,53,if1)
 an = rand(ka,0,1)
 as = dconv(an,ftlen(if2),if2)
 xout(as)
endop
```


This filter can be customized to generate other $1/f^{\nu}$ noise spectra. In addition to writing our own, we can also use one of the built-in generators in Csound. The opcode `pinkish` also implements both methods, a random number generator and a filter that produces pink noise if fed a white noise waveform. Another alternative is `pinker`, which employs a different generation algorithm.

Bandpass filters such as resonators can also be used to shape the spectrum of wideband distributed sources. Both white and pink noise sources may be appropriate for this, depending on the type of spectra being designed. The latter will produce a less harsh, more balanced source signal and may be preferred for most applications. Particularly interesting is the use of formant-modeling techniques such as the channel vocoder to shape wideband noisy sources.

9.2.5 Wavetables

Another means of producing noise for various synthesis applications is to record various different types of distributed-spectra waveforms and store them in wavetables. These can be accessed via table lookup, and mixed in various ways, through the use of amplitude envelopes. Dynamic spectra can be constructed by crossfades, moving from one source table to another. Depending on the shape of the recorded sound spectrum, heterodyning or single-sideband modulation can be used to shift/recenter it. In Csound, we can use GEN1 to put a recorded waveform from a soundfile onto a function table, for access via table readers or oscillators.

9.3 Grain

An important perceptual aspect of noise sounds, which also goes beyond the range of the standard spectral tools derived from the Fourier theorem, is *grain*. Often we can identify differences in sounds with uniformly distributed (long-time) spectra, which fail to appear in frequency analysis. These can be attributed to the way the sound mixture is composed in granular terms, rather than simply as a spectral distribution. A grain may be described as a time-frequency bundle, a waveform of some description wrapped in an envelope, generally of very short duration. In this description, we try to capture both spectral and temporal aspects of a signal.

In some ways, the analysis-synthesis model of frames we developed in Chapter 6 is based on a sort of granular model. Each bin of each frame can be thought of as grain, lasting for a short while and enveloped by the analysis window.[14] The waveform content of each grain is a bandlimited signal made up of a modulated sinusoid. This model is capable of capturing both deterministic and noisy sounds, and successfully processing them as a mixture. However, it only provides a clear and distinct description of the former, and is generally not capable of distinguishing between the two (unless some further processing is applied). We can also describe such a system, from the point of view of grain generation, as *synchronous*: all grains are perfectly aligned on a time frame.

From this perspective, we may see that for an effective time-frequency analysis approach, a consideration of the granular aspects of sound is very important. It can also be a very useful approach to take from a sound design perspective to synthesize dynamic noise spectra. Furthermore, considering sound as composed of grains gives us a slightly higher level, from a temporal perspective, to apply the ideas about stochastic processes we have developed earlier in the chapter. If we think of the sample as the lowest atomic level of a waveform, the grain gives us the next subdivision. Although the duration of a grain is undefined, we may set it using a perceptual threshold where individual instances are not long enough to have any defined spectral features. When summed up, though, they reveal a complex distributed spectrum.

9.3.1 Asynchronous Granular Synthesis

When dealing with stochastic processes, we need to consider the generation of grains as an asynchronous operation. Although we can control the overall trends, the actual time and duration of each instance is not defined. Other parameters of the grain, such as its spectral content and amplitude, are also not defined directly. We can think of the result as a cloud of individual particles, whose external shape can be manipulated, but the components are taken from a probability density function.

To provide a practical example, let's develop the code for an asynchronous granular synthesizer. We can base its stochastic processes on the Gaussian distribution, which has well-known properties that allow us to shape the spectral and temporal trends of the output very carefully. We have seen the probability cumulative and density functions for this distribution before; they tell us that with two parameters we can construct a random process that ensures values to lie within a given range. This is very suitable for creating a non-deterministic sound with certain overall characteristics.

While we know the trends of a Gaussian distribution, how do we actually realize them? We know of course how to generate uniformly distributed values, using the `rand` opcode, for instance. From this, we can apply the Box-Muller algorithm (Box and Muller, 1958), which is a known method to produce random normal deviates, or numbers in a Gaussian distribution with mean of zero and unity standard deviation. This is defined by the formulae,

$$\begin{aligned} Z_0 &= \sqrt{-2\log(U_0)}\cos(2\pi U_1) \\ Z_1 &= \sqrt{-2\log(U_0)}\sin(2\pi U_1) \end{aligned} \tag{9.26}$$

where U_0 and U_1 are two samples from a uniform distribution in the range 0 to 1. The following Python code implements this algorithm,

```
import random
class Gauss:
 def __init__(self):
  random.seed()
  self.flag = True

 def compute(self,mu=0,sigma=1):
   if self.flag:
     u0 = random.uniform(0,1)
     u1 = random.uniform(0,1)
     z = np.sqrt(-2*np.log(u0))*np.cos(2*np.pi*u1)
     self.z = np.sqrt(-2*np.log(u0))*np.sin(2*np.pi*u1)
     self.flag = False
     return z*sigma + mu
   else:
     self.flag = True
     return self.z*sigma + mu
```

where, in order to get different values from the pseudo-random sequence, we seed it (i.e., provide an initial value) using the system clock.

This algorithm is also implemented in one of Csound's `gauss` opcodes, which takes the mean and the standard deviation as parameters.[15] With this we can construct the noise cloud synthesis code. The approach is straightforward:

- We will design a grain as composed of a bandlimited noise waveform, with center frequency, bandwidth, and amplitude as spectral parameters.
- The grain envelope is given by a Hanning window. Other shapes may also be used, such as the Bartlett window or a truncated Gaussian function. GEN 20 can be used to create tables containing such windows.
- Start time and duration will be set for each grain.
- Each one of the five parameters will be defined by a Gaussian distribution, with separate means and standard deviations.

A total of ten parameters are needed to define the cloud. To generate the grains, we will employ again the very useful coding technique of temporal recursion, where each instrument instance schedules the next. In order not to have finite duration, we can add an extra parameter to determine an end to the process. With this recipe, we have the following grain generation instrument,

```
instr 1
 igs = gauss(p4,p5) // start time
 igd = gauss(p6,p7) // duration
 iga = gauss(p8,p9)*0dbfs // amp
 icf = gauss(p10,p11) // center freq
 ibw = gauss(p12,p13) // bandwidth
 ifn = p14 // envelope table

 ai = line:a(0,p3,ftlen(ifn))
 ae = table:a(ai,ifn)
 an = randi:a(ae,ibw)
 as = oscili(iga*an,icf)
 out(as)

 if times() < p15 then
  schedule(p1,igs,igd,p4,p5,p6,p7,
        p8,p9,p10,p11,p12,p13,p14,p15)
 endif
```


Now we need code to define the cloud parameters and initiate the synthesis by launching the first grain,

```
seed(0) // seed from system clock
igsm = 0.1 // start mu
igsd = 0.04 // start sigma
igdm = 0.2 // dur mu
igdd = 0.05 // dur sigma
igam = 0.2 // amp mu
igad = 0.08 // amp sigma
igfm = 5000  // cf mu
igfd = 300 // cf sigma
igbm = 1500 // bw mu
igbd = 500 // bw sigma
// env table and total duration
ifn = ftgen(0,0,1024,20,2,1)
idur = 10

schedule(1,0,igdm,
         igsm,igsd,
         igdm,igdd,
         igam,igad,
         igfm,igfd,
```

```
        igbm, igbd,
        ifn, idur)
```

This code can be placed in global space (instrument 0, in Csound parlance) for the generation of a single cloud, or inside another instrument for multiple instances.

Such synthesis design is a fairly elegant way to generate very complex non-deterministic dynamic spectra. A good variety of textures may be achieved, which could be further shaped on a macro level by a global envelopes controlling the trends of each parameter. We could have also derived the grain waveform from existing sound sources, for example by sampling a wavetable at different time points, and applying bandpass filtering to limit the spectral width of each grain and/or using variable-speed playback to transpose the sound.

9.3.2 Wavelets

As we noted earlier, the concept of grain as a time-frequency atom can be useful in spectral analysis. We can develop this idea a little further, and cast the short-time Fourier transform as a granular process. Let's first generalize a time-frequency transform as[16]

$$X(\gamma) = \int_{-\infty}^{\infty} x(t)\overline{\phi_{\gamma}}(t)dt, \tag{9.27}$$

where γ may be a single or a multiple parameter. For the (fixed-time) Fourier transform, we can think of it as a frequency index. The function $\phi_{\gamma}(t)$ represents the bases used in the transform, that is, a complex sinusoid in the case of the Fourier transform. Using this approach, the time-localized STFT, which we have used as the main component of the spectral analysis-synthesis system, can be defined as

$$X(u,f) = \int_{-\infty}^{\infty} x(t)\overline{\phi_{u,f}}(t)dt, \tag{9.28}$$

where we use as bases a time-frequency grain made up of an enveloped complex sinusoid, with frequency f and centered at time u,

$$\phi_{u,f}(t) = w(t-u)e^{j2\pi ft}. \tag{9.29}$$

This then yields

$$X(u,f) = \int_{-\infty}^{\infty} x(t)w(t-u)e^{-j2\pi ft}dt. \tag{9.30}$$

This is the transform we saw in Chapter 6 in a continuous-time continuous-frequency form. We know how to get to its discrete form, and similar principles apply to that case. So its application can be seen as a form of granular analysis, where the grains all have the same duration, determined by the window size. We also know we can reconstruct the original waveform by summing up all the grains,

$$x(t) = \int_{-\infty}^{\infty}\int_{-\infty}^{\infty} X(u,f)w(t-u)e^{j2\pi ft}dfdu. \tag{9.31}$$

Note that the double integral represents, in continuous time, the overlap-add operation we employed in the synthesis stage. As we have seen, this analysis system works well, but built into it there is a compromise between the precision of analysis in time and frequency. As we know, this is determined by the uncertainty principle. Given that we use a single size grain for all frequencies, often we have to consider the characteristics of the spectral content before we can decide on the time-frequency resolution.

The problem can be approached from another perspective, if we decide that $\phi_\gamma(t)$ can have varying lengths. This is given by the *wavelet* transform. The principle is to employ bases that are derived from a *mother* function, or a mother wavelet, by scaling and translating (e.g., dilating and shifting) it to create a family, which is then employed in the transform. A wavelet is a function $\psi(t)$ with zero average, and normalized. The set of wavelets derived from $\psi(t)$ is given by

$$\psi_{u,s}(t) = \frac{1}{\sqrt{s}}\psi\left(\frac{t-u}{s}\right) \tag{9.32}$$

where $s \geq 0$. As in the STFT case, u is the time center of the wavelet. The new parameter s is the *scale*, which defines its dilation.

With these definitions, the wavelet transform is given by

$$X(u,s) = \int_{-\infty}^{\infty} x(t)\frac{1}{\sqrt{s}}\overline{\psi}\left(\frac{t-u}{s}\right)dt. \tag{9.33}$$

This can also be interpreted as filtering operation,

$$X(u,s) = x(u) * h(u) \tag{9.34}$$

where the impulse response $h(t)$ is

$$h(t) = \frac{1}{\sqrt{s}}\overline{\psi}\left(\frac{-t}{s}\right). \tag{9.35}$$

Just as in the case of the STFT-based analysis, the time-domain signal can be reconstructed, for example,

$$x(t) = \frac{1}{C_\psi}\int_{-\infty}^{\infty}\int_{-\infty}^{\infty} X(u,s)\frac{1}{\sqrt{s}}\psi\left(\frac{t-u}{s}\right)\frac{ds}{s^2}du, \tag{9.36}$$

where C_ψ is a scaling factor based on the energy of the wavelet.

The wavelet may be purely real (as, for instance, in the case of cosine transforms), or complex (analytic). For example, we can base a mother wavelet on the windowed complex sinusoid, used in STFT analysis. The *Gabor* wavelet is one such example, based on a Gaussian window shape modulated by a sinusoid.[17] The family of wavelets derived from it in different scales can be called a *dictionary* of grains or atoms, with which the transform correlates an input signal. The result, as with the Fourier transform, is a coefficient of how well a given atom describes a component of that signal in time and frequency. Note that here, as in the case of the STFT, the analysis is the representation of a one-dimensional waveform as a two-dimensional function.

For discrete-time signals, the definition of the wavelet transform has to be constrained to scales in the range of $f_s < s < 1$, where f_s is the sampling rate. In this case, we can set $s = a^i$, where i is an integer. If $a = 2^{1/v}$, then it is possible to put the analysis in terms of v scales per octave. The discrete wavelet is defined to have K samples, in the interval $[-K/2, K/2]$, and a family of such wavelets is created for $1 \leq a^i \leq N/K$,

$$\psi_i(n) = \frac{1}{\sqrt{a^i}}\psi\left(\frac{n}{a^i}\right) \tag{9.37}$$

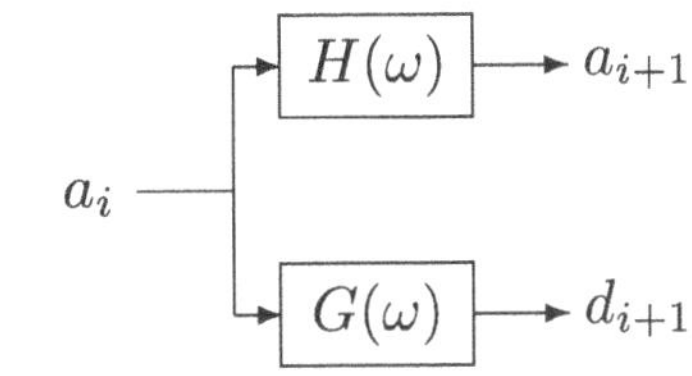

Figure 9.5 Dyadic discrete wavelet transform, single-level, as a pair of filters applied to an input a_i, producing the next level approximation and detail coefficients.

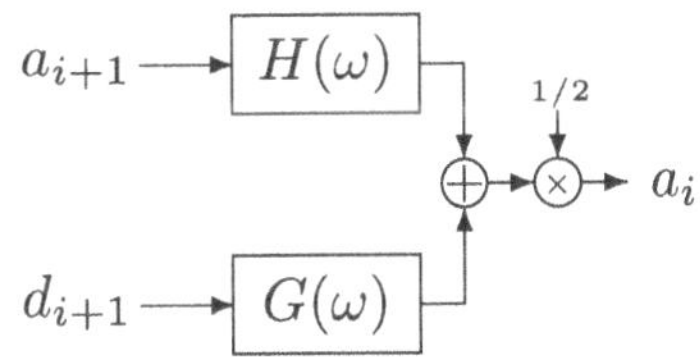

Figure 9.6 Inverse dyadic discrete wavelet transform, single-level, as a pair of filters applied to the approximation a_{i+1} and detail d_{i+1} coefficients, producing the scale signal a_i .

The discrete wavelet transform can then be computed as a convolution of $x(n)$ by $\overline{\psi}(-n)$,

$$X(n,a^i) = \sum_{m=0}^{N-1} x(m)\overline{\psi}(n-m), \tag{9.38}$$

for which a fast Fourier transform algorithm can be employed. The analysis is not complete, as larger scales, corresponding to low frequencies, are not included. However, if energies at these frequencies are negligible, then we can reconstruct the signal using a discrete form of Eq. 9.36 (Mallat, 2008, 115),

$$x(n) \approx \frac{2\log(a)}{C_\psi}\Re\left\{\sum_{i=1}^{I}\frac{1}{a^i}X(.,a^i) * \psi_i(n)\right\} \tag{9.39}$$

A special case exists where $s = 2^i$, called a *dyadic* wavelet transform. In this case, the transform can be computed using a cascade of pairs of highpass and lowpass filters. The outputs of the highpass filter provides the scale, or detail, coefficient, and the lowpass the residual low-frequency information, or approximation, which is downsampled and then reapplied to the filters. The process is repeated for each scales up to 2^I. At every level of the analysis, the number of coefficients is reduced, due to the decimation. The size of the output depends on the length of the filter corresponding to the wavelet employed. This method is called the *fast wavelet transform*.

To implement this algorithm, the two filters are designed according to a corresponding wavelet function. The highpass filter has an impulse response $h(n)$ defined by Eq. 9.35 and the lowpass IR $g(n)$ is obtained from its quadrature mirror filter.[18] A signal scale of scale a_i, which is the downsampled approximation output of the previous stage (or the transform input if at first level), is put through the filters, producing the input to the next level (a_{i+1}) and the detail coefficients for this level (d_{i+1}), as shown in Fig. 9.5. The inverse transform can be obtained recursively from the detail and approximation coefficients. Starting with the highest level, the next finer scale signal is the sum of the filter outputs, scaled by 1/2 (Fig. 9.6) (Mallat, 2008, 177). This is an exact inverse transform.

Probably the simplest example is given by the Haar wavelet, which is defined as (Mallat, 2008, 3)

$$\psi(t) = \begin{cases} 1, & 0 \leq t < \frac{1}{2} \\ -1, & \frac{1}{2} \leq t < 1 \\ 0, & \text{elsewhere,} \end{cases} \tag{9.40}$$

which is similar to one period of a continuous square wave. For this wavelet, the highpass and lowpass decomposition filter impulse responses are

$$\begin{aligned} h(n) &= \{-1, 1\} \\ g(n) &= \{1, 1\} \end{aligned} \tag{9.41}$$

normalized by $1/\sqrt{2}$.

There are several different wavelet bases, which can be chosen according to their characteristics. A Python implementation of the discrete wavelet transform is given by the PyWavelets package (Lee et al., 2019). With this it is possible to perform single- or multilevel analyses using a variety of wavelets. The following code shows an example, implementing a Haar wavelet transform of a sequence. It calculates all scales up to $2^I = N$, and in order to do so, it truncates N to a power-of-two length no greater than the input size. This code works for all length-2 wavelet filters. The output consists of the detail coefficients for all scales, with the finest (lower-order) last in the array. The first pair of coefficients are the approximation and detail at the highest level.


```
import pywt

def haarlet(x):
  len = 1
  while len <= x.size:
    len *= 2
  len //= 2
  out = np.array(x[:len])
  w = pywt.Wavelet('haar')
  while len > 1:
    a,d = pywt.dwt(out[:len],w)
    len //= 2
    out[:len] = a
    out[len:len*2] = d
  return out
```

The original N-samples of the time-domain signal can be reconstructed using the inverse transform. Starting with the coefficient array arranged as before, with the highest scale first, we have

```
def ihaarlet(x):
  len = 1
  out = np.array(x)
  w = pywt.Wavelet('haar')
```

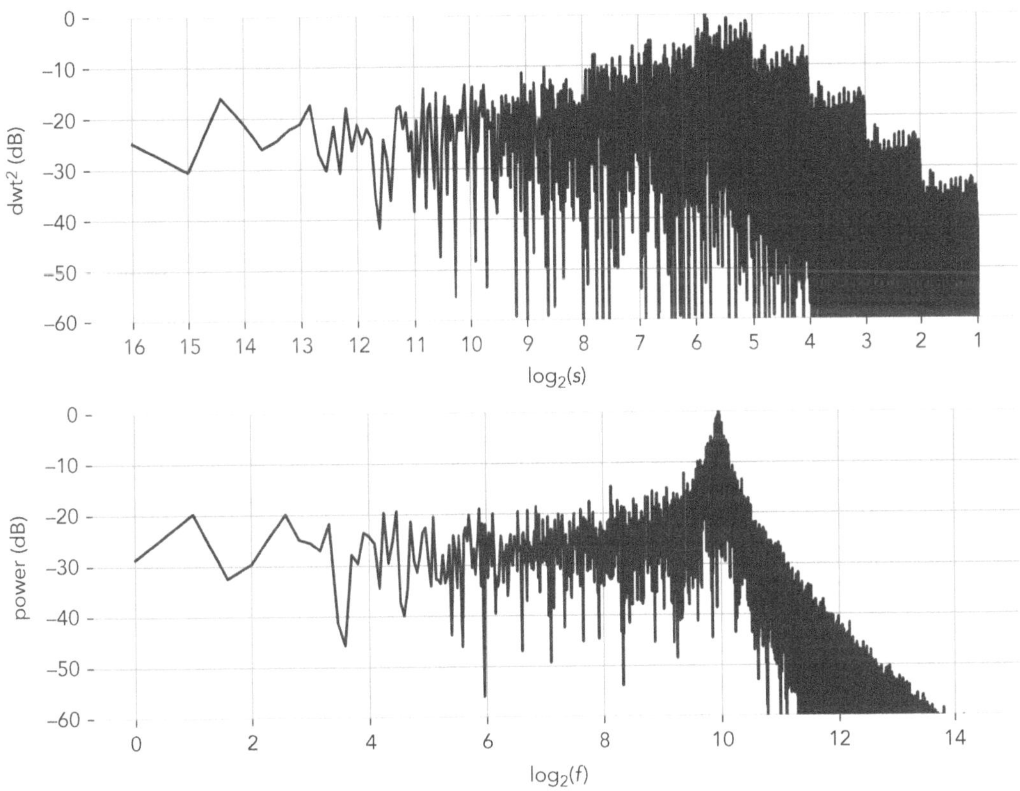

Figure 9.7
Two plots of the spectrum of a filtered noise waveform, showing the squared discrete Haar wavelet transform at different scales (*top*), and the corresponding power spectrum from a DFT analysis (*bottom*) in log frequency.


```
while len < x.size:
  print(len)
  a = pywt.idwt(out[:len],out[len:len*2],w)
  len *= 2
  out[:len] = a
return out
```

The discrete wavelet transform coefficients are not as straightforward to interpret as the output of a DFT. We have the concept of scale to grapple with, and the coefficients are the analysis of the waveform at different scales. Figure 9.7 shows a plot of the Haar dwt as applied to a filtered noise waveform, compared to a straight power spectrum from a DFT. As can be seen, the analysis picks up on the fact that the peak of the spectrum is around 1,000 Hz, corresponding to level-2^5 coefficients. So we can associate the higher-order scales with low frequencies and vice-versa quite closely in a log frequency plot. The Haar wavelet may be a too blunt instrument for the types of analysis that could prove useful here, but that is not to say that the wavelet method in general does not have interesting applications in audio signal processing.[19]

9.3.3 Matching Pursuit

A related granular method of analysis is given by the *Matching Pursuit* algorithm (Mallat and Zhifeng Zhang, 1993), where signals are decomposed into grains or atoms from a given dictionary (e.g., Gabor wavelets of various scales, translations, and center frequencies). The approach is to compute the projection of the grain vector and a residual, then repeating the decomposition, trying to minimize the residue. The signal then is going to be the mix of all the grains weighted by their projection coefficients and the remaining residue. For a grain $g_m(t)$, we have

$$c_m = \int_{-\infty}^{\infty} x(t)\overline{g_m}(t) \tag{9.42}$$

and the signal is first decomposed as

$$x(t) = c_0 g_0(t) + R_x^1(t), \tag{9.43}$$

where R_x^m is the residual at order m, so that $R_x^0(t) = x(t)$. The M-level matching pursuit decomposition is then

$$x(t) = \sum_{m=0}^{M-1} c_m g_m(t) + R_x^M(t). \tag{9.44}$$

We can view Eq. 9.44 as a synthesis equation, which may, for instance, describe the asynchronous granular synthesis method we developed earlier on. As an analysis process, we can view it as a form of mosaicking, where we attempt to represent a signal with a series of small blocks (as it is done in imaging and visual art). Computing the coefficients for a coefficients taken from a given dictionary may be performed as an offline operation.[20] With these and the corresponding grains, resynthesis is possible.

9.4 The Spectral Envelope Revisited

As we noted at the start, when it comes to noise waveforms, we are particularly interested in the spectral envelope. Understanding the different shapes, and the time evolution of these, for the various types of noisy sources is extremely useful in sound design. We have already developed a couple of methods to extract and manipulate spectral envelopes. The first one, as part of a spectral analysis-synthesis approach, used the cepstrum transform to smooth the amplitude spectrum and produce an approximation to the spectral envelope. The other one was the channel vocoder, a multiband analyzer that produced energy measurements at selected center frequencies. These could then be used to approximate the spectral envelope via a matching parallel filter bank. The complete analysis and synthesis filters constituted a form of adaptive equalization. Besides these, we can look at other ways to encode the spectral envelope, which are neither based on Fourier analysis nor in splitting the spectrum into bands, but on observation models. These can take advantage of the ideas we have developed earlier in this chapter in terms of random distributions of unknown processes.

9.4.1 Linear Prediction

Let's suppose we have a digital signal $s(n)$, of which we know nothing to start with, but assume that it behaves in ways that can be predicted. In that case, we could approach the problem of modeling $s(n)$ by trying to find its value at time n as a linear combination of past samples, plus some error or residual value. This is how such an idea can be expressed,

$$s(n) = \sum_{m=1}^{M} \beta_m + \varepsilon(n), \tag{9.45}$$

where the $s(n-m)$ etc are the past samples of the signal and β_m are weights attached to each one of them. The magnitude of the error $\varepsilon(n)$ is dependent on how well the signal can be predicted. Let's take the example of a steady-amplitude sinusoid of frequency f. It is possible to predict $s(n)$ with the use of only two past samples, with scaling factors set to

$$\beta_1 = 2\cos(2\pi f/f_s) \qquad \beta_2 = -1, \tag{9.46}$$

which can be used as recursive formula for a fast sinusoid oscillator. Furthermore, we can vary these two weights to give a rising or falling amplitude sinusoid.

At this point, maybe a sense of déjà vu is setting in. We have seen this before in some other shape. This prediction is defined by a Chebyshev polynomial of the second kind, which, as we have studied, also describes the impulse response of a resonator. Yes, what we have here is the resonator equation, with $x(n) = \delta_{0n}$ in place of $\varepsilon(n)$,

$$y(n) = x(n) - b_1 y(n-1) - b_2 y(n-2), \tag{9.47}$$

with $b_1 = -\beta_1$ and $b_2 = -\beta_2$. This means that the linear prediction equation (Eq. 9.45) can be re-written as an M-order all-pole[21] IIR filter,

$$y(n) = a_0 x(n) - \sum_{m=1}^{M} b_m y(n-m). \tag{9.48}$$

This reinterpretation of the filter coefficients as predictors allows us now to model the spectral envelope. Once we have these, we can use these to construct an all-pole filter to process audio as usual. At the synthesis stage, we can consider the error or residual as a undetermined source signal. A single impulse will produce the filter impulse response, as we know; this, scaled by a_0, is the approximation of $s(n)$. In the case of noisy spectra, resynthesizing the modeled spectrum can be done with a broadband source such as white or pink noise. However, we can use any waveform we like, so this method can be applied to cross-synthesis. Since the method works for deterministic and non-deterministic signals, it is a good approach for an all-round spectral envelope modeling. It has been particularly successful in extracting the formants of speech sounds.

9.4.2 Computing Prediction Coefficients

What we need to do now is to figure out a method of computing the prediction coefficients. Once we have done this, we have a complete analysis-synthesis approach. This may also be able to generate dynamic spectra as we can change the filter coefficients over time. Starting with the all-pole equation, and replacing $y(n)$ by $s(n)$, we consider the predicted value $\tilde{s}(n)$ as consisting solely of a linear sum of weighted past samples (Makhoul, 1975),

$$\tilde{s}(n) = -\sum_{m=1}^{M} b_m s(n-m). \tag{9.49}$$

The error $\varepsilon(n)$ is now $s(n) - \tilde{s}(n)$, and its square magnitude is

$$E = \sum_{n=0}^{M} \varepsilon(n)^2, \tag{9.50}$$

which also corresponds to the squared value of the filter scaling factor a_0.

The goal is to minimize this error so that the prediction coefficients can provide a faithful description of the signal. For this, we can now use some of the theory developed earlier in the modeling of stochastic processes. In particular, we saw that, for stationary signals, the power spectrum density is equivalent to the cosine transform of the autocorrelation sequence,

$$S_{xx}(f) = \int_{-\infty}^{\infty} r_{xx}(\tau)\cos(2\pi f\tau)d\tau. \tag{9.51}$$

Therefore we can estimate the signal power at different frequencies by taking its z-transform, since the cosine transform is a special case of it. The next step is to define the autocorrelation of $s(n)$. From Eq. 9.10, a finite length sequence $r(n)$ may be defined as

$$r(n) = \sum_{m=0}^{N-1-n} s(m)s(m+n), \tag{9.52}$$

for $0 \leq n < N$. Now we can call the power spectrum density of $s(n)$ the z-transform $S(z)$ of $r(n)$. The next assumption we now make is that the all-pole filter power spectrum $|H(z)|^2$ approximates the power spectrum density of the predicted signal,

$$S(z) \approx |H(z)|^2. \tag{9.53}$$

From this we can also infer that $s(n)$ can be replaced by its autocorrelation on both sides in Eq. 9.49. The effect of that is to minimize the error E, since it equates the power spectrum on both sides of the equation. This gives us a system of equations,

$$\sum_{m=1}^{M} b_m r(n-m) = -r(n), \tag{9.54}$$

for $1 \leq n \leq M$, which can be used to compute the prediction coefficients b_m. The total squared error is

$$E_M = r(0) + \sum_{m=1}^{M} b_m r(n-m). \tag{9.55}$$

It turns out that the system of equations in Eq. 9.54 can be computed by an efficient recursive algorithm[22]

$$\begin{aligned}
E_0 &= r(0) \\
k_m &= -\frac{r(m) + \sum_{i=1}^{m-1} b_i(m-1)r(m-i)}{E_{m-1}} \\
b_m(m) &= k_m \\
b_i(m) &= b_i(m-1) + k_m b_{m-i}(m-1), \qquad 1 \leq i \leq m-1 \\
E_m &= (1-k_m^2)E_{m-1}.
\end{aligned} \tag{9.56}$$

This recursive method computes the solutions for coefficients for all lower-order predictors before the final set is found. This is given by $b_i = b_i(M)$, for $1 \leq i \leq M$. The autocorrelation input can also be normalized by $r(0)$ to scale the solution to Eq. 9.54. Given an autocorrelation signal $r(n)$, the algorithm in Eq. 9.54 has $O(N)$ complexity.[23] We can now translate this algorithm into a Python function that takes an input signal, a predictor order, and outputs the all-pole coefficients and squared error. To smooth the analysis, the function can optionally apply a window to the autocorrelation sequence,

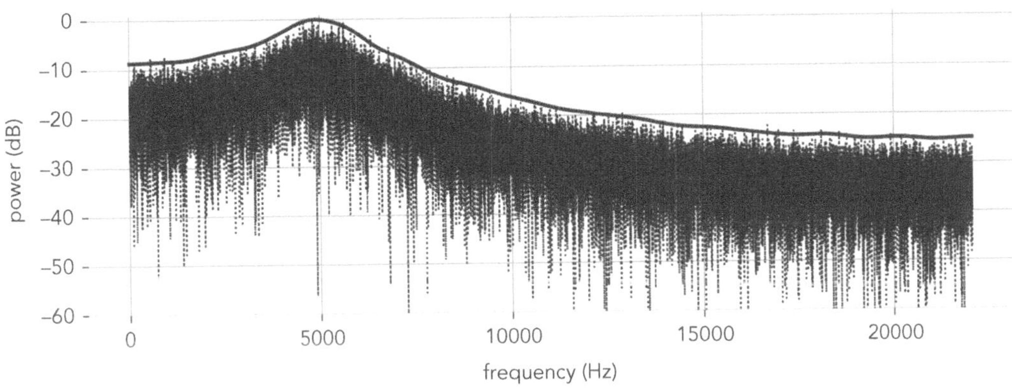

Figure 9.8 Power spectra (*normalized*) of a filtered noise waveform (*dots*) and its linear prediction with an order-30 filter (*solid line*).


```
# autocorrelation of x
def acr(x):
    r = np.correlate(x,x,mode='full')
    return r[r.size//2:]

# Linear Prediction function
# takes an input x and predictor order M
# and optional window w
# outputs M+1 coefficients as an array
# and squared error
def lpred(x,M,w=None):
 E = np.zeros(M+1)
 b = np.zeros((M+1,M+1))
 k = np.zeros(M+1)
 if w is None: r = acr(x)[:M+1]
 else: r = acr(x[:w.size]*w)[:M+1]
 r /= r[0]
 E[0] = r[0]
 b[0,0] = 1
 for m in range(1,M+1):
  s = sum(b[m-1,:m]*r[m:0:-1])
  k[m] = -(s)/E[m-1]
  b[m,m] = k[m]
  b[m,1:m] = b[m-1,1:m] + k[m]*b[m-1,m-1:0:-1]
  E[m] = (1 - k[m]**2)*E[m-1]
  b[m,0] = 1
 return b[M], E[M]
```

To demonstrate the process of extracting a spectral envelope from an input signal, we use his function to compute predictors for a filtered noise waveform. The resulting spectral envelope is plotted in Fig. 9.8 against the measured power spectrum of the signal.

Depending on the signal being analyzed, employing a window of some description may be useful. We can see an example of its application in Fig. 9.9, where a single sinusoid at 2,500 Hz is being analyzed. While the truncated (rectangular) input results in a

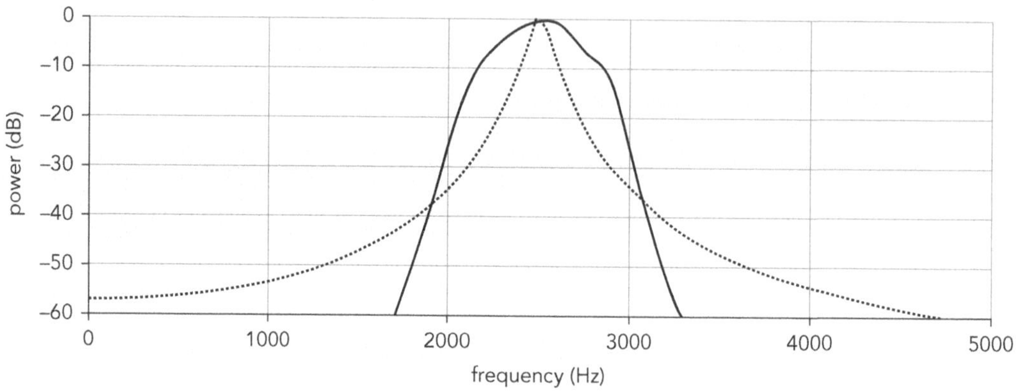

Figure 9.9 Linear prediction of a sinusoidal input using a Hanning (*solid*) and rectangular window (*dots*) input, with order-30 filter extracted from 120 samples at $f_s = 44,100$.

narrower peak but gentler rolloff, the spectrum of a windowed input has a more steep curve, although a wider −3 dB bandwidth. For generally stable signals, the rectangular window should be sufficient, but for more fast-changing spectra, the application of other windows may give a cleaner analysis (Makhoul, 1975, 569).

Since the linear prediction model is based on an IIR filter, we should consider the question of filter stability. If the autocorrelation function is computed from a non-zero signal, Eq. 9.54 should provide coefficients that make the filter poles lie inside the unit circle (Makhoul, 1975, 567). Furthermore, if we are computing the coefficients for short sections of input, and replacing these regularly (e.g., on a streaming basis), then it is possible that any unstable pole might not be present for long enough to cause the filter to blow up (Moore, 1990, 299). Nevertheless, it is also possible to employ routines to check the roots of the polynomial, making sure none of them leads to a singularity. However, root checking may be computationally expensive, and if the analysis is to be effected in realtime, then we may need to assume that the resulting polynomial has all of its poles inside the unit circle.

9.4.3 Synthesis

The following Python class implements a linear prediction filter as an example. Using the analysis code, it takes in an input and extracts filter coefficients. The processing code then implements the all-pole filter using a convolution of the prediction coefficients and the filter output:

```
class lpfil:
  def __init__(self,x,M):
    b,E = lpred(x,M)
    self.g = np.sqrt(E)
    self.b = b[1:]
    self.N = self.b.size
    self.yn = np.zeros(self.N)
    self.rp = 0

  def process(self,x):
```

```
n = 0
rp = self.rp
y = x*self.g
N = self.N
while n < N:
  y -= self.b[N - n - 1]*self.yn[rp]
  n += 1
  if rp == N - 1: rp = 0
  else: rp = rp + 1
self.yn[self.rp] = y
if self.rp == N - 1: self.rp = 0
else: self.rp = self.rp + 1
return y
```

A higher value of M will of course match the spectrum more finely, but it will also increase computation. We should also consider the length of the signal N being analyzed. Since we assume it is stable over that duration, similar considerations regarding time versus frequency resolution, which have been present in all our spectral methods, apply here. While the code above performs a single analysis when the object is constructed, it may be extended to perform multiple regular predictor computations so that we can model time-varying spectra.

9.4.4 Spectral Representations

A linear predictor may be represented in alternative forms. To implement the synthesis code, we have of course employed its filter description,

$$H(z) = \frac{a_0}{1 + \sum_{m=1}^{M} b_m z^{-m}}, \tag{9.57}$$

but a variety of other representations are possible.

For example, for even orders M, the all-pole model can be factored out as cascade of second-order resonators, with pole magnitudes R_k and frequencies ω_k,

$$H(z) = a_0 \prod_{k=1}^{M/2} \frac{1}{1 + 2R_k \cos(\omega_k) z^{-1} - R_k^2 z^{-2}} \tag{9.58}$$

which can be obtained from center frequencies f_k and bandwidths B_k, as we have seen before. This model is particularly useful if we want to track formants, as each resonator can attempt to reproduce a formant region, and the spectrum is then matched that way. In this case, instead of using the direct method of coefficient estimation, some probabilistic model is employed, and an iterative approach is used, successively correcting the error and approximating the spectral envelope. An example of this is found in the use of extended Kalman filters as a formant tracking method.[24]

The spectral description of linear prediction can also be very useful, as we may intermix it with the other forms of analysis-synthesis developed throughout this book. We can represent the relationship between signal, prediction, and error in a spectral form as

$$S(z) = E(z) \frac{|H(z)|^2}{E_M} \tag{9.59}$$

where $E(z)$ is the power spectrum density of the error signal $\varepsilon(n)$. The denominator of $H(z)$ (Eq. 9.57) can also be described as an inverse filter $H_I(z)$, whose power spectrum is denoted by $B(z)$

$$S(z) = \frac{E(z)}{B(z)} \tag{9.60}$$

From this we may have yet another representation for the linear predictor. Since $B(z)$ is the z-transform of the autocorrelation of the filter coefficient sequence b_m, we have

$$\rho(n) = \sum_{m}^{M-n} b_m b_{m+n}, \tag{9.61}$$

for $0 \leq n \leq M$, and $b_0 = 1$.

Furthermore, since $S(z)$ is the power spectrum of the observed signal, we can use its inverse DFT to get the autocorrelation sequence for computing the predictor coefficients. This might be useful in spectral analysis-synthesis context, since we can get $S(k)$ by squaring the amplitude of bin k in an analysis frame. From the computed coefficient sequence b_m we have two basic options to get a spectral representation of the linear predictor. The first one is to compute the amplitude spectrum of the filter, and the second is to extract the cepstrum coefficients from which a spectral envelope can be derived.

9.4.4.1 Amplitude Spectrum

The sequence of filter coefficients b_m is the impulse response $h_I(n)$, of the inverse filter $H_I(z)$. Therefore, we can proceed by zero-padding this signal to the transform size N, then take its DFT, using a real FFT algorithm, which will give use the complex spectrum $H_I(k)$. The next step is to compute the amplitudes $A(k)$. The spectral envelope is then given by

$$|H(k)| = \frac{a_0}{A(k)} \tag{9.62}$$

Alternatively, we can look to obtain $B(z)$ instead from $\rho(n)$, which is computed from b_m by Eq. 9.61. Since the autocorrelation is even, its spectrum is real-valued,

$$B(\omega) = \rho(0) + 2\sum_{k=1}^{M} \rho(k)\cos(\omega k). \tag{9.63}$$

We can now take the DFT $P(k)$ of the sequence

$$p(n) = \begin{cases} \rho(n), & n = 0 \\ 2\rho(n), & 1 \leq n \leq M \\ 0, & n > M \end{cases} \tag{9.64}$$

and get $B(k) = \Re\{P(k)\}$. This is the power spectrum of the inverse filter, therefore

$$|H(k)| = \frac{a_0}{\sqrt{B(k)}} \tag{9.65}$$

This second method is more appropriate in cases where the power rather than the amplitude spectrum is required. It can also be computed using a Discrete Cosine Transform instead of a DFT.

9.4.4.2 Cepstral Coefficients

A third representation of linear predictors can be given in the form of cepstral coefficients. These of course could be computed from the amplitude or power spectrum forms, but in this case it would be more work than necessary. However, there is much simpler relationship between b_m and the cepstrum, which can be useful in some applications. We may remember that in Chapter 6, we defined the cepstrum as the Fourier transform of the log amplitude spectrum. Since we have been employing the power spectrum density to get the linear prediction coefficients, we will instead take the logarithm of the squared magnitudes as an input to the transform. From this, the continuous cepstrum of the power spectrum of the all-pole filter is given by

$$\begin{aligned} c(\nu) &= \int_{-\infty}^{\infty} \log(|H(f)|^2) e^{-j2\pi f \nu} df \\ &= \log(a_0^2) - \int_{-\infty}^{\infty} \log(B(f)) e^{-j2\pi f \nu} df. \end{aligned} \tag{9.66}$$

For $\nu = 0$, the integral on the left vanishes and we have,

$$c(0) = log(E_M). \tag{9.67}$$

To compute the cepstrum coefficients of $B(z)$, we use the fact that this is assumed to be a minimum-phase filter.[25] This gives a recursive expression to compute all cepstral coefficients, $1 \leq n \leq M$ (Makhoul, 1975, 575),

$$c(n) = \begin{cases} b_n, & n-1 \\ b_n - \sum_{m=1}^{n-1} \frac{m}{n} c(m) b_{n-m}, & 1 < n \leq M. \\ 0, & n > M. \end{cases} \tag{9.68}$$

These are the cepstral coefficients for the inverse filter. For the all-pole case, they are given by $-c(n)$. To represent the linear predictor, these should be all we need. The spectral envelope is extracted by taking the DFT $C(k)$ of this sequence,

$$|H(k)| = |e^{C(k)}|, \tag{9.69}$$

where as before $H(k)$ is the spectrum of the all-pole filter. The following Python code produces an N-point cepstrum from the linear prediction parameters:

```
# lpc to cepstrum
# Takes coefficients b and
# squared error E, and
# cepstrum size N
# returns cepstral coefficients
def lpceps(b,E,N):
    c = np.zeros(N)
    M = b.size - 1
    c[0] = -np.log(E)
    c[1] = b[1]
    for n in range(2,N):
     if n > M:
       c[n] = 0
```

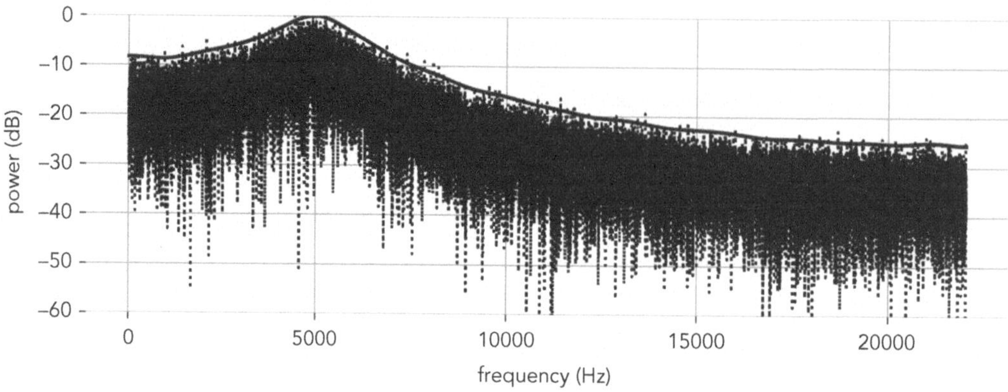

Figure 9.10 Power spectra (normalized) of a filtered noise waveform (*dots*) and the spectral envelope obtained by the inverse cepstrum from an order-30 predictor.

```
    else:
        s = 0.
        for m in range(1, n):
            s += (m/n)*c[m]*b[n-m]
        c[n] = b[n] - s
    return -c
```


Note that we have to be careful to reverse the sign of $c(0)$ before we perform the computation since we output the negative coefficients as per Eq. 9.68. In Fig. 9.10, we show the power spectrum of a filtered noise waveform and its spectral envelope from linear prediction cepstral coefficients.

By examining Eq. 9.68, we can define its inverse. With this, we can compute the M all-pole filter coefficients, $0 \leq m \leq M$ from a real-valued cepstrum,

$$b_m = \begin{cases} 1, & m = 0, \\ c(1), & m = 1, \\ -c(m) - \sum_{n=1}^{m-1} \frac{m-n}{m} b_m c(m-n), & 1 < m \leq M. \end{cases} \tag{9.70}$$

We can then set $E = e^{c(0)}$. The following code implements this equation, returning the all-pole coefficients and squared error,

```
# cepstrum to lpc
# Takes cepstrum b and
# predictor order M
# returns lpc coefficients and
# squared error
def cepslp(c,M):
    M += 1
    b = np.zeros(M)
    b[0] = 1
    b[1] = -c[1]
    for m in range(2,M):
        s = 0.
        for n in range(1, m):
```

```
            s -= (m-n)*b[n]*c[m-n]
        b[m] = -c[m] + s/m
    E = np.exp(c[0])
    return b,E
```

We may remember from Chapter 6 that in order to extract a smooth spectral envelope, we normally keep only a small number of low-order coefficients. It may be possible, under certain circumstances, to construct an all-pole representation of the cepstrum spectral envelope. However, we should note that in order to be converted using this algorithm, the cepstrum needs to be computed from the power rather than the amplitude spectrum of a waveform. This is because the assumption built into Eq. 9.70, which was derived from Eq. 9.66, is that we are dealing with the spectrum of an autocorrelation sequence, or the power spectrum density. For example, the following code produces a filter HC from a cepstral analysis of an arbitrary input signal, alongside another, HL, from a predictor of the same order. Note that since the cepstrum coefficients are real-valued, we can use a complex-to-real inverse FFT to compute them. The fact that we have swapped the forward for an inverse transform to compute the cepstrum has no bearing on the result.[26].


```
S = np.abs(np.fft.rfft(sig))**2
c = np.fft.irfft(np.log(S))
b,E = cepslp(c,30)
bb = np.zeros(c.size)
bb[:b.size] = b
HC = np.sqrt(E)/(abs(np.fft.rfft(bb)))
# lpc filter
b,E = lpred(sig,30)
bb = np.zeros(c.size)
bb[:b.size] = b
HL = np.sqrt(E)/(abs(np.fft.rfft(bb)))
```

We can apply this to model the spectrum envelope of a filtered noise signal such as the one in Fig 9.8. The normalized amplitude responses for the resulting cepstrum-analysis and linear-prediction all-pole filters are shown in Fig. 9.11. As can be seen, in this example the cepstrum-based filter approximates the linear-prediction model fairly well. However, there is a snag, as the resulting filter is not guaranteed to be stable if it is simply derived from spectral and cepstral analyses, and numerical errors may cause slight discrepancies in the two estimates of the spectral envelope. Checks for stability may be required.

In any case, the two approaches (cepstrum or linear prediction) provide different alternatives for spectral envelope design, depending on the circumstances and the type of input data to be analyzed. If we happen to have only time-domain signals, the linear prediction approach is a very good choice, but if the signals have already been put into a spectral form, then cepstral analysis might be more easily implemented.

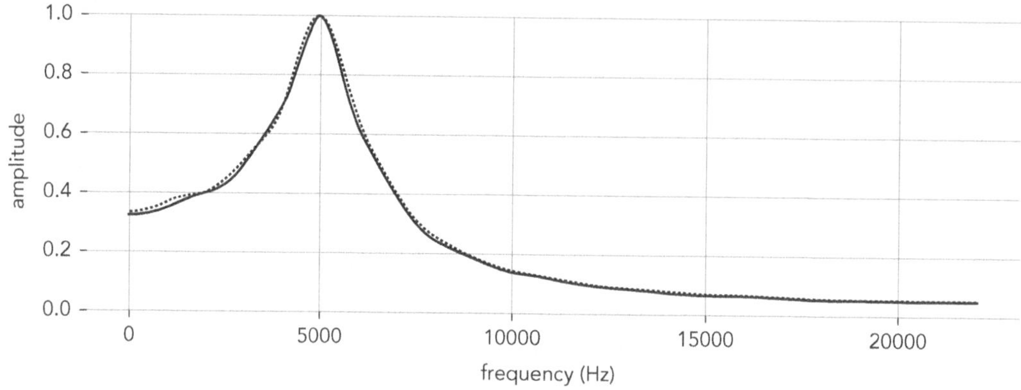

Figure 9.11 Normalized amplitude responses of two filters generated by cepstral analysis (*solid line*) and linear prediction (*dots*) of an arbitrary broadband noise source.

9.4.5 Streaming Linear Prediction

As noted in Sec. 9.4.3, a streaming implementation of the analysis-synthesis methods developed here is possible. In this case, we can take an arbitrary signal as input, apply linear prediction to estimate the filter coefficients, then use these coefficients in an all-pole filter to process another signal. This approach can be likened to the vocoder operation. There are three important parameters in this process:

- The linear predictor order.
- The autocorrelation size.
- The analysis frequency.

We have seen the first two of these. The linear predictor order determines the all-pole filter length and has an effect on the spectral envelope detail. Too few coefficients and the shape of the spectrum is only captured in general lines, too many make the computation less efficient. The autocorrelation size determines how much of the input we are analyzing, and that balances frequency and time precision. Longer frames average out time precision, and are also more expensive to compute, although they provide better-quality analysis. The final parameter determines how often we want to replace the coefficients. The minimum time spacing may be a single sample, but the maximum is undefined (e.g., we can have a single frozen set of coefficients over a given duration).

To reproduce the original spectral envelope more faithfully, we would want to perform overlapped analyses. That is, we want an analysis period that is shorter than the autocorrelation length. This will make the new coefficients not too dissimilar to the previous ones, minimizing artifacts. Another helpful idea is to employ analysis windows, which tend to produce smoother results. Note that there are no particular restrictions to the analysis period, filter length, and autocorrelation length, so these can be modified to suit different types of inputs.

Finally, it is also important to consider the filter input scaling. In the earlier synthesis example, we used the analysis error as an input gain, which is the correct approach. However, we can also employ the analysis input RMS to complement it. We can compute it as

$$s_{rms} = \sqrt{\frac{r(0)}{N}}, \tag{9.71}$$

where, as before $r()$, is the autocorrelation signal of size N. The filter input can then be scaled by $s_{rms}\sqrt{E}$, which should keep its output within the normal range.

Finally, since we have to compute the autocorrelation, we may as well use it to estimate the pitch of the input signal. The fundamental frequency appears as a peak in the autocorrelation function at a given lag, defined by the fundamental period. Dividing the sampling rate by this yields the estimated pitch in Hz. This method is not the most precise, particularly as it is prone to octave errors. We may apply a peak picking procedure (e.g., for instance, the one described in Sec. 9.5.2) to isolate the most prominent peaks in the autocorrelation function, and then determine which one is the correct to use employing some heuristics (e.g., the highest peak, or the one within a certain range, etc.).

In Csound, linear prediction is implemented in two ways, through the use of a utility program, `lpanal`, based on the original Lansky program (Boulanger, 2000), and by a suite of streaming opcodes. The latter include both analysis, synthesis, and conversion operations in various different combinations. For example, the following opcode, provided in two overloaded versions, implements linear prediction analysis:


```
kCfs[],krms,kerr,kcps lpcanal asrc,kflg,kprd,kN,kM[,ifw]
kCfs[],krms,kerr,kcps lpcanal koff,kflg,ifn,kN,kM[,ifw]
```

The first of these takes an input signal as the linear prediction source, and the second a function table. In the former case, a new analysis is taken every `kprd` samples, provided that `kflg` is non-zero. In the other opcode, linear predicition is always performed if `kflg` is non-zero, and the input samples are taken from the function table starting at the `koff` position (in samples). The input size `kN` and predictor order `kM` are as discussed before, and an optional window function stored in table `ifw` may be applied. The output is an array of M coefficients (b_1 to b_M), and the estimates of the input RMS, the error, and the fundamental frequency (from autocorrelation).

The filter coefficients may be used for resynthesis with

```
a3 = allpole(asig,kCfs[])
```

which is an all-pole filter implementation using direct convolution (as per the code in Sec. 9.4.3). It takes a filter input, and the all-pole coefficients b_1 to b_M, which can actually be obtained by any means, including linear prediction, but are assumed to be stable (no checks are performed). The input is not scaled, but if `lpcanal` was used, we can apply the RMS and error values as multipliers to keep levels within a reasonable range.

In addition to these, the `lpcfilter` opcode, also in two versions, can be used as a single analysis-synthesis processor,

```
aout = lpcfilter(asig,asrc,kflg,kprd,kN,kM[,ifw])
aout = lpcfilter(asig,koff,kflg,ifn,kN,kM[,ifw])
```

where the estimate of the spectral enveloped is applied directly using an all-pole filter.

Alternatively, it is possible to produce a streaming phase vocoder signal from linear prediction analysis with the `pvslpc` opcode, and then apply it as a filter in the spectral domain,

```
fenv = pvslpc(asrc,1024,128,64,gifw)
```

```
fexc = pvsanal(asig,1024,128,1024,1)
fsig = pvsfilter(fexc,fenv,1,0.5)
```

and, from the other direction, starting with an fsig, we can get all-pole coefficients via the cepstrum transform as outlined earlier, and apply these to time-domain filtering,

```
fenv  = pvsanal(asrc,1024,256,1024,1)
kcfs[],krms,kerr pvscfs fenv,30
a3 = allpole(asig*krms*kerr,kcfs)
```

These opcodes provide a bridge between streaming linear prediction and streaming spectral processing. The first one, `pvslpc`, produces an fsig consisting of the spectral envelope (encoded as the bin amplitudes) and the autocorrelation frequency (as a harmonics of the fundamental frequency, encoded as the bin frequencies). While this signal can be resynthesized directly, frequency artifacts will be present. These will vary depending on the fundamental estimation and the number of coefficients used in the analysis. Its most typical application is, however, as a filtering signal (that is, discarding its frequency data). The other opcode, `pvscfs`, extracts the spectral envelope of an fsig, and then converts it to all-pole filter coefficients. To guarantee stability, coefficients are put in polar representation, and checks are run. Any poles outside the unit circle can be either reflected back inside or simply limited to provide a stable filter.

9.5 Spectral Models

While we have been concentrating on noise and distributed spectra, it is often the case that musical sounds will actually contain a combination of deterministic and non-deterministic elements. Therefore, it is also important to consider the sonic object as a whole and try to model its spectrum as a mixture of various components, some with a defined description, others with stochastic characteristics. For example, if we take an instrumental tone of the pitched variety, say from a violin or a flute, it is quite clear to ear that there are stable components in it, carrying the pitch and various other fleeting elements (e.g., caused by friction, wind/breath turbulence, etc.) that are also very important to the total spectrum of that particular waveform. A naive approach to mimicking these instruments is generally focused on the stable spectrum, which is derived from a Helmholtzian view of what constitutes a musical tone. We of course know better and understand that there is much more to it, accepting that the picture is much more complex.

Thus it should be useful to try to model the spectra of musical sounds in a way that allows us to accommodate deterministic and non-deterministic components of sound. At this stage, we have develop a good set of tools to deal with both the stable elements (from the Fourier theory and its extensions), and the stochastic (developed in this chapter). What we need is a method of splitting a waveform into two sets of components, corresponding to the deterministic and non-deterministic categories. Note that the elements we are characterizing as stable here are not limited to static waveforms, but, more generally, spectral components that can be modeled as individual partials. Dynamic spectra then can be thought of a mix of time-varying sinusoids and time-varying stochastic spectra.[27]

There are two basic routes to solve this particular separation problem. We can look at the frequency-domain data that is produced by a time-varying analysis process, which may be for instance the STFT or perhaps a wavelet transform. Then we determine what are the stable components according to some metric and isolate them from the rest of the spectrum. Some sort of comparative or subtractive method can then be applied to separate the non-stable part, which we can call the *residual*.[28] This method can be made to be fairly efficient. It assumes that for the duration of an analysis frame, the waveform is stationary. The other approach, which does not make any such assumptions, is a time-domain method. In this case, the stable components are determined from the STFT analysis, but then resynthesized using a precise method of additive synthesis, which models the inter-frame changes in amplitude and phase of each partial. The residual is then extracted through a sample-by-sample subtraction of the original waveform and the deterministic part.

This method, although somewhat involved, is particularly interesting and worth exploring in detail. The idea is to analyze a waveform into a set of partial *tracks*, which are then sent to be resynthesized. Each track is composed of a set of amplitudes, frequencies, and phases. This data format models the spectrum in a different way from what we have been used to. Instead of rectangular frames made of bins, each one corresponding to a fixed frequency band, we have tracks following the trajectory of a partial. If a partial moves across bins, the analysis will continue to pick up a single sinusoid, whereas in the STFT, the result is that the component will show up at different bands, as it moves in the spectrum. This model is often called a *sinusoidal model* (McAulay and Quatieri, 1986), defined by

$$s(t) = \sum_{k=1}^{K} a_k(t)\cos(\phi_k(t)), \tag{9.72}$$

where K is the number of partials, $a_k(t)$ are their corresponding instantaneous amplitudes, and $\phi_k(t)$ their instantaneous phases. The latter can be computed from the phases and frequencies of each track, and of course, in practice, all functions are sampled at time points $t = nT$.

9.5.1 Partial Tracking

We can break the partial analysis process down into a number of steps, namely:

- STFT data analysis: take windowed DFT analysis at defined time intervals.
- Peak extraction: determining the peaks in the spectrum that may correspond to stable components, from a DFT frame.
- Peak interpolation: approximating the peak locations and their amplitudes.
- Track formation: matching peaks in successive analysis frames to make a spectral tracks modeling partials.
- Frequency and phase allocation: from the peak and frame data, estimate the frequency and phase for each partial.

At one end, we get the audio waveform split into DFT frames taken at regularly spaced intervals. At the other we come up with an array of tracks containing parameters that can be used to resynthesize it. With this description of what we need to achieve an each step of the process, we can now attempt to program it. One of the important code design decisions we need to make at an early stage is how to structure track data. This determines the sequencing of operations and also influences how the resynthesis process is programmed. Since the data format does not follow naturally from the process, as for instance is the case with the straight STFT, we need to consider our options.

If we are designing this as a strictly off-line, non-streaming, process, we could possibly arrange each track as an array or list of analysis parameters indexed by time. Then the resynthesis could proceed from the beginning of each track until its end, summing successive tracks to produced the output signal. However, this would prevent the code from being reused in a streaming fashion, as we did before in the spectral analysis-synthesis system. A better approach is to think of a way to structure tracks as if they were bins, in frames, so that at each analysis point we have a complete description that may be resynthesized. From this, it is also possible to reformat all of the data into a track-oriented structure fairly easily if we need to. The code can then be used in off-line or streaming applications.

Following that decision, we have to solve one small but significant issue: how to define what a track is across successive frames. This is not a problem with bins, since they are always found in the same order in successive analyses. With tracks, the situation is fairly different: they come and go, as partials may appear or disappear when the spectrum changes over time. We need to find a way of connecting tracks from one frame to another, as they may not be ordered in any particular way, so that the data can be used for further modification, resynthesis, or display. One way we could do this is to attach a unique ID element to each track data, so that at each frame we can look for these tags. A track is then easily identified.

There is however an even a simpler way to connect tracks in adjacent frames. We can begin by assuming a track at a given frame index position is a continuation of the track from the previous frame at same index. The only time this is not going to be the case is when the track is dead. So we transformed the problem in a simpler one: how to kill a track. We can do this by using a convention of some kind, like setting one of the parameters, such as the amplitude to zero, or to be negative (since amplitudes, magnitudes, are always non-negative). Continuation can then be determined by checking this parameter. This allows us to reuse the storage space of dead tracks when new ones are found.

We now have the structure of a frame of tracks, and we can define the format of each element of this frame. We know the three parameters needed for resynthesis are amplitude, frequency, and phase. Besides these we can build some redundancy, which may be useful at later stages by adding a time stamp and the peak position. We can then use the time element as a dead track marker: dead tracks have negative time stamp. The track format becomes

```
[time,peak,amplitude,frequency,phase]
```

This can be defined as a standard NumPy floating-point array, so that for instance if we need the frequency of a track, we can just look at `track[4]`. Furthermore, as the number of tracks is not predictable before the analysis start, we can start with an empty track frame and make it grow as more storage is needed. Now that we know the format of the input (a DFT frame) and the output (a frame of tracks), we can look at the intermediate steps.

9.5.2 Peak Identification

In order to identify the peaks in the spectrum, we need the amplitude data in a form where there is good spacing between troughs and crests. We can achieve this by placing the data in a logarithmic scale. This also makes the information more perceptually relevant, as we are attuned to ratios of intensities rather their differences. With this, we then look to see if a given DFT bin contains more energy than its neighbors. If so, we mark it as a peak. We also need to discount peaks with very small amplitudes, so we use a thresholding approach to confirm the finding.

The resulting code is shown below as a Python function. The input to it is a frame of amplitudes in logarithmic form, and the output is an array containing the indexes of the bins where we found a peak,

```
# pick peaks
# takes a DFT and threshold
def pkpick(X,thr):
 N = X.size
 pk = np.zeros(N).astype(int)
 n = 0
 for i in range(0,N-1):
  if i > 0:
   if X[i] > X[i-1]: t1 = True
   else: t1 = False
  else: t1 = False
  if X[i] >= X[i+1]: t2 = True
  else: t2 = False
  if (X[i] > thr) and (t1 and t2):
      pk[n] = i
      n += 1
 return pk[:n]
```

This code is fairly straightforward; we apply two tests to check if a bin has a higher value than its neighbors. If these are passed and the threshold is cleared we have a peak. Otherwise we move to look at the next bin. Note that this code excludes peaks at DC or Nyquist frequency bins. The code return an array whose length is the number of peaks detected.

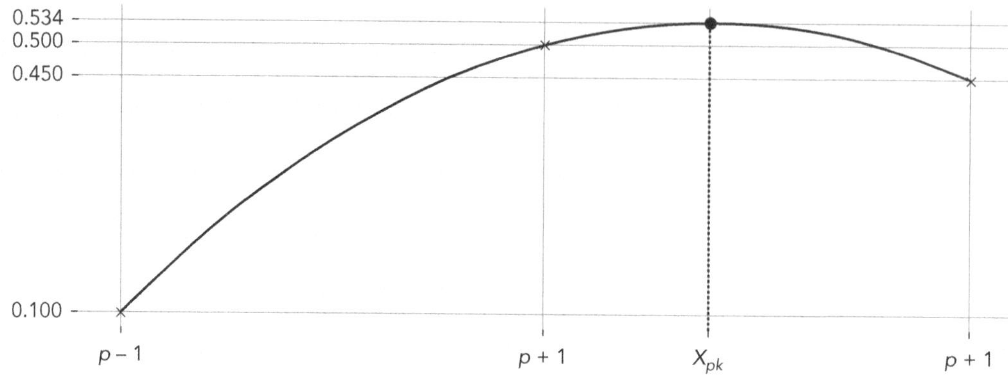

Figure 9.12 Peak interpolation: true peak position x_k computed by fitting a second-order polynomial to three points around a detected peak in bin p.

9.5.3 Peak Interpolation

Once we have the bin indices corresponding to identified peaks, we need to home into a more precise location for these. We can do this by interpolating the log amplitude spectrum from the values found at the analysis bins. Since the peaks were found around three bins, it makes sense to use the values of three adjacent bins in the process. For this we need a second-order interpolator, which takes data from three bins around the central one and estimates the fractional position of the peak. The idea is to fit a polynomial function

$$f(x) = -ax^2 + bx + c, \tag{9.73}$$

so that for a peak bin p we have $f(p-1) = \log(|X(p-1)|)$, $f(p) = \log(|X(p)|)$, and $f(p+1) = \log(|X(p+1)|)$. We first need to compute the coefficients for this polynomial. Solving for $p = 0$ gives us the three coefficients,

$$\begin{aligned} c &= \log(|X(p-1)|) \\ a &= \log(|X(p)|) - \frac{\log(|X(p+1)|) + \log(|X(p-1)|)}{2} \\ b &= a + \log(|X(p)|) - \log(|X(p-1)|) \end{aligned} \tag{9.74}$$

We are interested in the peak of this function, that is, where its derivative is zero,

$$-2ax + b = 0. \tag{9.75}$$

This is the point $x_k = \frac{b}{2a}$ marking the maximum point of the polynomial. The peak amplitude is then given by $f(x_k)$. The interpolation procedure is shown in Fig. 9.12, where the true peak position is located at the vertex of the parabolic curve.

```
# quadratic interpolation
# returns interpolated peaks
# and amplitudes
def pkinterp(pk,X):
 N = X.size
 A = np.zeros(pk.size)
 P = np.zeros(pk.size)
 n = 0
```

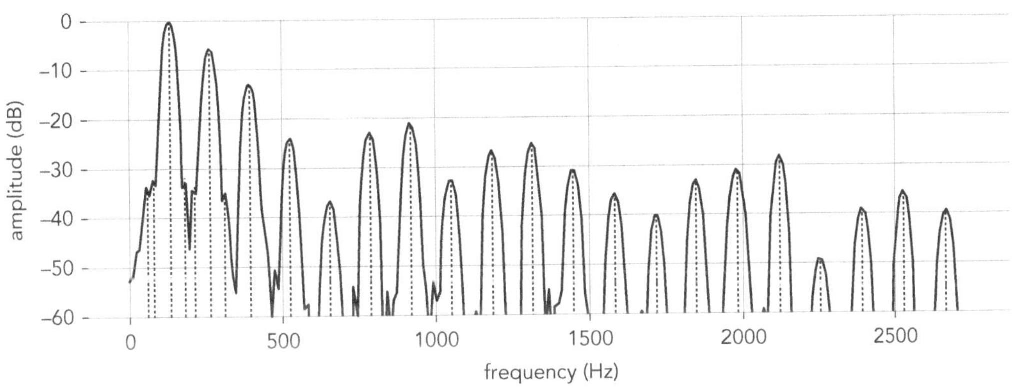

Figure 9.13
Detected peaks (*dotted vertical lines*) from an amplitude spectrum of a piano tone (*solid line*).


```
for p in pk:
  if p != 0: c = X[p-1]
  else: c = X[p+1]
  dy = X[p] - c
  if p < N-1:
    ddy = X[p+1] - c
  else:
    ddy = X[p] - c
  a = (ddy-2*dy)/2
  b = 1-dy/a
  P[n] = p-1+b/2
  A[n] = np.exp(y-a*b*b/4)
  n += 1
return np.array([P,A])
```

The code computes not only the interpolated peak position but also its interpolated log amplitude. A 2-dimensional array containing a row of peaks and a row of amplitudes is returned. The size of each row is determined by the number of peaks detected. In Fig. 9.13, a plot of an amplitude spectrum of a piano tone is shown against the extracted and interpolated peak amplitudes. Although a few low-amplitude spurious *false positives* are picked up, the detection routine generally does a good job of finding partials in a predominantly deterministic spectrum such as this one.

9.5.4 Track Formation

Track formation is very much dependent on program logic, using checks for matching and continuation as the means to create and kill tracks. It works on a process of elimination: first we find what continues, then we deal with creating new tracks. It is also entirely dependent on the previous peak identification steps, the previously found peaks, and the existing tracks. It is recursive in nature, we can only produce new data on the basis of past outputs.

An array of peaks does not constitute a track, as we know. We need to check for inter-frame continuation to determine if a track exists. The key to this is a process called peak matching. We take a currently found peak and look for the closest one in a list of existing peaks. The following Python function implements this process, given a peak and an array of peaks to search,

```
# matches a peak to an existing
# array of peaks
# returns matched peak index
# or -1 if not found
def pkmatch(pk,pks):
    N = pks.size
    if N == 0: return -1
    dif = np.zeros(N)
    for i in range(0,N):
       dif[i] = abs(pks[i] - pk)
    ndx = np.argmin(dif)
    if dif[ndx] <= 1:
        return ndx
    return -1
```


Once we found the closest match, we check that the absolute distance between the two is less than one bin bandwidth. If that is the case, we have a match and we return the corresponding array index. The search is always performed against previously found peaks.

If a peak is not matched between two adjacent analyses, it is too short-lived and we dispose of it. However, if a match is found, a track is created, which has a minimum of two time points. This track is continued in subsequent analyses if peaks continue to match against it. Otherwise they are killed. Once a peak is found we need to allocate it a track. This is either an existing (continuing) or a new track. A matched peak only makes it into an existing track if

- It corresponds to an existing track peak.
- Its amplitude does not suddenly jump.

Therefore we need to check first that the peak we have matched to is present in an existing track. Then we need to check that the amplitudes have not changed so much to cause this to be a new track, instead of continuation. Once these conditions have been met, we can copy the peak position and amplitude to the track. At this stage, we can mark this as a continuation track and flag that it has been used. All existing tracks with no continuation are marked as dead by setting their time stamp to a negative value and flagged as used. The storage for these tracks may be reused in future analyses.

We now have non-allocated peaks, that is, the ones where a match was found but could not be placed in an existing track. That happens if a peak had only just appeared in the previous analysis so that it has not made into a track yet. For these, we need to look for free tracks, the ones marked as unused. These correspond to the ones that had been previously killed and their storage can now be reused. If we run out of free tracks, we can expand the output track frame.

A Python function implementing track formation is shown below. It takes as input the current time, current array of peaks (as output by the peak interpolation function), previous peaks, and previous track frame. From this, it produces the current track frame based on the process outlined earlier. Since this is quite involved, comments indicate the significant steps of the computation,

```
# make tracks, take time,
# current array of peaks (two rows with N columns,
# row 0 holds peaks, row 1 holds peak amplitudes),
# previous arrays of peaks and track frame
# returns track frame
def mktracks(time,pks,prev=None,trks=None):
 N = pks[0].size
 # if no tracks or prev peaks exist
 if prev is None and trks is None:
   trks = np.zeros((N,6))
   for i in range(0,N):
    trks[i,0] = time
    trks[i,1] = pks[0,i]
    trks[i,2] = pks[1,i]
    trks[i,3] = 40
   return trks
 M = trks.shape[0]
 ndx = np.zeros(N).astype(int)
 flag = np.zeros(M).astype(int)
 # find all matching peak ndxs
 for i in range(0,N):
  ndx[i] = pkmatch(pks[0,i],prev[0])
 # find corresponding tracks
 for k in range(0,M):
  for i in range(0,N):
   if ndx[i] >= 0: # a matched peak
     n = ndx[i]
     # found a track that is not dead
     if prev[0,n] == trks[k,1] and trks[k,0] >= 0:
      # check for amp continuity in dB
      A  =  trks[k,2]
      pA = pks[1,i]
      dbs = 20*np.log10(A/pA)
      adt = trks[k,3]
      # track matches
      if dbs <= adt:
       trks[k,0] = time # cur time
       trks[k,1] = pks[0,i] # peak
       trks[k,2] = pks[1,i] # amp
       # update adt
       a1 = dbs*1.5
```

```
      a2 = adt - (adt - 1.5)*0.049
      if a1 > a2: trks[k,3] = a1
      else: trks[k,3] = a2
      flag[k] = 1 # mark as used
      ndx[i] = -1 # marked as tracked
# kill unmatched tracks
for k in range(0,M):
  # unused and not dead, kill it
  if flag[k] == 0 and trks[k,0] >= 0:
     trks[k,0] = -time # marked as dead
     trks[k,1] = trks[k,1] # same peak
     trks[k,2] = 0 # zero amplitude
     trks[k,3] = 40 # reset adt
     flag[k] = 1 # mark as used
# new tracks
m = 0
for i in range(0,N):
 if ndx[i] >= 0: # matched and untracked
  # reuse old dead track
  if m < M:
   j = m
   for k in range(j,M):
    if flag[k] == 0:
     trks[k,0] = time
     # use cur peak data (not previous)
     # so it can be matched later
     # for continuation
     trks[k,1] = pks[0,i]
     trks[k,2] = pks[1,i]
     flag[k] = 1 # mark as used
     ndx[i] = -1 # marked as tracked
     break
    m = k+1 # update search index
# or resize
for i in range(0,N):
 if ndx[i] >= 0: # matched and untracked
    ntrks = np.resize(trks, (M+1,6))
    ntrks[M,0] = time
    ntrks[M,1] = pks[0,i]
    ntrks[M,2] = pks[1,i]
    ntrks[M,3] = 40 # reset adt
    return ntrks
return trks
```

This function outputs a partial set of track parameters, since it has only been supplied peaks, amplitudes, and time. Frequency and phase are added once the track frame has been constructed.

9.5.5 Frequency and Phase

With the individual steps of partial tracking broken as functions, we have only a few extra elements to add. As before, it is a good idea to encapsulate the complete analysis code as a class so that it can be easily manipulated in a program. So we can now add the STFT as part of a driving function inside this analysis class, and complete the track data by adding frequencies and phases. In Chapter 6, we reviewed a series of standard methods of instantaneous frequency analysis, from which we can pick the most suitable to use here. The characteristics of the partial tracking process are such that

- We need to obtain the instantaneous frequency from a single frame at a given time point.
- We already have a windowed DFT from which we have extracted the amplitudes.

To meet these conditions, we have to discard the phase differences method, as it depends on two adjacent windows. We are left with instantaneous frequency distribution, the one-sample hopsize, and the sliding DFT as possible techniques. The second and third options, although more efficient, require a straight rectangular-window analysis, which we do not have. Therefore, the most suitable approach is compute the IFD. Additionally, to get a better approximation of both frequency and phase, we apply a linear interpolation according to the true peak position. Note that for this to be correct as far as the phase is concerned, we need to interpolate the complex DFT output first, then compute the relevant arctangent. However, we only need to do this for each track. Also, if the window size is the same as the DFT size, phase precision is increased by rotating the frame by half, which realigns the phase spectrum, but does not affect the magnitudes. This also has the effect of shifting the phase reference; therefore it needs to be accounted for when trying to determine the resynthesis phase offset.[29] A longer transform size (e.g., twice the window size) provides better phase resolution and does not require any rotation.

The following Python code defines the track analysis class and its processing function:

```
# Track Analysis
# takes DFT size N
# and sampling rate fs
class TrackAn:
  def __init__(self,N,fs=44100):
    self.N = N
    self.fs = fs
    self.fc = np.arange(0,N//2+1)*fs/N
    self.fac = fs/(2*np.pi)
    self.s1 = np.zeros(N)
    self.s2 = np.zeros(N)
```

```
    self.pks = None
    self.trks = None

  # Analysis
  # takes time t
  # signal sig
  # and optional threshold thr
  def process(self,t,sig,thr=.003):
    N = sig.size
    if N > self.N: N = self.N
    w = np.hanning(N)
    dw = fod(w)
    scl = sum(w)/2
    if self.N == N:
      self.s1 = sig*w
      self.s2[:self.N//2] = self.s1[self.N//2:]
      self.s2[self.N//2:] = self.s1[:self.N//2]
      S = np.fft.rfft(self.s2)/scl
      self.s1 = sig*dw
      self.s2[:self.N//2] = self.s1[self.N//2:]
      self.s2[self.N//2:] = self.s1[:self.N//2]
      Sd = np.fft.rfft(self.s2)/scl
    else:
     self.s1[:N] = sig[:N]*w
     self.s2[:N] = sig[:N]*dw
     S = np.fft.rfft(self.s1)/scl
     Sd = np.fft.rfft(self.s2)/scl
    a = abs(S)
    fr = self.fc + self.fac*np.imag(Sd*S.conjugate())/(a*a)
    thr = np.log(thr*max(a)/5.)
    la = np.log(a)
    pks = pkinterp(pkpick(la,thr),la)
    trks = mktracks(t,pks,self.pks,self.trks)
    for k in range(0,trks.shape[0]):
      ndx = int(trks[k][1])
      frac = trks[k][1] - ndx
      if ndx < fr.size:
       trks[k][4] = fr[ndx] + frac*(fr[ndx+1] - fr[ndx])
      else: trks[k][4] = fr[ndx]
      Si = S[ndx] + frac*(S[ndx+1] - S[ndx])
      trks[k][5] = np.angle(Si)
    self.pks = pks
    self.trks = trks
    return trks
```

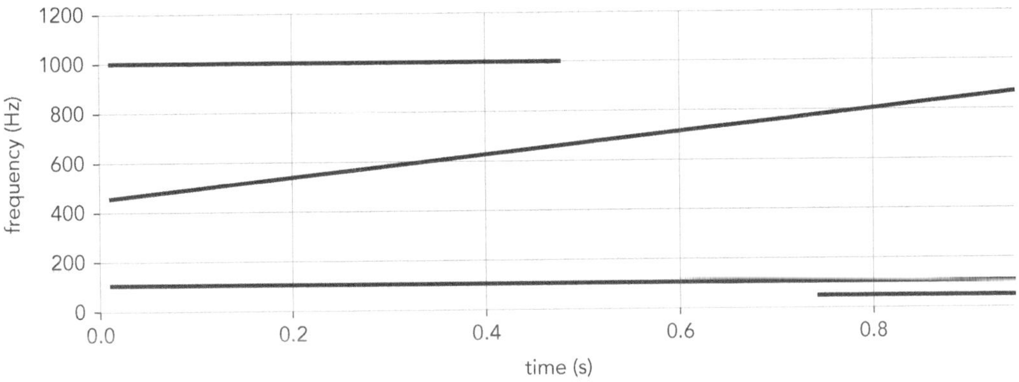

Figure 9.14
Plot of four tracks produced by the partial analysis program.

The track class holds the current track frame and previous peaks. From these, it drives the track formation function as expected. A new output track frame is produced at each call to the analysis process, and it can be fed directly into a synthesis function, or further modified. As intended, this approach can be used directly in a streaming processing context.

In order to demonstrate this code, a synthetic waveform consisting of four partials was analyzed. It starts with three sinusoids, at 105, 440, and 1,000 Hz. The middle partial glides an octave upward over the duration of the sound, and the higher one stops midway through. At about 3/4 of the way, a fourth sinusoid is added, at 50 Hz. A plot of track frequencies generated by the program is shown in Fig. 9.14. As can be seen, the program isolates the partials into separate tracks, which may start and stop and different times, and it follows the trajectory of any component changing in frequency. While this example is somewhat contrived, it shows the basic functionality of the method.

9.5.6 Synthesis

While it is possible to convert a track frame back into an STFT frame for overlap-add resynthesis, we will instead concentrate on an additive synthesis approach. Since we have not yet discussed this in any detail, it is appropriate to turn our attention to it at this moment. In conceptual terms, additive synthesis is far simpler, and more forgiving, than the inverse STFT. It is also fairly flexible, as partials can be left out of resynthesis without compromising the whole process. However, there are a number of technical details that are very important to note. These need to be correct and precise in order to enable a perfect reconstruction of the track data, particularly in terms of accurate phase detail. Earlier on in Chapter 6, we used a windowing method to interpolate between two additive-synthesis generated frames. Here we require a different method, since there are no frames to speak of, only track data.

For this, we compute the synthesis parameters of Eq. 9.72 on a sample-by-sample basis. That is, the functions $a_k(n)$ and $\phi_k(n)$. The amplitudes can be computed by linearly interpolating between two adjacent time positions for which we have track data, h and $h+H$, over the hopsize H,

$$a_k(h+m) = a_k(h) + m\frac{a_k(h+H) - a_k(h)}{H} \tag{9.76}$$

for $0 \leq m < H$.

If we are not concerned with phase offset accuracy, the instantaneous phases $\phi_k(n)$ can be integrated directly from time-varying instantaneous frequencies. We can generate these in a similar fashion, from the data obtained in adjacent bins, linearly interpolated. In this case, we generally do not employ the phase data stored in each track.[30] On the other hand, if we want to precisely match the phases of the detected partials, we need to use a higher-order polynomial interpolation. Since frequencies encode the phase difference between two time positions, we can use a combination of these and the detected phases to implement four-point or cubic interpolation.

This involves determining the coefficients of a polynomial, which defines the phase curve between the two time points,[31]

$$\phi_k(x) = c_0 + c_1 x + c_2 x^2 + c_3 x^3 \tag{9.77}$$

for $0 \leq x < H/fs$. The zeroth- and first-order coefficients are given by the current phase $\phi_k(h)$ and radian frequency $\omega_k(h)$. The other two coefficients are given by a combination of the phase difference between the two points, and the frequencies. We first define

$$\Delta_k = \phi_k(h+H) - \phi_k(h). \tag{9.78}$$

Then we compute a correction offset to make this fit with the estimated frequencies,

$$\varepsilon = \frac{1}{2\pi}\left[\frac{H}{f_s}\frac{\omega_k(h)+\omega_k(h+H)}{2} - \Delta_k\right], \tag{9.79}$$

$$\tilde{\Delta}_k = \Delta_k + 2\pi\left\lfloor \varepsilon + \frac{1}{2} \right\rfloor. \tag{9.80}$$

From these corrected phase differences and the radian frequencies at the two points, we can compute the higher-order polynomial coefficients,

$$\begin{aligned} c_2 &= 3\left(\frac{f_s}{H}\right)^2\left[\tilde{\Delta}_k - \frac{H}{f_s}\frac{2\omega_k(h)+\omega_k(h+H)}{3}\right] \\ c_3 &= \frac{1}{3}\left(\frac{f_s}{H}\right)^2\left[\omega_k(h+H) - \omega_k(h) - 2c_2\frac{H}{f_s}\right] \end{aligned} \tag{9.81}$$

With either of these methods (cubic phase or linear interpolation), an additive synthesizer can be constructed to reproduce track data. Again, here we find that the most complex parts of the resynthesis code have to do with the correct program logic to deal with track startup, continuation, and end. Once this is defined, the synthesis code is just slotted in to process the track data. As in the analysis case, it is useful to encapsulate the process in a class so that state can be kept and managed properly.

The synthesizer will need to keep data pertaining to the previous track frame so that the interpolation procedures can be executed. Therefore the class will provide storage for amplitudes, frequencies, and phases, as well as for continuation flags. Since the number of tracks will vary in time depending on the input waveform, there is no point in predicting how many of these we will have. So we can start the process with no tracks, that is, no stored data. If a new track is seen at the input,

- Storage is added/increased for amplitudes, frequencies, phases, and state (continuation).

- An onset is created, by setting current amplitude to 0, and rewinding phase and frequency.

This means that the actual track phase and amplitude are reached at the end of the current hopsize period. This imposes a delay of 1 hopsize in relation to the input track data, which is important to note if we are trying to match signal phases. For any existing stored tracks we need to perform a couple of checks,

- Check that the input track has not been killed by the analysis.
- If the track is alive, check state for continuation.

If a track passes these tests, we take the stored track data as the current time point parameters and the input data as the next time point parameters. If a track has been killed, we set the next amplitude to zero, keep the frequency steady and extrapolate the phase. That will create a natural track decay, and we flag its state as discontinued. Tracks that are dead and discontinued are skipped. The final case is when a discontinued track is alive. In this case we proceed as if it were a new track, without adding any new storage.

That completes the summary of the program logic. The following Python class implements cubic phase interpolation using the code structure outlined above. We comment it copiously so that its structure can be followed without much difficulty.

```
# Track Synthesis
# takes hopsize h and
# sampling rate fs
class TrackSyn:
 def __init__(self,h,fs):
  self.h = h
  self.trn = 0
  self.fac = h/fs
  self.fs = fs

 # resynthesis
 # takes an array of tracks
 # and an optional amp threshold
 def process(self, trks, thr=0):
  fac = self.fac
  facsq = fac*fac
  fs = self.fs
  h = self.h
  out = np.zeros(h)
  twopi = 2*np.pi
  K = trks.shape[0]
  n = 0
  for k in range(0, K):
   trk = trks[k]
   if k > self.trn-1:
     # new track, expand storage
```

```
    self.trn += 1
    if self.trn > 1:
     self.freq = np.resize(self.freq,self.trn)
     self.ph = np.resize(self.ph,self.trn)
     self.amp = np.resize(self.amp,self.trn)
     self.state = np.resize(self.state,self.trn)
     self.state[k] = 1
    else:
     self.freq = np.zeros(1)
     self.ph = np.zeros(1)
     self.amp = np.zeros(1)
     self.state = np.ones(1).astype(int)
    # get track data, do onset
    frnxt =  trk[4]
    phnxt = trk[5]
    ampnxt = trk[2]
    fr = frnxt
    ph = phnxt - fr*fac*twopi
    a = 0
# get data from existing track
else:
 if trk[0] > 0:  # track is not dead
  frnxt = trk[4]
  phnxt = trk[5]
  ampnxt = trk[2]
 else:
  frnxt = self.freq[k]
  phnxt = self.ph[k] + frnxt*fac
  ampnxt = 0
 if self.state[k] > 0: # continuation
  fr = self.freq[k]
  ph = self.ph[k]
  a = self.amp[k]
 else: # restart a track
  fr = frnxt
  ph = phnxt - twopi*fr*fac
  a = 0

# track is dead and discontinued
# move on to the next track
if trk[0] < 0 and self.state[k] == 0:
   continue
else:
 # track is dead, discontinue it
 # carry on to synthesise decay
```

```
        # else confirm it continuing
        if trk[0] < 0:
         self.state[k] = 0
        else:
         self.state[k] = 1

      if ampnxt > thr:
        # cubic phase interpolation
        delta = modpi(phnxt - ph)
        # scale freq
        # fac = h/fs
        frnxt *= twopi
        cph = ((frnxt + fr)*fac*0.5 - delta)/twopi
        delta += int(cph + 0.5)*twopi
        a2 = 3/facsq*(delta - fac/3*(2*fr+frnxt))
        a3 = 1/(3*facsq)*(frnxt - fr - 2*a2*fac)
        # linear amplitude interpolation
        an = (ampnxt - a)/h
        cnt = 0
        pn = 1/fs
        phas = ph
        # synthesis
        for n in range(0,h):
         out[n] += a*np.cos(phas)
         cnt += pn
         phas = ph + cnt*(fr + cnt*(a2+a3*cnt))
         a += an
       # store frs, phs, amps
       self.freq[k] = frnxt
       self.ph[k] = phnxt
       self.amp[k] = ampnxt
     # output frame
     return out
```


The processing method includes an optional argument defining an amplitude threshold, which can be used to filter out tracks whose amplitude is below a certain value. Note that this acts after the tracks have been picked and so it does not affect the continuity logic. However, this also means that the track can appear or disappear at specific hop periods depending on whether it falls below or above the threshold.

The linear frequency interpolation method is far simpler, although it does not guarantee that the resynthesis phases are going to reproduce the original partial phases,

```
        # linear frequency interpolation
         an = (ampnxt - a)/h
         fn = (frnxt - fr)/fr
         for n in range(0,h):
```

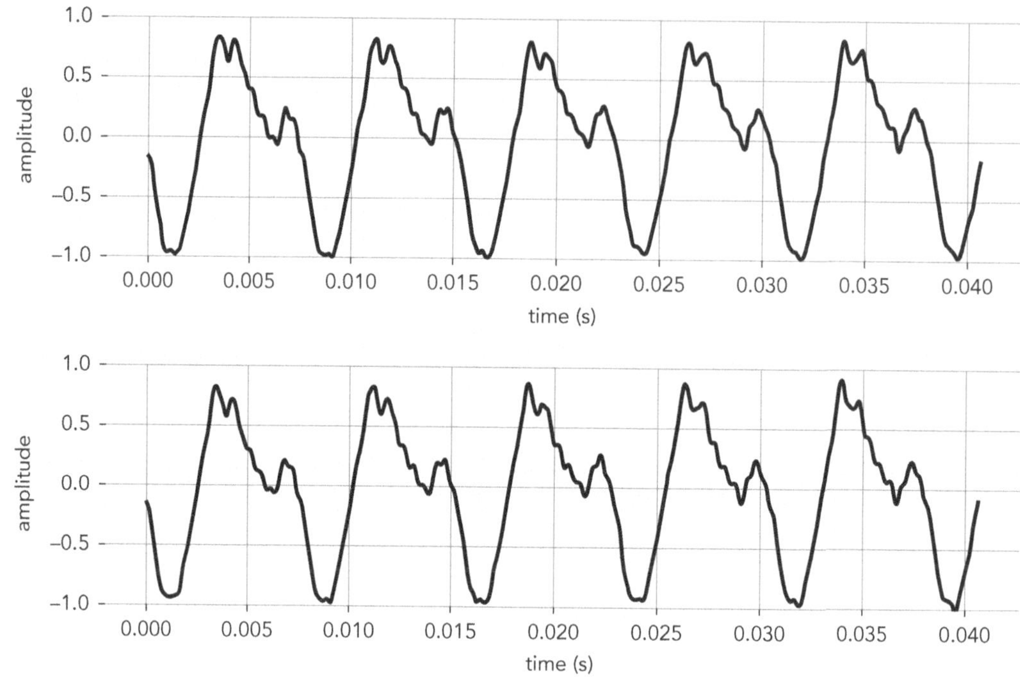

Figure 9.15 Original (*top*) and resynthesized (*bottom*) waveforms corresponding to a piano tone.


```
out[n] += a*np.cos(ph)
ph += fr*2*np.pi/fs
a += an
fr += fn
```

In this case we keep the integrated phases for next time as the track continues. For starting tracks we can either reset the phase to zero, or use the first track phase value as the starting point.

In Fig. 9.15, a demonstration of the analysis-synthesis process is given. A piano tone was analyzed using the partial tracking code and then resynthesized using cubic phase interpolation. The result shows a very good match of time-domain features, which is enabled by the careful phase analysis and reconstruction achieved by the methods developed above.

9.5.7 Residual Extraction

Now that we have a good reconstruction of phases and amplitudes, it is possible to attempt to extract the residual from the analysis-synthesis output. The idea is that the tracking does not capture the more fleeting elements of a sound, which do not make it into tracks, and only the stable partials are present in the resynthesis. This is true to some extent, but it depends very much on the type of input and on the analysis parameters. If the conditions are right, a simple time-domain subtraction of the two waveforms with the correct offset and optional scaling,

$$e(n) = as_{in}(n-d) - s_{out}(n), \tag{9.82}$$

should be enough to yield a residual signal $e(n)$, off the input $s_{in}(n)$ and output $s_{out}(n)$ of partial tracking analysis-synthesis. The following program uses the classes developed earlier to implement partial resynthesis and residual extraction:


```
import sys
import wave
import numpy as np
import sinus

# input file
fname = sys.argv[1]
wf = wave.open(fname, 'rb')
fs = wf.getframerate()
L = wf.getnframes()
# io buffers
inp = np.zeros(L)
output = np.zeros(L)
residual = np.zeros(L)
# read input
norm = 2**(wf.getsampwidth()*8)
chn = wf.getnchannels()
s = np.frombuffer(wf.readframes(L)[::chn],dtype=np.int16)
inp[:s.size] = s/norm
wf.close()
# analysis -- synthesis
N = 2048
H = N//8
analysis = sinus.TrackAn(2*N,fs)
synthesis = sinus.TrackSyn(H,fs)
# delay buffer
delay = np.zeros(H) + 0.001

# processing
for i in range(0,L-N,H):
  sig = inp[i:i+N]
  tracks = analysis.process(i,sig)
  out = synthesis.process(tracks)
  res = out - delay
  delay = np.array(inp[i:i+H])
  output[i:i+H] = out
  residual[i:i+H] = res

# write output data, partial resynthesis
wf = wave.open('syn-%s' % fname, 'wb')
wf.setnchannels(1)
```

```
wf.setsampwidth(2)
wf.setframerate(fs)
s = np.ndarray.tobytes((output*(2**16)).astype(np.int16))
wf.writeframes(s)
wf.close()
# write output data, residual
wf = wave.open('res-%s' % fname, 'wb')
wf.setnchannels(1)
wf.setsampwidth(2)
wf.setframerate(fs)
s = np.ndarray.tobytes((residual*(2**16)).astype(np.int16))
wf.writeframes(s)
wf.close()
```


To demonstrate it, we can apply this program to an instrumental tone with a certain amount of stochastic components and observe the amount of separation that can be achieved. In Fig. 9.16, we plot three waveforms, corresponding to the recording of a flute note, its partial resynthesis, and the residual. As we should have expected, the partial resynthesis captures most of the original waveform fairly well, since this is dominated by deterministic components. In addition to this, at a low level there is a certain amount of residual, which was not modeled by partials.

The nature of this residual can be understood by looking at how the energy is distributed over the frequencies. In Fig. 9.17, we plot the spectra corresponding to these three waveforms, the original flute recording, its partial track resynthesis, and the analysis residual. As can be seen, there is a significant amount of broadband noise in the high register, caused by the turbulent air flow. This is by and large suppressed in the partial track resynthesis. The residual shows a fairly distributed spectrum, with an envelope marked by the most prominent harmonic peaks, which is expected due to the resonance effects of the flute pipe. It shows the presence of noise throughout the spectrum, albeit at a low level. On audition, it is very clear that the breath sounds dominate the residual, with some remnants of pitch caused by the resonance created by the low-frequency spectral peaks.

In order to get these results, each DFT was padded to twice the input frame size. This allowed a precise phase match of each detected partial. However, it is possible to achieve an even better separation. In fact, the partial tracking method we developed is capable of capturing a significant part of the stochastic element as short tracks; therefore, it is not specifically designed for partial separation. By placing more constraints on continuity and stability, it is possible to remove more of the non-sinusoidal components. We could, for instance, enforce a higher minimum number of time points for track formation; the current code only assumes a track can be made of a minimum of two points. Other methods may attempt to model the spectrum as made of harmonics, and only allow partials whose frequencies correspond to multiples of a fundamental. In any case, the system developed here provides the basic functionality for residual extraction.

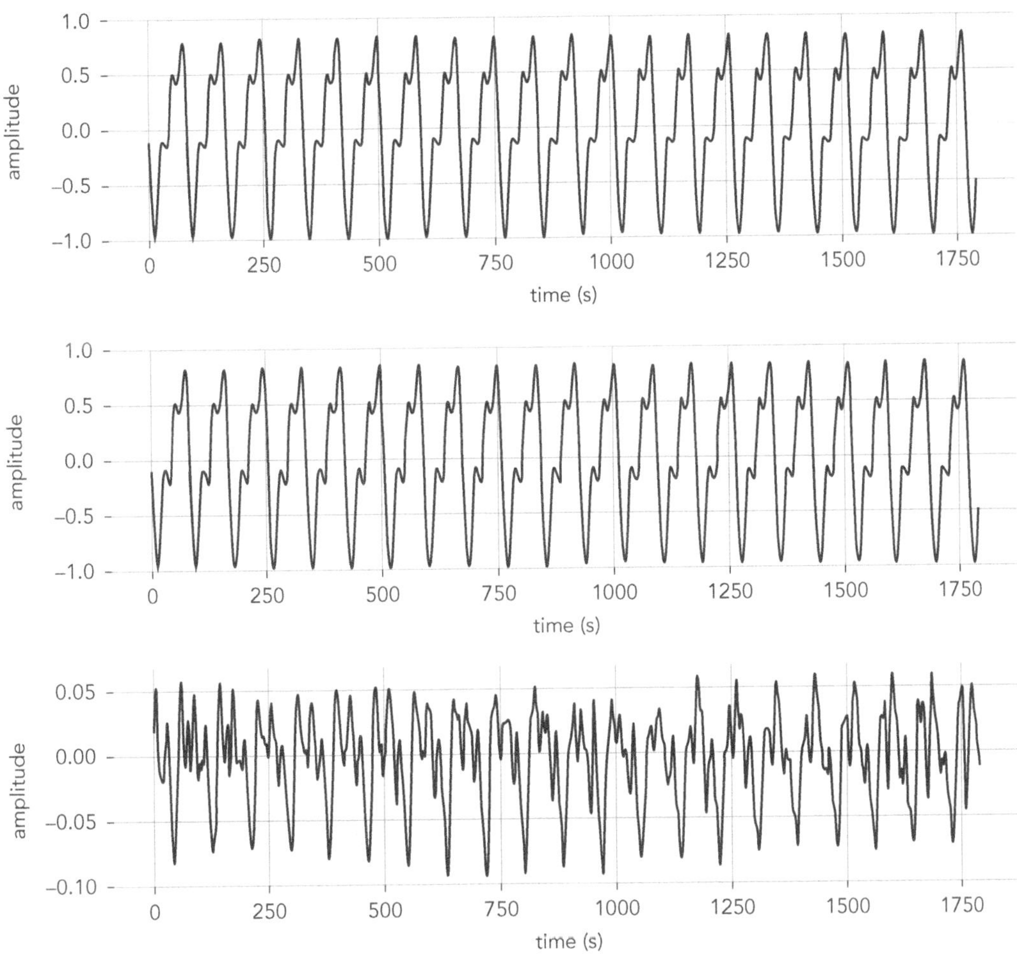

Figure 9.16 The original (*top*), resynthesized (*middle*), and residual (*bottom*) waveforms of a flute note.

9.5.8 Modeling the Residual

The residual extraction program is designed in such a way that at every analysis point we get both the partial resynthesis and the residual as a block of hopsize samples. It is possible to apply some form of modeling to the residual data, which may allow us to resynthesize it later. The model parameters can, for example, be stored alongside the track data to complement it. At a minimum, we can extract the total residual energy at every hopsize, and use that as a representation of the non-deterministic part of the sound,

$$\hat{e} = \sqrt{\frac{\sum_{m=0}^{H-1} e(m)^2}{H}}, \tag{9.83}$$

its RMS measurement. For a more complete, but still compact, description of the residual, we could add, for example, the first four spectral moments: centroid, bandwidth, skewness, and kurtosis.

This model is dynamic, since regular measurements are taken at each hopsize. It can drive a resynthesis engine that uses one of the bandlimited noise generation approaches discussed earlier. Alternatively, we can expand the model to hold a more complete

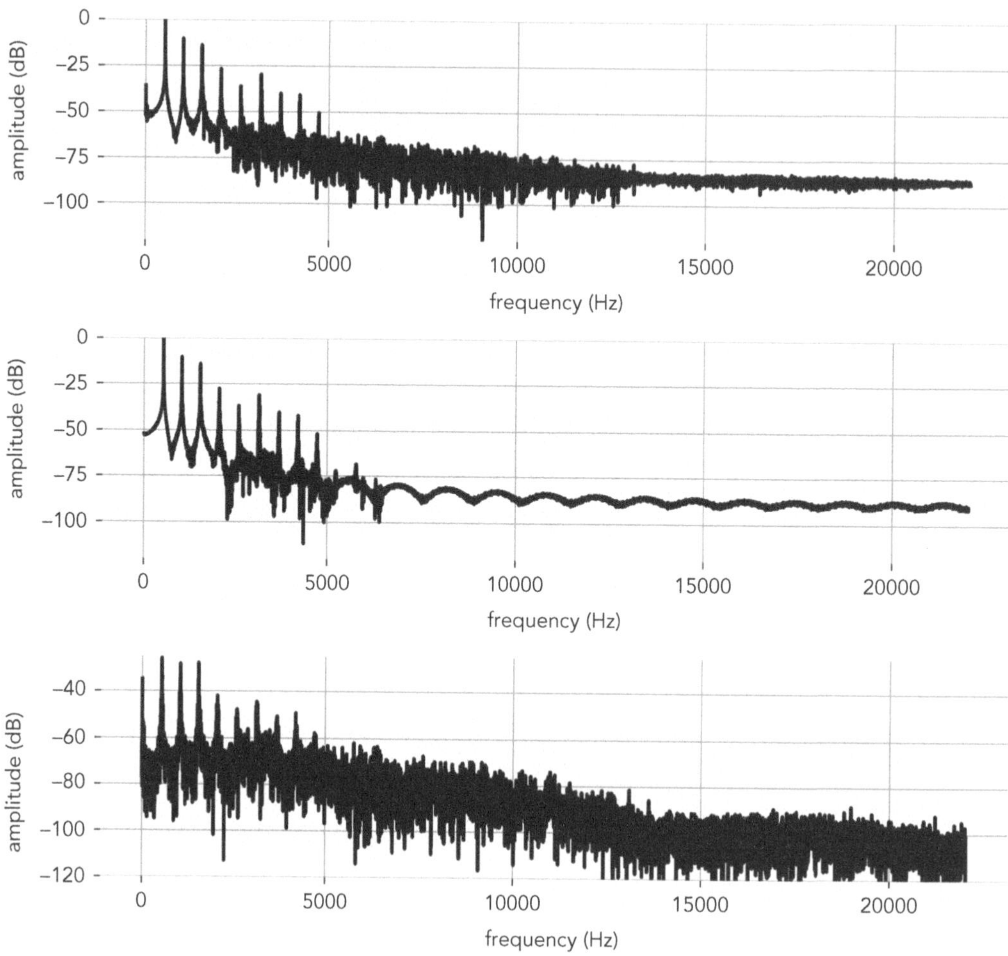

Figure 9.17 The original (*top*), resynthesized (*middle*), and residual (*bottom*) spectra of a flute tone.

description of the spectral envelope of the residual. Each residual frame of hopsize samples can be analysed with a DFT, its power or amplitude spectrum taken and stored alongside track data. The following Python code excerpt demonstrates the idea:

```
spec = []
for i in range(0,L-N,H):
  sig = inp[i:i+N]
  tracks = analysis.process(i,sig)
  out = synthesis.process(tracks)
  res = abs(np.fft.rfft(out - delay))
  delay = np.array(inp[i:i+H])
  output[i:i+H] = out
  spec.append((np.array(tracks),np.array(res)))
```

where the spectral track data is augmented by the residual amplitude envelope. The Python list contains a time-ordered sequence of tuples consisting of the track frame and the residual. This can be stored in a data file for later resynthesis of both partials and noise.

Other spectral envelope representations can be used, for instance, a coarser filter-bank or spectral analysis, lower-order cepstrum coefficients, etc. We can also use a series of all-pole coefficients taken from a linear predictor. Using the code we have developed earlier, we can replace the noise spectrum with $M+1$ coefficients and the residual energy,

```
spec = []
for i in range(0,L-N,H):
  sig = inp[i:i+N]
  tracks = analysis.process(i,sig)
  out = synthesis.process(tracks)
  res, E = lpred(out - delay, M)
  delay = np.array(inp[i:i+H])
  output[i:i+H] = out
  spec.append((np.array(tracks),np.array(res), E))
```


where `lpred()` is the linear prediction function developed earlier in the chapter. The resynthesis can then use an all-pole filter fed with a pink or white noise generator.

9.5.9 Transients

Alongside with noise, the phenomena of transients is also an important part of musical sounds. These refer to the short-lived components that appear and disappear when a sound starts and stops abruptly. It is possible to bundle transients with either the sinusoidal partials or the stochastic element when these are separated, but a more complete treatment could try to isolate them as a separate component on their own. Given that they are linked to changes in the amplitude (time) envelope, the extraction of clear functions defining the evolution of the sound amplitude over time is probably a good starting point. Quick changes in sound energy would signal the presence of transients, and a time window can be applied to isolate them. In general, similar approaches to the ones discussed earlier can be used for separation, either in the time or in the frequency domain.

9.5.10 Streaming Partial Track Processing

As we noted before, the ideas developed here can be applied to a streaming processing context. In Csound, we find opcodes to perform partial track analysis, manipulation, and resynthesis, which use a specially defined form of the fsig data type (called `PVS_TRACKS`). This is designed to work fairly transparently, in the same fashion the streaming phase vocoder system. We begin by producing amplitude, frequency, and phase data using the instantaneous frequency distribution,

```
ffr,fphs pvsifd asig,iN,iH,iwin
```

which produces a `PVS_AMP_FREQ` and a `PVS_AMP_PHASE` output. These are then fed into the partial tracking opcode,

```
ftrk = partials(ffr,fphs,kthr,iminpts,imaxgap,500)
```

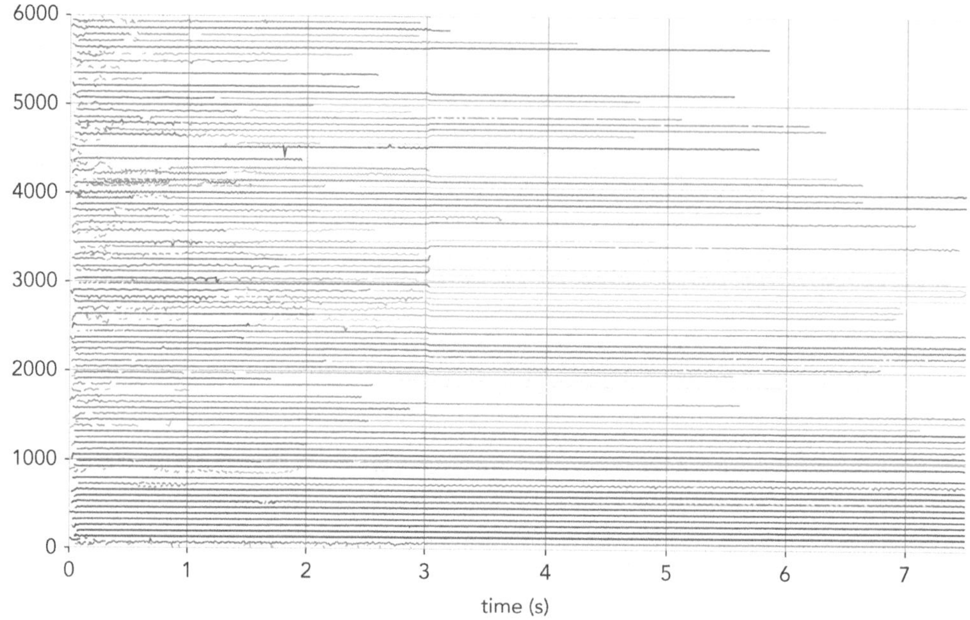

Figure 9.18 A plot of partial tracks from a harpsichord note, as analysed by `partials`.

which takes a threshold, a minimum number of time points, a maximum allowed continuation gap, and a maximum number of tracks. Figure 9.18 shows a plot of partial tracks from a low-register harpsichord note, as analyzed by this opcode.

A partial tracks signal can be resynthesised in various ways:

- Cubic phase interpolation additive synthesis, using phase data from analysis,

  ```
  asig = sinsyn(ftrk,kscl,kparts,ifn)
  ```

  which takes a scaling gain, the maximum number of partials to use, and a function table containing a cosine wave.
- Cubic phase interpolation additive synthesis, using extrapolated phase data,

  ```
  asig = resyn(ftrk,kscal,kpitch,kparts,ifn)
  ```

  which allows pitch shifting to be used.
- Linear frequency interpolation additive synthesis,

  ```
  asig = tradsyn(ftrk,kscal,kpitch,kparts,ifn)
  ```

  which does not use any phase information from the partial analysis.
- Inverse STFT, by constructing a PVS_AMP_FREQ signal from tracks and then using the appropriate opcode,

  ```
  fsig = binit(ftrk,iN)
  asig = pvsynth(fsig)
  ```

For example, a residual extraction opcode can be written as

```
/**************************************************
ares,asin Residual ain,kthresh,isize,ifcos
```

```
ares - residual output
asin - sinusoidal output
kthr - analysis threshold
isize - DFT size
ihop - hopsize
ifcos - function table containing a cosine wave
*********************************************/
opcode Residual, aa, akiii
 ain,kthr,isiz,ihsiz,ifcos  xin
 idel = isiz/2 + ihsiz
 ffr,fphs pvsifd ain,isiz,ihsiz,1
 ftrk = partials(ffr,fphs,kthr,1,1,500)
 aout = sinsyn(ftrk,1,500,ifcos)
 asd delayr idel/sr
 asig = deltapn(idel)
      delayw ain
 aenv = linsegr(0,idel/sr,0,1/sr,1,1,1)
 aout = balance(aout,asig)
 xout(aout*aenv-asig,aout)
endop
```


Complementing the analysis and resynthesis opcodes, as in the phase vocoder system, there are several unit generators for manipulating partial tracks in Csound,

- `trcross`: cross-synthesis of two partial tracks.
- `trfilter`: filtering of partial tracks.
- `trhighest`: extracts the highest-frequency track.
- `trlowest`: extracts the lowest-frequency track.
- `trmix`: partial track mixing.
- `trscale`: track frequency scaling (pitch shifting).
- `trshift`: partial track frequency shifting.
- `trsplit`: partial track splitting by frequency.

Since partial tracks can be obtained from phase vocoder signals and vice-versa, these two frequency domain processing approaches can be used side by side as a complete spectral analysis-synthesis system.

9.5.11 ATS

In addition to the streaming partial tracking, Csound also implements offline analysis with resynthesis opcodes based on the Analysis, Transformation, and Synthesis (ATS) system (Pampin, 2004). It provides a utility called `atsa`, which is used to produce analysis files for use with a suite of opcodes. The ATS system employs a track analysis and residual modeling designed along similar lines to the methods discussed earlier. The following opcodes are provided to read and manipulate ATS data:

- `ATSadd`: additive synthesis from analysis file data.

- `ATSaddnz`: noise resynthesis.
- `ATSbufread`: reads data from file and stores it in an internal data table of frequency, amplitude pairs.
- `ATScross`: cross synthesis from analysis files.
- `ATSinfo`: reads data out of the header of an analysis file.
- `ATSinterpread`: extracts spectral envelope from an `ATSbufread` buffer.
- `ATSread`: reads partial data from an analysis file.
- `ATSreadnz`: reads noise data.
- `ATSpartialtap` returns a frequency, amplitude pair from an `ATSbufread` buffer.
- `ATSsinnoi`: resynthesis of partials and noise.

All these systems make Csound quite a formidable synthesis engine for spectral processing. It is a fully featured environment to explore the ideas developed in this book for creative, as well as research, purposes.

9.6 The Non-Deterministic Spectrum

In this chapter, we focused specifically on the fleeting and less clearly defined components of a musical sound. These were shown to be better modeled by stochastic processes, and we went on to study their nature and various ways to capture, model, and generate such signals. As part of this, we managed to expand our evolving definition of spectrum to encompass random signals and the processes underlying them. We have discovered that

> It is possible to extend the notion of spectrum to represent stochastic processes. In this case, we have a more abstract concept than what we have previously used for signals. A measure of the spectrum a random process is given by its power spectrum density, which can be approximated by the power spectrum of its realization (or an observation of it). These principles allow us to model the non-deterministic aspects of musical sound, so they can be captured, analyzed, manipulated, and resynthesized along with the more stable and individually defined sinusoidal partials.

At the end of the chapter, we attempted to develop models that would describe a musical sound as a whole, consisting of well-defined partials, which are nicely described by the Fourier theory and its extensions, and distributed spectra, covered by the principles of stochastic processes explored in this chapter. This completes our discussion of spectral audio from a technical, signal-processing and computational, perspective. We can now move our attention to creative and aesthetical aspects of sound design and music making. Armed with the knowledge built up in these first three sections, we can embark in the final part of this volume to explore the various approaches that are available for spectral music composition.

Notes

1 We have used the word *energy* in a loose way until now, which has worked for us. More strictly, we can define it using *Parseval's theorem*, where the total energy of a signal is

$$E = \int_{-\infty}^{\infty} |x(t)|^2 dt = \int_{-\infty}^{\infty} |X(f)|^2 df. \tag{9.84}$$

From this, we can also consider energy as a density function of frequency,

$$|X(f)|^2, \tag{9.85}$$

which is equivalent to the squared amplitude spectrum. In this case, it turns out to have a similar definition to what we have called the power spectrum.

2 In the present case, we normally expect this to be normalized by $1/N$.

3 This is a different measurement from, although proportional to, the half-power bandwidth, which we used when looking at the spectra of filters.

4 Depending on the process, we may need to consider $E[x(t)]$, that is, the expected value at time t. For sake of simplicity, we will assume, however, that this is not necessary here.

5 This is not to be confused with a noise waveform, whose spectrum can be analyzed with, for instance, a DFT. A random sequence does not have a Fourier transform because it is said to have infinite energy (Oppenheim et al., 1999, 376–377), and so it cannot be represented in such a way.

6 For real-valued x.

7 This relationship is given by the *Wiener-Kinchin* theorem (Champeney, 1987, Chapter 11), whose applicability is not universal, as it assumes signals to be stationary.

8 Note that for real signals, the complex conjugate is an identity.

9 In some places, the power spectrum as derived from a DFT analysis is called a *periodogram* (Oppenheim et al., 1999, 542). The extra terminology is not particularly helpful for our applications, so it has not been adopted here.

10 We need to note the difficulties with such a description at 0 Hz, where a singularity may emerge. This is a particular problem with such a spectral description, which nonetheless characterizes many important random processes (Watkins, 2017).

11 Remembering that we are referring to a 1/f power spectrum curve, and not as we did in other chapters, to the amplitude or magnitude spectrum.

12 The variance has also been characterised as the instantaneous bandwidth of the signal (Loughlin and Davidson, 2000).

13 This is a similar shape to that of a single triangular waveform cycle, which we have seen in Chapter 4, or a Bartlett window (Chapter 5).

14 This has been one of the original interpretations of Gabor's STFT method, which he called *acoustic quanta*, describing single bins in single DFT frames, localized in frequency and time (Gabor, 1947).

15 The other, a single-parameter opcode, uses a different algorithm with zero mean.

16 The following definitions and notation are adapted from Mallat (2008, Chapter 4).

17 The name is an homage to Gabor's pioneering time-frequency analysis work, in which he employed a Gaussian-window STFT.

18 A digital quadrature filter pair has amplitude responses that are symmetric around $f_s/4$, or $\omega = \pi/2$. For example, the first-order difference and first-order sum are quadrature mirror filters, which correspond to the *Haar* wavelet.
19 See Mallat (2008) for a very detailed look at possible applications.
20 See Mallat (2008, 645) for an optimized algorithm.
21 This means the filter has no zeros, only poles.
22 The system of equations defined by Eq. 9.54 is represented by the following equation:

$$\begin{bmatrix} r(0) & r(1) & r(2) & \dots & r(M-1) \\ r(1) & r(0) & r(1) & \dots & r(M-2) \\ r(2) & r(1) & r(0) & \dots & r(M-3) \\ . & . & . & & . \\ . & . & . & & . \\ . & . & . & & . \\ r(M-1) & r(M-2) & r(M-3) & \dots & r(0) \end{bmatrix} \times \begin{bmatrix} a_1 \\ a_2 \\ a_3 \\ . \\ . \\ . \\ a_M \end{bmatrix} = - \begin{bmatrix} r(1) \\ r(2) \\ r(3) \\ . \\ . \\ . \\ r(M) \end{bmatrix}$$

The big matrix on the left-hand side is called a *covariance* matrix, which is symmetric and has identical elements along its diagonals. Furthermore, the vector on the right-hand side has the same elements found in the covariance matrix. Combined, these two characteristics lead to an efficient recursive procedure first discovered by Durbin (1960) (Makhoul, 1975, 566). This method is therefore called *Durbin's* algorithm. There are other ways of solving these equations, such as the Gauss elimination method (used in Csound's `lpanal` utility), but these are nowhere as efficient.
23 As opposed to $O(N^2)$ in the case of the other known solution methods (Makhoul, 1975, 565–6)
24 The Kalman filter is an algorithm that uses a stochastic model of a system that is iteratively adjusted based on observation data. The extended Kalman filter approach is a non-linear version of it, which is more suited to the problem of fitting formant frequencies and bandwidths to an all-pole representation (Lazzarini and Timoney, 2009).
25 Since this is the inverse filter of an all-pole expected to be stable, all of its zeros are necessarily inside the unit circle.
26 Note that here, by computing the IDFT of a real-valued function (the log of the power spectrum), the result is an even function, as expected (since it is only made of cosine partials). In order to recover a real-valued power spectrum, we then need to apply a real-to-complex DFT to an even cepstrum function. However, using this method, if the cepstrum is not an even function, then we have to take the magnitude of the result to recover the spectral envelope. For example, if we have computed the cepstrum from the linear prediction coefficients, we will have an odd cepstrum function (since the coefficients are only in first positions of the array, the rest is padded with zeros), and therefore the following code applies,

```
b,E = lpred(out,M)
ceps = lpceps(b,E,N)
spe = abs(np.exp(np.fft.rfft(ceps)))
```

if we are using the forward (real-to-complex) DFT to compute the spectral envelope from the cepstrum. On the other hand, we can, as we did before in Chapter 6, use the inverse (complex-to-real) DFT. In that case, since we are assuming a real-valued result, we get the spectral envelope directly,

```
b,E = lpred(out,M)
ceps = lpceps(b,E,N//2+1)
spe = np.exp(np.fft.irfft(ceps*N//2))
```

but we have to employ an array size that is half the transform size (N) plus the Nyquist point. The assumption is that the cepstrum is Hermitian symmetrical, which is correct in this case (in fact, it is just even, because it is purely real). The confusion over using forward or inverse DFTs to compute the cepstrum is universal, so it is important to make sure we are absolutely clear about what we are doing.

27 The idea of modeling the spectrum as a combination of sinusoids and noise is originally due to X.Serra (1989), who built on the original work of McAulay and Quatieri (1986), but it was later expanded and reworked by many others in various ways (Bonada et al., 2011).

28 In the spectral domain we can look at each DFT frame and pick the most prominent amplitude spectrum peaks. We assume these to be the individual partials, and assign them to sinusoids with the measured amplitude and phase. Everything else is removed and defined as noise. The stenciling method discussed in Chapter 6 is an example of this approach. We could also apply the methods developed here for the time-domain approach, but use spectral methods for reconstruction and residual extraction. Meurisse et al. (2006) also showed an interesting alternative approach, which models each bin as a mixture of stable and stochastic elements, using probability distribution methods to do the separation of sinusoids and noise.

29 This imposes a negative shift of N/2 samples relative to a non-rotated frame.

30 We can, alternatively, use these to set an initial phase offset at the start of each track.

31 A proof of this method is given by McAulay and Quatieri (1986).

PART III

Design

10

Spectral Design in Music

We can now return to some of the threads started in Chapter 2. In particular, it will be interesting to trace the origins and development of a spectral approach to composing and listening, which is arguably a key characteristic of today's music. We should be at this stage well acquainted with a wide notion of spectrum that encompasses various definitions and concepts. With these in mind, we can proceed to examine their importance to music-making and appreciation. It is clear that attention to spectral details in composition and performance did not start at any specific point in time, in that composers and listeners were always attuned to nuances and expressivity of timbre. This is a logical conclusion from the fact that a large range of instrumental families were developed along the years to bring about spectral variety in music-making. However, at the same time, the notion of timbre, or spectral color, has traditionally been a fleeting one, very seldom notated, and never as precisely defined as other parameters of music composition.

It is possible to argue that this lack of precision in terms of notation has made timbre a secondary concern for music structuring, behind pitch, duration, and intensity. While pitch and intensity are parameters that can describe a spectrum, they do so in an incomplete way; therefore, a complete spectral characterization should include these in combination with timbral information. Even as a secondary consideration for composers, spectral color was never displaced in importance when it came to performance expression: the ability to shape timbre has always been one of the most important interpretative qualities of a good player. So while notated aspects of music may not have consistently incorporated color as an element, we have to suspect that it has always been the case to a large extent that the realization of music as sound has always been concerned with manipulating timbre in one way or another.

With this out of the way, we can now focus on music that is composed with what we may call *spectral shaping* as one of its structural characteristics, which treats with equal importance the different parameters of the spectrum. In this chapter, we attempt to show how this became a central aspect of today's music. Part of this development has been devoted to make spectral color a more central concern in the minds of composers. Many authors have already written extensively on the emergence of timbre as a musical structuring element. Erickson notes that

> (...) much of the music of modern times is more about timbre than it is about pitch. Pitch is secondary to timbre in most electronic and tape music, where various aspects of the timbral dimension are often quite precisely controlled. (Erickson, 1975, 18)

Spectral Music Design. Victor Lazzarini, Oxford University Press. © Oxford University Press 2021.
DOI: 10.1093/oso/9780197524015.003.0010

While this was most likely written with concert music[1] in mind, we can argue that is largely true of many other different genres of music made today.

If the spectral qualities, in particular color, of sound are now on equal footing with the other traditional parameters of musical texture, we propose that this is due foremost to one particular factor. This is the development of capture, processing, and reproduction of sound, in particular in its electronic form, which superseded the early mechanical methods. Since the lack of precise notation inhibited the development of timbre as a structural concern for composition, sound recording provided a means of registering it, with increasing precision. Through this medium, we are able to communicate spectrally oriented musical ideas without any difficulties, and these in turn have been accepted as an essential part of the discourse. This includes both music that is composed purely for instrumental[2] forces, as well as for electronically generated sounds, or for a combination thereof. Within the recorded medium, a distinction between these two is completely artificial, for reasons that will become clear in this chapter.

The tools developed in earlier parts of this book provide key resources for the composition of spectral music through computational means. While an electronic music approach is the focus of this book, it is by no means the only one leading to the manipulation of spectra as a primary aspect of composition. It should be clear from the outset that there are many ways to write music, including some not involving any sort of electronic devices or computer technology, which may be considered as spectral. Hopefully, as a more developed idea of what this means emerges at the end of this chapter, the principles of a spectral approach will be able to inform our future music-making activities, whichever our choice of medium and tools may be.

10.1 The Emergence of Spectral Color as a Structuring Device

As we have noted, timbre has always been an element of music. However, it was only toward the end of the 19th century that it began to feature properly as a structural element in composition. As the tonal-functional approach to harmonic organization developed and finally unraveled, the notion of a chord as a timbral bundle started to emerge. For many theorists, a chord has to be considered a whole, rather than as individual notes sounding together, due to the spectral relationships between its constituent sounds. This principle is associated with the idea of *Klang*, brought forward by German theorists in the late 19th century. Hugo Riemann used it to describe a triad:

> The ear perceives a tone with its direct relatives (third and fifth or their octaves) on the same side, therefore prime, over-third, and over-fifth, or prime, under-third, and under-fifth, to be connected together as one Klang (. . .), (H.Riemann, 1893, 7)

Helmholtz also employed the word to denote similar ideas. Although this was originally connected to the theory of tonal harmony, the notion of a chord as a single *sound*, or as a *timbre*, to be taken by its color, rather than constituent tones, is a powerful one. At the dawn of the 20th century, this principle becomes part of the compositional language of various composers. It is observed particular in many of Debussy's works, but also seen in the music of Schoenberg, Webern, Stravinsky, and many others. The dissolution of the tonal-functional structuring principle was probably one of the most important causes of this change in perspective. Devoid of a clear function within a harmonic sequence, the role of a chord needed a redefinition. It was an obvious step to take it for what it was, for its spectral characteristics.

10.1.1 Chords and Spectra

New chordal qualities began to be created by the inclusion of higher harmonic tones as integral components of a *Klang*. Messiaen noted the technique of added notes used by Debussy in some of his works:

> With the advent of Claude Debussy, one spoke of appoggiaturas without resolution, of passing notes with no issue, etc. In fact, one found them in his early works. In *Pélleas et Mélisande, les Estampes, les Préludes, les Images* for the piano, it is a question of foreign notes, with neither preparation nor resolution, without particular expressive accent, which tranquilly make part of the chord, changing its color, giving it a spice, a new perfume. These notes keep a character of intrusion, of supplement: the bee in the flower! They have, nevertheless, a certain citizenship in the chord, either because they have the same sonority as some classified appoggiatura or because they issue from the resonance of the fundamental. They are *added* notes. (Messiaen, 1956a, 49)

Messiaen picked up on the clear intention by the composer to look for chordal choices that were based on the quality of the *Klang*, rather than for their function in a sequence. Variations in color were now replacing the tension-resolution of common-practice music. It is interesting to see the mention of *resonance* as a key element in chord construction. This idea had of course been around since Helmholtz, where it was used as a basis for his harmonic theory. Here, it takes a more autonomous form. At first, it provides a method for adding higher notes to color the triad, then it is extended as a way to emulate more complex inharmonic timbres, such as bells.[3]

Such ideas are fully realized in Messiaen's own music, where visual metaphors are constantly used to evoke the sense of color that permeates it. He speaks of "those flows of blue-orange lavas, those planets of turquoises, those violet shades, those garnets of long-haired arborescence, those wheelings of sounds and colors in a jumble of rainbows (...)" (Messiaen, 1956a, 52). In order to describe his approach to harmonic composition. A typical example of the realization of such ideas is found in his *accord de la résonance*, which is based on a (theoretical) low fundamental, and composed of the fourth harmonic followed by upper odd partials (Fig. 10.1). While this betrays a naive approach to acoustics, since the chord notes may not correspond exactly to harmonic frequencies depending on the instrumentation and tuning employed, it nevertheless shows a focus on the spectral qualities of musical materials. Messiaen himself was probably aware of the difference between the intention of resonance and acoustical reality, as he mentions "effects of resonance,"

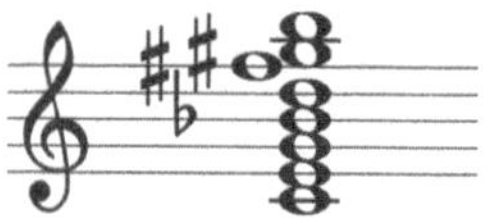

Figure 10.1 Messiaen's *accord de la résonance* (Messiaen, 1956b, 37).

attributing the idea to Dukas, but qualifies that these are "[e]ffects of pure fantasy, similar by a very distant analogy to the phenomenon of natural resonance" (Messiaen, 1956a, 51).

Approaches such as these demonstrate that composers were employing spectrally informed methods to organize pitch and timbral material in their music. Erickson (1975, 19) speaks of three different categories of such musical materials, a single pitch (spectrally colored), chord, and *a sound*. He organizes the continuous link between these as a triangle in which each one of these perceptual qualities occupies a vertex. The first one of these categories emerges when a single pitch among a tonal complex is clearly recognized as a *note* we may associate with it. A chord is taken as a bundle of tones organised in some way, and the third category, *a sound*, is mostly connected to more complex unpitched percepts. Composers started navigating the space between these categories in the early part of the 20th century. In the period following the end of World War II, much of the musical material used in concert music was derived from relationships found in this triangle.

10.1.2 Instrumentation and Spectra

Another spectral shaping element at the composer's disposal has been the variety of sonorities found in the modern orchestra. From the mid-part of the 19th century onward, a less formulaic approach to orchestration emerged, with composers starting to pay careful attention at the combination of instrumental timbres as a structural concern. While we may be able to trace this back to Beethoven, and point to the early example of Berlioz's *Symphonie Fantastique* (1830), the exploration of orchestral color only came to full fruition in the 20th century. We can find many early examples of spectral shaping in the music of Debussy, Stravinsky, Bartók, and Schoenberg, among others. In the practice of rearranging piano music for orchestral forces, we find another form of this, most famously employed by Ravel, Rimsky-Korsakov, and Webern. Composers began to employ the instrumental palette consistently to create various spectral color nuances, which became an essential part of these re-created works.

In particular, we find that this careful use of orchestration brought about the notion of *composing* timbre. Debussy, in works such as *Nuages* (1899), makes the listener keenly aware that the focus is not so much on the sounds as carriers of tonal-harmonic relationships, but on tonal colors, and the timbral mix of orchestral sonorities. Schoenberg goes one step further in inventing the term *Klangfarbenmelodie*, or melody of tone colors, for his approach of timbral composition:

> Now, if it is possible to create patterns out of tone colors that are differentiated according to pitch, patterns we call "melodies," progressions, whose coherence evokes an effect analogous to thought processes, then it must be possible to create such progresssions out of tone colors of the other dimension, out of what

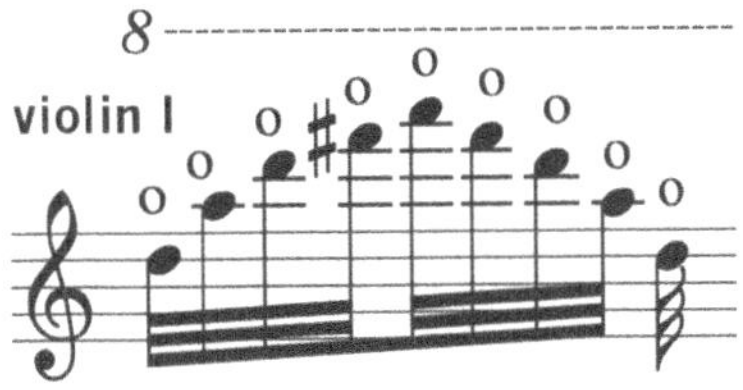

Figure 10.2 Harmonic glissando from *Firebird*, violin I.

which we call simply "tone color," progressions whose relations with one another work with a kind of logic entirely equivalent to that logic which satisfies us in the melody of pitches. That has the appearance of a futuristic fantasy and is probably just that. But it is one which, I firmly believe, will be realized. I firmly believe it is capable of heightening in an unprecedented manner the sensory, intellectual, and spiritual pleasures offered by art. I firmly believe that it will bring us close to the illusory stuff of our dreams; that it will expand our relationships to that which seems to us today inanimate as we give life from our life to that which is temporarily dead for us, but dead only by virtue of the slight connection we have with it.

Tone-color melodies! How acute the senses that would be able to perceive them! How high the development of spirit that could find pleasure in such subtle things! (Schoenberg, 1983, 421–422)[4]

In the third movement of his Op. 16 (1909), which he calls *Farben*, we have an example of such manipulation of timbre, in particular with chords being deliberately explored for their color rather than as a traditional means of harmonization, with careful use of orchestration. The type of technique used here was further explored by Webern, in, for instance, his 1934 orchestration of Bach's *Ricercar a 6* from the Musical Offering, where there is a subversion of the linearity of the counterpoint in splitting the fugue subject and other melodic structures into several instruments, "illuminating," or "coloring" them. In his own serial pieces for ensembles, such as his Symphony Op. 21 and Concerto Op. 24, Webern would also employ this approach as an integral part of the composition.

In the introduction to Stravinsky's *Firebird* we find another typical instance of the kind of coloristic orchestration employed by composers of the period, where a pure spectral effect is used in the strings (measure 17 in the first edition of the score). Over a low D, sustained by the basses, a harmonic glissando is placed in the cellos and violins, also supported by three sustained harmonic tones in the violas. The sound is of an arpeggio of harmonics (although it is played as a glissando, by sliding the finger over the strings). Played on the D string of cellos and second violins, it requires a retuning of the first string in the first violins from E to D (Fig. 10.2).

A further, particularly striking, example is also found in the same piece, the *tutti* chord from the *Infernal Dance* (score mark 133, first edition). This is an open fifth chord preceded by a short appoggiatura figure and spread through the whole orchestra range from contrabassoon to piccolo: a short hit, whose function is also purely spectral, punctuating the main melodic idea. It is of such a distinct effect that it became a detached sound object on its own, over eighty years later at the dawn of the digital sampling era.

This short hit was captured from an orchestral recording as a *sample* and reused in various musical settings (Fink, 2005) for its spectral qualities, completely removed from its original setting. Early 20th century music is full of such sonic gems.

More widely, the notion of a musical tone, or what sounds are fit for music, gets to be challenged by many composers. Nowhere this is more evident than in the music of Futurism movement, launched by the 1913 manifesto *The Art of Noises*, by Luigi Russolo (1967). In this work, we find the description of a new musical aesthetic based on the sounds of everyday life, in particular, the ones stemming from modern industry. Along with this, we find descriptions and instructions for the creation of instruments that would be capable of mimicking these sounds, called *intonarumori*, tuned noises. While the terminology was not quite there yet, considering that these attempts anticipated a more theoretical study of the spectra of these new sounds, we can observe the will to conquer all sounds for the implementation of a new music.

Following this path, the work of Edgar Varèse is probably the most celebrated pre-electronic music effort to create a musical aesthetic that is centered on the spectra of sounds. Varèse had a consistent approach to creating structures based on their spectral quality, one that could not be described in terms of the typical methods of orchestration, such as doubling.[5] A most extreme example of this type of instrumental writing is provided by his *Ionisation* (1931), which employed no instruments that would be able to produce *musical tones* in Helmholtz's sense.[6] While rhythmic structuring is an important aspect of the piece, the timbral colors of the instruments are also brought to the fore. By abstaining from using pitch as part of his musical materials, the composer is setting himself the task of looking for other sound attributes with which to define musical texture. Varèse would go on to employ electronic instruments, such as the Ondes Martenot,[7] as he tried to reach out for sounds not possible through mechanical means.[8]

Percussion was also regarded by John Cage as a means of conquering the spectrum. His *First Construction (in Metal)* (1939) for various percussion instruments, similarly to Varèse's work, explored a universe of sound where pitch was not a major concern. Cage also pioneered the use of pre-recorded sound in his *Imaginary Landscapes No. 1* (1939), where test recordings of steady tones were played on variable-speed turntables, alongside a cymbal and piano (played both on keys and directly on the strings). Although recordings had been used before (e.g., in Respighi's *Pini di Roma* from 1924), this was a much more purposeful use of recorded sound, as an electric instrument, fully integrated into the score. Cage's introduction of the *prepared piano*,[9] for which he composed several works, was also indicative of his intent in exploring spectral colors as an element of the composition. In 1939, Cage gave a lecture on the possibilities of these new sounds, which was later reprinted as *The Future of Music: Credo* (Cage, 1961, 3-6). This included many visionary ideas, which would become reality in the next few decades. At the outset he stated

> Wherever We are, what we hear is mostly noise. When we ignore it, it disturbs us. When we listen to it, we find it fascinating. The sound of a truck at fifty miles per hour. Static between the stations. Rain. we want to capture and control these sounds, to use them not as sound effects but as musical instruments. (Cage, 1961, 3)

He also predicted that electric instruments would soon be developed to provide "complete control of the overtone structure of tones (. . .) and to make these tones available in any frequency, amplitude, and duration" (Cage, 1961, 4). In the 1950s, Cage will move on to become one of first composers to work directly with tape recordings, by which time he was also interested in the use of chance operations to control the creative process (also an early example of *algorithmic composition*).

These early compositional approaches to shaping timbre, which have been extremely influential in 20th-century concert music, were developed from mostly an intuitive understanding of spectral qualities of sound. As the century developed, a better comprehension of the acoustic and psychoacoustics allowed composers to explore these ideas further. In particular, the advent of electronic means of sound production brought composers into contact with more precise and technical information, as well as with direct experimental means. These were key, as they allowed a hands-on approach to sound manipulation. In time, the electronic studio experience would feed back into instrumental composition, pushing it toward more fully fledged forms of spectral music.

10.2 Audio Technology

From a 21st-century perspective, we take audio technology for granted, as it is just another element of contemporary life. However, it is important to note how this has transformed our hearing, as it highlights the spectral detail in music in ways that would have been impossible without it. From a music consumption angle, access to high-quality reproduction of musical performances is in many cases instantaneous, and this allows us to discern nuances of sound that are beyond the simple musical parameters of pitch, duration, and intensity. Finally, from the music composition and production points of view, careful consideration of such spectral nuances is an integral element of professional and artistic practice. The musical work as an audio product is a complex spectral bundle composed of many essential elements,[10] and it is significantly different from what existed before the advent of electronic sound technology. While we may want to concentrate on its implications for the development of a spectrally informed composition, it is equally important to note how it transformed music creation and appreciation in general.

10.2.1 Recording and Broadcasting as Carriers of Spectral Information

The appearance of a means to record, process, broadcast, and reproduce sound was, as noted earlier, the single most important development that made timbre, and the spectrum, a central concern of composers and audiences. This follows from the fact that, for the first time, there was a practical method of directly and precisely registering the spectrum of the sound, as opposed to capturing only specific parameters such as pitch, duration, and intensity. As the pianist Glenn Gould put it:

> If we were to take an inventory of those musical predilections most characteristic of our generation, we would discover that almost every item on such list could be attributed to the influence of the recording. (...) [L]isteners have come to associate musical performance with sounds possessed of characteristics which two generations ago were neither available to the profession nor wanted by the

public—characteristics such as analytic clarity, immediacy, and indeed almost tactile proximity (Gould, 2006, 115-6).

Of course, as we know, this took place slowly as technology developed. Early mechanical recording was too crude to capture the nuances of instrumental timbre, but with the advent of amplification, and electrical registration methods, significant advances were made. From the 1940s, the quality of audio recording entered a phase of great development, supported by an expanding mass market for music. This raised the tolerance and expectations of both consumers and producers, which demanded a higher level of spectral detail in sound capture and reproduction. Recording and broadcasting became a precise means of disseminating spectral audio information.

The extent to which spectral qualities became an important and essential part of music was shown by the race toward frequency high fidelity in recording and reproduction.[11] The only reason for spending more technical resources to capture frequencies well above 5 kHz is to make timbral nuances clearer and more pronounced, since high frequencies are not good carriers of pitch information. The same argument applies for the extensive technical development in microphone technology and use, which was portrayed as bringing more *realism* into recordings. We may argue that such qualities are spectral ones, which are being placed on a par with the other traditional musical parameters. The attention of audiences, and the interest of composers, was positively turned toward the sound as a whole through the improvement in the quality of recordings. This was observed in many genres of music, from contemporary (*new*) music (where timbre is of course a central concern), to film music, as well as jazz and the various early examples of pop music, from the 1940s onward.

10.2.2 Changes in Instrumental Sound

Interesting examples of the influence of recording in shaping the spectra of instrumental sounds over a historical timeline are found in jazz performance. If we take, for instance, the evolution of the saxophone timbre from early recordings, we will see significant changes, which were supported by the fact that sound, rather than notated music,[12] becomes the immediate mode of musical communication. Recording provides the means in which the timbral nuances of particular players are disseminated. Much attention has been paid to the pitch content of jazz improvisation, but equally important has been the way instrumental timbre is shaped by players. The sound of the jazz saxophone deviated from its classical counterpart largely via the dissemination of recordings (and radio broadcasts) by great players from the 1930s onward. As the timbral idiosyncrasies of each performer could be captured and reproduced, a direct sonic bridge was established with up-and-coming musicians (who in turn adopted and transformed the instrument tone further).

An early example of this is seen in the recordings of Lester Young's saxophone performances, which were widely influential. The spectral features of his *sound* may be described as mellow, that is, lacking in strong high-frequency components, but also with a significant amount of air noise, which was used by him as an important expressive element. The various timbral nuances he created when articulating a melody, and the subtle combination of deterministic and non-deterministic spectral elements in his performances, could only be disseminated through audio recordings, particularly as the quality and frequency fidelity of the medium improved. They were integral elements to his *sound*, which could not be captured in mere transcriptions of his solos. Likewise, other individual approaches to shaping the sound of the saxophone that emerged in jazz were communicated widely, transforming the sound of the instrument.[13]

Many other interesting examples of the influence of recording on the instrumental sound (and stylistic) development of jazz can be found. The increasing prominence of ride cymbals as carriers of pulse in jazz performance is one. This, again, was only possible by the fact that recording and reproduction technologies were sufficiently developed to capture the high-frequency spectrum. A symbiosis between the performer and the recording engineer shaped what by the early 1960s became the distinct sound of jazz percussion. Equally, progress in amplification and in the spectral range of recordings allowed the rise of the double bass as a solo instrument, beyond its basic functions of accompaniment. The microphone, and its capacity of highlighting nuances, also played an essential role in the development of vocal techniques in jazz and pop music genres.[14]

Similar examples of timbral evolution can be given for other instruments that are as malleable as the saxophone, or even less so as in the case of the piano, which are attributable to the effects of recording and the dissemination of music as sound. Beyond jazz, the changes in the spectral quality of the electric guitar sound provide another important case. We can trace a remarkable timbral transformation, which in fact was even more abrupt: in the space of around ten years, from the early 1960s, changes in its sound make it almost unrecognizable as the same instrument. Of course, in this case, electronic processing in the form of non-linear distortion and other effects also played an important part. Nevertheless, without the direct means of spectral communication provided by recording, any such changes would have been much less pronounced, gradual, and not so widespread.

10.2.3 The Mechano-Acoustic and the Electro-Acoustic

The development of audio technology has made the distinction between the mechano-acoustic and the electro-acoustic virtually meaningless. Since the most typical contact with musical material is via some sort of loudspeaker delivery, the fact that the source of the sound might have originated at some point from a mechanical means of production is largely moot. Therefore any categorization that we may want to impose along these lines is artificial. There may be a vestigial separation between the *instrumental*, as in performed by physical actions on an identified instrument (mechanic or electronic), and the *non-instrumental*, as a product of synthesis or sound manipulation, but that is also on the way out, if the listening experience is as usual via an electro-acoustic transducer.[15] While it may be important for some to know whether a piece was *played* or created through other means, from the purely sonic, or spectral, point of view, that is an external consideration.

In many cases, it is impossible to define the boundaries of the instrumental and the electronic, and similarly, of the mechano-acoustic or the electro-acoustic. Are the recorded, digitally encoded sounds of a piano, played back through keyboard triggering, a form of electronic music, instrumental music, or an undefined territory between these? Are spliced-up recordings of an existing piano piece[16] a registration of a performance of concert music, or do they make up a specific electronic work based on the said piece? We would have to consider these as open questions, if it were not for the perspective given by electro-acoustic listening. Since all we care about is the *sound*, which we characterized earlier as a complex spectral bundle, all of these contrasting, and sometimes paradoxical, viewpoints are resolved into one single answer: these are all instances of musical audio products. We have moved into a territory that is occupied by *sound*-based music.[17] That leads us to take a an approach to music listening (and by extension, production) that is of a spectral nature.

10.3 Electronic Music

Besides the more widely significant element of sound recording and broadcasting, audio technology also gave us the means of electronically processing, and synthesizing, spectra. This has had a much more obvious impact in the development of music that is directly concerned with timbre, and spectrum, than anything else. As we already noted in Chapter 2, decisive steps for the establishment of a musical practice that was based on sound alone, and through electronic means, were first made in the immediate postwar period by Schaeffer in his *studio d'essai* (research studio) at Radio France.

Schaeffer was particularly interested in examining the nature of sounds as captured by his recording equipment, and how these could be transformed and assembled to make pieces of music. Some of the early pieces betray the influence of the Futurists, where we hear the sounds of modern-world machines, such as locomotives, now played through the lens provided by the microphone. From one perspective, this elevates everyday sounds into a category of musical sounds to which they did not originally belong. Taken as for what they are, they represent, or stand for, or their source cause, they are *mimetic*. This is a characteristic that we do not normally assign to instrumental sounds; we do not listen to a piano mimetically, but generally, *spectrally*, for the *aural* qualities of its sound. With the use of recorded sounds, the added dimension of mimesis is fully integrated into the practice of music composition.[18]

Complementarily, Schaeffer's use of everyday sounds also allowed us to approach these as we normally do instrumental tones. This is quite revolutionary in that it provides a change of perspective, which employs a spectral way of hearing sounds that disregards their source or causation. Such new modes of listening would later form a central tenet of his musical object theory. Besides exploring everyday sounds, Schaeffer also turned his attention to instrumental spectra, and their transformation. In one of his first studies, *Étude Violette* (1948), we hear various spectral evolutions created from piano sound, whose attack has been removed, forming different textures that were mixed and sequenced to make the final result.

The recording and manipulation of different sound sources led to the development of an approach that put spectral concerns at the fore of the composition process. This was combined with an experimental method based on the *primacy of the ear*. Structure could be derived directly from the sound matter itself, and the organization of a piece was oriented mainly by the sonic result. In many ways, that is how many examples of electronic music of all genres and styles have been, and continue to be, composed. It can be considered as the opposite of applying some external, and abstract, structuring scheme to provide the scaffolding, or in many cases the complete material, for a musical work.

Emmerson (1986) delineated the various intermediate approaches to structuring a composition that lie in between these poles. He puts these in the form of two axial concepts: *syntax* and *discourse*. The latter has to do with the sound material of the composition and how it is used, and the former with how it is organised into a musical piece. While in terms of discourse, we may treat sounds from an spectral (or aural) or from a mimetic perspective, it is possible to apply a syntax that is abstracted from the sonic material, or impose one that is external to it. Approaches to electronic music composition may use aural and/or mimetic discourse to different degrees, combined with abstracted and/or abstract syntax. It is clear that works employing an aural discourse, whose structure is derived from its materials, are intentionally (and intrinsically) linked to a spectrally oriented way of composing.

However, it is also obvious that any work may be appreciated from a spectral hearing perspective, regardless of the composer's intentions, as this is the listener's prerogative. To describe this, Schaeffer proposed the idea of *acousmatic* hearing (Schaeffer, 2006). This term has an origin on the practices of the Pythagorean school, where lectures were delivered through a curtain or a veil, and the speaker was out of sight.[19] Similarly, when listening to sound through a loudspeaker, the focus may be turned onto its spectral qualities rather than its cause or source. The musical discourse is then constituted of *sound objects*, which are listened to for themselves, for their own particular characteristics and nuances. The listening method involved in this process is called *reduced listening* (Schaeffer and Chion, 1983). It is not a natural mode of listening, as our hearing and perceptual mechanism tends to be directed to the concrete rather than the abstract aspects of the sound world. However we can and do develop this for listening to music of all genres. We may argue that closely following various parametric aspects of a composition is a form of reduced listening.[20]

The musical practice that emerged following Schaeffer's Musique Concrète (and to some extent from the German Elektronische Musik school) has been associated to a very specialized form of concert music, which is often called *electro-acoustic* music. This became a catch-all term, which often is meant to include computer music and other subgenres. As it developed, the aesthetics of electro-acoustic composition favored careful exploration of sound synthesis and processing. This was fundamental for the development of many novel techniques of spectral manipulation. However, we may argue that distinctions between electro-acoustic music and various other forms of electronic music may not be relevant today.

10.3.1 The Feedback on Instrumental Writing

Within one decade of Schaeffer's early works, electronic music became an important element of the contemporary concert music scene in Europe and North America. Considerable interest in developing new technologies for sound manipulation led to a number of interesting inventions in the magnetic tape and analogue processing studio, which were used in creative, and sometimes disruptive ways, by composers. Backed by public institutions such as the national radios, or educational bodies such as colleges and universities, facilities of various sizes and technical means became the catalyst for a flurry of activity in experimental music. Much of this was fundamentally oriented toward the techniques of spectral manipulation.

The sounds and methods of electronic music, in time, fed back into the scores of instrumental music, and a primarily spectral approach to writing emerged in the works of Ligeti, Penderecki, Lutosławski, and Xenakis, among others. Classic examples are given by Ligeti's *Apparitions* (1959) and *Atmosphères* (1961), and Penderecki's *Threnody* (1960). The earlier work *Metastaseis* (1954) by Xenakis, while not directly influenced by electronic music, shares with it principles of spectral manipulation, such as the shaping of sound masses. Informed by his own practice of tape music composition, Stockhausen also employed similar spectrally oriented methods, in works such as *Stimmung* (1968), where the manipulation of vocal formants is used as a main structuring principle.

The influence of electronic music on instrumental composition continued to be felt in later works of the late 1960s and following decades. The iterative music of Steve Reich, for example, based on phasing effects, was heavily influenced by some of the typical magnetic tape studio methods, which the composer himself had explored in earlier works. Spectral concerns become central to other American musicians of this generation, such as La Monte Young and Terry Riley. In Europe, a number of composers started to transcribe electronic effects into score notation to emulate certain types of spectral transformations. An example of this is Murail's *Mémoire—Érosion* (1976), which attempts to mimic the various artifacts created by feeding back signals in tape echo arrangement. Due to use of a specific and deliberate approach to shaping musical materials from a spectral perspective, the music of this group of composers received the label of *Spectral Music* (La Musique Spectrale). However, we cannot say this music is any more *spectral* than several other examples of music composed from 1945 onwards.[21] Toward the end of the 20th century, as the conditions to process audio in realtime with computers became a more practical proposition, the blurring of the boundaries between instrumental and electronic music helped embed a spectral approach to composition more widely.

10.3.2 Electric Jazz, Rock, and Pop

Jazz, rock, and pop may be considered electro-acoustic native genres, since they developed hand in hand with audio technology, at first primarily with recording, broadcasting, then later through the use of amplification, sound processing, and synthesis. This allowed a smooth adoption of electronic music methods, which started to appear more consistently midway through the 1960s, and then became somewhat integral to their musical practice. It is possible to single out some early electronics-oriented approaches to composition in both rock and jazz: Jimi Hendrix's use of guitar and tape effects in *Electric Ladyland* (1969), and Miles Davis's extensive use of tape-editing in *Bitches Brew* (1970) provide some standout moments. A number of other similar examples may be noted.[22] The rise of the synthesizer as a performance instrument, alongside other sources of distinct timbres (e.g., the mellotron, the processed tone-wheel and divide-down organs, various types of electric guitar and the transformations applied to them), and the use of the multitrack recording studio as a composition tool, allowed rock (and jazz-rock/fusion) to incorporate fully fledged electronic music sub-genres.

Starting with progressive and experimental rock, electronic music practices began to filter more consistently into mainstream music. By the 1980s, many of the techniques that emerged from the early electronic music studios became essential parts of the music production methods in pop. The most obvious example of this is *sampling*, which first appeared in its modern recognizable form[23] in Schaeffer's studio of the early 1950s, provided by specially designed tape playback machines, the *phonogènes* (Teruggi, 2007), and then was carried over to commercial instruments such as the Mellotron. With the introduction of digital audio technologies, sampling became so central to pop music production that it permeates all of what is done within today's recording environment (to a point where it has gone into the background, integrated into the typical studio routine). This supports the idea that making distinctions between various forms of music-making in terms of methods, sound sources, and materials is problematic and unhelpful. Furthermore, we continue to observe that the more experimental end of pop music (e.g., some of electronic dance and ambient music sub-genres) approaches and converses with the experimental end of contemporary concert music.[24] A process of hybridization (Waters, 2000), where boundaries of various kinds are broken, was put in train by the development of audio and electronic music technologies, and is being fully realized in the continuum of today's music. As we have been arguing consistently, it is increasingly hard to make distinctions in an age where music appreciation and consumption is fundamentally oriented toward the *sound* (and the musical product as a spectral complex).

10.3.3 Spectromorphology

As we have already noted in Chapter 2, Schaeffer's ideas on structural aspects of electronic music were extended significantly into the concept of *spectromorphology* (Smalley, 1986). Introducing it, Smalley speaks of a split in the development of Western music, in which one of the branches can be characterized by spectromorphology:

> The development of Western music in the twentieth century is dominated by a historical bifurcation in musical language: tonality with its metrically-organised harmonic and melodic relationships has continued to be the vernacular language, (...) while the other fork, in its most recent guise, is represented by spectromorphology. (Smalley, 1986, 61)

While we may doubt that this assertion, simply put, actually represented the reality on the ground forty years ago, and is certainly questionable now,[25] the principles of a theory for composing spectrally oriented music (which is what we are talking about here) are certainly useful. Some of the concepts developed in spectromorphology can be applied to describe the various aspects of spectral sound design we have been studying from a technical perspective. There are three elements to this theory, namely,

- Spectral typology: the various types of objects classified from a spectral perspective.
- Morphology: the dynamic shapes that such objects can assume in time.

- Motion typology: the patterns and types of motion, which may be constructed from morphologies.

The typology of spectral objects is close to Erickson's triangle, discussed earlier, but is organized in a linear fashion, with three separate categories that are placed within a continuum: *note*, where the focus is on definite pitch attributes; *node*, where pitch is diffuse, and the focus moves to the overall spectral shape; and *noise*, which is fundamentally based on non-deterministic spectra. Such a model, while it ignores the multi-dimensionality of spectral perception, has an advantage that is simple and applicable to the techniques we have developed in this book. Thus it can provide a language with which we can discuss general aspects of spectral design. In complement, the morphology of spectral objects attempts to describe their dynamic aspects, in a small scale, through a set of envelope *archetypes*, which may be combined or stringed. Again, some of these may prove useful as sound object descriptors, although the simplicity observed in the typology is somewhat lost here. While three basic envelope archetypes are given, these get multiplied into several morphological models and string, yielding a much larger set of elements to describe this aspect of spectromorphology. Finally, motion typology is concerned with a longer timeframe, and seems to provide support for specific principles of music composition, which may be less applicable to our concerns. Here, the theory becomes increasingly more complex, and less elegant, with an enlarged set of terminologies and relationships. Nevertheless, as a whole, the principles of spectromorphology may provide useful insights into the structuring of spectrally informed music.

10.3.4 Spectral Hearing

From a sound design point of view, regardless of whether we are going to employ a sound object for their mimetic or aural (spectral) characteristics, *spectral hearing* skills are fundamental. These are quite separate from, and much more objective than, the listening *attitude* we may have toward a piece of music. Being able to detect the nuances and specific characteristics of different types of sounds, and how to achieve them through synthesis and/or manipulation, is key. Spectral hearing requires a specific kind of ear training that is directed at the spectrum as a whole rather than to specific parameters such as pitch and duration (rhythm). Tuning our ear to the technical means of realizing specific sounds through digital signal processing, as discussed in this book, may be achieved by an aural understanding of the methods employed, as well as the practice of applying these in diverse situations. It also involves an acknowledgment of psychoacoustic effects, such as, for instance, spectral masking, tonal fusion/separation, and a consideration of listening environment conditions. Spectral hearing is an extensive topic, which although central to the subject of this book, is beyond the scope of our discussion.[26]

10.4 Computer Music

As we enter the third decade of the 21st century, it is becoming increasingly hard to find examples of music that, from the perspective of production tools, are not computer-based. As audio computing technology becomes ubiquitous and disappears in the background, only very few examples of musical products are untouched by it. However, to simplify our discussion, we may keep ourselves to music that is intrinsically computer-based, in the sense that it emerged directly from the application of the technology.[27] In Chapter 2, we traced the key technologies involved in the development of computer music. Here, we will attempt to focus mostly on the aesthetics and musical techniques that emerged from it.

We should think of computer music, as it emerged in the 1960s, as a branch of electronic music with very particular features, which make it quite unique. In special, its methods made it particularly suitable for the manipulation of the spectrum, and we can consider the computer as a natural environment for spectral music. Beginning with Tenney, and then Risset and Chowning, we have composers who were exploring the precision of computational approaches to carefully sculpt and manipulate the spectrum. In this space, we see the development of key techniques that would be later employed in all genres of electronic composition. The two classic examples are of additive synthesis, within the work of Risset, and frequency modulation synthesis, introduced by Chowning. Since these ideas were being developed under a music programming environment, their formulation and dissemination became much more stable, if compared, for instance, with the routines of the tape studio. The MUSIC N-family of computer music systems was an essential element of early computer music, providing the environment for a lot of the direct experimentation with sound synthesis and manipulation.

10.4.1 Risset's *Catalog*

Starting with Risset, the idea of creating collections of sound synthesis recipes as a means of documentation is a noteworthy aspect of computer music. Very early on, composers understood that in order to learn the craft of electronic music, it was important to understand both the means of sound generation and the mechanisms of auditory perception. In his introductory notes to the *Catalog of Computer Synthesized Sounds*, Risset notes:

> It has long been recognized that in order to take advantage of the unlimited resources of computer synthesis of sound, one had to develop a body of psychoacoustical knowledge, enabling him to specify the physical parameters corresponding to a desired type of sound. Experiments with the seemingly well-known sounds of some musical instruments have shown that such knowledge was still very poor, but that computer synthesis of sound was an invaluable tool to remedy this situation. (Risset, 1969, 2)

Risset was quite clear that the examples provided in his catalogue were not meant as models, and in particular, not as efficient recipes to synthesize a particular sound, and that no attempt at closely imitating real instruments sounds was made. The different items in his catalogue were more or less organized around sound types, which could be employed as a starting point for systematic investigations based on the modification of synthesis parameters. As part of Risset's work at Bell Labs a number of interesting spectral synthesis ideas were developed. These were duly registered in some of his catalogue examples; number 513 is one such case, describing the Risset-Shepard tone synthesis. The MUSIC V instrument for this sound is given as (Risset, 1969, A93)

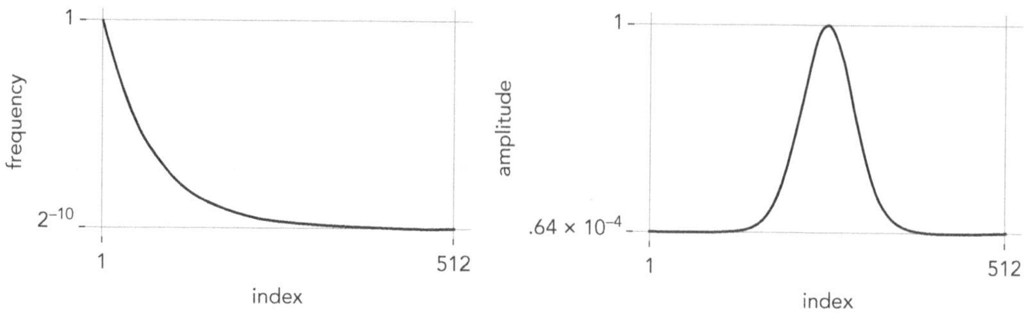

Figure 10.3
Function tables from Risset's example 513.

```
INS 0 1;
IOS P5 P7 B3 F2 P8;
IOS P6 P7 B4 F3 P9;
IOS P3 B4 B5 F1 P25;
OUT B5 B1;
END;
```


which is simply an instrument made up of three interpolating oscillators (implemented by the IOS unit generator), the first two reading from function tables defining frequency and amplitude curves, and the third defining sine wave controlled by these parameters. It is effectively one partial of an additive synthesis design, and as such, very straightforward. What makes this example special are the way this instrument is employed and the frequency and amplitude tables used by it, which are shown in Fig. 10.3.

Each instance of the instrument is used to produce a glissando whose amplitude is shaped by a bell-shaped curve. This means that we never hear where it begins or ends; it just fades in and out smoothly. If we use a number of these at the same time, we can make the fade in/out imperceptible by offsetting the glissandos so that they do not start and end at the same time. If the glissando has an N-octave width, we can use N oscillators, spaced in octaves, whose glissando starts by $1/N$ of the total cycle time, an illusion of an endless glissando may be created. In example 513, Risset uses 10 oscillators, descending 10 octaves in 120 seconds; as each oscillator reaches the lowest frequency, it starts again at the highest, but since this happens when the amplitude is zero (at the edges of the amplitude window), then we do not hear this jump. Effectively, the perceptual effect is of a constantly sliding sound, much like an Escher staircase that goes forever up (or down). Variations on these principles of auditory illusion, as well as careful spectral synthesis methods, are the basis of his classic compositions *Little Boy Suite* (1968) and *Mutations* (1969): chord—timbre—melody transitions, smooth changes of octaves, glissandos that simultaneously go up and down in frequency, etc.

Risset hoped that his catalogue would provide a model and inspire other people working in computer sound synthesis to document their findings similarly. It was in fact very influential, followed, for example, within the Csound community by collections of instruments from Richard Boulanger and others (e.g., *Csound TOOTs*, the *Amsterdam Catalogue*, etc.), which continue to be useful and influential to this day. With the advent of the internet and the world-wide web, the sharing of synthesis recipes became common practice within computer music communities.

10.4.2 Case Studies

In order to understand spectral composition in computer music, it is worth looking at specific cases, and in isolation, making an attempt to dissect the methods employed. It is a somewhat unnatural approach in that what we may study in these specific cases may not be always heard separately from other dimensions of compositions. However, with that caveat in mind, we can proceed to explore certain aspects analytically, and then attempt some sort of mental synthesis at the end. The case studies will not be particularly work-oriented, that is, dedicated to specific compositions, but will aim to explore certain approaches that are common to a number of them.

We may organize spectral composition in two categories, which can be generally described as synthesis-oriented and transformation-based. The former is quite clearly concerned with pieces whose sound material is created from scratch, via one or more methods of sound generation, and the composer attempts to shape the spectrum via synthesis parameters. The latter is related to the use of pre-existing recorded material, and employing various forms of processing to transform[28] their spectra. Extreme cases exist in which a distinction between these categories may not be possible: through the various processes of transformation, a single short and non-distinct sound may be the source for several different types of spectra; is this synthesis-oriented or transformation-based? Do extensive processing techniques applied to synthetically generated material constitute a form of transformation or synthesis? However, if we take the view that such examples are the exception rather than the norm, it is possible to explore these two categories separately. The two case studies that follow are thus organized.

10.4.2.1 Case Study I: The Synthesis of Spectra

From a chronological perspective, spectral manipulation via synthesis came first to computer music. As we have already noted in Chapter 2, the development of the MUSIC N languages allowed an increasing number of approaches to sound generation, which was explored by composers, starting with Tenney, to create spectrally oriented music. This was done with a surprisingly small number of unit generators; MUSIC IV had only a handful, and their immediate successors did not expand on that very significantly. As we have shown with an example from Risset's catalogue, it can be fairly straightforward, from a sound synthesis instrument perspective, to construct quite complex sounds through additive synthesis. Most of the spectral effort design goes into the specification of the time functions for frequency and amplitude, and the parametric control of the instruments.

Very early on, composers latched on to the idea of composing from the ground up: creating sounds from the most basic building blocks. As a practical approach to electronic music composition, this actually precedes computer music by over ten years. It was already the mode of operation at Eimert's studio in Cologne, although their methods were entangled with the somewhat limited horizons of serialism. Risset talks about discovering spectral composition in his first work of computer music:

> The [Little Boy] *Suite* introduced two important concepts that depended upon computer synthesis for their implementation. The first of these concepts was the composition of sound and the inharmonic structuring of timbre. Synthesis

> allows us to compose sounds themselves, just as one composes a chord; composing spectra helps impart to timbre a functional role and not only a coloristic one. Harmony can be prolongated into synthetic timbres—such as certain bell-like tones or the slowly gliding tones that evoke the flight of planes—or it can be the root of melodic textures, through spectral analyses of chords, such as in the harmonic arpeggios opening of the piece. (Risset, 2009, 27)

Faced with a parametric approach to designing timbre, it was soon the case that composers found means of structuring compositions through different control schemes. Again, we see this in the way Risset handles continuous and discrete scales of frequency in *Mutations*, where an overall idea (transitions/mutations) is imposed on the various aspects of sound generation. The discovery of frequency modulation as a synthesis method by Chowning propelled such methods of spectral design, or spectral composition, into a new level. In an early work using the technique, *Turenas* (1972), he explores careful exchanges between harmonic and inharmonic spectra, which fall naturally within the scope of simple FM. Both Chowning's and Risset's methods point to the idea of a *metacompositional* practice in spectral music synthesis, which may or may not be due to the fact that computers offer a programming interface to mediate music-making.

Certainly the fact that constructing sonic material from the ground up favors a *meta* perspective for composing: we end up creating structures, routines, and programs, which are not themselves the musical matter, but direct the composition process. These can be seen as *metathemes*, themselves not the actual musical themes,[29] but the generators of the spectral elements that will make up the thematic material. This metathematic approach appears to be a feature of spectral music that is composed primarily using synthesis methods (but not exclusive to it). While we can say that the structuralist/serialist approach of the 1950s is also based on such an approach, the main difference when it comes to spectral music composition is that perceptual considerations play a key guiding role in the process. This is observed, for instance, in the music of Xenakis, whose metathematic concerns were extensive but always served the final sonic/aural result.

We can illustrate the metathematic principle in computer music through another classic FM synthesis work by Chowning, *Stria*, composed in 1977. In this piece, the metatheme is based on the *golden ratio* (Meneghini, 2007), and this determines the spectral design of the work. The golden ratio is defined as a ratio between two numbers that is equivalent to the ratio of their sum and the larger of the two numbers (Herz-Fischler, 1987, 1). We can translate this as

$$\frac{m+n}{m} = \frac{m}{n}, \tag{10.1}$$

which puts the numbers m and n are in a golden ratio relationship. If this is the case, then, it follows algebraically that

$$\frac{m+n}{m} = 1 + \frac{n}{m}. \tag{10.2}$$

Calling the golden ratio $\phi = m{:}n$ we have

$$\phi = 1 + \frac{1}{\phi}, \tag{10.3}$$

which defines a second-order polynomial equation

$$\phi^2 - \phi - 1 = 0. \qquad (10.4)$$

This polynomial has two roots, which are found through Bhaskara's (quadratic) formula, $(1 \pm \sqrt{5})/2$. One of these is positive, $1.618\ldots$, and the other negative, $-0.618\ldots$. The golden ratio ϕ:1 is defined by the former. If we instead, use the inverse ratio, 1:ϕ, we will get $0.618\ldots$ and $-1.618\ldots$ as the roots of the polynomial $\phi^2 + \phi - 1$, and so $\phi^{-1} = 0.618\ldots$.

From Eq. 10.3, we have the following unique relationship

$$\phi^{-1} = \phi - 1, \qquad (10.5)$$

which can be used to create a recursive formula to determine any power of ϕ,

$$\phi^{n+1} = \phi^n + \phi^{n-1} \qquad (10.6)$$

The exponents in this expression make up three consecutive terms of the *Fibonacci* series F_n,

$$F_n = \begin{cases} 1, & n \leq 0 \\ F_n + F_{n-1}, & n > 0, \end{cases} \qquad (10.7)$$

where we find the next term for $n \leq 0$ as the sum of the previous two. In fact the Fibonacci sequence and the golden ratio are intimately linked. The ratio of successive terms in that series approximates the golden ratio with increased precision, and at the limit it is equal to it,

$$\lim_{n \to \infty} \frac{F_{n+1}}{F_n} = \phi \qquad (10.8)$$

Finally, with Eqs. 10.6 and 10.7 together, we have a means of transforming a polynomial in ϕ directly into a first-order expression $a\phi + b$. Examining this a little more closely, we can also conclude that the coefficients a and b are also consecutive terms in the Fibonacci series,

$$\phi^n = F_n\phi + F_{n-1} \qquad (10.9)$$

The choice of the golden ratio as the basis for the construction of the spectra in *Stria* is not casual. The relationships shown here are very important for the synthesis method employed. One of the motivations was to find a method to organize materials based on inharmonic partials, which may bear a methodological resemblance to how harmonic spectra are constituted. In this case, the golden ratio becomes an *ersatz* octave, and pitch scales are constructed using it as the repeating interval. A division of nine steps (also subdivided again in 18 half-steps) is imposed on the golden ratio, to create the basic frequency scale,

$$p_n = \phi^{\frac{n}{9}} \qquad (10.10)$$

which has an origin at the base frequency of 1,000 Hz. This is extended downward using the reciprocals of powers of ϕ (to ϕ^{-7}, equivalent to 34 Hz). The scale has an upward limit of ϕ^3, or 4,236 Hz). The complete pitch space defined by this scale is shown in Fig. 10.4.

Since the golden ratio is not too far from a major sixth ($5{:}3 = 1.666\ldots$),[30] a 9-step scale makes the step interval only slightly shorter than a semitone by a ratio of about 0.995 (or 7 cents). This allows the construction of melody-like segments that have a

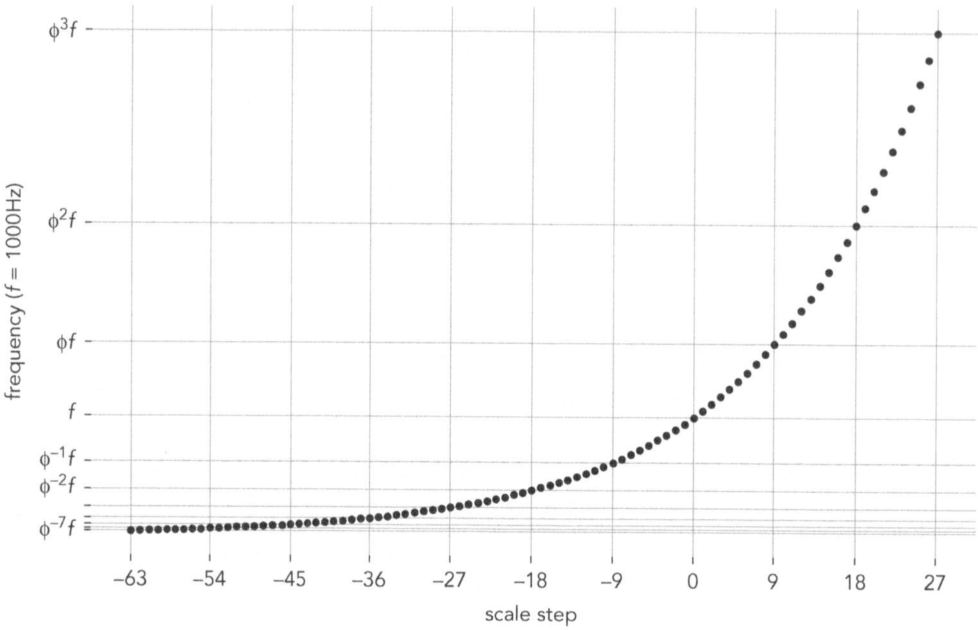

Figure 10.4 The golden-ratio based scale used in *Stria*.

natural feel, although somewhat disrupted from the usual forms of temperament. Due to the shorter step, it takes about 13 steps for the pitch to be close to an octave (an interval ratio equivalent to 2.003).

In *Stria*, a single method of synthesis is employed, a double-modulator form of FM synthesis. The synthesis instrument in its re-creation using Csound code[31] is fairly straightforward,

```
instr 1
 iamp = p4  // amp
 isk = p14 // mod skew
 idis = p15 // distance
 ifc  = p5  // carrier freq
 ifma  =  p6 // modulator A freq
 indxa1 = p7 // mod index A1
 indxa2 =  p8 // mod index A2
 idifa  = p8-p7
 ifmb  = p9  // modulator B freq
 indxb1  =  p10 // mod index B1
 indxb2  =  p11  // mod index B2
 idifb  = p11-p10

  // get correct index values for end.mem
  if (p32 == 1)  then
   indxb2 =  indxa2
  endif
  imag = 1 // fun to test
  ifactor  =  (p3-(p12+p13))/2
```

```
// generate global envelope pointer
 kndx = linseg(0,p12,0.25,ifactor,0.5,ifactor,0.75,p13,1)

 // generate first envelope (env)
 ifn1 =  p28
 ixmode  =  1
 ktable = table(kndx,ifn1,ixmode)
 kenv  = ktable*iamp

  // generate second envelope (skew)
 ifn2  = p29
 ktable2 = table(kndx,ifn2,ixmode)
 ksk = ktable2*isk
 kskew  = 1+ksk

 // generate third envelope (fda)
 ifn3  = p30
 ktable3  = table(kndx,ifn3,ixmode)
 kfda  =  ktable3*(ifma*idifa*imag)

 // generate fourth  envelope (fdb)
 ifn4  = p31
 ktable4 = table(kndx, ifn4, ixmode)
 kfdb = ktable4*(ifmb*idifb*imag)

 // double-carrier FM
 amoda = oscili((ifma*indxa1*imag)+kfda, (ifma*imag)*kskew,1)
 amodb = oscili((ifma*indxb1*imag)+kfdb, (ifmb*imag)*kskew,1)
 asig = oscili(kenv, ((ifc*imag)*kskew)+amoda+amodb,1)

 // spatialisation and output bus routing
 gaudio = gaudio+(asig*p16)  // sends reverb signal
 asig = asig*idis   // adjust distance
 aout1 =  asig*p22  // FL
 aout2  =  asig*p23  // FR
 aout3  =  asig*p24  // RR
 aout4  =  asig*p25  // RL
 ga1 =  ga1+aout1
 ga2 =  ga2+aout2
 ga3 =  ga3+aout3
 ga4 =  ga4+aout4
endin
```

As we have seen in Chapter 8, the use of complex modulation allows a smoother spectral transitions as many more partials are created with reasonably small amounts of modulation. Chowning employs different envelopes for timbre and amplitude, also applying a small deviation, called a *skew* in the code, to the modulation frequency in order to produce less static results.

Furthermore, as FM synthesis is used to generate materials, setting the carrier to modulator ratio as powers of ϕ creates a partials whose frequencies are always powers of ϕ (and so fall on frequency grid created by the golden ratio). As we have seen, these are the sums of differences of powers of ϕ, and can also be described by the linear expression $a\phi + b$. Such a spectral plan produces a series of hidden connections (Dahan, 2007) that makes the inharmonic material sound cohesive in a similar fashion to how harmonic spectra seems to match tonal music. The composer has indicated that he believes this is due to the "unified structure of spectral formation" (Dahan, 2007, 68).

Interestingly, a precedent in the use of non-octave scales is found in Stockhausen's *Studie II*. In that piece, we find a scale with steps of $5^{1/25}$ ratio. The spectra of the sounds used in the work is also composed (by additive synthesis) of partials taken from that pitch space. However, the result is nowhere as cohesive and natural sounding as in *Stria*, perhaps due to two factors: the more or less arbitrary choice of interval in which to base the scale, and the static quality of the sounds used. Chowning, on the other hand, was very careful in creating sounds whose spectra changed over time. Also as evidenced by the Csound code of its re-creation, he employed some small deviations in the frequency modulation to introduce a lively quality to the resulting sounds. The importance of dynamic spectra cannot be understated.

In addition to determining the spectral shape at the synthesis level (through the carrier to modulator ratios), and at the event level (through the nine-step scale), the golden ratio is also employed to define in broad terms the overall shape of the piece. The climax of the work is at 585 seconds from the start, and the complete piece is 948 seconds long,[32] giving a ratio of 1.617..., which is only slightly shorter than ϕ. In Fig 10.5, we can see the overall frequency arch given by changes in tessitura over time, where the two parts of the work can be clearly identified. This use of the golden ratio at various levels has been noted to give the piece a fractal-like structure (Dahan, 2007), where macro features (such as the timing of sections) are also carried over to the lower levels of its fabric. A metathematic approach supports this type of construction very well, which can be helpful to provide an overall result that is perceived as cohesive.

The use of FM synthesis as the sole method of sound generation in *Stria* and other pieces by Chowning is a significant factor in his search for a unified approach to composing the spectrum. It is often the case that composers will concentrate on particular method and try to get the maximum variety of spectra out of it, while still preserving strong connections between the different sonic material used. Another classic example of this is found in Barry Truax's *Riverrun* (1986), a piece composed solely using granular synthesis. In this work, there is a one-to-one match of sound synthesis technique and compositional intention: the main idea behind it is to depict how a complex phenomenon can emerge from the accumulation of small, nondescript, elements. Granular methods are exactly based on that principle, as separate grains may not have any striking spectral features, but when collected in large numbers, can create very complex sounds

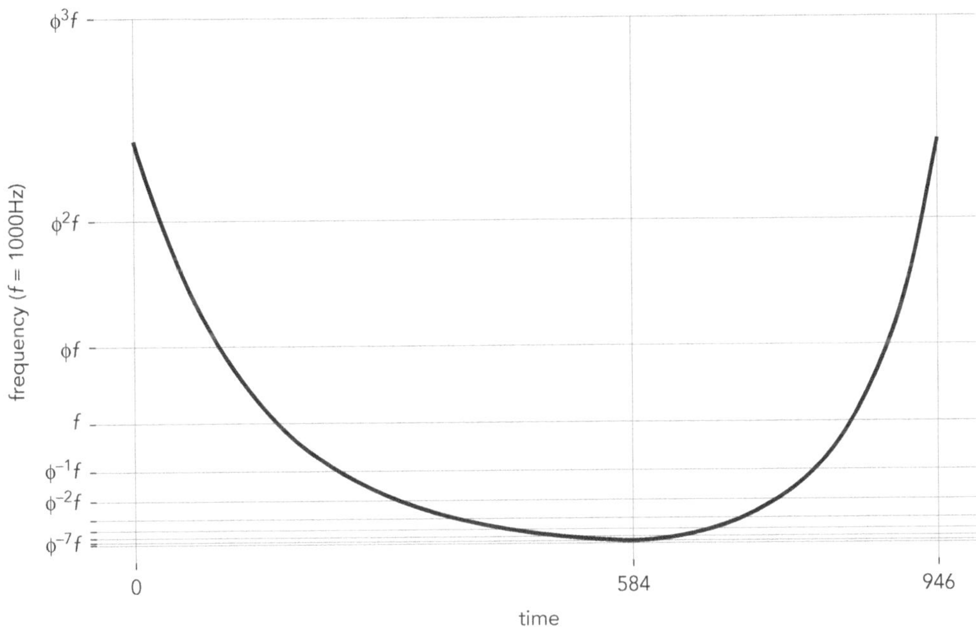

Figure 10.5 Average tessitura and main sections of *Stria*.

with intricate spectral structures. Truax notes the link of the synthesis method to the musical idea:

> The fundamental paradox of granular synthesis—that the enormously rich and powerful textures it produces result from its being based on the most "trivial" grains of sound—suggested a metaphoric relation to the river whose power is based on the accumulation of countless "powerless" droplets of water. (Truax, 1988, 25)

Also important to note is that the overall design and programming approach may be heavily oriented by the synthesis method. In the case of *Riverrun*, there are sections where a density of over 2,000 grains per second was used. It is impossible for the composer to specify the parameters for each individual instance, so a principle of guiding the overall trends of parameters needs to be applied. Truax employed ramps as functions of time to define the dynamic aspects of these trends, but also mentions the possibilities of using tendency masks to define ranges of underlying stochastic behaviors (as, for instance, discussed in Chapter 9). This, in the particular case of granular synthesis, is very appropriate, as it allows control over the macrostructure, leaving the microstructure to be defined by random processes. While the overall spectral shape and its emergent features can be sculpted carefully in this way, a very natural-sounding quality is retained due to the non-deterministic behavior of the grain generation process.

These and other classic examples have an important common aspect: the possibility of a *pure* form of computer music practice supporting spectral composition comprehensively. It is pure in the sense that no other means of music-making can reach the same levels of precision, as it is not possible to recreate this music using any other electronic studio routines. Of course, this is nothing new: there are plenty of examples in today's music production practices (of all genres) that would be impossible to achieve before computers became an integral part of them. However, the point we are attempting to make here is a different one: in spectral (or spectrally oriented) music composition, the power of the computational approach can be harnessed in structural ways to help the composer shape a piece. This is an intentional, and technically-informed, approach, which goes beyond the simple application of synthesis methods in arbitrary ways. A parallel may be made with tonal music practice: while it is possible to reach a particular harmonic/polyphonic structure intuitively, *by ear*, there is a much more straightforward, and possibly more cohesive path, to do this through the known methods of function harmony, voice-leading, and contrapuntal techniques. Equally, here, we can arrive at interesting sonorities and musical ideas purely by experimentation, but a planned and technically oriented approach may be more appropriate to reach the desired result.

10.4.2.2 Case Study II: The Transformation of Spectra

A very wide range of possibilities is provided by applying spectral processing techniques to recorded sounds. Composers have taken advantage of this in many ways, including using the frequency-domain structure of sounds as a metatheme for the work. This is the method employed by Jonathan Harvey in his piece *Mortuos Plango, Vivos Voco* (1980) (Harvey, 1981), where the recording of a bell is used as a starting point. The spectra of this sound were analyzed and the most prominent partials were extracted, from which eight frequencies were selected to form the structure of the work. Each section is based on one of these, which are the basis for a hierarchical system of frequencies. They are used, for instance, in the transpositions of the bell spectra, and in organizing the pitch material of the other sonic component of the piece, the voice of a boy soprano. Various transformations from harmonic vocal spectra to inharmonic bell spectra are used, as well as cross-synthesis of the two. Similarly to the examples discussed before, the use of a unifying method (in this case, spectral analysis and additive synthesis), together with a metathematic approach, plays an important part in making this succeed as a coherent musical work.

Perhaps the most important single contribution to the development of a spectral music based on the process of transformation has been given by the composer Trevor Wishart. Starting in the tape studio environment, Wishart worked consistently to develop a language based on processing of recognizable sounds into various forms, informed mostly by an aim of making these aurally *credible* (in his words), having a naturalistic quality, even though they may be completely artificial. Early on, he found that a computational approach was very useful to enable his musical ideas:

> The computer opens up areas of compositional exploration that were previously inaccessible. The precision with which sound materials can be specified implies two things: (1) Given an understanding of acoustics, sounds can be transferred directly from the composer's imagination to the performance situation; (2) Areas of sonic organization previously inaccessible to composers through the existing media of notation can be explored, opening up a new world of dreamed of, but unsung possibilities. (Wishart, 1988, 27)

His environment of choice is given by the phase vocoder, which he augmented with various programs of his own design.[33] As with many composers of his generation, Wishart tends to devise new methods and compositional principles for each individual

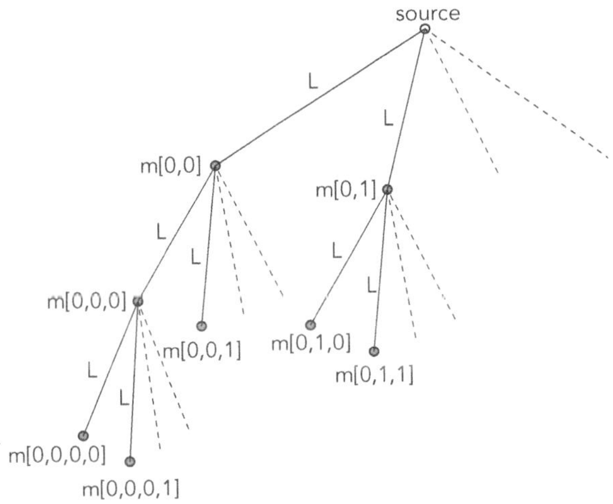

Figure 10.6
A tree of spectrally related sounds, where sonic materials at each level are linked by aurally recognizable relationships *L*, each one produced by some method of metamorphosis.

work. However, some of these appear to be common to many, transformations employing spectral interpolation, cross-synthesis, harmonic – inharmonic – noise transitions, and the use of various sounds derived from vocal sources. Another important aspect of his aesthetic is the emphasis on the manipulation of *sound-images*, where, in opposition to a pure spectromorphological view, the nature of source, attributed causation, and acoustic environment are considered essential. Wishart refers to this as the aural *landscape*. His compositional technique is based on transformations applied to landscapes, with the aim of constructing sonic metaphors, which work complementarily both at a purely aural level and in more visual/story-telling way (Wishart, 1996). This balance of concerns is a central aspect of Wishart's music.

A systematic use of spectral transformation forms the basis of most of his pieces. A particular example, which has been extensively analyzed by the composer himself, is *Tongues of Fire* (1994) (Wishart, 2000). Wishart actually makes a distinction between *transformation* proper and *metamorphosis*. While the former is related to dynamic changes in spectra over time, the latter defines the process of modifying sounds, making audibly related versions of these. Transformations build sonic relationships, which themselves support musical structures in time, and Wishart makes an analogy to the role of *modulation* in tonal music, although noting that any such sonic modulation is multi-dimensional in the case of spectra. The role of metamorphosis in his compositional practice is also important to note. Applied iteratively, this can create trees of interrelated sounds, which serve as the basic sound materials for the composition. This process is depicted in Fig 10.6, where at each consecutive metamorphosis level, sounds are linked to their origin by an audible relationship. Such a method of sonic material generation is in fact quite a common practice in electronic music composition, where libraries of sounds are constructed from source recordings. In Wishart's case, this is done primarily through the application of spectral processing methods in a computing environment.

In some ways, Wishart's music structuring methods can be fairly traditional. Once materials and transformations are created, the arrangement of these in sequential order, or in parallel streams is analogous to the construction of themes, melodies, and counterpoint. Although an extremely challenging composition from an aural point of view, *Tongues of Fire* is straightforward in structural terms, using the most traditional of the forms, and the theme and variations. In a later work, *The Division of Labour* (1997), the variation form is used even more explicitly. We are, however, in spectral music territory, and should expect the principle of a *theme* to be significantly expanded in sonic and conceptual terms. The main thematic idea for the piece was constructed with the two complementary considerations of spectral richness (as it will be the material for all transformations) and metaphor (it is a vocal utterance, with strong expressive characteristics). From this short musical phrase, the complete piece is derived, using a diverse set of spectral manipulations. Throughout the work, the composer attempts to emphasize recognizable connections between metamorphic spectra, often using intermediate steps to bridge a starting and an ending point. Also, unlike the examples of Chowning and Harvey, which worked with discretely defined spectra (*lattice-based* in his terminology), Wishart embraces the continuous and multidimensional characteristics of the spectral composition.

10.5 The Musical Spectrum

This chapter attempted to look at the spectrum from the perspective of musical practice. It was not intended to be a fully detailed description of all forms of spectral music, but it focused on some of the more salient aspects of it. We noted in particular the fundamental contribution of audio recording technologies to our modern perception of music, which is mediated by the spectral qualities of what we hear. This is as much a byproduct of the fact that the music is mostly communicated and transmitted through recorded sound rather than printed notation or live performances, as it is linked to the (intuitive) understanding that any realization of a musical work is a spectral object (rather than just notes on the page). We elected to discuss a few examples of the crucial influence of recording in the development of the *sound* of music, but there are many others. Through audio technology, a different form of hearing emerged, which was focused intensely on spectral detail.

We have noted that our contemporary approach to listening is therefore of a spectral nature, and that music production (of all kinds) has the shaping of spectral qualities as a main concern. When we turned our attention to music that is intentionally oriented toward the manipulation of the spectrum, we noted that a fundamentally new way of music making has emerged in the 20th century, which we may describe by the term spectral music. We have also noted that, from a contemporary perspective, this does not stand in isolation from other forms of music-making, but is part of a continuum. Attempts to make it distinct and separate are not satisfactory, both in terms of the on-the-ground reality, and from an all-inclusive aesthetic perspective. From this, we can conclude our continuously developing definition of the spectrum with a simple addition,

The spectrum is the natural space of musical imagination and realization.

At this stage, having fully explored the spectrum, progressing from an intuitive, through a historic, then a technical and finally a musical perspective, we can move on to explore some practical ideas of sound design and composition. These constitute the subject of the final two chapters of this book.

Notes

1 We use the term to mean music that is primarily aimed a concert performances, even though it may find use elsewhere. This relates to what is commonly called *classical* music in some places, which is in this case an inappropriate denomination. The words *new* and *contemporary* music have also been used, but, similarly, are not very meaningful, as some of the compositions encompassed are neither new nor contemporary.

2 Here we consider *instrumental* in reference to instruments producing sound by mechanic means. These include, for instance, a violin, a bell, and a voice.

3 An early example of this is heard in Debussy's tenth prelude (book one), *La Cathédrale Engloutie*, where chords and associated melodic figures are constructed so to simulate bells and their lower resonances.

4 With this, Schoenberg, closes this 1911 harmony treatise, looking toward a time when spectra can as easily be manipulated as harmony. Of course, an equivalent to tonal-functional progressions does not seem to exist, and any theory oriented toward finding one might be particularly artificial.

5 Erickson (1975, 21) notes that such use of vertical structures may be considered as the development of individual sound objects. It is clear that Varèse's methods of instrumental writing anticipated the advent of electronic means of sound manipulation.

6 A siren and a piano are also used, both of which of course produce defined pitches (a glissando in the case of the siren). However, they are never employed for their pitch qualities, but rather for their timbral or spectral color characteristics.

7 These were also used by Messiaen, but in a melodic, rather than timbral, role.

8 Eventually, toward the end of his career, he was finally given access to a fully fitted studio, in which he composed his tape work *Poème Electronique*.

9 A piano whose strings are *prepared* through the insertion of various materials, creating a variety of timbres accessible by the keyboard. This could be equated to a one-person percussion orchestra.

10 Some take the view that there are two separate and distinct elements, the musical composition and the audio production. A recording of a piece is, in this case, the latter applied to the former, but in any case they still remain distinct and the *piece* exists beyond the *production*. This is a very traditional view, which is incompatible with the music of today. In fact, it is a form of understanding that was already wrong by the late 1960s. The audio recording of *Strawberry Fields Forever* as released by the Beatles is one and the same as the musical piece that it contains. A pub rendition of it with guitar and voice is not it, but another piece of music altogether.

11 The improvement in audio capture, mastering, and record cutting in 1940s and 1950s is discussed by Putnam (1980).

12 Or on an educational setting, teacher-pupil transmission.

13 It may be worth noting the timbral difference of the *classical* saxophone sound, transmitted mostly through the teacher-pupil relationship, to the various jazz saxophone sound, more widely spread through recording. We can speculate that without the means of audio technology such a divergence may have been less evident.

14 This is very significant in the emergence of vocal styles that did not require strong powers of projection by the singer. Suddenly, the extreme spectral shaping of the *bel canto* technique became a stylistic choice rather than a practical necessity. Microphones allowed the development of intimate (and sometimes more natural) forms of singing, as, for instance, seen in the music of the *bossa nova* sub-genre of jazz.

15 In the cases where it is not, which are mostly reduced to the concert hall, then we have a mode of engagement with music that is absolutely atypical.

16 Such as, for instance, the construction of a Bach fugue track from two or more takes as described by Gould (2006).

17 Landy (2007) introduced this term in a much narrower sense, attempting a distinction between modes of music creation that are, traditionally, *note*-based (e.g., instrumental music), and the new phenomenon of music that is based on *a sound* (which may be considered an extension of Erickson's use for the term). However, if we follow this reasoning a bit further we will end up with the paradoxes and open questions that stem from the same artificial distinctions between *electronic*, *instrumental*, *mechano-acoustic*, *electro-acoustic*, etc., which we have discussed here. It makes sense to consider that the contemporary listener is attuned to different forms of *sound*-based music, which go beyond Landy's original use of the term.

18 Of course, composers have attempted to use mimetic approaches through instrumental means before the advent of recording. However, this constitutes a significant leap: instead of trying to reproduce the sounds of nature artificially, we have an accurate representation of these sounds themselves.

19 Schaeffer quotes the Larousse to define the acousmatic as "a noise that one hears without seeing the cause of it" (Schaeffer, 2006, 77).

20 Even though the term was particularly used to describe a listening attitude to electronic music, we cannot exclude that it can also be used more generally in the apprehension of any type of music.

21 The label was used, as these things are, for promotional purposes, as an attempt to portray this music as novel, distinct, and innovative.

22 The early account of the rise of electronic music by Manning (1986), dedicates a whole chapter to the subject, noting that "[d]uring the late 1960s rising expenditure on basic electronic mixing and amplification equipment by the commercial sector led to the acquisition of studio and stage facilities which eclipsed those offered by all but the most prestigious of the [experimental music] institutions" (Manning, 1986, 201).

23 One may argue that sampling started with the first methods of mechanical recording of Edison and others, but that is beside the point.

24 If that designation can still be given any sense today.

25 Why? For the reasons we have been putting forward throughout this chapter, the idea of such a straightforward bifurcation is too simplistic, and it does not consider the thought that pitch- or harmony-oriented music may also include features that transcend these. Tonality (strictly speaking) may not even be the most common mode of pitch structuring in many cases and certainly (as the art of music production may attest) the aural qualities of the musical product were, by the 1980s, a central concern. Besides, the absolutist tone of the quote betrays an attitude that places the *grand tradition* of *art music* as some superior experience.

26 This has been the subject of a research project whose results are reported in (Brandtsegg et al., 2012), which the interested reader may want to consult.

27 Even here we may be making artificial distinctions, since much of the sound we hear today (in all genres of music), in the exact way it has produced, would not be achievable without computer audio technology.

28 Note that the use of the word *transform* and *transformation* here is quite distinct from the previously employed mathematical sense (as in the Fourier transform).

29 Considered in a very expanded sense.

30 It is actually closer to a minor sixth (1.6).

31 By M. Meneghini and K. Dahan.

32 The question of the actual section proportions is confusing because more than one version exists, with different timings. Chowning's original *nominal* intention was to have an *a*:*b* form with a section length ratio 610:377 seconds (1.618). The *actual* intended version was longer and contained a different sectional organization (Zattra, 2007). The timings considered here relate to an analysis of the reconstructed version in Csound code.

33 Some of the ideas originated by the composer are explored in detail in Chapter 6 and documented in Wishart (1994).

11

Computer Sound Design

We can now turn our attention to practical aspects of spectral music, more specifically, the development of computer instruments for sound design. We will approach the topic without making any assumptions about the possible applications of these spectral design ideas: we do not expect them to be only useful for a specific genre of music composition, or form of sonic art. The idea is to help the development of a craft, which may be used in a variety of settings. The methods, however, will be, as expected, computational and oriented to the making of *instruments*. Such an approach was chosen because of its malleability (it is very straightforward to modify and adapt it to suit our needs) and the fact that it can be explored in a systematic way.

The subject of computer sound design is vast, and the techniques, even if we keep ourselves to spectral processing, are numerous. It is therefore not possible to cover all of them in a single chapter. However, we can explore some of the key areas with a view to developing a methodology, which may then be applied elsewhere. So, beyond their existence as recipes for synthesis and processing, the examples discussed here show a way of looking at sound design which is computational in nature. Similarly to other computer instrument and code catalogues, we expect that each design in this chapter will serve as a launch point for further exploration of spectral audio, rather than being a finished product or preset to be used directly.

11.1 Additive Synthesis

We begin by looking at techniques that can fall within the large category of additive synthesis. Taken from a macroscopic perspective, any method that tries to create a complex and dynamic spectrum by summing signals of a simple nature can be classed as additive. However, we might narrow it down to the use of sinusoidal sources as the building blocks for additive synthesis. Conceptually, this is very straightforward, but of course, a certain amount of control complexity may arise in cases where a large number of components are required. However, there are ways to tackle these issues, as we will see.

As we discussed in Chapter 10, we can model the additive synthesis element as a three-oscillator instrument. A straight translation to Csound of the MUSIC V code for this would be

```
instr 1
 a1 = oscili(p4,1/p3,p6)
 a2 = oscili(p5,1/p3,p7)
```

Spectral Music Design. Victor Lazzarini, Oxford University Press. © Oxford University Press 2021.
DOI: 10.1093/oso/9780197524015.003.0011

```
 out(oscili(a1,a2))
endin
```

that is, we have one oscillator to control amplitude and another one to control frequency, reading two separate function tables (taken from parameters 6 and 7, denoted by p6 and p7 respectively). Their frequency is set to the reciprocal of the sound duration (p3); this means the oscillators will run once through the table, a one-shot operation. Their output is scaled by parameters 4 and 5 (p4 and p5), respectively, so a different base amplitude and frequency can be used for each component. Finally, the signal they generate is used to control a sine wave produced by the third oscillator. A more up-to-date version of the instrument could be made using a single time signal generator and table lookups,

```
instr 1
 at = line:a(0,p3,1)
 out(oscili(p4*tablei:a(at,p6,1,0,1),
                p5*tablei:a(at,p7,1,0,1)))
endin
```

since the first two oscillators are only used as function table readers.

This instrument can be used as the basis for the realization of various types of additive synthesis. For example, we can employ it to re-create the Risset-Shepard tone synthesis (instrument 513 of Risset's catalogue, as discussed in Chapter 10). For this we need to define specific amplitude and frequency functions. The amplitude should be a bell curve (e.g., a Gaussian envelope),[1] and the frequency, an exponential covering ten octaves. In Csound, we can use GEN 20 to create the former, and GEN 5 for the latter. In addition, we need to offset the start of the table reading in each component, so that the oscillators are tuned in octaves and they enter and exit in a staggered fashion.

Let's code this example step by step. We start by defining the function tables,

```
isd = 0.75
ioct = -10
ifa = ftgen(0,0,16384,20,6,1,isd)
iff = ftgen(0,0,16384,-5,1,16384,2^(ioct))
```

where isd defines the standard deviation of the Gaussian curve, and can be used to adjust how narrow or broad we want it to be. The second table is an exponential from 1 to 2^{-10}, making the oscillators glide by 10 octaves over the course of the full glissando. We define it as a negative GEN number to switch off normalization in table generation. This allows us to make exponentials that are either rising or falling, depending on the sign of ioct. As we have noted, these functions will need to be offset, which is done at the lookup stage. To implement this, we make a small modification to the first line of the instrument

```
at = line:a(0,p3,1)+p8
```

with parameter 8 determining the offset.

The next step is to create the 10 instances of the oscillator. We could do this by hand, but it is much better to employ a loop,

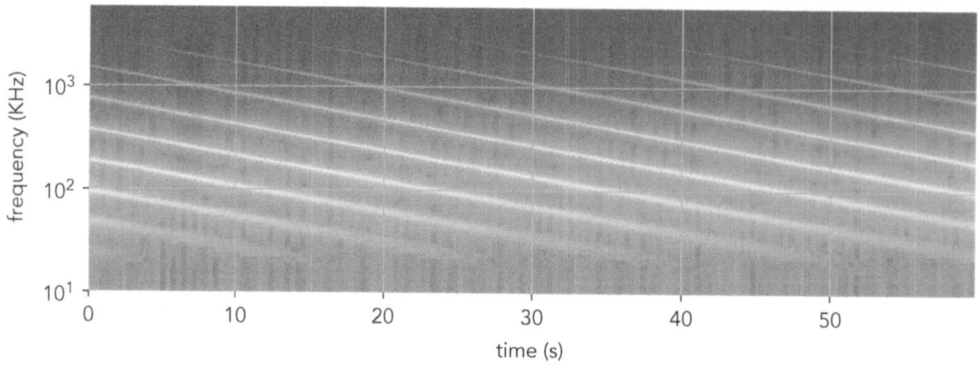

Figure 11.1
Spectrogram of an endless glissando sound.

```
idur = 120   // cycle duration
ifmx = 6000 // top frequency
icnt = 0
insts = abs(ioct)
while icnt < insts do
 schedule(1,0,idur,0dbfs/2,ifmx,ifa,iff,icnt/insts)
 icnt += 1
od
```


Note that we have parametrized almost everything. So if we want to extend, limit, or change the direction of the glissando, we can change the `ioct` and `ifmx` values, and if we want to change its rate, we can modify `idur`. We now have an extended design, based on Risset's instrument, to realize a variety of endless glissandos. The perception of these will depend on the parameters used: a short/fast glissando will probably make us focus on the repetitive aspect of it, destroying the auditory illusion. A spectrogram of the endless glissando sound is shown in Fig. 11.1, as synthesized by the Csound code.[2]

The same approach can be employed to design percussive inharmonic sounds, such as bells, gongs, and barphones. In this case, we do not expect the frequencies to glide or modulate, so we can simplify the instrument somewhat,

```
instr 1
 at = line:a(0,p3,1)
 out(oscili(p4*table:a(at,p6,1),p5))
endin
```

using only one time function, for amplitude. A bell-like sound can be designed using an exponential decaying envelope. Individual partial durations, amplitudes, and frequencies can be defined in arrays, and we can retain more or less the same control code,

```
imin = .001
ifa = ftgen(0,0,16384,5,1,16384,imin)
idur[] fillarray 1,.9,.7,.6,.2,.1,.07,.05,.03,.01
ifr[] fillarray 120,95,97,176,250,300,610,755,933,1020
iamp[] fillarray 1,0.5,0.9,0.3,0.4,0.6,0.2,0.1,0.15,0.05
itot = 20
icnt = 0
```

```
insts = lenarray(idur)
while icnt < insts do
 schedule(1,0,idur[icnt]*itot,iamp[icnt]*0dbfs/insts,ifr[icnt],ifa)
 icnt += 1
od
```

This design can be used to re-create Risset's bell synthesis (example 430 of his catalogue (Risset, 1969, A64–A66)). In his description of the instrument, Risset notes that "in actual bells, it is attempted to approximate the following ratios for the first five components: .5, 1, 1.2, 1.4, 2 (corresponding for example to the following succession of notes: G, G, B flat, D, G, called respectively hum notes, fundamental, minor third, fifth, nominal)" (Risset, 1969, A64). From this, Risset derives a sequence of partial ratios: .56, .92, 1.19, 1.71, 2, 2.74, 3, 3.76, 4.07. In order to make the sound more natural, decay times are set to be approximately proportional to their frequencies (in our first example above, decay times also generally become shorter as the frequency increases). In addition to this, he splits the two lower partials in two, to generate a small amount of beating, which is also common in bell sounds. The transcription of Risset's parameters is shown below, with the addition of an amplitude scaling factor to accommodate the different range employed in Csound:


```
imin = .001
ifa = ftgen(0,0,16384,5,1,16384,imin)
idur[] fillarray 20,18,13,11,6.5,7,5,4,3,2,1,1.5
ifr[] fillarray  224,225,368,369.7,476,680,800,1096,1200,
                 1504,1628
iamp[] fillarray 150,100,150,270,400,250,220,200,200,150,200
iscal = 2000 // amp scaling
itot = 20
icnt = 0
insts = lenarray(idur)
while icnt < insts do
 schedule(1,0,idur[icnt],iamp[icnt]*0dbfs/iscal,ifr[icnt],ifa)
 icnt+=1
od
```

The control code itself can be put into an instrument and used to trigger any number of separate, additive synthesis instances. For example, let's create the code for a barphone synthesis. We can use exactly the same synthesis code, but change the parameters slightly. The sound of a marimba or vibraphone tends to have few significant partials, some of which are close to the harmonic series, and some of which are more inharmonic. We could actually base the re-creation on the analysis of real sound, but it can also be interesting to arrive at the parameters by ear. In this example we use only four partials, and the higher tones start with the fifth harmonic and are all slightly deviated from the exact ratios. Higher partials decay fast and have low amplitude, and we also set the total duration of the sound to be inversely proportional to the fundamental frequency. The complete code is

```
imin = .001
gifa = ftgen(0,0,16384,5,1,16384,imin)

instr 1
 idur[] fillarray 1,.2,.15,.05
 ifr[] fillarray 1,5.1,6.95,13.3
 iamp[] fillarray 1,0.1,0.05,0.01
 itot = 800/p5
 icnt = 0
 insts = lenarray(idur)
 while icnt < insts do
   schedule(2,0,idur[icnt]*itot,iamp[icnt]*p4,ifr[icnt]*p5,
            gifa)
   icnt+=1
 od
endin

instr 2
 at = line:a(0,p3,1)
 out(oscili(p4*table:a(at,p6,1),p5))
endin
```


where we have moved the synthesis instrument up to a higher number. This is because execution order in Csound is controlled by the numbering of instruments. So in this case, the dispatching instrument is run first, and the synthesis instrument starts straight away. If the order was reversed, we would wait a complete k-cycle before sound was started. In realtime situations, this might make a difference. Also note that we stored the function table number in a global variable so that it can be referred to inside an instrument. This code can be used for instance as a MIDI synthesizer, directly from a controller, by mapping note velocity and note number to amplitude and frequency in parameters 4 and 5.

11.1.1 Recursion

Although loops are very useful to create several instances of the same instrument in parallel, an alternative programmatic approach can be used for similar applications. This is known as *recursion*, and is demonstrated by the following example:

```
instr 1
 iamp = ampdbfs(p4)/p6
 ifreq = cpsmidinn(p5)*p6
 if ifreq <= sr/2 then
   iatt=p6*0.005
   ka = expseg:k(0.001,iatt,iamp,
                 p3-iatt,0.001)
   asig = oscili(ka,ifreq)
   out(asig)
```

```
    schedule(1,0,p3,p4,p5,p6+1)
  endif
endin
schedule(1,0,5,-6,60,1)
```

This method is based on making each instance of the instrument create another one, by placing a `schedule()` call inside it. We also might want to protect this inside a conditional section, to avoid infinite recursion, although there are applications for that case, which we will see later. While the instrument example above produces a harmonic spectrum, it is possible to modify it to create different types of sounds. Since each harmonic is created on the basis of successive values of parameter 6, the following line can make the spectrum stretch out inharmonically:

```
schedule(1,0,p3,p4,p5,p6+sqrt(2))
```

since all partials will be separated by multiples of $\sqrt(2)$. Following the example of *Stria*, we could also plug in the golden ratio and obtain spectra that is based on that interval.

We can also easily add an envelope to the frequency,

```
kf = expseg:k(ifreq*1.1,iatt,ifreq, p3-iatt,ifreq*.95)
asig = oscili(ka,kf)
```


in this case creating a slight downward glissando in each partial. These are going to happen in parallel, at the same rate; to make them variable, we can apply some randomization using the `birnd()` function, which draws random numbers from a random distribution within a given bipolar range,

```
kf = expseg:k(ifreq*(1+birnd(.3)),p3,ifreq*(1+birnd(.5)))
```

Partials also need not start at the same time and have the same duration, so we can randomize that too. For example, start times may be arpeggiated and durations set to be distributed within a certain positive range using `rnd()`,

```
schedule(1,rnd(.05),rnd(20),p4,p5,p6+(1+sqrt(5))/2)
```

Here's the fully modified example, which creates a texture of randomly moving partials:

```
seed(0)
instr 1
 idur = 20
 ist = .05
 iamp = ampdbfs(p4)/p6
 ifreq = cpsmidinn(p5)*p6
 if ifreq <= sr/4 then
   iatt = p6*0.005
   ka = expseg:k(0.001,iatt,iamp,
                p3-iatt,0.001)
   kf = expseg:k(ifreq*(1+birnd(.3)),p3,ifreq*(1+birnd(.5)))
```

```
    asig = oscili(ka,kf)
    out(asig)
    schedule(1,rnd(ist),rnd(idur),p4,p5,p6+(1+sqrt(5))/2)
  endif
endin
schedule(1,0,20,-6,60,1)
```

where a call to seed() initializes all random sequences to start with a value taken from the system time. This guarantees a different sequence each time the instrument is run.

If we take the case to space the start times of each new instrument instance by more than one block size (ksmps), then we may be able to use infinite recursion to create a spectral texture where partials enter and exit at different times, continuously. Using random values for most parameters can make the result very lively. Let's take the *Stria* scale as the basis and create partials whose frequency starts in one of these steps, then glides out of it. Setting gir to the golden ratio, we can use the following frequency scaling factor

```
iff = gir^(floor(birnd(28))/9)
```


which produces values within the range of $\phi^{\pm 3}$. We can then set the minimum start times to a small value, larger than sr/ksmps. The start times determine the separation between the entry of each partial and, together with duration, the overall spectral density. Amplitudes can also be set randomly to create a sculpted result. In the following example: we set the minimum start time to 10 milliseconds and maximum partial duration to 5 seconds. The scale origin (its center point) is set to 4,000 Hz,

```
seed 0
gir = (1+sqrt(5))/2 // golden ratio
icf = 4000        // scale centre
instr 1
 idur = 5      // max dur
 ist = .01     // min start time
 iff = gir^(floor(birnd(28))/9)
 iamp = p4
 ifreq = p5*iff
 iatt = rnd(0.05)
 ka = expseg:k(0.001,iatt,iamp,
                    p3-iatt,0.001)
 kf = expseg:k(ifreq,iatt,ifreq,p3-iatt,
                  ifreq*(1+birnd(.2)))
 asig = oscili(ka,kf)
 out(asig)
 schedule(1,ist+rnd(ist),rnd(idur),rnd(0.1),p5)
endin
schedule(1,0,5,0.1,icf)
```

The result is a cascading set of partials produced at quite a fast rate, and creating a lively and bright spectrum (Fig. 11.2).

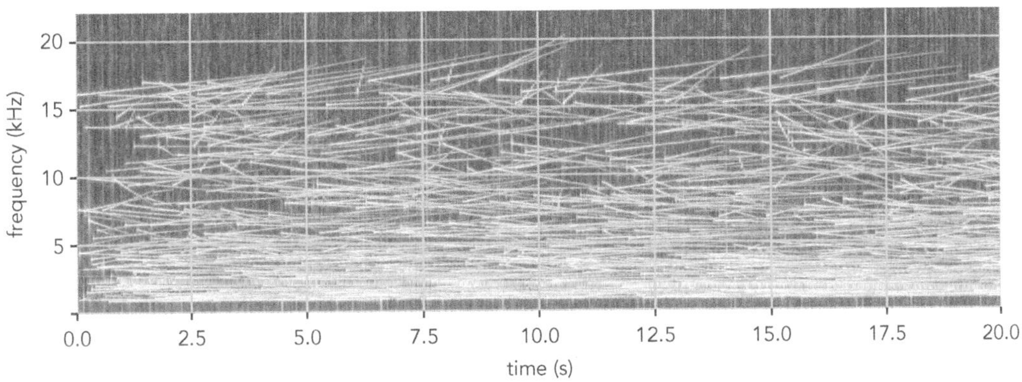

Figure 11.2
Spectrogram of a recursively generated additive synthesis texture.

We can easily make the parameters that define this texture change over time, by, for instance, making them depend on global variables and use another instrument to modify these. At this stage, we are moving from designing additive synthesis spectra to designing processes, and we start to approach the principles of granular synthesis, which we discussed in Chapter 9, and in fact, used a similar recursive method as an example. The boundaries between synthesis techniques can be quite fluid, and we can benefit from this when designing our sounds.

Finally, we should note that additive synthesis can be characterized by a data explosion; while it is very straightforward conceptually, as the number of partials increases there are substantially more parameters to deal with. We have done well with a small set of partials, on one hand, and with programmatic ways to control the amplitudes and frequencies. However, if we try to model more complex spectra, it may be more difficult to control all parameters. In this case, the best approach is to use some analysis method, from which we can extract and modify the amplitude and frequency data. Such an approach to computer sound design will be discussed later in this chapter.

11.2 Non-Linear Distortion

In many ways, non-linear distortion methods are the opposite of additive synthesis. In place of a multitude of parameters, we have only a few; less computation is required to generate equally complex spectra. On the other hand, the conceptual simplicity of additive synthesis is lost, and we have, in some situations, a more complicated relationship between controls and resulting sound. However, that can be overcome with good instrument design, and a change of mindset. Once the principles of non-linear distortion are understood, programming it does not represent too much of a challenge. The design ideas we will develop in this section stem from the theory and examples explored in Chapter 8.

11.2.1 Operator FM

Frequency modulation synthesis can be implemented and integrated into spectral instrument design in various ways. We may use the original simple FM formulation, whose theory actually describes phase modulation, by connecting two oscillators together, and using envelopes for amplitude and timbre. As we saw, however, there are other ways

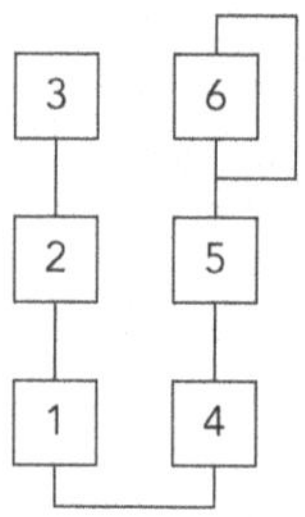

Figure 11.3 Algorithm 3 in the Yamaha DX7 synthesizer.

of connecting components to realize FM, employing complex (or more than one) modulators and carriers, as well as feedback. A great number of modulation *topologies* are possible, if we also consider the stacking of modulators, and multiple parallel carriers. It is worth therefore considering how such designs can be implemented in a systematic fashion. The solution was given in Yamaha's DX line of FM synthesizers, where the different topologies were organized in what they called *algorithms*.

The basic building block for an algorithm is the FM *operator*, which is a sinusoidal oscillator whose phase may be modulated by an input signal. Thus, instead of frequency modulation, what is actually implemented is phase modulation. This is because the latter is much more malleable to multiple topologies, whereas the former, due to the implicit integration of frequency built into an oscillator, only really works with simpler modulation arrangements. The operator also included an amplitude envelope, which shapes the output of the oscillator. The core of the operator can therefore be resumed to a single line of code,

```
asig = tablei:a(phasor:a(kf)+afm,-1,1,0,1)
        *madsr(iat,idc,isus,irl)
```

with which we can construct an operator as a user-defined opcode in Csound.

Operators can be connected in various topologies, and the Yamaha DX7 synthesizer had a large number of algorithms that defined the linking of its six operators.[3] In Fig 11.3, we show one such topology, algorithm 3, which is made of two parallel stacks of operators (two stacked modulators driving a carrier). We should also note the feedback connection in operator 6, indicated by the output signal being redirected to its input. Therefore the system has three types of operators, defined by their inputs. An operator may

- Take an FM input from another operator.
- Take no inputs, produce a sinusoid.
- Take a feedback input from itself.

We now have the basic principles with which we can start design FM instruments. It is important to emulate not the exact details of the DX line, but the approach. We can also expand on these ideas and allow for further possibilities that were not anticipated (more operators, more use of feedback, etc.). We can begin by designing three UDOs to match the types discussed above. Csound allows the *overloading* of opcodes, so we can use the same name but include different parameters to distinguish between the types. This is very useful in this programming context, as we can define three different operators, but use the same opcode name for them.

The first of these, an FM-input operator, is implemented as

```
/* FM operator
 asig = Operator(afm,kamp,kfc,iatt,idec,isus,irel)
 afm - freq modulation signal
 kamp - operator amplitude
 kfc - carrier freq
 iatt, idec, isus, irel - ADSR envelope
*/
opcode Operator,a,akkiiii
 afm,kamp,kfc,iatt,idec,isus,irel xin
 afm /= (2*$M_PI)
 acar = tablei:a(phasor:a(kfc)+afm,-1,1,0,1)*kamp
 xout(acar*mxadsr:a(iatt,idec,isus+.0001,irel))
endop
```

which is effectively the one-line code wrapped up in an opcode, using an audio-rate exponential envelope instead (mxadsr).[4] Note that we have to scale the FM signal by $1/2\pi$ because the table lookup index is normalized to the range $[-1,1]$. This allows us to use the normalized peak amplitude of the input to define the maximum index of modulation directly.

The next operator type, producing a sinusoid, is much simpler, just an oscillator with an envelope,

```
/* Sine operator
 asig = Operator(kamp,kfc,iatt,idec,isus,irel)
 kamp - operator amplitude
 kfc - carrier freq
 iatt, idec, isus, irel - ADSR envelope
*/
opcode Operator,a,kkiiii
 kamp,kfc,iatt,idec,isus,irel xin
 acar = oscili(kamp,kfc)
 xout(acar*mxadsr:a(iatt,idec,isus+.0001,irel))
endop
```

The final operator type uses a feedback signal. In order to implement this properly, we need to set the computation block to a single sample. This allows a unit sample delay between input and output, which is required by the feedback FM equations,[5]

```
/* FM feedback operator
 asig = Operator(kamp,kfc,kfdb,iatt,idec,isus,irel)
 kamp - operator amplitude
 kfc - carrier freq
 kfdb - feedback amount
 iatt, idec, isus, irel - ADSR envelope
*/
```

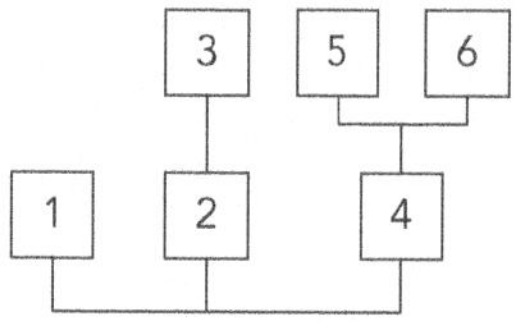

Figure 11.4 Algorithm 26, as used in the electric piano instrument.

```
opcode Operator,a,kkkiiii
 setksmps 1
 asig init 0
 kamp,kfc,kfdb,iatt,idec,isus,irel xin
 asig *= kfdb/(2*$M_PI)
 acar = tablei:a(phasor:a(kfc)+asig,-1,1,0,1)*kamp
 asig = acar*mxadsr:a(iatt,idec,isus+.0001,irel)
 xout(asig)
endop
```


In order to allow the parser to distinguish this operator from the previous overload, we need to make sure that the third parameter is provided as a k-rate signal, instead of an i-var.

With these we can now implement the DX7 algorithms (and more, since we have all the power of a music programming language to support us). For example, we can look at the typical FM emulation of a tine-based electric piano, which uses algorithm 26 (Tomlyn and Aikin, 1985), shown in Fig 11.4. The arrangement is made up of a single sine wave, mixed with two FM signals, one with a single and the other with a double modulator. The electric piano sound can be decomposed into three elements:

- Hammer strike.
- High frequency inharmonic components.
- Fundamental plus a few low-amplitude harmonics

So in this arrangement, the idea is to use the single sine wave to create a fixed low-frequency sound corresponding to the hammer strike, the double modulation component to create the short-lived high inharmonic components, and the simple FM arrangement to produce the sustained sound (actually slowly decaying) consisting of the fundamental and some harmonics. Furthermore, the inharmonic contents are more apparent at the higher notes, so we can use some sort of scaling based on the fundamental frequency to control the amount of modulation and the level of these components. Also, in order to simulate changes in brightness in response to dynamics, we can scale the amount of modulation produced by operator 3 by the amplitude. Here's the resulting Csound code:

```
/* FM electric piano
   algorithm 26
          op3   op5   op6
           |      \   /
     op1  op2      op4
*/
instr 1
```

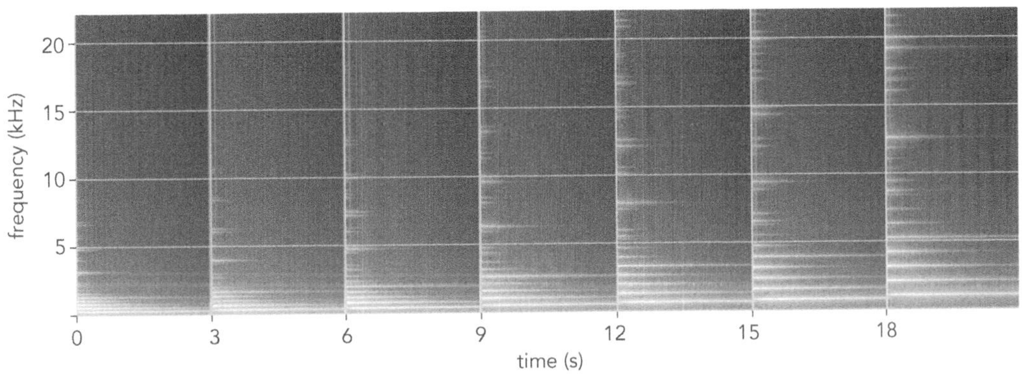

Figure 11.5
Spectrogram of an arpeggio played on the FM electric piano instrument.

```
 iscl = p5/5000 // frequency scaling
 iatt = .005    // attack
 idec = 6       // decay
 irel = .1
 ia = p4
 ifr = p5
 a6 Operator iscl,ifr*13.3,iatt,idec/8,0,irel // op6
 a5 Operator iscl+.3,ifr*7,iatt,idec/4,0,irel  // op5
 a4 Operator a5+a6,ia*iscl*.5,ifr*5,iatt,idec/2,0,irel // op4
 a3 Operator ia*4,ifr,iatt,idec,0,irel // op3
 a2 Operator a3,ia,ifr,iatt,idec,0,irel  // op2
 a1 Operator ia*.1,97,iatt,0.01,0,0.01 // op1
  out((a1+a2+a4)*0.5)
endin
```

In Fig. 11.5, we see the spectrogram of a two-octave arpeggio from middle C played on this instrument. As the pitch gets higher, there are increasingly more significant high-frequency components in the spectrum. This is a typical characteristic of the electric piano sound. As in the barphone example discussed earlier, we can map MIDI key velocity to parameter 4 (amplitude), and MIDI key number to parameter 5 (frequency, in Hz), and this code becomes a fully functional realtime polyphonic FM synthesizer.

The approach we took here can be extended to cover other examples of the FM repertory. It can also be modified to create various types of evolving textures, similarly to what we did with additive synthesis. The FM operators can also be complemented with frequency envelopes, which are not present in the original, but may be a useful addition. The modular aspect of this type of design is very scalable: starting with a simple topology, it is possible to develop it to become quite complex while still keeping the synthesis code reasonably succinct and elegant.

11.2.2 Synthesis of Resonance

An alternative to FM, which can be used as a drop-in replacement since it relies on similar parameters, is modified frequency modulation. Besides realizing similar sounds to simple FM, it can also be used successfully in other applications. One of these is the synthesis of resonance, which can re-create the effects of filtering directly in one compact package.

We will concentrate on this aspect of ModFM. As in the case of FM, it is useful to attempt to describe it in an modular form. Since this is a more specialized application, the design of a ModFM operator for resonance synthesis is slightly more involved, and less general.

We can begin by defining the core of the method, which is based on synchronous ModFM. The following code outlines it,

```
aph = phasor:a(kfo)
amod = tablei:a(aph,-1,1,.25,1)
aexp = exp(kndx*(mod-1))
afm = tablei:a(aph*kn,-1,1,.25,1)*aexp
```

that is, we use a single phase source (aph) to drive both the sinusoid modulator and the cosine carrier.[6] This is run at an integer multiple (kn) of the fundamental (kfo), depending on the resonance center frequency. In order to allow fine-tuning of this frequency, we use two carriers and derive the output by linearly interpolating their outputs,

```
kfn = kfc/kfo
kn = int(kfn)
ka = kfn - kn
aexp = exp(kndx*(amod-1))
acar1 = tablei:a(aph*kn,-1,1,.25,1)
acar2 = tablei:a(aph*(kn+1),-1,1,.25,1)
afm = (ka*acar2 + (1-ka)*acar1)*aexp
```


With this we have the core of the phase synchronous ModFM operator. We only need to add the code to compute the modulation index (kndx) from the resonance bandwidth,

```
kg = 2^(-kfo/(.29*kbw))
kndx = 2*kg/(1-kg)^2
```

When synthesizing more than one resonance region, we will probably want to use a single phase source for all operators, and since the sinusoid modulator is common to all of them, we can include in the design a phase and sinusoid inputs. Finally, we may also use a Q control instead of defining the bandwidth directly. This completes the phase-synchronous ModFM operator design,[7]

```
/* ModFM operator
   kamp - amplitude
   aph - phase signal (kfo Hz)
   amod - sinusoid mod (kfo Hz)
   kfo - fundamental
   kfc - resonance centre
   kQ - resonance Q
*/
opcode ModFMOp,a,kaakkk
 kamp,aph,amod,kfo,kfc,kQ xin
 kbw = kfc/kQ
 kg = 2^(-kfo/(.29*kbw))
```

```
 kndx = 2*kg/(1-kg)^2
 kfn = kfc/kfo
 kn = int(kfn)
 ka = kfn - kn
 aexp = exp(kndx*(amod-1))
 acos1 = tablei:a(aph*kn,-1,1,.25,1)
 acos2 = tablei:a(aph*(kn+1),-1,1,.25,1)
 acos = (ka*acos2 + (1-ka)*acos1)*aexp
    xout kamp*acos
endop
```

This opcode can be used in several resonance synthesis applications. A typical example is in the emulation of singing voice formants. These are characterized by three parameters:

- Center frequency.
- Relative amplitude.
- Bandwidth (or Q).

Generally, a minimum of three formants are used to create a vocal sound, but we can use more. Each ModFM operator can be employed to re-create one formant region, so an opcode can be designed to synthesize a voice from formant data by mixing three to five of these. Here is an example:

```
/* Voice synthesis
   asig1 = Voice(ka,kf,kA[],kF[],kQ[])
   ka - amplitude
   kf - fundamental
   kA[] - formant amplitudes
   kF[] - formant frequencies
   kQ[] - formant Q factors
*/
opcode Voice,a,kkk[]k[]k[]
 ka,kf,kA[],kF[],kQ[] xin
 aph = phasor:a(kf)
 asin = tablei:a(aph,-1,1,0.25,1)
 a1 = ModFMOp(kA[0],aph,asin,kf,kF[0],kQ[0])
 a2 = ModFMOp(kA[1],aph,asin,kf,kF[1],kQ[1])
 a3 = ModFMOp(kA[2],aph,asin,kf,kF[2],kQ[2])
 a4 = ModFMOp(kA[3],aph,asin,kf,kF[3],kQ[3])
 a5 = ModFMOp(kA[4],aph,asin,kf,kF[4],kQ[4])
 xout((a1+a2+a3+a4+a5)*ka)
endop
```

The `Voice` opcode generates a single set of formants, which has quite a raw synthetic nature. To make it more lively, we can double it up, and add some random detuning and vibrato. This creates a chorus effect, which will result in a dynamic spectrum and contribute to making the instrument sound more natural. An example instrument implement these ideas is shown below, using a simple linear trapezoidal envelope (linenr):

```
instr 1
 ia = p4/3
 ifr = p5
 kA[] fillarray 1,0.5,0.42,0.38,0.08
 kF[] fillarray 650,1080,2650,2900,3250
 kQ[] fillarray 8,10,12,14,18
 kvib = oscili(0.004,4)
 krnd = randi(0.004+kvib,7)
 asig1 = Voice(ia,ifr*(1+kvib),kA,kF,kQ)
 asig2 = Voice(ia,ifr*(1+krnd),kA,kF,kQ)
 asig3 = Voice(ia,ifr*(1-krnd),kA,kF,kQ)
 amix = linenr:a((asig1+asig2+asig3),0.1,0.1,0.01)
 out(amix)
endin
```


Similarly to the FM example, this is driven by two parameters, amplitude and frequency, which can be easily mapped from a MIDI controller. We are using a fixed vowel, but the design can be easily modified to use time-varying formant data (from function tables, for example).

Another application of the ModFM operator is in simulating an oscillator–filter arrangement. In this case, we can use two operators, the first one creating a peak at the fundamental with a wide bandwidth, and the second providing a resonant peak at a higher frequency. Such an opcode can produce an effect that is similar to a complex periodic waveform being fed into a resonant low-pass filter,

```
/* Resonant synthesis
   asig = Resonans(kamp,kf,kc,kQ)
   kamp - amplitude
   kf - fundamental
   kc - resonant frequency
   kQ - Q factor
*/
opcode Resonans,a,kkkk
 ka,kf,kc,kQ xin
 aph = phasor:a(kf)
 amod = tablei:a(aph,-1,1,0.25,1)
 asig2 = ModFMOp(ka,aph,amod,kf,kc,kQ)
 asig1 = ModFMOp(ka,aph,amod,kf,kf,1)
 xout(asig2+asig1)
endop
```

In Fig. 11.6, we show the spectrogram produced by the following code:

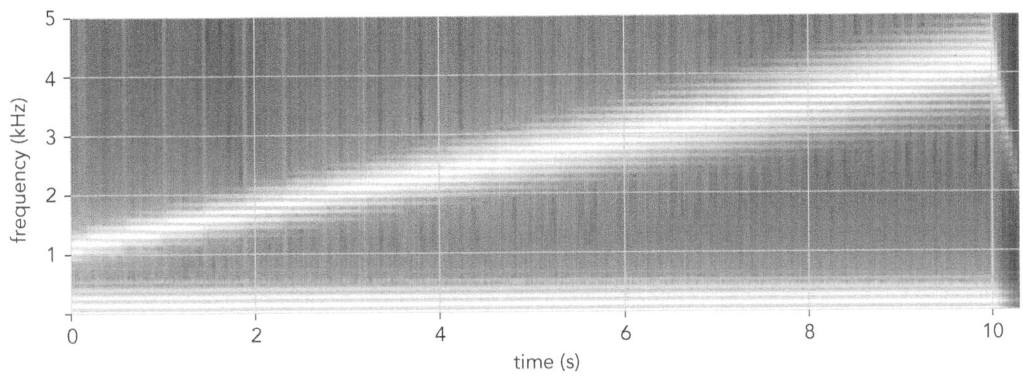

Figure 11.6
Spectrogram of a resonant synthesis instrument.


```
instr 1
 ia = p4
 ifr = p5
 kc = linsegr(1000+ifr,10,4000+ifr,0.3,2000)
 kQ = 15
 asig1 = Resonans(ia,ifr,kc,kQ)
 amix = linenr:a(asig1,0.1,0.1,0.01)
 out(amix)
endin
schedule(1,0,10,0.5,cpspch(7.00))
```

which produces an filter-like sweep of the spectrum, picking up harmonics at the resonance center frequency.

One of the advantages of this method over an oscillator + filter arrangement, as discussed in the next section, is that we have a phase-coherent spectrum, where all components are in cosine phase. Therefore any mixing of components will not have unexpected dips or peaks due to partial phase interactions. Furthermore, if implemented using table lookup waveshaping, it can be quite an efficient method of resonance synthesis.

11.3 Source-Modifier Techniques

While non-linear methods are designed to produce dynamic spectra directly, a common alternative is to split the design into a complex sound source and a spectral shaper or modifier, which sculpts it. This may be a filter of some description, in the most general case, but it can also be another transformation process. Source-modifier synthesis methods are also commonly described as *subtractive*, but as we noted earlier in the book, there is no subtraction involved, so we may as well use a more suitable term for them. These techniques are very typical of hardware synthesis instruments (and analogue synthesizers in particular).

11.3.1 String Machines

We can begin by looking at the typical sound sources of analogue synthesizers and try to emulate them in a digital computing environment. A very common approach to generate notes electronically is provided by the so-called divide-down schemes. These are quite economical means of producing full polyphony from a very small number of source signals. The idea is based on having 12 oscillators, one for each of the chromatic steps of the 12-tone scale, at a very high frequency, then divide the frequency of each one of these successively to generate the various lower octave sounds. It is a very ingenious way to have any number of octaves produced at the same time. This method was originally conceived to produce organ sounds, but later on it was also employed in synthesizers, particularly to produce string ensemble sounds. One of the characteristics of the sound of these *string machines* was the in-sync octaves, which were a by-product of the divide-down scheme.

If we want to emulate these sounds in software, we might need a high sampling rate (through oversampling)[8] for a true divide-down oscillator. However, we can use a different approach to generate the phase-synchronized sounds. Instead of starting with a top octave, we can create a slowly varying phase signal and multiply it to get the higher octaves. These signals can then be used to look up tables containing bandlimited waveforms (sawtooth waves), which will produce the notes. This is enabled by the fact that table lookup can be done in a wrap-around way, so that phase signals are read modulo 1. For this idea to be implemented we need three components:

- One bandlimited wave table per octave.
- Twelve phase signals.
- The actual note generator.

Starting with the wave tables, to enable a full bandwidth sawtooth without aliasing, we create a different one for each octave. Let's say we want eight octaves of non-aliased notes. The following code creates these, numbering them according to the actual octave they belong to:

```
// lowest oct
gilo init 4
ioct init gilo
// highest oct
gihi init 12
// partial arrays
iarr[] init sr/(2*cpspch(ioct+1))

// sawtooth table generation
while ioct < gihi do
 // reset partial table
 icnt =  0
 while icnt < lenarray(iarr) do
   iarr[icnt] = 0
   icnt += 1
 od
```

```
 // fill partial table
 icnt = 1
 while icnt <= sr/(2*cpspch(ioct+1)) do
  iarr[icnt-1] = 1/icnt
  icnt += 1
 od
 // create bandlimited table
 ifn ftgen ioct-gilo,0,16384,10,iarr
 ioct += 1
od
```

Note that we set it so that the full octave is non-aliased, which means that the lower notes of a set may not have harmonics up to the Nyquist. Since these are possibly beyond most people's hearing range, it should not be a problem.[9] In order to define the lowest frequencies, we use the Csound octave notation, where 8 defines middle C. Therefore we have four octaves on each side of this pitch.

The next is step is to create the low octave frequency signals. We can use a global array to store these so that the note generator can easily access them. To instantiate the phase sources we can use a loop, or more elegantly, recursion,


```
// 12 phase signals
gaphs[] init 12
instr 1
 inst = p4/100
 gaphs[p4] = phasor:a(cpspch(gilo+inst))
 if p4 < 11 then
   inst += .01
   schedule(1+inst,0,p3,p4+1)
  endif
endin
schedule(1,0,-1,0)
```

where the frequency for each of the twelve notes is determined using pitch class notation (0-11) and the lowest octave: 4.00, 4.01, ..., 4.11. This is the octave.pitch-class notation (`pch`) used in Csound, which is converted to Hz (`cps`) for the phase generator. Passing a −1 as an instrument duration (parameter 3) makes it run continuously for an indeterminate length of time.

The final step is to produce the sawtooth signals for each note. This is the actual sound-generating part of the design. Again, it is fairly straightforward: we take a pitch in `pch` notation, extract its octave to get the correct wave table, and the pitch class gives the index into the phase signal array. This is scaled up to the right octave and used in the table lookup,

```
// note generator
gasig init 0
instr 2
 indx = round(frac(p5)*100)
```

```
  ioct = int(p5)
  ioct = ioct < gilo ? gilo : (ioct > gihi ? gihi : ioct)
  ioct -= gilo
  a1 = tablei:a(gaphs[indx]*2^ioct,ioct,1,0,1)
  gasig += a1*linenr:a(p4,0,.2,.01)
endin
```

Pitch is given by parameter 5 and amplitude by parameter 4. The instrument uses an envelope to create a short release (0.2 seconds), and it puts its signal into a global variable that acts like a bus. This is used instead of the straight output so we can apply some simple modifications to the sound before it is sent out.

String machines typically used two forms of modifiers. The first one is a simple amplitude envelope, which is applied using a voltage-controlled amplifier (VCA), with attack and decay (release) controls. There is only one envelope that is globally applied to all of the sounds, and this is triggered as the first key is pressed and sustained while any other notes are played. When all keys are released, the decay is applied. We can model this *paraphonic*[10] VCA with another instrument that takes the bus signal and shapes it,


```
// VCA envelope
instr 3
  kact active 2
  kgate = kact > 0 ? 1 : 0
  if kgate > 0 then
    kht = .2
   else
    kht = .1
   endif
  kenv = portk(kgate,kht)
  gasig *= kenv
endin
schedule(3,0,-1)
```

This instrument derives a gate signal from checking if there are any active instances of instrument two. If there any active instances, it produces a high signal, if there are none, we get a low signal. The gate is used directly in a portamento generator, `portk`, which is a first-order lowpass filter with a half-time setting.[11] That creates the envelope, smoothing out the gate transitions. We employ separate times for attack and decay (0.1 and 0.2 seconds, respectively). The envelope scales the global signal bus, emulating the single VCA.

The second modifier is also applied globally to the sound to create an ensemble effect, without which the string sound would be very dull and synthetic. This is based on a stereo *chorus* design, which transforms the spectrum dynamically by mixing delayed and frequency-modulated copies of the signal to the left and right channels,

```
// chorus
nchnls = 2
instr 4
```

```
 amod1 = randi:a(3,.75)+oscili(2,.45)+17
 amod2 = randi:a(2,.65)+oscili(3,.55)+13
 a1 = vdelay(gasig*.5,amod1,25)
 a2 = vdelay(gasig*.5,amod2,25)
 out(.5*(gasig + a1-a2), .5* (gasig + a2-a1))
 gasig = 0
endin
schedule(4,0,-1)
```

As we have seen in Chapter 8, varying the delay time of a delay line has the effect of modifying its pitch. If we modulate it, we get a frequency-modulated signal. A slow variation makes it a form of a vibrato effect, but if it is also delayed slightly and combined with the original signal, some phase interactions will cause the spectrum to change periodically. That is similar to instruments playing together as an ensemble, not exactly synchronized in time (and sometimes pitch). The chorus effect used here applies a periodic (sinusoidal) and aperiodic (noise-based) modulation to the delay time. This does not to be very wide, as only a few milliseconds are needed (a maximum of 5 ms of modulation is used in the code). The chorus signal is then sent to the output buffer.

This string machine emulation can be used as a MIDI instrument by mapping note velocity to amplitude and note number to octave.pitch-class. Some further alterations may be made to this. Filters can be used to shape the sawtooth waves, emulating slightly different "string" instruments; other waveforms may be added (e.g., square waves for clarinet-like spectra); a low-frequency oscillator can be applied to the low-octave generation to create a vibrato effect; and a resonant low-pass filter can be applied to the signal path, paraphonically, with a single envelope.

11.3.2 The Vocoder

Another classic source-modifier design is the channel vocoder, which we have already explored in some detail in Chapter 7. We can now use the ideas developed there to implement an instrument in Csound. The fundamental principle of the vocoder is to use a bank of filters to detect the energy at different bands in the spectrum, then apply this to impose the analyzed spectral envelope on an excitation signal (the complex audio source). This is normally done via a matching filterbank modifying an impulse train or another component-rich input.

We can concentrate first on the filterbank design. Ideally we would like to program it in such a way to allow an adjustable number of bands, spread over a given frequency interval. To do this we can employ a different form of recursion, now applied to an opcode. This is a very useful approach to create parallel or serial filter connections. The principle is to instantiate the requested number of opcodes and then distribute the signal to them, summing their outputs. In order to have a steep rolloff, we can employ fourth-order bandpass filters. The Butterworth design is a suitable type of filter for this application, since it has the characteristic of a good stopband attenuation coupled with a maximally flat passband.

The vocoder band design is therefore

```
kfreq = kmin*(kmax/kmin)^(icnt/ibnd)
kbw = kfreq/kQ
abnd = balance(butterbp(butterbp(asig,kfreq,kbw),kfreq,kbw),
               butterbp(butterbp(anal,kfreq,kbw),kfreq,kbw))
```

where ibnd center frequencies are exponentially spaced over an interval defined by kmin and kmax. Now we can wrap this band in an opcode, and call it recursively ibnd-1 times (the first band is given by the first call from an instrument),

```
opcode VBands,a,aakkkio
 anal,asig,kmin,kmax,kQ,ibnd,icnt  xin
 amix init 0
 kfreq = kmin*(kmax/kmin)^(icnt/ibnd)
 kbw = kfreq/kQ
 abnd = balance(butterbp(butterbp(asig,kfreq,kbw),kfreq,kbw),
                butterbp(butterbp(anal,kfreq,kbw),kfreq,kbw))
 if (icnt < ibnd-1) then
   amix = VBands(anal,asig,kmin,kmax,kQ,ibnd,icnt+1)
 endif
 xout(amix+abnd)
endop
```


Note that the type of the last parameter (the band count) is given as o, which means an i-time optional parameter, defaulting to 0. Next we can take an input, extract its pitch and RMS,[12] then use that to control a source signal, which in this case is a bandlimited pulse oscillator,

```
opcode Vocoder,a,akkki
 anal,kmin,kmax,kq,ibnd xin
 anal = dcblock2(anal)
 afun,ars plltrack anal,0.1
 kamp = port(rms(anal),0.01)
 kfun = port(k(afun),0.01)
 kfun = (kfun > 20 ? kfun : 20)
 asig = buzz(kamp,kfun,sr/(2*kfun),-1)
 aout = VBands(anal,asig,kmin,kmax,kq,ibnd)
     xout(aout)
endop
```

The vocoder opcode can then be used in an instrument. For example, we can take the input from a sound file and resynthesize it,

```
instr 1
 Sfile = "/Users/victor/audio/paisley-full.ogg"
 av,ab diskin Sfile
 asig = dam(av+ab,0.05,0.001,1,0.01,0.1)
 p3 filelen Sfile
```

```
 aout = Vocoder(asig,100,8000,40,32)
 out(aout*linenr:a(0dbfs,0.1,0.1,0.01))
endin
schedule(1,0,0)
```

Various modifications can be made to this design. We can, for instance, modify the spectral envelope by stretching/compressing formants, using different frequencies and bandwidths in the synthesis filterbank. It is possible to use various other input (excitation) sources, pitch or unpitched, and to realize cross-synthesis of two spectra. We can also replace the synthesis filterbank and source signal by a bank of ModFM operators, using the opcode designed earlier, which have center frequency and bandwidth/Q parameters. The vocoder was one of the first practical examples of spectral analysis-synthesis techniques; although somewhat primitive, it can be used in very sonically interesting and musical ways.

11.4 Granular Processing

In Chapter 9, we introduced the concept of grain as a characteristic of noise signals. We used that idea to develop a grain texture synthesizer using parameter controls that followed Gaussian distributions. The basic principle was to generate grains based on bandlimited noise waveforms, and have the duration, time and frequency densities, amplitude, etc., defined as time-varying trends, with their values drawn from random number generators. The instrument design used a recursive approach,

```
instr 1
 igs = gauss(p4,p5) // start time
 igd = gauss(p6,p7) // duration
 iga = gauss(p8,p9)*0dbfs // amp
 icf = gauss(p10,p11) // center freq
 ibw = gauss(p12,p13) // bandwidth
 ifn = p14 // envelope table

 ai = line:a(0,p3,ftlen(ifn))
 ae = table:a(ai,ifn)
 an = randi:a(ae,ibw)
 as = oscili(iga*an,icf)
 out(as)

 if times() < p15 then
  schedule(p1,igs,igd,p4,p5,p6,p7,
        p8,p9,p10,p11,p12,p13,p14,p15)
 endif
```

and once we set the values for the mean and standard deviation for each parameter, stochastic generation would take control of the process.

This is a very interesting approach to generate different types of textures, particularly if we make it into a more general-purpose signal processing method. We can modify the instrument above to take an audio wave table as a source for grains, instead of a single period of a sinusoid, and then apply the same sound generation principle. This introduces a new parameter, defining the location on the table from where the grain waveform is extracted. Following stochastic principles, we can also control this using a random variable. This time, however we can accumulate the value so that the reading may progress forward (or backward). This can be used as a sort of timescaling control, compressing or stretching the original sound.

We also need to repurpose the center frequency and bandwidth controls, since we are not generating noise grains anymore. These may be reused as controls for a bandpass filter that is applied to the grain. This can create a spectral mosaicking result, as different grains from different bands are combined together to make a reconstruction of the original sound source. We just need to be careful to employ the absolute values of frequency and bandwidth in case these become negative at some point. With these ideas in mind, here's a commented redesign of the original granular synthesis instrument,


```
instr 1
 igs = gauss(p4,p5) // start time
 igd = gauss(p6,p7) // duration
 iga = gauss(p8,p9)*0dbfs // amp
 icf = abs(gauss(p10,p11)) // center freq
 ibw = abs(gauss(p12,p13)) // bandwidth
 ist = gauss(p14,p15)+p18 // start pos
 ifn = p16 // envelope table
 ifw = p17 // wave table

 ai = line:a(0,p3,ftlen(ifn))
 ae = table:a(ai,ifn)
 aph  = phasor:a(sr/(ftlen(ifw)))
 as = tablei:a(aph,ifw,1,ist,1)
 as = butterbp(butterbp(as,icf,ibw),icf,ibw)
 out(ae*as*iga)

 // keep s
 ist = ist > 1 ? ist - 1: (ist < 0 ? ist + 1 : ist)
 if times() < p19 then
  schedule(p1,igs,igd,p4,p5,p6,p7,
        p8,p9,p10,p11,p12,p13,p14,p15,p16,p17,ist,p19)
 endif
endin
```

This instrument can now be driven by passing some initial parameters to it. There is quite a significant range of processing effects that can be achieved if we used different settings. The grain duration is particularly important now, as it will determine the size of the chopped-up segments of sound. Larger grains will render the original audio more intelligible, but of course depending on the grain rate, defined by the successive grain start times, it can create very dense textures. Here's an example that granulates a source sound, but keeps it within some level of intelligibility:

```
seed(0) // seed from system clock
igsm = 0.02 // start mu
igsd = 0.01 // start sigma
ism = 0.002 // pos mu
ist = 0.001 // pos sigma
igdm = 0.25// dur mu
igdd = 0.05 // dur sigma
igam = 0.7 // amp mu
igad = 0.05 // amp sigma
igfm = 1500  // cf mu
igfd = 3000 // cf sigma
igbm = 500 // bw mu
igbd = 250 // bw sigma

// env table and total duration
ifw = ftgen(0,0,0,1,"fox.wav",0,0,0,1)
ifn = ftgen(0,0,1024,20,3,1)
idur = 60

schedule(1,0,igdm,igsm,igsd,
         igdm,igdd,igam,igad,
         igfm,igfd,igbm,igbd,
         ism,ist,ifn,ifw,0,idur)
```

The first ten seconds of the granular processing above are plotted in Fig 11.7, corresponding to a single repeat of the phrase "the quick brown fox jumps over the lazy dog." The original recording lasts just short of three seconds; through granular reconstruction, we have stretched it to around ten seconds. The effect of the grain segmentation can be seen clearly in the spectrogram.

In particular, the filter settings can make a significant difference, especially if the bandwidth is set to be fairly narrow. This design can be the starting point for further exploration of granular processing. In particular, we may want to make the spectrum evolve dynamically according to some pattern. While the parameters are stochastically generated, their trends are fixed over time. Further spectral changes can be created if we apply some form of time varying control over these, for instance, with the use of envelopes. The parametrization approach used in the design of this instrument should make any such code modifications fairly straightforward.

11.5 Analysis-Synthesis

Analysis-synthesis methods represent an important category of computer sound synthesis and processing. Within this category, we find various techniques, from the phase

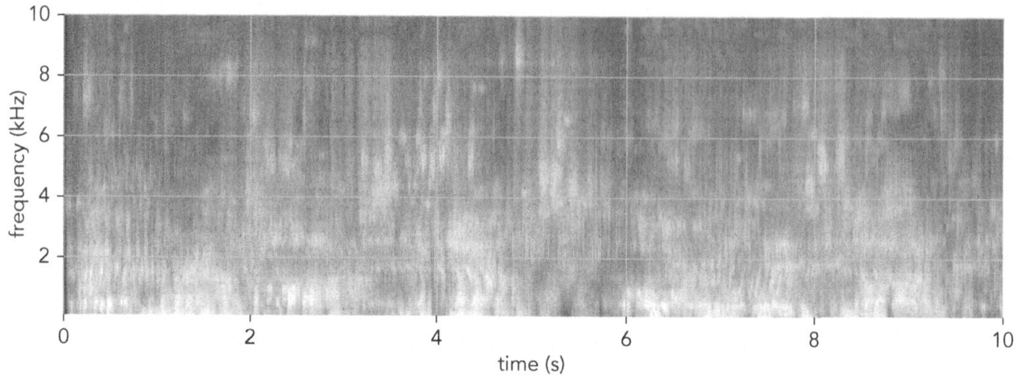

Figure 11.7 Spectrogram of a granular-processed reading of "the quick brown fox jumps over the lazy dog."

vocoder to linear prediction to partial tracking and other approaches. The principle behind all of these methods is to start with an existing waveform, breaking it down according to some model, and then recomposing it with some modifications. Some of the techniques allow a transparent process, that is, if no changes are made to the analysis data, the reconstruction is faithful. In general, we will want to apply transformations to it, since the whole point of these methods is to expose certain parameters and characteristics of sound spectra which cannot be manipulated directly.

11.5.1 Spectral Envelopes

A typical application of analysis-synthesis is to allow access to the spectral envelope, independently of the frequency/partials content of a sound. We have seen earlier on that a primitive (yet quite musical) method of extracting the spectral envelope is provided by the channel vocoder. More sophisticated alternatives are represented by linear prediction and cepstral analysis, which may be used interchangeably. In Chapter 6, we noted that a fundamental application for spectral envelope processing is given by formant preservation in pitch transposition, which attempts to maintain intelligibility of vocal sounds by keeping the resonant peaks in place while moving the frequency of each partial.

We can construct a simple vocal harmonizer using pitch transposition and formant preservation. The following code adds three voices above the input signal, using cepstral analysis to extract the spectral envelope and replace the transposed amplitudes with it (`pvscale` formant mode 1, third parameter). We also scale up the gain (`pvscale` fourth parameter) of each added voice to counteract any reductions caused by the spectral envelope replacement,

```
instr 1
 isiz = 2048
 a1,ad diskin "paisley-full.ogg"
 fs1 = pvsanal(dcblock2(a1),isiz,isiz/8,isiz,1)
 a2 = pvsynth(pvscale(fs1,5/4,1,2)) // major 3
 a3 = pvsynth(pvscale(fs1,3/2,1,2)) // perf 5
 a4 = pvsynth(pvscale(fs1,7/4,1,2)) // just min 7th
 a1 = delay(a1,(isiz*9/8)/sr)
```

```
  out((a2+a3+a4+a1)/4)
endin
```

It is possible to stretch or compress the spectral envelope, and then reapply it to the signal. This type of warping can be used to modify the gender characteristics of a voice. For example, we may independently scale the spectral envelope and the pitch of a signal,

```
instr 1
 isiz = 2048
 as = diskin2:a("fox.wav",1,0,1)
 fs1 = pvsanal(dcblock2(as),isiz,isiz/8,isiz,1)
 fs2 = pvscale(pvswarp(fs1,0.6,0),2) // warp and transpose
 a2 pvsynth fs2
 outs a2
endin
```

where we first warp the spectral envelope downward by the frequency ratio 0.6, and then scale up the pitch by an octave without preserving the formants. We then change the male voice characteristic of the recording into a female one. This is because the vocal formants get slightly scaled up under transposition but are still recognisable. Without the pre-warping, the result would just be the typical *Donald Duck* effect resulting from scaling the formants by a factor of 2. On the other hand, preserving the formants while transposing would also not necessarily impart a gender change effect.

Spectral envelope warping can be used for various other applications. It can be scaled and/or shifted in frequency (in the example above, there is zero shifting) to create different types of shapes. These parameters can also be varied over time, for dynamic modifications. We can, for instance, apply these to individual notes of instrumental sounds to create various timbral modifications. Starting with recorded wave tables, a spectral shaping synthesizer using spectral envelope warping is an interesting possibility.

Spectral envelope extraction can also be used in cross-synthesis applications, where we apply the spectral envelope of one sound into another,

```
instr 1
 isiz = 2048
 a1 = diskin2:a("fox.wav",1,0,1)
 a2 = diskin2:a("pianoc2.wav",1,0,1)
 fs1 = pvsanal(dcblock2(a1),isiz,isiz/8,isiz,1)
 fs2 = pvsanal(dcblock2(a2),isiz,isiz/8,isiz,1)
 outs(pvsynth(pvsvoc(fs1,fs2,1,1)))
endin
```

where the voice filters a piano note, producing a hybrid of the two. Other component-rich sounds can be used as the source excitation signal, including music excerpts and synthesized tones.

Cross-synthesis is a territory in which linear prediction excels. So as an alternative to spectral analysis and cepstral processing, we might want to use an all-pole filterbank to impose the spectral envelope obtained from a linear predictor. While it may produce similar results, it is a fundamentally different approach, which is not conditioned by modeling the spectrum in equally spaced bins. Here is a streaming linear prediction example, based on a 30th-order filter:

```
gifw  = ftgen(0,0,1000,20,2,1)
instr 1
 a1 = diskin:a("mu_brahms.aif",1,0,1)
 a2 = diskin:a("fox.wav",1,0,1)
 out(lpcfilter(a1,dcblock2(a2),1,100,1000,30,gifw))
endin
```

where we now use a recording of Brahms's lullaby[13] as the excitation signal (first input). The filter is computed from the second input every 100 samples, using a 1,000-sample autocorrelation sequence. The function table holds a Hanning window, which is applied to the input before analysis. It is also possible to get the all-pole coefficients as a k-rate array using `lpcanal`, and apply these with the `allpole` filter (effectively breaking the process into two separate components, analysis and synthesis). The spectral envelope can be extracted an fsig using `pvslpc`, and all-pole coefficients can be obtained from an fsig (via the cepstrum) using `pvslpc` (also as a k-rate array).

11.5.2 Morphing

Spectral morphing was first introduced by Trevor Wishart in his phase vocoder manipulation programs and used extensively to create dynamic changes in sounds from one recognizable character to another. This is implemented by interpolating the amplitude and frequency data of two spectral signals. For example, we may want to create a transition from two instrumental notes, starting with one and ending with the other. Ideally, we will start the morphing sometime after the attack and end it before the note stops, so that at the beginning and end we have the full spectra of the two sounds we are working with. This example morphs a flute sound (transposed down two octaves, and a piano note,

```
instr 1
 ka = linseg(0,.5,0,5,1,1,1)
 kf = linseg(0,.5,0,5,1,1,1)
 a1 = diskin:a("flutec4.wav",0.25)
 a2 = diskin:a("pianoc2.wav")
 fs1 = pvsanal(a1,2048,256,2048,1)
 fs2 = pvsanal(a2,2048,256,2048,1)
  out(pvsynth(pvsmorph(fs1,fs2,ka,kf)))
endin
```

It is important to try and match the two spectra so that it does not result in a simple crossfading. This makes partials from one sound move seamlessly into the other, as in the example above, where we needed to transpose the flute sound to match the piano spectrum.

In this example, we have used the same rate of transition for both amplitude and frequency data, but since these are independent sometimes a faster transition in one parameter may result in a better morphing effect. In this particular flute to piano morph, having a faster amplitude change (e.g., 1 second instead of 5) actually produces an interesting result, bringing out the air noise of the flute over the piano spectrum as the start of the morphing. Experimentation is key, as different spectral characteristics may require fine adjustments in the interpolation parameters.

11.5.3 Timescaling

Since we can change the pitch of a signal without affecting the waveform duration, it follows that we can modify the duration of a sound without changing its pitch. There are several ways to demonstrate this. The simplest is to use a variable rate playback to reproduce a recording at different speeds, like a tape recorder varispeed control. This alone has the effect of changing the pitch of the sound, lowering it when the speed is less than the original and raising it if we play the recording at a faster rate. To counteract this, we can apply frequency scaling, bringing the sound back to the correct pitch,

```
instr 1
 kspd = p4
 kscl = 1/kspd
 a1 = diskin:a("fox.wav",kspd)
 fs1 = pvsanal(a1,4096,256,4096,1)
 out(pvsynth(pvscale(fs1,kscl)))
endin
```


where the timescale (and playback) ratio is defined by the instrument parameter 4.

In fact, frequency and time scales are independent, so we could also scale the pitch if we need to, multiplying kscl by the desired frequency ratio. A more flexible approach is to load the sound to memory, using a function table and then use an all-in-one pitch and time scaling opcode to do the reading,

```
gifn = ftgen(0,0,0,1,"fox.wav",0,0,1)
instr 2
 out(temposcal(p4,1,p5,gifn,1))
endin
```

which has the advantage of using phase locking to reduce any artifacts that may arise with extreme time stretching. We now use parameter 5 as the pitch ratio. The only issue with this approach is that transposing vocal sounds is of course going to shift formants, as these are not preserved. To fix this, we can, instead, read the table using pvstanal, which produces an fsig, whose frequency envelope can be warped as required,

```
gifn = ftgen(0,0,0,1,"fox.wav",0,0,1)
instr 3
 fs1 = pvstanal(p4,1,p5,gifn)
 out(pvsynth(pvswarp(fs1,1/p5,0)))
endin
```

but of course, since the phase data has been modeled as instantaneous frequencies, phase locking is not available. However, this may not be a concern in most applications.

Using function tables may also allow a random access approach to sound playback. We can navigate freely to play different parts of the recorded sound as we wish, with whatever speed and pitch we want, possibly stopping (*freezing* the spectrum) at some points, etc. The following code shows an example of this, using a noise generator to determine the reading position:

```
gifn = ftgen(0,0,0,1,"fox.wav",0,0,1)
instr 1
 iftd = ftlen(gifn)/sr
 kt = timeinsts:k()%iftd
 apos = randi(iftd*p4,2)
 out(mincer(apos+kt,1,1,gifn,1))
endin
```

using the `mincer` opcode, which is similar to `temposcal`, except that it takes a position in seconds to read from. This defines where the samples are taken from the function table. In this example, the bandlimited noise generator produces values that are added to a time reference (`kt`), which proceeds forward at normal pace. If parameter 4 is zero, there is no added noise and the playback is normal; as that parameter is increased, a wider range of random variation is used as the reading position. We can also link the position to a *scrub* control of some sort (e.g., from a graphical user interface), and this can be used to navigate the playback.

For any kind of timescaling application, we need to store audio in memory. In the examples above we used a soundfile and a function table for this, which were accessed read-only. If we want to apply time stretching or compression to streaming audio, we need to fill a memory buffer with data and then read from it. This is fairly easy to do with a function table. We can create an empty table, write the input signal to it, then read from it at a slower (or faster) rate,

```
isiz = 10 // table holds 10 seconds
gifn = ftgen(0,0,isiz*sr,7,0,isiz*sr,0)
instr 2
 asig = inch(1)
 iftd = ftlen(gifn)/sr
 arp = line:a(0,iftd*p3,p3)
 tablew(asig,arp,gifn,1,0,1)
 out(temposcal(.5,1,1,gifn,1))
endin
```

There will of course be some latency between input and output. Also, in the case of time stretching, as we run out of memory, the table will be overwritten, so there will always be some loss of input. This is inevitable, as we are reading at a slower rate than we are writing. Alternatively, we can use spectral delay lines, which we will explore in the next section.

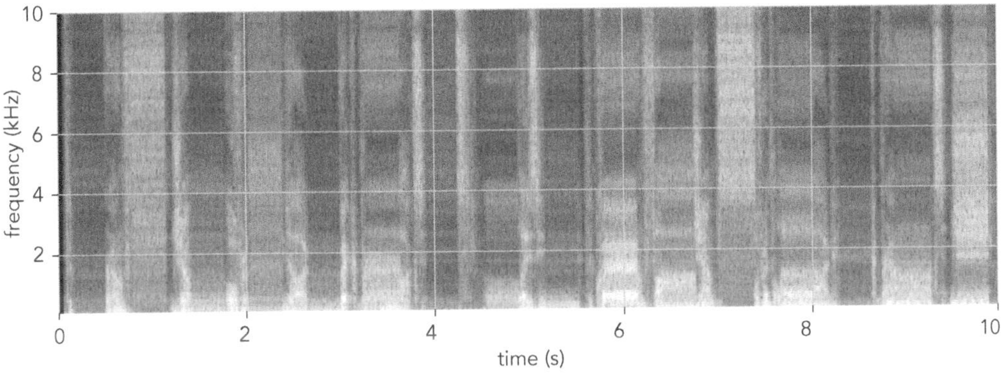

Figure 11.8
Spectral freezing.

In some cases, however, we might want to just freeze the spectrum of an input sound at some points. In this case, we can use `pvsfreeze`, which can hold steady the amplitudes or frequencies of an input,

```
instr 1
 isiz = 2048
 kfa = abs(oscili(1.5, 0.8))
 kff = abs(oscili(1.5, 0.7))
 as = diskin2:a("fox.wav",1,0,1)
 fs1 = pvsanal(as,isiz,isiz/8,isiz,1)
 out(pvsynth(pvsfreeze(fs1,kfa,kff)))
endin
```

In this example, whenever the values of `kfa` or `kff` are one or above, the amplitudes or frequencies, respectively, are frozen. The output of this code is plotted In Fig. 11.8, where we can clearly see the regular freezing of the spectrum produced by this process.

11.5.4 Spectral Delays

The streaming timescaling example discussed in the previous section employed a function table as a buffer in the time domain, from which data could be read at different rates. In this case, the table could be considered as a sort of a delay line, holding a portion of an audio signal to be released after a certain time. A similar approach can be applied to the frequency domain directly, using a spectral delay line,

```
instr 1
 isiz = 2048
 as = diskin2:a("fox.wav",1,0,1)
 fs1 = pvsanal(as,isiz,isiz/8,isiz,1)
 ihn,kt pvsbuffer fs1,10
 fs2 = pvsbufread(kt*p4,ihn)
 out(pvsynth(fs2))
endin
```

Here, the signal frames are placed in a buffer set up by `pvsbuffer`, then read by another opcode, `pvsbufread`. This uses the `ihn` variable as a buffer handle to define the correct memory location for access. The time can be dilated or compressed through scaling the time reference returned by the buffer opcode (`kt`), which is done via parameter 4.

More than one reader can access a single buffer, and the reading can proceed in any direction, without necessarily using the buffer reference. Conversely, a single reader can hop between multiple buffers (by making the buffer handle a k-rate variable). Optionally, the readers can exclude frequencies below and above a certain value.

Many applications besides timescaling can be implemented with this arrangement. For example, we can create complex textures out of a single source, by using several readers, at different frequency bands. The example below spawns 10 reader instruments, each bandlimited to an octave, spaced in half-octave intervals, and with different delay times, which are further modulated by a random signal,

```
instr 1
 SFile = "fox.wav"
 asig = diskin:a(SFile,1,0,1)
 fs1 = pvsanal(asig,2048,256,2048,1)
 gih,gkt pvsbuffer fs1, filelen(SFile)
endin

instr 2
 kmin = p5
 kmax = p6
 kdel = p4 + randi:k(p4,p7)
 fsig = pvsbufread(gkt*kdel,gih,kmin,kmax)
 out(pvsynth(fsig))
endin

idur = 60
schedule(1,0,idur)
imin = 400
imax = 800
icnt = 0
while icnt < 10 do
 intv = 2^(icnt/2)
 schedule(2,0,idur,1/(icnt+1),imin*intv,imax*intv,2/(icnt+1))
 icnt += 1
od
```


The spectrogram of the output produced by this code is shown in Fig. 11.9.

Different types of textures and moving spectra can be created using bandlimited spectral delays. It is possible to consider these design ideas as similar to granular processing, where each reader is modeling a continuous stream of grains. There is fertile ground for further experimentation. Since streaming spectral processing can be readily applied to any input signal, live signal processing is an obvious application to this and other analysis-synthesis designs.

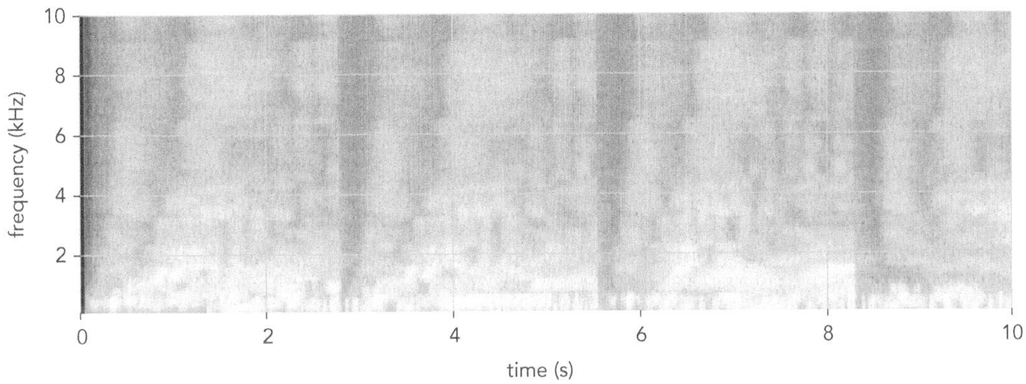

Figure 11.9
Spectral delays applied to "the quick brown fox...".

11.6 Design Methods

As we noted at the outset, sound design is a very wide area, with many techniques and approaches, which makes it difficult to tackle in a single chapter, even if we focus specifically on spectral techniques. Nevertheless, if we concentrate on the methodology and the thinking that underpins it, we may be more successful in conveying useful ideas. Therefore what we have studied here should not be taken simply as a collection of recipes for a particular computer music environment, but as examples of design approaches. Recipes are good, but methods are much better. What we learn from these can be transferred to other particular tasks we may need to execute.

So what are these design methods? There are many of them in the collection of code we built here, but we can highlight a few of these. The first one of them can be described as *modularity*, which we have seen from the start. We take a complex sound or spectral description and break it down into its constituents, then we try to design components, or modules, to represent these. An instrumental approach to computer music is modular. So when we look, for instance, at the endless glissando example, and figure out it is made of sinusoids that slide in frequency and have tapered ends, we can then model each one of these as an instrument. Similarly, when we try to process the spectrum at sub-band level, it is useful to design each band separately as a module, then combine several of these together. High-level programming languages such as Python and Csound have plenty of support for this approach.

Alongside modularity, another important design idea is *parametrization*. Once we create our modules, we might want to take a step back and see what parameters we want to expose, and what we want to keep constant or use some automatic method of control. So when we think about a sound, we may want to design the parametrization of its synthesis in such a way that it becomes malleable. Again, back to the endless glissando example, by exposing a few parameters we can create different instances of this textural idea that have varied characteristics: direction, length/rate, frequency width, etc. Good attention to parametrization also allows designs to be more easily modifiable. Note that it is not only a matter of allowing access to every possible parameter, but a balance between flexibility and ease of application (since parameter explosion can also pose difficulties in design).

Many of the examples in this chapter took advantage of iteration, through loops or recursion, which made the code compact and more manageable. This is another very important design method, which we should take advantage of whenever possible, if the programming environment allows it.[14] So when thinking about how to realize a sound or spectrum, we should try to see how its synthesis can be designed from an iterative perspective. This of course goes hand in hand with the other ideas of modularity and parametrization. Many processes in music are iterative or recursive by nature, and therefore this design principle is extremely useful here. Recursion can also been seen in the application of a single principle at different levels of structure, which was examined closely in Chapter 10. This is such an elegant idea that is worth serious consideration at the sound design stage.

Finally, it is also important to note that underlying all of these methods is a computational approach. We have had just about sixty years of computer music languages, since Mathews introduced the idea of an acoustic compiler in MUSIC III. There are many such systems, open source and proprietary, which we can use for sound design and music composition. We should take full advantage of these and the wide possibilities they offer. It is also worth noting how they might help the documenting of our work, since computer programs may be preserved quite well, or at least in a recoverable form.[15] In that sense, open source systems have an advantage, as, unlike proprietary software, they need not necessarily be discontinued or lost when the original maintainers, owners, vendors, or suppliers move on to other projects. For example, Csound allows the performance of works written over forty years ago for its predecessor (MUSIC 11), and has the principle of backward compatibility and work-preservation as one of its development strategies. This is quite unique even within open source software, and no proprietary system can ever reach this level of preservation of computer music works. Documentation, and support for it, is an essential aspect of sound design, which may have been overlooked by practitioners.

Notes

1 In the original formulation, Risset used a raised-inverted cosine with a small DC offset, similar to a Hamming window, shaped by an exponential function. A Gaussian probability density function provides a more flexible way to achieve a similar result.

2 We employ a Python script to create the spectrogram, running the Csound code and passing the output to the spectrogram plot function,

```
import ctcsound
import numpy as np
from matplotlib import pyplot as pl
import sys
f = open(sys.argv[1],'r')
code = f.read()
cs = ctcsound.Csound()
cs.setOption('-n')
cs.compileOrc(code)
sr = cs.sr()
ksmps = cs.ksmps()
```

```
dur = int(sys.argv[2])
end = dur*sr
buffer = np.zeros(int(end))
cs.start()
time = 0
while time < end:
    cs.performKsmps()
    buffer[time:time+ksmps] = cs.spout()
    time += ksmps
fig,axs = pl.subplots(1,1,figsize=[8,3])
axs.specgram(buffer,16384,sr)
pl.tight\_layout()
pl.show()
```

which can be run by passing the Csound code as a text file to the first argument and the spectrogram duration as the second. Note that in the case of the endless glissando, for clarity we set the plot to be in a logarithmic frequency scale, and limited the max frequency to 6,000 Hz.

3 Simpler models had only four operators, and some models like the TX81Z also allowed other operator waveforms besides a sinusoid.

4 Since an exponential curve will not reach 0, we add a minimum value, at −80 dB from peak amplitude, to the sustain parameter in case it is set to 0. We also make sure the audio rate version is selected so that any zipper noise may be avoided.

5 Some algorithms, however, employ feedback between operators, which is only possible by setting the global ksmps to 1. Since this incurs a significant performance penalty, it is possible to ignore this detail and use a ksmps-long delay between input and output. While not matching the feedback equations exactly, a frequency modulation effect will still be present.

6 The table lookup uses a sine wave table (−1) and offsets it by 1/4 of a cycle to get a cosine.

7 It is also possible to avoid the direct exponential function call by defining a function containing a certain range of the exponential function, then looking it up

```
   ifn = ftgen(3,0,131072,"exp",0,-14,1)
...
aexp = table(kndx*(1-acos)/14,3,1)
```

with the exponential drawn from 0 to its minimum value, which is chosen according to the expected maximum modulation index. Note that in order to correspond to the call to `exp()`, we need to change the sign of the argument and scale it. This implements waveshaping via table lookup.

8 As, for instance, described in Wiltshire (2016).

9 A solution would be to have more than one table per octave, but we do not require that much complexity.

10 As opposed to *polyphonic*, where each note has a separate envelope articulation.

11 The portamento opcodes use the same filter design as the 6 dB/octave filter introduced in Chapter 7 to compute the RMS of a signal. The difference is that instead of a

cutoff frequency parameter, they have a half-time control. This determines the time the signal takes to reach halfway toward the target input value.

12 A DC blocking filter is used, to help with the pitch and RMS tracking.

13 Thanks to Richard Boulanger for the source sound suggestion.

14 It is the case that some systems are less capable in this regard, and that tends to lead to code bloat. An example is Pure Data (PD), which is very useful for small projects, but large designs make it extremely difficult to program and debug, given its graphical nature.

15 The recreation of *Stria* thirty years after its composition is a good example of this.

12

Composing the Spectrum

This final chapter builds on the ideas developed in all sections of the book to discuss approaches to compose the spectrum from a musical perspective, using computational methods. The focus here is on practical aspects of the theories and techniques explored earlier on. The text first discusses some key aspects of spectral music, extending some of the ideas introduced in Chapter 10 toward the craft of music composition. These concepts are then put in the context of the creation of a musical work that is an example of a piece of *pure* computer music developed using a strictly spectral method.

12.1 Spectral Music-Making

One of the important arguments of Chapter 10 was that all music is spectral,[1] and that the means of making it cannot be reduced solely to computational or other methods.[2] Nevertheless, we can speak of a way of crafting and performing music that treats all the parameters of the spectrum at the same level, and which is informed by an understanding of how these can be manipulated in an extended fashion. While a computational approach is still not a requirement for this (it can be created purely with instrumental means), it can provide powerful supports for its composition and performance practice. We can summarize this by saying that computer music is *spectral native*.

We should also be more clear about the meaning of spectrally informed, which we are using to characterize this native form of spectral music. Let's start with an extreme example, Bach's *The Art of Fugue*. At the outset, we cannot say that this is not informed by spectral considerations, since it is strictly based on frequency relationships between different notes that are determined by their spectra. That is, it is carefully constructed using methods whose rules are derived from spectral aspects of the sounds employed.[3] However, it is not fully spectrally informed because the actual realization is undefined, we have no indication of instrumentation or any other performance instructions. Granted, it might have been composed with specific spectral shapes in mind, based on the kind of ideal instrumental tone described by Helmholtz. In any case, this is something that has to be imposed externally, after the fact, which will of course have a significant impact on the actual sonic result. The piece as realized by a brass ensemble is very different from a harpsichord rendition. In the limit case, if it were performed by a percussion group, it could begin to lose its sonic shape altogether, and the carefully constructed sequence of consonances and dissonances could become meaningless. Therefore, such a piece cannot be characterized as spectrally informed in the sense meant here.

Spectral Music Design. Victor Lazzarini, Oxford University Press. © Oxford University Press 2021.
DOI: 10.1093/oso/9780197524015.003.0012

From the other direction, and keeping to classic works to facilitate the discussion, we can name *Mutations* by Risset, as a fully spectral-native work. In this piece, all considerations are dictated by the spectrum, and its different parameters are given a level playing field. We can characterize this work as composed in accordance with a strictly spectral approach. Between these extremes, represented by these two examples, lies a continuum of possibilities, and musical works can be located in it according to how they manipulate spectral parameters. We should reiterate that being spectrally informed says nothing about the genre or style of music, it is a mostly technical denomination, relating to the process of composition. It should also be stressed that it does not also follow that certain aspects, such as the use of pitch constructions, melodies, or strongly enforced rhythmic patterns (repetitive or otherwise), are alien features of spectral music. As should have been expected, a great part of electronic music composition is fundamentally spectrally informed. Nonetheless, as we also argued in Chapter 10, many works from all sorts of music-making practices share different levels of spectral orientation. The word *spectral* is not a trademark of any one composer or group of composers.

12.1.1 Metaphors

The use of metaphors as ways of describing musical traits is very common, particularly in music criticism, but also in musicology and in the words of composers themselves. In spectral composition, this may be helpful at times, but also a hindrance. Metaphors, particularly visual ones, are often considered as the basis for interfaces for manipulating audio. However, there are limits to their usefulness as a general method for sound design and composition. A metaphor is helpful when it clarifies a sonic idea that is hard to put in objective terms, and often this is the case with musical attributes. It stops being of value if it obscures the technical means that lead to the realization of an idea. For example, by employing analogies with color in composition, it may be possible to describe a certain transformation or another musical device that is employed in a piece. This, however, tells us little about the technique used to achieve it, as the same metaphor can be translated to sound in different ways. Besides, its interpretation is by definition very personal and subjective. Therefore, it is very hard to extend the use of metaphors to provide universal interfaces for spectral composition. On the other hand, it is possible to facilitate, via a computational approach, the development of such metaphors for proprietary (or personal) use, alongside their translation into very precise sound synthesis routines. Open systems that allow such interfaces to be constructed can be very flexible tools for spectral music-making.

Metaphors may take various forms, beyond the obvious and most common types that are generally of visual nature. Some programming environments provide strong support for visual/graphic metaphors in detriment to other representations. This is very typical of graphical object and flowchart systems. On the other hand, text-based environments are less imposing and support the construction of interfaces dealing with any metaphorical connotations. Therefore, although perhaps less immediate, these constitute on the whole a far better development platform for spectral music-making. We should also consider the level of a particular programming system: lower-level languages may enable more sophistication in the development of programming metaphors, at a higher cost of programming. Higher-level languages, on the other hand, may only allow a narrower range of metaphors, but in a more immediate way.

12.1.2 Terminology

At this stage, it is worth considering an appropriate terminology for discussing spectral music. As noted in Chapter 10, a full consideration of electronic music may require an extension of traditional musical concepts. Since these were developed to deal with music where the pitch aspects of the spectrum were the primary composition materials, they clearly cannot be comprehensive enough to describe more extended ways of making music. However, we should stress that traditional terminology does not need to be completely discarded, as it may be repurposed and augmented by a wider dictionary. The principles introduced by spectromorphology attempt to address this gap in terminology, but suffer the problems of over-specification, and so are not always useful.

Theorists of electronic music have been exploring new terminologies since its inception. Some of these terms may be very useful for us, as, for instance, the concept of *sound* (or *sonic*) *object*, which, as discussed elsewhere, was first introduced by Schaeffer. It can be particularly enlightening if we use it as an augmentation of the concept of *musical note*. We can then describe a variety of music that transitions from very definitely and well-defined *notes* to more extended building blocks that may not be so easily characterized. The sound object encompasses all of these different materials. As we have seen, the term *note* is also used in spectromorphology, but in a more limited way, denoting a harmonic spectrum, which is not necessarily the case here.[4] We may go on and define various terms characterizing sound objects, but that may lead to an explosion of terminologies, which is not helpful here.

Terms referring to musical syntax, or the organization of notes in a composition, may also be extended to encompass more complex structuring of spectra. We can start by considering the word *harmony* as describing simultaneous and mutually dependent sound objects, and *counterpoint* as sequentially organized independent elements. To these, we can add a repurposed sense of the term *texture*, and the concept of *gesture*. The first one of them characterizes the overall shape and the permanence aspects of a sound object or a combination of these. Thus we can speak of granular textures, continuous textures, which can describe such characteristics of a musical work. Gestures are also related to shape, but more in terms of its time-sequential aspects. In this case, it is an extension of the concept of a melodic line, or a motif. In a setting where pitch is manipulated at the same level as spectra color, it is not appropriate to use the term *melodic*, so we can speak of the *gestural* instead. The latter term can encompass the former, just as a sound object is a superset of a note.

12.1.3 Realtime Systems and Performance

While the focus here is on composition aspects of spectral music, it is worth also discussing its performance, in particular when it involves live processing of instrumental sounds. With the increase in computer power over the past twenty years, demanding tasks such as the computation of a series of Fourier transforms of realtime input data have been enabled. This allows us to consider how the spectral manipulation of an instrumental sound can complement or augment it. In some cases, where the energy levels are sufficient, we need to consider how a blend between the mechano-acoustic (i.e., the direct instrument output) and the electro-acoustic (from the loudspeaker) can be achieved. It often requires careful consideration of source placement and location. In situations where the electro-acoustic source is dominant, with little direct sound, this is less of an issue.

Another key consideration when designing instruments for live electronics work is latency. Frequency-domain processing will most likely lead to some level of audible delay between input and output, which will depend on transform sizes and hop intervals. In addition, to regulate peaks in processor use and other operating system requirements, some amount of analogue-to-digital and digital-to-analogue buffering may be required, which will add to other implicit latencies. This needs to be carefully considered by the composer, especially in cases where time alignment is a requirement. Incorporating the delays as part of the fabric of the work is often a successful strategy. If negligible latencies are required, then only strictly time-domain processing is possible.

The design of a robust performance interface is also an important aspect of composing a live electronics work. This is something that is beyond the scope of our considerations here, but can be nevertheless be crucial to the success of a composition. A good gestural control over sound transformation parameters is key to the realization of a live spectral music performance. The advent of simple microprocessor platforms for sensing and control has simplified significantly the design of instruments for live electronics. These can be easily interfaced with systems such as Csound running on dedicated synthesis system or even on the devices themselves.

12.1.4 Physical and Virtual Space

A separate, although somewhat related, concern in composition is the question of the physical space of performance. Before the introduction and development of audio technology, music listening was always mediated by physical space, but with the now common use of headphone and earphone listening, this is not the case anymore. Nevertheless, we have continue to consider space as an important characteristic of musical performance, and the simulation of real spaces through reverberation design is very common. This of course is superimposed onto the real space of performance if loudspeakers are used, which may be problematic in the case of environments with non-negligible reverberation characteristics. So if some form of virtual acoustical space design is employed in a work, it needs to be designed taking in consideration the intended performance conditions.

If earphone or headphone delivery is the target, then the composer may create a completely virtualized environment for a composition. This is what is described as *binaural* audio, and the virtual space characteristics can be modeled by applying position-dependent filters to sound sources. Such filters re-create the effect of the head and outer ear geometry on incident sound, which can be described by head-related transfer functions. The principle can be applied to both direct and reverberation signals, artificially modeling a virtual acoustic space around the listener. We can consider any such type of virtualization an aspect of spectral music, since it is achieved through the manipulation of a source sound spectrum. Other forms of space design in composition are dependent on specialized set-ups consisting of multiple channels and loudspeakers, and are less relevant to our discussion here.

12.1.5 Approaches to Composing the Spectrum

The discourse of spectral music is predominantly aural, which does not mean that mimetic images are not welcome. In fact, in some cases, as evidenced by Wishart's music, they can take a first-class status. However, we have to call it an aural approach because it is mediated by the qualities of the spectrum. Even if we want to portray a recognizable source, we will bear in mind its sonic qualities and how they connect with the other sound objects around it. That is what characterizes a recognizable spectrally informed way of composing. Therefore an analytical ear is an essential tool in this process.

As a help to shaping the spectrum, external structures may be imposed on it. A simplest case of these is a frequency lattice such as a scale. As we saw in Chowning's *Stria*, an extended concept such as the golden-ratio scale can be particularly useful as guiding device. Or we can derive such a structure from a source sound, as in Harvey's *Vivos voco*. Artificially imposing restrictions may not be successful, though. We can think of various permutation schemes used in serial music that make very little sense from an aural perspective. A sound reason (in both senses of the word) is needed for the use of spectral structures.

On the other hand, lattice-like structures may be limiting, as they cut into the continuum aspects of music. So, we may choose an approach that favors the in-between, the time between the notes,[5] the gliding and sliding aspects of sound objects. As we learned early on from Xenakis's music, we can construct very intricate and geometrically informed shapes that are continuous in frequency, in opposition to segmented and fixed-frequency sounds. A non-lattice approach may also help disperse any regular and square patterns that may arise as part of the composition. It allows a translation of the concept of *swing* to the spectrum, via, for instance, the use of parameter modulation, which may be periodic, random, or a combination of the two.

Perhaps the essential aspect of spectral composition is how it can enable the articulation of transitions and the dynamic, time-varying aspects, of sound objects. While this goes hand in hand with the non-lattice ideas of smoothness and continuity, it can be applied more generally. For example, we may have step-wise, discrete transformations that can articulate transitions in a successful way. Gestures that bring out the dynamic qualities of the spectrum are a good starting point for the construction of transitions. They can be created with devices as simple as the application of cross-fading envelopes. The technique of morphing may, for instance, be used to create time-localized forms of transition articulation. Time-varying filters can also be employed to make very successful gradual transformations from one sound object to another. Of course, careful control of frequency and amplitude envelopes in additive synthesis is another prime method of articulating transitions. It has been noted that such spectral transformations can be used in a structurally analogous way to modulation in tonal-functional music. The extent of the relevance of this assertion is of course dependent on the intentions of composers, and their applications of such devices.

12.2 The Composition of *Mouvements*

We will now explore *Mouvements* (Hommage à Risset) (2003)[6] as an example of a strictly spectral composition approach. In this work, Csound was used to generate all the synthesized sound objects, and the rough mix for the complete sound objects. The final version was assembled using a multitrack mix of individual sections. The fact that Csound is a quite flexible tool for creating sound objects was crucial in the composition process. This section will outline a working method for spectral music and the most salient aspects of the composition process.

12.2.1 The Generative Principle

This piece is structured around a single generative principle, a meta-theme, which governs the synthesis procedures used, resulting in music elements that are linked together by a certain audible trait. This generative principle (and its different implementations) was developed stepwise from an original design into a set of variations, which dictated the construction of the instruments used in the piece. The basic form for it was elaborated by modifying the Risset-Shepard idea, the acoustic illusion of a glissando that continuously descends or ascends without ever reaching an endpoint. The starting point was thus given by an instrument design that creates such glissandos by using 10 sine wave oscillators tuned in octaves, constantly sliding down 10 octaves from the topmost frequency, as discussed in Chapter 11.

The illusion is created by using an overall spectral envelope, in the shape of a bell window, which attenuates the lower and higher ends of the glissando. However, the actual endless glide was not intended to be used directly. Instead, a number of modifications were made to create the actual sounds for the piece. The first of these was the elimination of the frequency glissando. The effect of this was to create a pedal tone (a drone), constantly changing in octave. In order to make the instrument more dynamic and flexible, the interval between each oscillator was allowed to vary. With this modification, they could be separated by any frequency ratio, not only octaves. This design became the basis on which all other instruments were created for the piece, the seed for all the various spectral shapes employed. Since all sound generators share the same structure, their sonic output will also contain similar gestural and textural traits.

This 10-part basic design was used as the starting point for the composition of *Mouvements*, from which further modifications were applied to yield the actual instruments used. The spectral shape was determined by a choice of four window types: Gaussian, Bartlett (triangle), Blackman-Harris, and Kaiser.

```
giblk = ftgen(21,0,4096,20,5)   // Blackman-Harris
gigau = ftgen(22,0,4096,20,6)   // Gaussian
gibar = ftgen(23,0,4096,20,3)   // Bartlett
gikai = ftgen(24,0,4096,20,7,1,5) // Kaiser
```

The duration (period) of the envelope cycle and the base frequency of the sound were made to vary in time. With this, moving gestures could be articulated. The sound output of each oscillator was also panned between pairs of speakers in parallel with their individual amplitude envelope. Finally, the direction of the envelope cycle (upward/downward) and a sound object fade in/out time could also be set. This became the first instrument implementation of the basic design, and determined the shared characteristics of all of the synthesis methods used in the piece:

- 10 parallel audio signal generators with evenly spaced, but variable, frequency intervals.
- An overall spectral envelope (determined by the window shape) that cyclically sweeps from one end of the spectrum to the other at different speeds, emphasizing/deemphasizing the sound of each signal generator.
- Individual spatial placement of the spectral components (also controlled by the envelope cycle).

This first design can be coded in Csound as a recursive UDO,

```
gipan = ftgen(3,0,512,7,0.00001,128,0.00001,256,1,128,1)
/**
  Parallel oscillators
  asig1,asig2 Oscs aph,kfreq,kint,iwin,ifn,imx
  aph - phase input
  kfreq - base frequency
  kint - interval
  iwin - spectral envelope window shape
  ifn - oscillator wave
  imx - number of components
*/
opcode Oscs,aa,akkiiio
 aph,kfreq,kint,iwin,ifn,imx,icnt xin
 amp = tablei:a(aph,iwin,1,icnt/10,1)
 apa = tablei:a(aph,gipan,1,icnt/10,1)
 aout = oscili(amp,kfreq*kint^icnt,ifn)
 aout1 = apa*aout
 aout2 = -(apa-1)*aout
 icnt += 1
 if icnt < imx then
  asig1,asig2 Oscs aph,kfreq,kint,iwin,ifn,imx,icnt
 endif
 xout(aout1+asig1,aout2+asig2)
endop
```

where the panning function table is provided by gipan (defined externally). This was employed in the first instrument as follows:

```
instr 1
 kfreq = expon:k(p5,p3,p6)
 kint = expon:k(p7,p3,p8)
 kgdur = line:k(p9,p3,p10)
 idir = p11
```

```
 iphs = p12
 iwin = p13
 aph = phasor:a(idir/kgdur,iphs)
 a1,a2 Oscs aph,kfreq,kint,iwin,-1,10
 kline = linen:k(ampdb(p4)*0dbfs,p14,p3,p14)
  out(a1*kline,a2*kline)
endin
```

This first instrument employs a sine wave (table number −1), as in the original Risset-Shepard tone instrument. Note that the phase signal determines the speed and direction of the movement within the overall spectral envelope. The frequency and interval vary exponentially over the duration of the sound object. The instrument uses a large number of parameters (14) to control its operation. An example of an instrument 1-generated sound object used in the piece is provided by the following line:

```
schedule(1,0,40,-16,1050,980,1.04,1.06,1,4,-1,.1,gibar,8)
```

whose spectrogram is shown in Fig. 12.1.

12.2.2 Variations

In addition to this first instrument, three major variations on the basic design were developed. They are based on source-modifier, formant wave, and frequency modulation synthesis methods. From these, a small number of modified instruments were derived for specific applications.

12.2.2.1 Filtered Noise

The first one of these replaces the oscillators by a filters, and partial frequencies are re-interpreted as center frequencies. A white noise generator is used as a source, resulting in a thickening of each component, controllable by a Q parameter,

```
/**
  Parallel filters
  asig1,asig2 Fils asig,aph,kfreq,kint,kq,iwin,imx
  asig - audio input
  aph - phase input
  kfreq - base (centre) frequency
  kint - interval
  kq - filter Q
  iwin - spectral envelope window shape
  ifn - oscillator wave
  imx - number of components
*/
opcode Fils,aa,aakkkiio
 asig,aph,kfreq,kint,kq,iwin,imx,icnt xin
 amp = tablei:a(aph,iwin,1,icnt/10,1)
 apa = tablei:a(aph,gipan,1,icnt/10,1)
```

```
 kcf = kfreq*kint^icnt
 aout = reson(asig*amp,kcf,kcf/kq,2)
 aout1 = apa*aout
 aout2 = -(apa-1)*aout
 icnt += 1
 if icnt < imx then
  asig1,asig2 Fils asig,aph,kfreq,kint,kq,iwin,imx,icnt
 endif
 xout(aout1+asig1,aout2+asig2)
endop
```

With the addition of the *Q* control, the instrument driving it requires an extra parameter (15). It also includes a white noise generator to produce the input signal,

```
instr 2
 kfreq = expon:k(p5,p3,p6)
 kint = expon:k(p7,p3,p8)
 kgdur = line:k(p9,p3,p10)
 idir = p11
 iphs = p12
 iwin = p13
 iq = p15
 asig = rand(1)
 aph = phasor:a(idir/kgdur, iphs)
 a1,a2 Fils asig,aph,kfreq,kint,iq,iwin,10
 kline = linen:k(ampdb(p4)*0dbfs,p14,p3,p14)
  out(a1*kline,a2*kline)
endin
```


This instrument is used in the opening of the piece to create a slow sweep over the spectrum (Fig. 12.2),

```
schedule(2,0,62,-12,1000,1100,1.3,1.1,18,.6,1,.75,gigau,2,500)
```

It produces sound objects that extend the original design, but are not too dissimilar to it.

12.2.2.2 Formant Synthesis

In contrast, the second variation employs a completely different synthesis method, formant wave function (FOF). This is implemented by `fof` opcode in Csound, and it can produce sounds that vary from vowel emulations to granular textures. It is based on another type of resonance synthesis, where a number of parallel synchronous grain streams are generated. Each grain models the impulse response of a bandpass filter, which is effectively a sinusoid shaped by an envelope. Therefore, when all streams are mixed together, the result is similar to a pulse wave being fed into a filter. Each `fof` is thus capable of modeling a formant region. In this variation, each component is produced by a FOF generator, and the base frequency is used as the fundamental, above which 10 formant frequencies are placed, following the interval spacing pattern. Thus this makes the highest[11] formant $k f_0$, where k is the interval and f_0 the fundamental frequency.

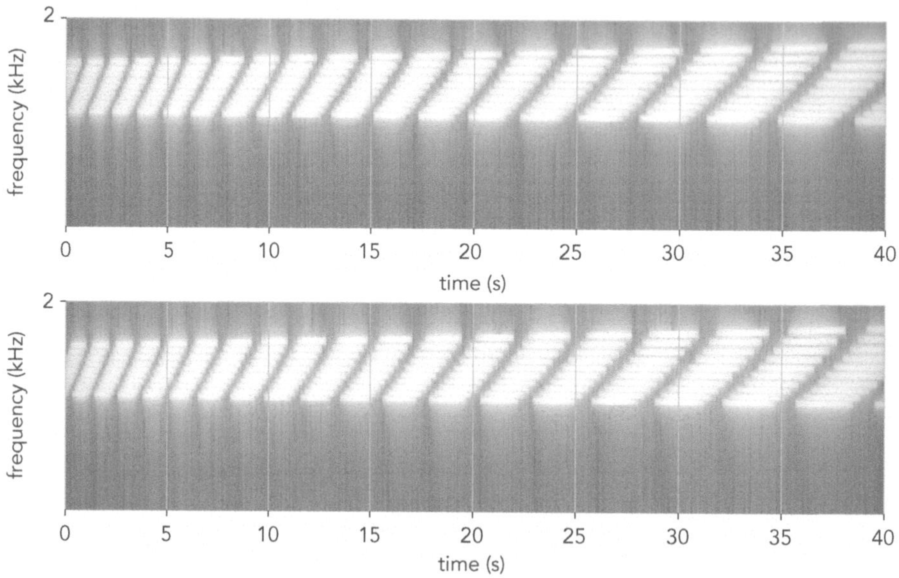

Figure 12.1 Spectrogram of instrument 1 example (left channel, *top*; right channel, *bottom*).

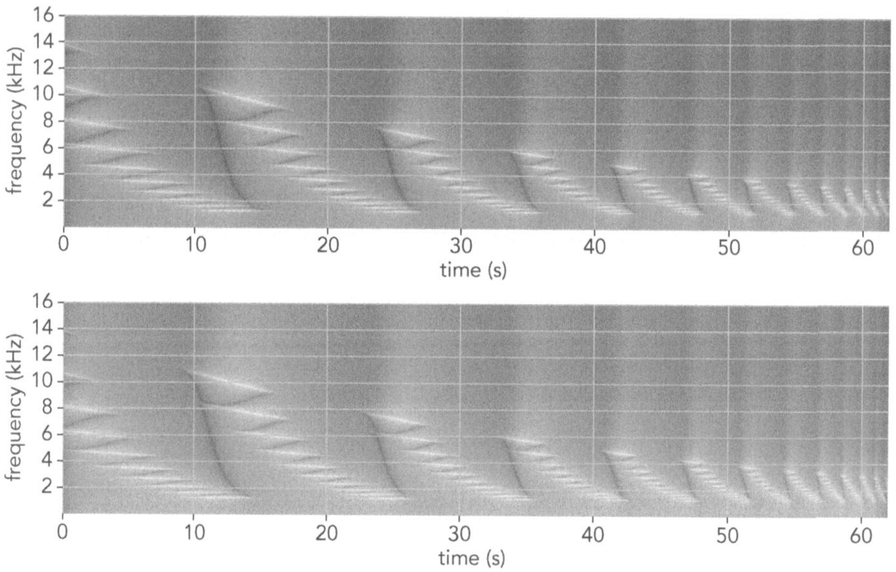

Figure 12.2 Spectrogram of instrument 2 example (left channel, *top*; right channel, *bottom*).

The code for the second variation is based on the following UDO,

```
gienv = ftgen(4,0,4097,5,.000001,4096,1)
/** Parallel fofs
    asig1,asig2 Fofs aph,kfun,kfreq,koct,kint,
                  kq,iris,kdur,idec,iwin,imx
    asig - audio input
    aph - phase input
    kfun - fundamental
    kfreq - base (centre) frequency
    koct - FOF octave control
    kint - interval
    kq - formant Q
    iris - FOF grain rise time
```

```
    kdur - FOF grain size
    idec - FOF grain decay
    iwin - spectral envelope window shape
    imx - number of components
*/
opcode Fofs,aa,akkkkkikiiio
 aph,kfun,kfreq,koct,kint,
     kq,iris,kdur,idec,iwin,imx,icnt xin
 amp = tablei:a(aph,iwin,1,icnt/10,1)
 apa = tablei:a(aph,gipan,1,icnt/10,1)
 kcf = kfreq*kint^icnt
 aout = amp*fof(1,kfun,kcf,koct,kcf/kq,iris,kdur,idec,
                1000,-1,gienv,p3)
 aout1 = apa*aout
 aout2 = -(apa-1)*aout
 icnt += 1
 if icnt < imx then
  asig1,asig2 Fofs aph,kfun,kfreq,koct,kint,
                   kq,iris,kdur,idec,iwin,imx,icnt
 endif
 xout(aout1+asig1,aout2+asig2)
endop
```


which has a few more parameters, which are relative to FOF synthesis, octave control, grain duration, and grain rise and decay. The first of these controls the perceived octave of the tone: as it increases, the fundamental moves gradually to lower octaves. This is done by fading out successive grains in the stream, creating also a granulation effect. The third instrument takes advantage of this and includes a time-varying octave control,

```
instr 3
 kfun = expon:k(p5,p3,p6)
 kint = expon:k(p7,p3,p8)
 kgdur = line:k(p9,p3,p10)
 koct = line:k(p16,p3,p17)
 idir = p11
 iphs = p12
 iwin = p13
 iq  = p15
 idur = .04
 iris = .007
 idec = .003
 aph = phasor:a(idir/kgdur,iphs)
 kran = randi:k(kfun*0.005,4)
 kfreq = kfun+kran
 a1,a2 Fofs aph,kfun,kfreq,koct,kint,
            iq,iris,idur,idec,iwin,10
```

```
 kline = linen:k(ampdb(p4)*0dbfs*.6,p14,p3,p14)
 out(a1*kline,a2*kline)
endin
```

which increases the parameter count to 17. Note also that the instrument adds a frequency jitter to the base frequency (but not to the fundamental), to create small deviations in the formant frequencies. An example of an instrument 3 sound object used in the first section of the piece is given by

```
schedule(3,0,20,-13,110,130,1.55,1.2,4,2,-1,.5,
         gibar,3,20,0,0)
```

which has quite a vocal-like character, but also shows some similarities to the original sonic idea. A spectrogram of this sound object is shown in Fig. 12.3.

12.2.2.3 Frequency Modulation

The third fundamental design variation employs FM synthesis. This uses a single modulator and each component is a carrier oscillator. The frequencies for this carrier oscillators can be derived in two ways:

- Directly as an interval of the base frequency.
- Indirectly as a harmonic of the base frequency.

For this, a parameter is added to define one of the two methods, as a frequency split: components above the value of the parameter use the first method, and the lower ones use the second method. The FM-based UDO is

```
/**
  FM synthesis, parallel carriers
  asig1,asig2 FMs aph,amod,kfreq,kint,iwin,ihr,imx
  aph - phase input
  amod - modulation input
  kfreq - base frequency
  kint - interval
  iwin - spectral envelope window shape
  ihr - split component
  imx - number of components
*/
opcode FMs,aa,aakkiiio
 aph,amod,kfreq,kint,iwin,ihr,imx,icnt xin
 amp = tablei:a(aph,iwin,1,icnt/10,1)
 apa = tablei:a(aph,gipan,1,icnt/10,1)
 kcf = kfreq*kint^icnt
 if icnt < ihr then
  khr = int((kcf/kfreq)+.5)
  kfc = kfreq*khr
 endif
 aout = oscili(amp,kfc+amod*icnt/10)
```

```
 aout1 = apa*aout
 aout2 = -(apa-1)*aout
 icnt += 1
 if icnt < imx then
  asig1,asig2 FMs aph,amod,kfreq,kint,iwin,ihr,imx,icnt
 endif
 xout(aout1+asig1,aout2+asig2)
endop
```

where the ihr parameter determines which is the component at which the carrier frequency is obtained directly. The difference between the two methods is that the first one can produce quite complex inharmonic spectra, which may not always be appropriate. By splitting the components into two groups, it is possible to produce a spectrum whose lower components are harmonically related, but the higher frequency partials may be inharmonic. In the instrument that uses this opcode, parameter 17 is used to determine the split,

```
instr 4
 kfreq = expon:k(p5,p3,p6)
 kint = expon:k(p7,p3,p8)
 kgdur = line:k(p9,p3,p10)
 idir = p11
 iphs = p12
 iwin = p13
 aph = phasor:a(idir/kgdur,iphs)
 kndx = line:k(p15,p3,p16)
 amod = oscili(kndx*kfreq,kfreq)
 a1,a2 FMs aph,amod,kfreq,kint,iwin,p17,10
 kline = linen:k(ampdb(p4)*0dbfs,p14,p3,p14)
  out(a1*kline,a2*kline)
endin
```

The following three examples of sound objects created by this instrument show how different levels of split (as determined by the last parameter in the list) are used in the composition:

```
schedule(4,0,40,-12,117,120,1.66,1.4,10,4,1,.75,
         giblk,4,5,0,5)
schedule(4,0,43,-12,482,467,1.38,1.43,2.3,7,1,.25,
         giblk,4,5,0,3)
schedule(4,0,38,-12,975,990,1.22,1.15,1.4,6,1,.75,
         gigau,4,5,1,0)
```

The spectrogram of the first of these three sound objects is shown in Fig. 12.4. The fact that it uses the first four components as harmonics of the fundamental gives it a vocal-like quality due to the presence of quasi-formant regions in the spectrum.

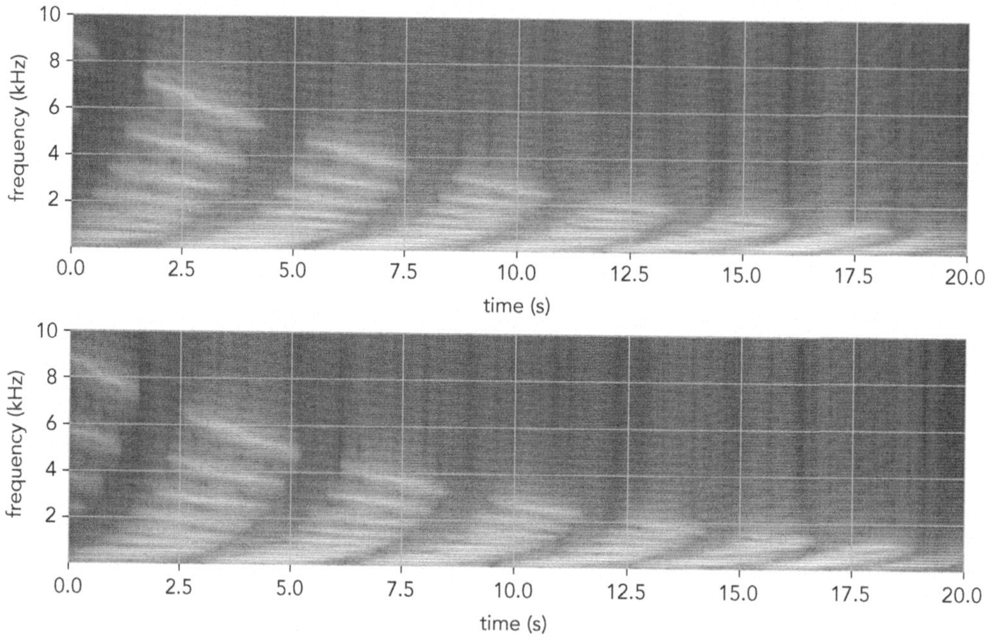

Figure 12.3 Spectrogram of instrument 3 example (left channel, *top*; right channel, *bottom*).

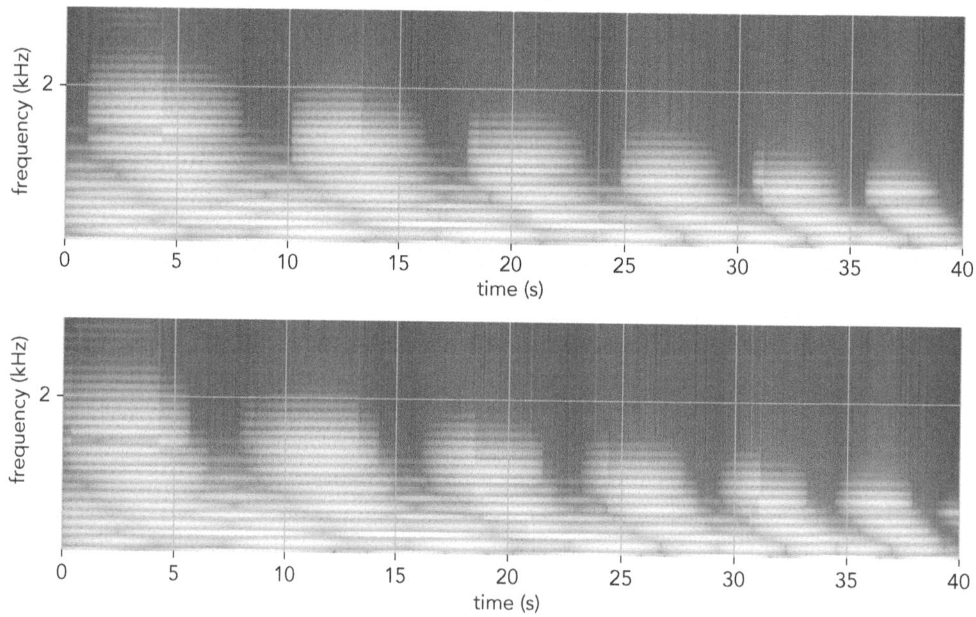

Figure 12.4 Spectrogram of instrument 4 example (left channel, *top*; right channel, *bottom*).

12.2.3 Other Variants

A number of minor variations on these basic sound generation methods were also developed. The first instrument has a variant where the interval ratio is made to fluctuate randomly 10 times a second and a wave with a few harmonics is employed:

```
giwav = ftgen(11,0,4096,10,1,1,0,1,0,1)
instr 11
  kfreq = expon:k(p5,p3,p6)
  kint = oscili:k(p7, randi:k(p8,10))
  kint += 1;
  kgdur = line:k(p9,p3,p10)
```

```
  idir = p11
  iphs = p12
  iwin = p13
  aph = phasor:a(idir/kgdur,iphs)
  a1,a2 Oscs aph,kfreq,kint,iwin,giwav,10
  kline = linen:k(ampdb(p4)*0dbfs,p14,p3,p14)
   out(a1*kline,a2*kline)
endin
```

As an example, the following parameters

```
schedule(11,0,30,-24,50,55,.08,.09,3,6,1,0,gikai,2)
```

create a low-frequency texture (Fig 12.5) used in the second section of the piece. This is fairly removed from the original sonic idea and provides some contrast to it.

The filtered noise design was also varied again by making the amplitude envelope table lookup position take its value from a random number generator. In instrument 2 (and the others) this was controlled by a constantly cycling variable going from start to end or end to start of the envelope (using phasor).


```
instr 21
 kfreq = expon:k(p5,p3,p6)
 kint = expon:k(p7,p3,p8)
 kgdur = line:k(p9,p3,p10)
 idir = p11
 iseed = p12
 iwin = p13
 kq = line:k(p15, p3, p16)
 asig = rand(1)
 aph = randi:a(1,1/kgdur,iseed)
 a1,a2 Fils asig,aph,kfreq,kint,kq,iwin,10
 kline = linen:k(ampdb(p4)*0dbfs,p14,p3,p14)
  out(a1*kline,a2*kline)
endin
```

Another example from the second section provides an idea of how this instrument is used,

```
schedule(21,0,48,-15,11,0,100,1.01,1.6,8,0.5,-1,0,
         gigau,3,50,100)
```

This example uses the instrument to create an ascending gesture based on noise bands, which starts low down in the spectrum and widens to sweep a large band of frequencies, as shown in Fig. 12.6.

Finally, the FOF-based instrument uses three further variants. The first one of these uses a time-varying Q, and takes the octave control from a random number generator,

```
instr 31
```

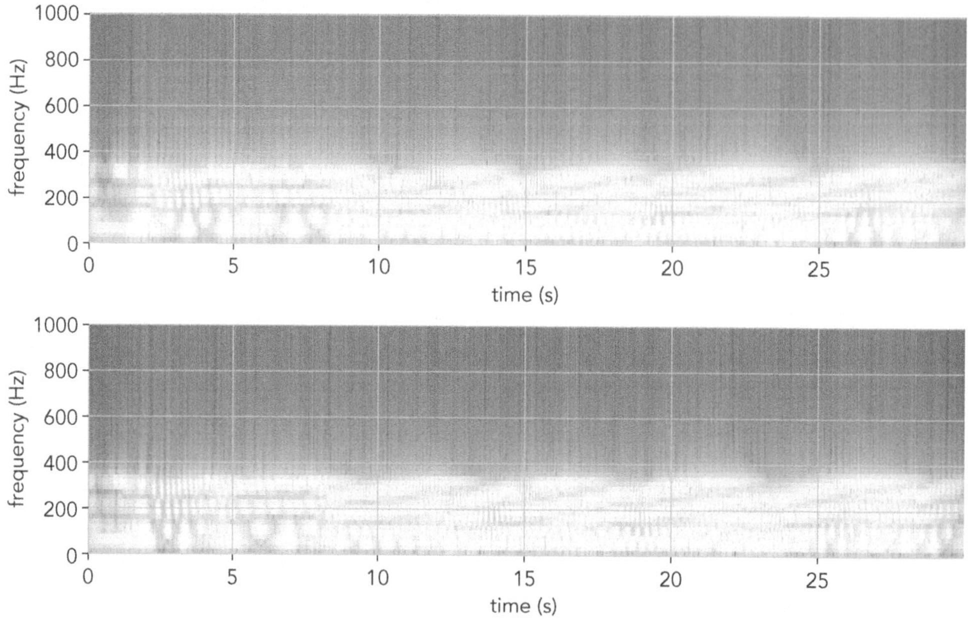

Figure 12.5 Spectrogram of instrument 11 example (left channel, *top*; right channel, *bottom*).

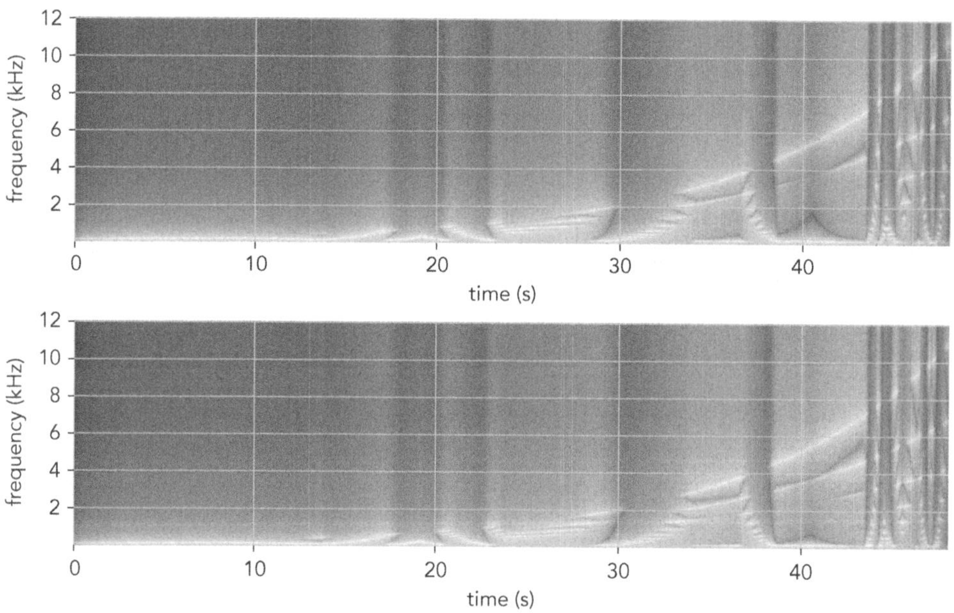

Figure 12.6 Spectrogram of instrument 21 example (left channel, *top*; right channel, *bottom*).

```
kfun = expon:k(p5,p3,p6)
kint = expon:k(p7,p3,p8)
kgdur = line:k(p9,p3,p10)
koct = abs(randi:k(line:k(p17,p3,p18),2.3,2))
idir = p11
iphs = p12
iwin = p13
kq = line:k(p15,p3,p16)
idur = 0.04
iris = .007
idec = .003
iolaps = 1000
```

```
 aph = phasor:a(idir/kgdur,iphs)
 kran = randi:k(kfun*0.005,4)
 kfreq = kfun+kran
 a1,a2 Fofs aph,kfun,kfreq,koct,kint,
            kq,iris,idur,idec,iwin,10
 kline = linen:k(ampdb(p4)*0dbfs*.6,p14,p3,p14)
 out(a1*kline,a2*kline)
endin
```

A slight variation on randomization of the octave control yields the next instrument,

```
instr 32
 kfun = expon:k(p5,p3,p6)
 kint = expon:k(p7,p3,p8)
 kgdur = line:k(p9,p3,p10)
 koct = abs(randi:k(p18-p17,p19,2))+p17
 idir = p11
 iphs = p12
 iwin = p13
 kq = line:k(p15,p3,p16)
 idur = 0.04
 iris = .007
 idec = .003
 iolaps = 1000
 aph = phasor:a(idir/kgdur,iphs)
 kran = randi:k(kfun*0.005,4)
 kfreq = kfun+kran
 a1,a2 Fofs aph,kfun,kfreq,koct,kint,
            kq,iris,idur,idec,iwin,10
 kline = linen:k(ampdb(p4)*0dbfs*.6,p14,p3,p14)
 out(a1*kline,a2*kline)
endin
```

These two variations were created to explore the granular characteristics of FOF,

```
schedule(31,0,45,-13,110,105,1.69,1.3,4,2,1,.75,giblk,
      .5,100,10,4,7)
schedule(32,0,12,-13,120,117,1.4,1.6,1.1,2,1,.5,giblk,
       1,50,500,1,4,1.5)
```

It is interesting how they blend the shape of the overall spectral envelope (and the basic principle behind the sonic construction of the piece) and a granular, sometimes vocal-like, texture (Figs. 12.7 and 12.8).

The final variation makes the FOF grain duration over time,

```
instr 33
 kfun = expon:k(p5,p3,p6)
```

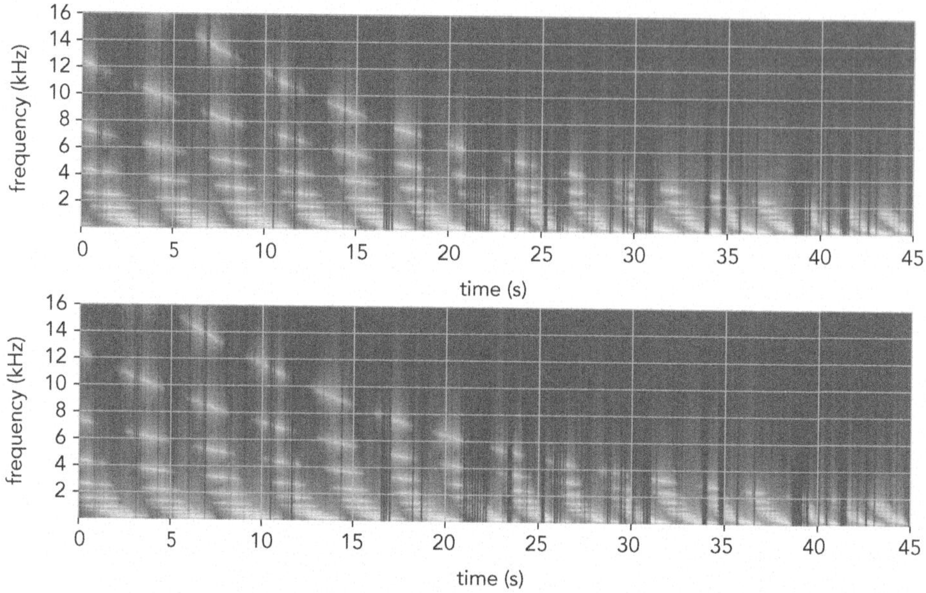

Figure 12.7 Spectrogram of instrument 31 example (left channel, *top*; right channel, *bottom*).

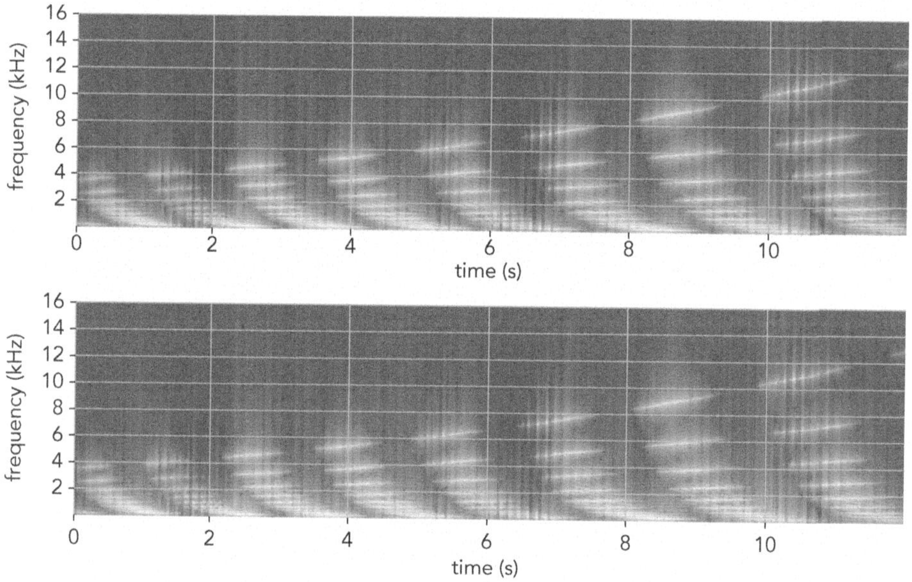

Figure 12.8 Spectrogram of instrument 32 example (left channel, *top*; right channel, *bottom*).


```
kint = expon:k(p7,p3,p8)
kgdur = line:k(p9,p3,p10)
koct = abs(randi:k(p18-p17,p19,2))+p17
idir = p11
iphs = p12
iwin = p13
kq = line:k(p15,p3,p16)
kdur = line:k(p20,p3,p21)
iris = .01
idec = .01
iolaps = 1000
aph = phasor:a(idir/kgdur,iphs)
kran = randi:k(kfun*0.005,4)
```

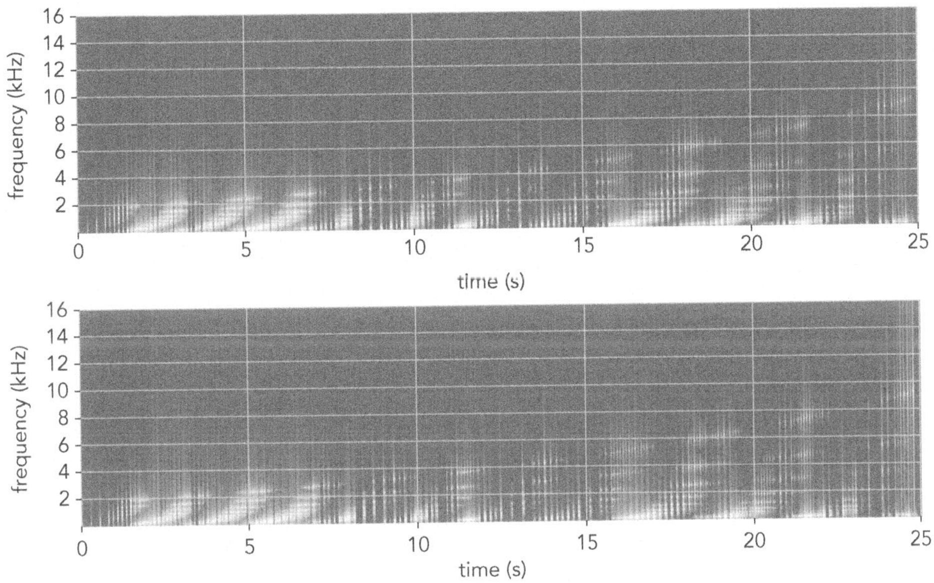

Figure 12.9 Spectrogram of instrument 33 example (left channel, *top*; right channel, *bottom*).


```
 kfreq = kfun+kran
 a1,a2 Fofs aph,kfun,kfreq,koct,kint,
            kq,iris,kdur,idec,iwin,10
 kline = linen:k(ampdb(p4)*0dbfs*.6,p14,p3,p14)
 out(a1*kline,a2*kline)
endin
```

This was also developed to explore the granular aspects of FOF, and to create transitions from continuous sounds and percussive, rhythmic, textures. It is used toward the end of the piece when the continuous characteristic of most of the sound objects used are broken down into a discrete rhythmic pattern (Fig. 12.9),

```
schedule(33,0,25,-13,123,80,1.3,1.61,1.6,3,-1,.5,giblk,
         1,150,15,3,6,1.2,.04,.08)
```

All of these different implementations were developed in response to musical needs. By keeping to the general design and varying the sound synthesis technique, a variety of interconnected musical elements were generated. Depending on the choice of parameters (spectral envelope shape, sweep rate, etc.) and instrument design, different types of gestures and textures can be created: spectral arpeggios, glissandos, drones, chords, timbral mutations, etc.

12.2.4 Macrostructure

The work is divided into four sections and a coda. An amplitude profile, outlining the sectional structure of the piece, is shown in Fig. 12.10. From an amplitude perspective the culminating point of the piece is placed toward the end of section 4, although the musical climax is set earlier in section 3, at 6′05″, at a moment of maximum spectral density.

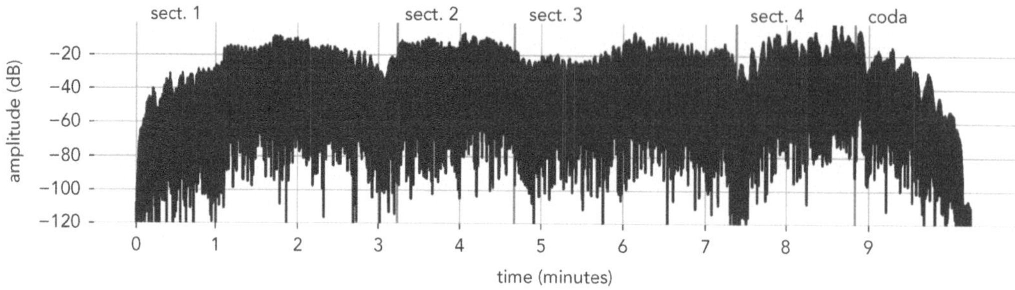

Figure 12.10 Amplitude profile and macrostructure of *Mouvements*.

For each part of the piece, the different implementations of the generative principle were developed along with the composition, as required. The opening was completed using only the first three instrument designs, with different performance parameters, gently introducing the spectral principle and its variations,

```
/* opening  */
schedule(2,.5,62,-12,1000,1100,1.3,1.1,18,.6,1,.75,22,2,500)
schedule(1,45,40,-16,1050,980,1.04,1.06,1,4,-1,.1,23,8)
schedule(2,50,30,-13,800,500,1.03,1.5,.3,10,1,0,21,2,250)
schedule(1,65,36,-13,100,110,1.6,1.01,2,4,-1,.5,22,1)
schedule(2,75,25,-18,800,950,1.5,1.15,.9,2,-1,.7,23,9,50)
schedule(1,87,20,-15,1200,1100,1.3,1.09,1,3,1, 25,21,2)
schedule(3,95,20,-13,110,130,1.55,1.2,4,2,-1,.5,23,3,20,0,0)
schedule(1,100,30,-18,5000,7000,1.004,1.09,2,1,-1,0,23,4)
schedule(2,105,20,-15,4000,4500,1.1,1.17,1.3,.8,1,0,21,6,100)
schedule(2,115,27,-24,2500,1000,1.1,1.3, 5,8,-1,.75,22,4,50)
schedule(1,123,22,-18,2000,2050,1.008,1.02,.3,3,1,0,21,2)
schedule(3,125,6,-20,800,850,1.1,1.3,6,6,-1,.5,22,3,500,0,0)
```

The idea here is to introduce different envelope cycle speeds and interval sizes. Various forms of movement are built up by cycling over the spectral envelope. The first section runs for 3′44″, and has a local (musical) culminating point at about 2′25″, after which it decays quite quickly with a small amount of granulation. The overall shape is designed as a slow condensation of texture leading to dissolution. The structure of the complete first section is depicted quite clearly in Fig. 12.11. Interestingly, spectrograms of musical works sometimes can be meaningless. However, in this case, these beautifully arranged figures tell us a lot about the sonic organization of *Mouvements*.

The first part introduces a quite wide spectral bandwidth to the listener. In contrast, the second section was devised to be much more limited in width, exploring low-frequency textures, which provide a darker sonic impression, as shown in the instrument 11 example above. Even though some higher-frequency sounds come into play midway through it, there is an overall predominance of energy in the lower parts of the spectrum. This is also clearly seen in its spectrogram (Fig. 12.12). This section employs instruments 11 and 21, which are variations on 1 and 2, providing not only a spectral contrast, but also different types of sonic gestures. It is significantly shorter, starting at 3′44″ and ending at 4′40″.

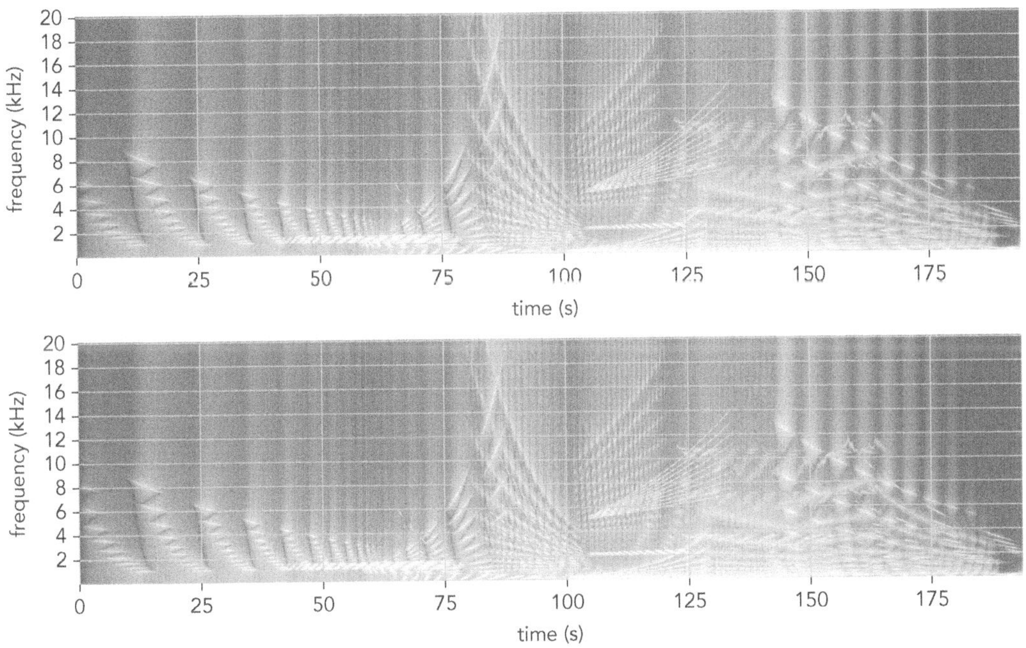

Figure 12.11
Spectrogram of the first section (left channel, *top*; right channel, *bottom*).

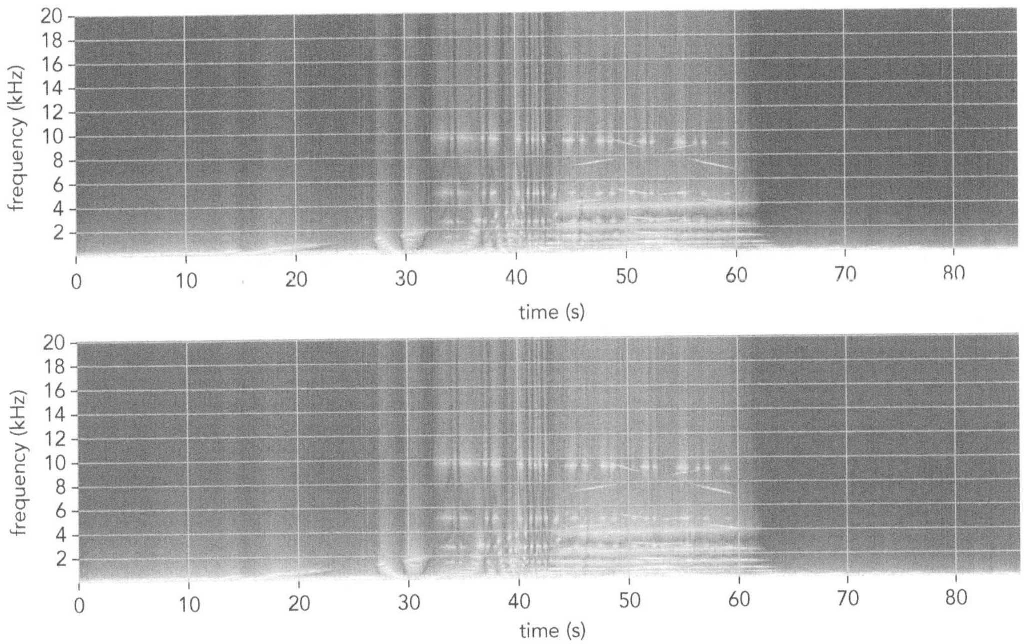

Figure 12.12
Spectrogram of the second section (left channel, *top*; right channel, *bottom*).

The following section expands the bandwidth and the dynamic range of the piece again, introducing more harsh inharmonic spectra courtesy of the FM synthesis instrument. This was designed to develop into the culminating point of the piece in terms of textural intensity and density, reaching a higher amplitude peak than in the first section, although more briefly. The continuous sounds are then dissolved granularly, more dramatically than in the first section, using the articulation possibilities given by the FOF generators as controlled by instruments 31 and 32. This section ends at 7′23″, and its spectrogram is shown in Fig. 12.13.

The last section was built from a contrasting granular texture, created by percussive sound objects made with the FOF versions of the design. It starts with a *fugatto* section,

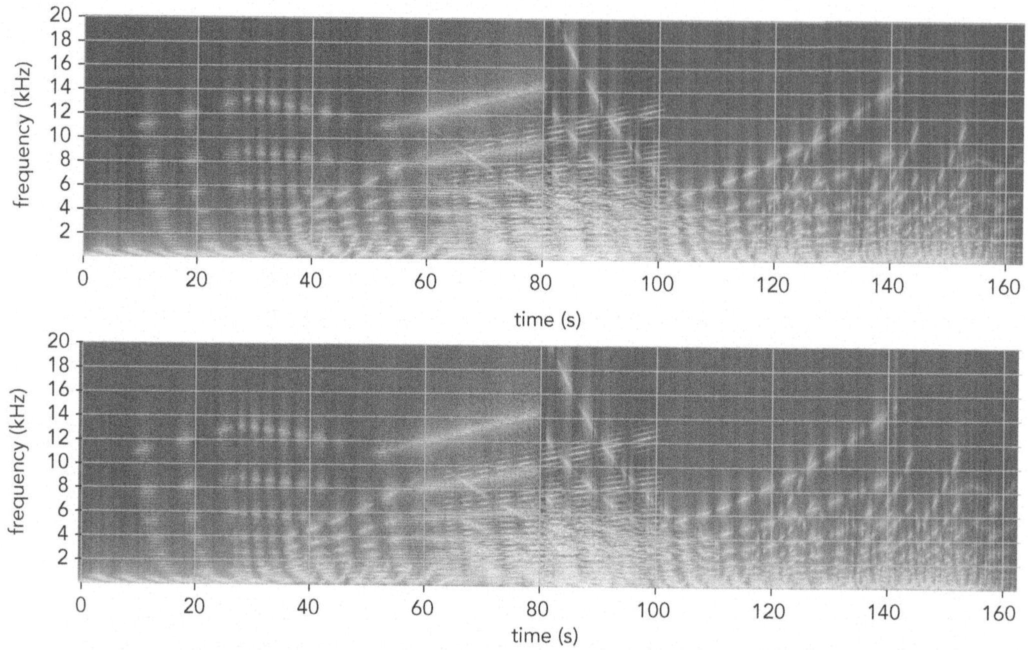

Figure 12.13
Spectrogram of the third section (left channel, *top*; right channel, *bottom*).

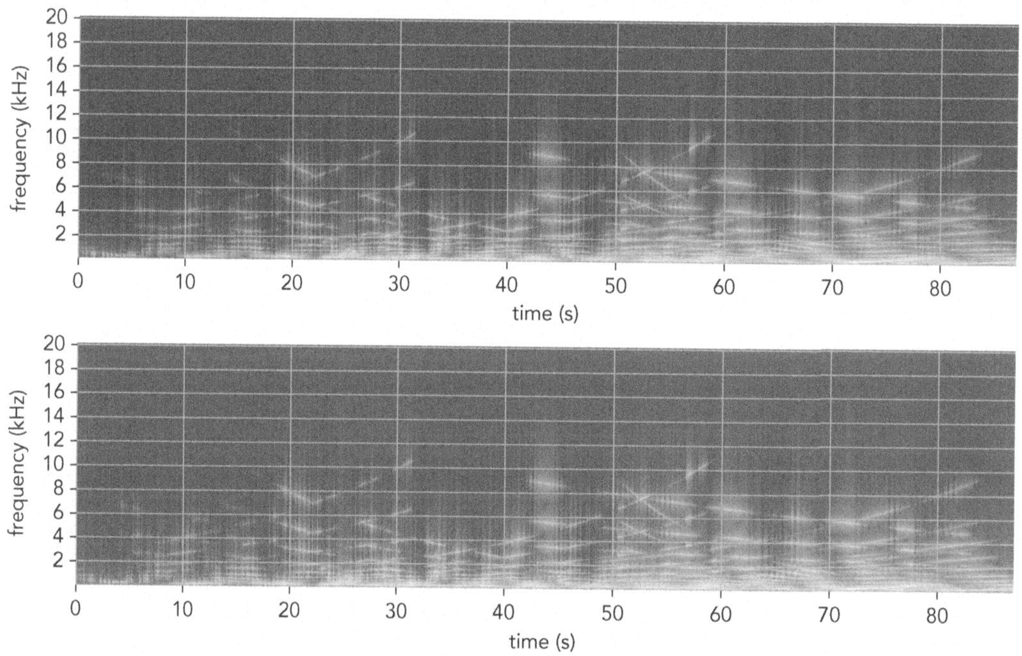

Figure 12.14
Spectrogram of the fourth section (left channel, *top*; right channel, *bottom*).

which develops randomly generated rhythms in non-literal imitation. In the spectrogram (Fig. 12.14), we can see how the grains are colored by a moving spectral envelope, which is created by basic generative principle. So each percussive sound object has a different timbre, and we can differentiate the various lines in the fugal counterpoint by their changes in colors. Eventually, the sonority of the piece gradually moves back to the continuous textures of early sections, in a slow condensation of the granular streams. As these become less discrete, spectral arpeggios begin to emerge, which evoke the sonic gestures characteristic of earlier parts of the work. This section makes it evident that we can move from well-defined *notes* to more generalized sound objects in a gradual and

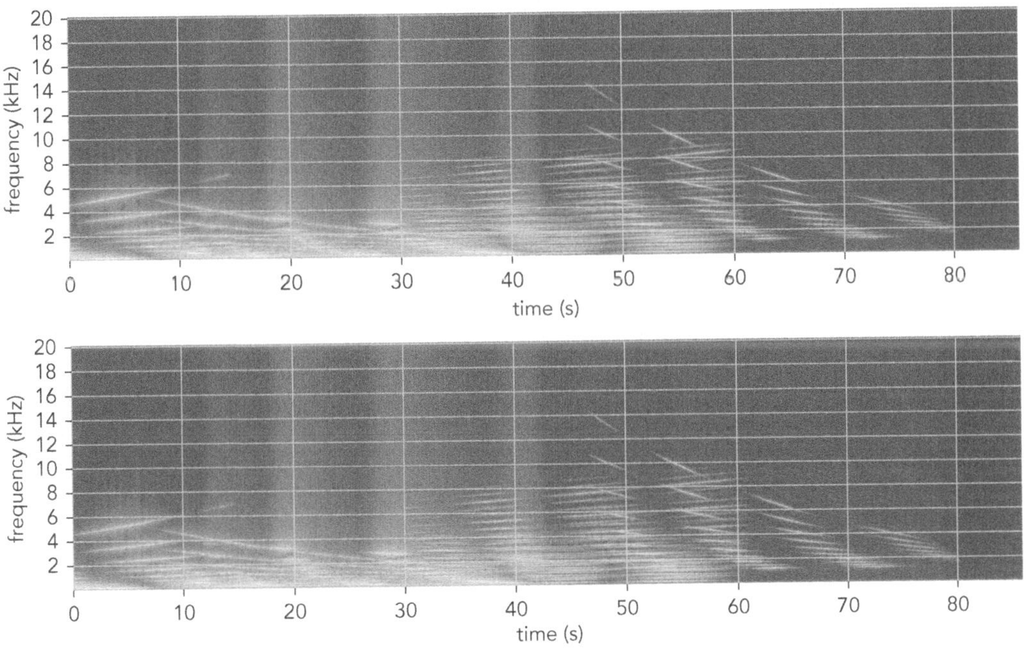

Figure 12.15
Spectrogram of the coda (left channel, *top*; right channel, *bottom*).

smooth way. It also indicates that the methods (and space) of spectral music allows such transitions to be articulated very precisely.

From 8′50″, a *coda* revisits the musical material used in the opening section. This brings us to a final gesture that is similar to the very first sound used in the piece, now synthesized with sinusoids rather than filtered noise. It reinterprets the beginning as a cleaned-up, pure, spectral phrase. This can be clearly seen in the spectrogram (Fig. 12.15), starting at about 46 seconds into the section, closing the work (at 10′16″).

12.2.5 Discussion

The macrostructure organization of *Mouvements* uses some traditional approaches, such as segmentation into sections, and the concept of a coda. It is not uncommon to use these to be employed in a more extended setting, as they provide good structural support for the development of longer forms. Working on a sectional basis can also help focus the attention to the local detail and how it may relate to the overall piece. An example of this is how the first section of the piece, which is in fact the longest, contains a time-compressed version of the complete formal arch. This is defined by a culminating point at about 2/3 of the length (in section 3), followed by an (extended) dissolution. In section 1, we also see a buildup to a climax, whose relative duration is close to the same proportion. The complete arch of the piece, excluding the final coda, starts with continuous sound objects and dissolves into granular streams, this is mirrored in section 1, where granulation is also used toward the end.

The use of contrasts is also a somewhat traditional device in composition. There is a clear change in sonic quality between sections 1 and 2, which is emphasized by the tessitura. The arpeggio gestures, which characterize the former, are condensed into a low-frequency mass in the latter, then are briefly transformed into noise-band sweeps midway through. Another use of contrasting material is heard in the continuous versus discrete aspects of the piece. The transitions between these two textures are used extensively to articulate the changes from one section to another.

Unity and cohesion of ideas are qualities that are often sought in the composition of a musical work. This can be achieved through various means; in traditional pitch-oriented approaches, themes, motives, collections of pitches (e.g., scales), harmonies, etc., can be structured in such a way that links between different sections of the work can be established. Here, as should be expected, there is no recognizable pitch-melodic theme or any form of grid that sustains the frequency material used. This is replaced by the fundamental working principle, the meta-theme, which is used to link all sounds, textures, and gestures in the piece. While there is no literal repetition of sound objects anywhere in the piece, a sense of continuity and integration is meant to be present throughout. This is carried by the way the instruments are constructed to create interrelated musical materials, reused in slightly different variations. The overall shape of the work is then more likely to be perceived as a united whole, even though the use of traditional structuring methods has been eschewed. Thus, unity and cohesion are achieved in a subtle manner, without recourse to straight recapitulation or exact reuse of sound gestures.

The composition of *Mouvements* featured a working method that is *instrumental* in nature, and the composer in this instance also takes the role of instrument maker. This is not an uncommon practice in electronic music, but it is made more evident by the working environment, the tools, and the computational approach involved in the process. Since the early days, the idea of the computer as a musical instrument has been discussed widely. Here, however, it is better to think of it as a platform for instruments, rather than a single instrument.

So, how could we describe the nature of such instruments? Borrowing from the terminology of computer languages, these are computational objects, which model something. This may be a process, or more directly, a spectrum, or, perhaps more completely, a sound object. So, the composition method can be described as led by instrument development, as the actual sonic material heard in the piece is mediated by these computational objects. We should have noticed that sounds are not composed directly, rather they are the result of variations on synthesis designs, and then the performance of instruments, given certain parameters. This is an key aspect of the composition process in *Mouvements*, and one that is fully supported by a system like Csound.

Finally, it is worth noting that while the piece was mixed using an external software package, this is actually not a necessity. At the time of the composition of this piece, common working methods included the use of graphical waveform editors and multi-tracking software, which provided a transition from analogue tape routines.[7] However, the complete work can still be recreated solely using Csound (as the opening code example shows), and in fact it may be preferable to preserve in that form. Pieces composed using similar methods to the ones describe here benefit from being fully defined in computer code, for a variety of reasons: documentation (as code may be seen as a form of self-documentation, a score for electronic music), compactness (only a few kilobytes of data are needed), readability, and, in the case of Csound, ubiquitousness of performance (since the software guarantees it can be played almost anywhere).

12.3 Conclusion: The Spectral Playground

We have now reached the end of a journey into spectral music, which began by a gentle introduction of the concept of spectrum, and continued through historical, technical, and finally musical and aesthetical perspectives. What we have seen is that instead of accepting a single definition, the spectrum can be approached from a variety of viewpoints. Historically, these developed through a combination of scientific discoveries in tandem with artistic interventions, leading to the development of a spectrally informed form of music-making, which emerged in the 20th century. The tools given to us by the Fourier theorem and the methods arising from it were shown to be very useful, within a computational environment, for a detailed composition of the spectrum. What was left out of these techniques was complemented by various other methods that dealt with non-linear systems and noise. In combination, these ideas provided a formidable set of spectral manipulation principles, which were shown to have practical computer-based implementations.

Since the first digital sound synthesis program by Max Mathews in 1957, many of these principles were put into practice by musicians pursuing spectral methods to compose music. At this stage, it is hoped that a comprehensive understanding of this form of music-making has been developed. The picture we would like to convey now is that of a spectral playground, a space for musical activities, enabled by the use of computing platforms. Such an environment has been dreamed of by many musical visionaries who sought extended ways to create music. It is now at our disposal, available as computing devices become ubiquitous.

Conquering the spectrum requires a change of mindset. Firstly, we need to expand our hearing, developing an analytical ear to all the sounds around us. We should be able to observe and consider how their spectra develop over time. It is key to recognize the musical potential of all sound sources, from the ones we learn to synthesize, to the ones created mechanically, which we can record and analyze. Learning such an analytical approach to hearing is both a technical and an aesthetical development. It is also helpful in educating us about the multitude of ways in which music-making is possible.

Parallel to this, developing the discipline and skills of computer music programming is essential. This is something that may be considered akin to learning an instrument. As noted elsewhere:

> Computer music as a field has been likened to a building with a sign on it saying "Best Eats in Town." Many people go into this building expecting to find an elegant restaurant with a parchment menu, formidable wine list, and pleasant, efficient, even charming service. What they find instead, to their surprise, is a shiny, enormous, extremely modern kitchen, with abundant supplies of every kind of foodstuff in voluminous, refrigerated storage. Indeed, the "Best Eats in Town" are available here, but only to those willing to learn to cook.[8]

It should be evident by now that such skills are a central element of the *métier* of spectral music, in its computational form, along with the rest of the theoretical foundations that formed the core of this book.

Above all, it is important to leave behind any unhelpful distinctions of genre and style that are fundamentally outdated in the 21st century. Although we have looked in detail at works that may fall under the aesthetics of an electro-acoustic concert music, we did not at any point imply that spectral music is limited to it, or even attempt to limit what the aesthetics of such music might encompass. While it might have been felt to be appropriate, thirty years ago, to distinguish a grand form of electro-acoustic music against *vernacular* music practice, this is hopelessly anachronistic now. With the multiplication of ways of creating music, mostly enabled by audio technology, it is not possible to make claims of legitimacy or of a special status for one particular niche approach to music-making. The spectral playground is open to all: anyone is welcome to share its delights and exercise their creativity within it.

Notes

1 While we may speculate on music as an abstract or intellectual construct, with no actual sound involved, this does not correspond to anything within the scope of this discussion.

2 The only requirement is a means of making sound.

3 It is possible to develop a spectral theory of harmony and counterpoint; in fact, since at least Helmholtz, we have had music theorists who have proposed ideas to that effect.

4 Notes may have non-pitched spectra. We may characterize a well-defined note more as a self-contained sound with a defined amplitude envelope shape, and with no splits in its spectral development.

5 Which relates the color to the scenes.

6 A stereo recording of this piece is available on the CD, *Voices Inside*, Farpoint Recordings fp024 (https://www.farpointrecordings.com/product-page/victor-lazzarini-voices-inside). The concert version of the piece uses 8 channels, and is available from the author.

7 They are still heavily used by many composers, but in the case of music composed in environments such as Csound, are always going to be an extra piece of software needed to complete the work. Since Csound offers all mixing and post-production facilities needed (Lazzarini et al., 2016, Chapter 22), use of external software is not a requirement.

8 Pope (1993) quotes this from the CARL software package release notes by F. R. Moore, D. G. Loy and others. It is a fit description for the computational approach taken in this book.

REFERENCES

Abe, T., T. Kobayashi, and S. Imai (1996). Robust pitch estimation with harmonics enhancement in noisy environments based on instantaneous frequency. In *Proceeding of Fourth International Conference on Spoken Language Processing. ICSLP '96*, Volume 2, pp. 1277–1280.

ANSI (1994). *American National Standard Acoustical Terminology*. New York: American National Standards Institute.

Auger, F. and P. Flandrin (1994). Generalization of the reassignment method to all bilinear time-frequency and time-scale representations. In *Proceedings of ICASSP '94. IEEE International Conference on Acoustics, Speech and Signal Processing*, Volume iv, pp. 317–320.

Barker, A. (1989). *Greek Musical Writings: Volume 2, Harmonic and Acoustic Theory*. Cambridge: Cambridge Univ. Press.

Barker, A. (2004). *Scientific Method in Ptolemy's Harmonics*. Cambridge: Cambridge Univ. Press.

Barry, D., B. Lawlor, and E. Coyle (2004). Sound source separation: Azimuth discrimination and resynthesis. In *Proc. of 7th International Conference on Digital Audio Effects, DAFX?04*, pp. 1–5.

Boethius (1971). Concerning the principles of music. In R. Lindsay (Ed.), *Acoustics: Historical and Philosophical Development*, pp. 35–40. Stroudsburg, PA: Dowden, Hutchinson and Ross.

Bogert, B., M. Healy, and J. Tukey (1963). The quefrency alanysis of time series for echoes: cepstrum, pseudo-autocovariance, cross-cepstrum and saphe cracking. In *Proceedings Symposium on Time Series Analysis*, pp. 209–243.

Bonada, J., X. Serra, X. Amatriain, and A. Loscos (2011). Spectral processing. In U. Zoelzer (Ed.), *DAFx Digital Audio Effects*, pp. 394–445. London: Wiley and Sons.

Boulanger, R. (2000). *The Csound Book*. Cambridge, MA: MIT Press.

Box, G. E. P. and M. E. Muller (1958). A note on the generation of random normal deviates. *The Annals of Mathematical Statistics 29*(2), 610–611.

Bracewell, R. (2000). *The Fourier Transform and Its Applications*. Singapore: McGraw Hill.

Bradford, R., R. Dobson, and J. ffitch (2005). Sliding is smoother than jumping. In *Proceedings of the ICMC 2005*, pp. 287–290.

Bradford, R., R. Dobson, and J. ffitch (2007). The sliding phase vocoder. In *Proceedings of the 2007 International Computer Music Conference*, pp. 449–452. ICMA and Re:New.

Bradford, R., J. ffitch, and R. Dobson (2011). Real-time sliding phase vocoder using a commodity GPU. In *Proceedings of ICMC 2011*, ICMC, pp. 587–590. University of Huddersfield and ICMA.

Brandtsegg, O., S. Saue, and V. Lazzarini (2018). Live convolution with time-varying filters. *Appl. Sci. 8*(1), 1–29.

Brandtsegg, O., S. Saue, V. Lazzarini, J. Inderberg, A. Tidemann, H. Kvidal, J. Tro, J. Rudi, and N. Thelle (2012). The development of an online course in DSP eartraining. In *15th International Conference on Digital Audio Effects, DAFx 2012 Proceedings*, York, UK, pp. 1–8.

Cage, J. (1961). *Silence*. Middletown, CT: Wesleyan University Press.

Carson, J. (1922). Notes on the theory of modulation. *Proceedings Institute of Radio Engineers 10*(1), 57–64.

Champeney, D. C. (1987). *A Handbook of Fourier Theorems*. Cambridge: Cambridge University Press.

Charpentier, F. (1986). Pitch detection using the short-term phase spectrum. In *ICASSP '86. IEEE International Conference on Acoustics, Speech, and Signal Processing*, Volume 11, pp. 113–116.

Cherniakov, M. (2003). *An Introduction to Parametric Digital Filters and Oscillators*. New York: John Wiley & Sons.

Chladni, E. F. F. (1787). *Entdeckungen über die Theorie des Klanges*. Leipzig: Weidmanns Erben und Reich.

Chladni, E. F. F. (1971). Discoveries in the theory of sound. In R. Lindsay (Ed.), *Acoustics: Historical and Philosophical Development*, pp. 119–124. Stroudsburg, PA: Dowden, Hutchinson and Ross.

Chowning, J. (1973). The synthesis of complex audio spectra by means of frequency modulation. *Journal of the Audio Engineering Society 21*(7), 527–534.

Chu, E. and A. George (1999). *Inside the FFT Black Box: Serial and Parallel Fast Fourier Transform Algorithms*. Computational Mathematics. Boca Raton, FL: Taylor & Francis.

Cohen, L. (1989). Time-frequency distributions—A review. *Proceedings of the IEEE 77*(7), 941–981.

Cooley, J. W. and J. W. Tukey (1965). An algorithm for the machine calculation of complex Fourier series. *Mathematics of Computation 19*(90), 297–301.

Dahan, K. (2007). Surface tensions: The dynamics of Stria. *Computer Music Journal 31*(3), 65–74.

d'Alembert, J. (1971). Investigations on the curve formed by a vibrating string. In R. Lindsay (Ed.), *Acoustics: Historical and Philosophical Development*, pp. 156–165. Stroudsburg, PA: Dowden, Hutchinson and Ross.

Danielson, G. C. and C. Lanczos (1942). Some improvements in practical Fourier analysis and their application to X-ray scattering from liquids. *J. Franklin Inst. 233*(4 and 5), 365–380, 435–452.

Dedekind, R. (2005). Essays on the theory of numbers. In S. Hawking (Ed.), *God Created the Integers*, pp. 906–964. London: Penguin Books.

Dolson, M. (1986). The phase vocoder: A tutorial. *Computer Music Journal 10*(4), 14–27.

Dudley, H. (1940). The carrier nature of speech. *Bell System Technical Journal 19*(4), 495–515.

Durbin, J. (1960). The fitting of time-series models. *Revue de l'Institut International de Statistique / Review of the International Statistical Institute 28*(3), 233–244.

Elster, A. C. (1998). Fast bit-reversal algorithms. In *Proc. ICASSP 89*, Glasgow, UK, pp. 1099–1102.

Emmerson, S. (1986). The relation of language to materials. In S. Emmerson (Ed.), *The Language of Electronic Music*, pp. 17–40. London: Macmillan Press.

Erickson, R. (1975). *Sound Structure in Music*. Berkeley: University of California Press.

Euler, L. (1768). *Institutionum calculi integralis*. Number v. 1 in Institutionum calculi integralis. imp. Acad. imp. Saènt.

Fink, R. (2005). The story of ORCH5 or, the classical ghost in the hip–hop machine. *Popular Music 24*(3), 339–356.

Flanagan, F. and R. Golden (1966). Phase vocoder. *Bell System Technical Journal 45*(9), 1493–1509.

Fletcher, H. and W. Munson (1933). Loudness, its measurement and calculation. *The Journal of the Acoustical Society of America 5*, 82–108.

Fourier, J. (1822). *La Théorie analytique de la chaleur*. Paris: F Didot.

Fourier, J. (2005). Propagation of heat in an infinite rectangular solid. In S. Hawking (Ed.), *God Created the Integers*, pp. 500–562. London: Penguin Books.

Friedman, D. (1985). Instantaneous-frequency distribution vs. time: An interpretation of the phase structure of speech. In *ICASSP '85. IEEE International Conference on Acoustics, Speech, and Signal Processing*, Volume 10, pp. 1121–1124.

Gabor, D. (1940). Theory of communication. *Journal of the Institution of Electrical Engineers—Part III: Radio and Communication Engineering 93*(26), 429–457.

Gabor, D. (1947). Acoustical quanta and the theory of hearing. *Nature 159*(4044), 591–594.

Gardner, W. G. (1995). Efficient convolution without input-output delay. *Journal of the Audio Engineering Society 43*(3), 127–136.

Goulart, A., M. Queiroz, J. Timoney, and V. Lazzarini (2017). Am/fm dafx. In *Proceedings of the 2017 DAFx Conference*, Trondheim, Norway, pp. 1–8.

Gould, G. (2006). The prospects of recording. In C. Cox and D. Warner (Eds.), *Audio Culture, Readings in Modern Music*, pp. 115–136. London: Continuum Press.

Grey, J. (1978). Timbre discrimination in musical patterns. *Journal of Acoustical Society of America 64*(2), 467–472.

Griffiths, D. (1995). *Introduction to Quantum Mechanics*. Upper Saddle River, NJ: Prentice-Hall.

Harvey, J. (1981). Mortuos Plango, Vivos Voco: A realization at IRCAM. *Computer Music Journal 5*(4), 22–24.

Hawking, S. (2005). Jean Batiste Joseph Fourier: His life and work. In S. Hawking (Ed.), *God Created the Integers*, pp. 491–499. London: Penguin Books.

Heideman, M. T., D. H. Johnson, and C. S. Burrus (1985). Gauss and the history of the fast Fourier transform. *Archive for History of Exact Sciences 34*(3), 265–277.

Helmholtz, H. (1954). *On the sensations of tone*. New York Dover.

Herz-Fischler, R. (1987). *A Mathematical History of Division in Extreme and Mean Ratio*. Waterloo, Ontario: Wilfried Laurier Univ. Press.

Hochstrasser, U. (1972). Orthogonal polynomials. In M. Abramowitz and I. Stegun (Eds.), *Handbook of Mathematical Functions*, pp. 771–792. Washington, DC: US Department of Commerce.

Howard, D. and J. Angus (2009). *Acoustics and Psychoacoustics*. Oxford: Focal Press.

Hunt, F. (1978). *Origins in Acoustics*. New Haven, CT: Yale Univ. Press.

Hutchins, B. (1975). The frequency modulation spectrum of an exponential voltage-controlled oscillator. *Journal of the Audio Engineering Society 23*(3), 200–207.

ISO 226:2003 (2003). *Acoustics: Normal Equal-Loudness Level Contours*. Geneva: International Organization for Standardization.

Jensen, K. (2002). The timbre model. *The Journal of the Acoustical Society of America 112*(5), 2238–2238.

Johnson, D. H. (1980). The relationship between spike rate and synchrony in responses of auditory-nerve fibers to single tones. *The Journal of the Acoustical Society of America 68*(4), 1115–1122.

Johnstone, B. and A. Boyle (1967). Basilar membrane vibration examined with the Moessbauer effect. *Science 158*(3799), 389–390.

Jolley, L. (1961). *Summation of Series*. New York: Dover.

Kleimola, J., V. Lazzarini, J. Timoney, and V. Valimaki (2009). Vector phaseshaping synthesis. In *Proceedings of the 14th Conference on Digital Audio Effects*, Paris, pp. 233–240.

Kleimola, J., V. Lazzarini, V. Valimaki, and J. Timoney (2011). Feedback amplitude modulation synthesis. *EURASIP J. Adv. Sig. Proc. 2011*, 1–18.

Knuth, D. (1997). *The Art of Computer Programming, 3rd Edition*. Reading, PA: Addison-Wesley.

Landy, L. (2007). *Understanding the Art of Sound Organization*. Cambridge, MA: MIT Press.

Lazzarini, V. (2013). The development of computer music programming systems. *Journal of New Music Research 4*(42), 97–110.

Lazzarini, V. (2017). *Computer Music Instruments*. Berlin: Springer.

Lazzarini, V. (2020). Parallel computation of time-varying convolution. *Journal of New Music Research 49*(5), 403–415.

Lazzarini, V. and J. Timoney (2009). New methods of formant analysis-synthesis for musical applications. In *Proceedings of the ICMC 2009*, Montreal, pp. 239–242.

Lazzarini, V. and J. Timoney (2010). New perspectives on distortion synthesis for virtual analog oscillators. *Computer Music Journal 31*(1), 28–40.

Lazzarini, V. and J. Timoney (2010). Theory and practice of modified frequency modulation synthesis. *Journal of the Audio Engineering Society 58*(6), 459–471.

Lazzarini, V. and J. Timoney (2013). Synthesis of resonance by nonlinear distortion methods. *Computer Music Journal 37*(1), 35–43.

Lazzarini, V., J. Timoney, and T. Lysaght (2008). The generation of natural-synthetic spectra by means of adaptive frequency modulation. *Computer Music Journal 32*, 9–22.

Lazzarini, V., S. Yi, J. ffitch, J. Heintz, Ø. Brandtsegg, and I. McCurdy (2016). *Csound: A Sound and Music Computing System*. Berlin: Springer.

LeBrun, M. (1977). A derivation of the spectrum of FM with a complex modulating wave. *Computer Music Journal 1*(4), 51–52.

LeBrun, M. (1979). Digital waveshaping synthesis. *Journal of the Audio Engineering Society 27*(4), 250–266.

Lee, G., R. Gommers, F. Waselewski, K. Wohlfahrt, and A. O'Leary (2019). Pywavelets: A Python package for wavelet analysis. *Journal of Open Source Software 4*(36), 1237.

Lindsay, R. (1945). Historical introduction. In R. Lindsay (Ed.), *J W S Rayleigh, The Theory of Sound*. New York: Dover.

Lindsay, R. B. (1966). The story of acoustics. *The Journal of the Acoustical Society of America 39*(4), 629–644.

Lissajous, M. J. (1857). Mémoire sur l'étude optique des mouvements vibratoires. *Annales de chimie et de physique LI*, 147–231.

Loughlin, P. J. and K. L. Davidson (2000, Aug). Instantaneous spectral skew and kurtosis. In *Proceedings of the Tenth IEEE Workshop on Statistical Signal and Array Processing (Cat. No.00TH8496)*, pp. 574–578.

Makhoul, J. (1975). Linear prediction: A tutorial review. *Proceedings of the IEEE 63*(4), 561–580.

Mallat, S. (2008). *A Wavelet Tour of Signal Processing, Third Edition: The Sparse Way* (3rd ed.). San Diego: Academic Press, Inc.

Mallat, S. G. and Zhifeng Zhang (1993). Matching pursuits with time-frequency dictionaries. *IEEE Transactions on Signal Processing 41*(12), 3397–3415.

Mandelbrot, B. (2006, September). Some noises with i/f spectrum, a bridge between direct current and white noise. *IEEE Trans. Inf. Theor. 13*(2), 289–298.

Mandelbrot, B. B. and J. W. van Ness (1968). Fractional Brownian motions, fractional noises and applications. *SIAM Review 10*, 422–437.

Manning, P. (1986). *Electronic and Computer Music*. Oxford: Oxford Univ. Press.

Mathews, M. (1961). An acoustical compiler for music and psychological stimuli. *Bell System Technical Journal 40*(3), 553–557.

Mathews, M. (1963). The digital computer as a musical instrument. *Science 183*(3592), 553–557.

Mathews, M. and J. E. Miller (1964). *MUSIC IV Programmer's Manual*. Murray Hill, NJ: Bell Telephone Lab.

Mathews, M., J. E. Miller, F. R. Moore, and J. R. Pierce (1969). *The Technology of Computer Music*. Cambridge, MA: MIT Press.

McAulay, R. and T. Quatieri (1986). Speech analysis/synthesis based on a sinusoidal representation. *IEEE Transactions on Acoustics, Speech, and Signal Processing 34*(4), 744–754.

Meneghini, M. (2007). An analysis of the compositional techniques in John Chowning's Stria. *Computer Music Journal 31*(3), 26–37.

Messiaen, O. (1956a). *The Technique of My Musical Language, volume one*. Paris: Alphonse Leduc.

Messiaen, O. (1956b). *The Technique of My Musical Language, volume two*. Paris: Alphonse Leduc.

Meurisse, G., P. Hanna, and S. Marchand (2006). A new analysis method for sinusoids+noise spectral models. In *Proceedings of the 9th International Conference on Digital Audio Effects, DAFx 2006*, Montreal, pp. 139–144.

Moog, R. (1967). Voltage controlled electronic music modules. *AES Preprint 346*, 1–19.

Moore, B. and B. Glassberg (1983). Suggested formulae for calculating auditory filters bandwidths and excitation patterns. *The Journal of the Acoustical Society of America 74*(3), 750–753.

Moore, F. R. (1990). *Elements of Computer Music*. Upper Saddle River, NJ: Prentice-Hall, Inc.

Moorer, J. (1977). Signal processing aspects of computer music: A survey. *Proceedings of the IEEE 65*(8), 1108–1137.

Moorer, J. (1978). The use of the phase vocoder in computer music applications. *Journal of the Audio Engineering Society 26*(1/2), 42–45.

Moorer, J. A. (1976). The synthesis of complex audio spectra by means of discrete summation formulas. *Journal of the Audio Engineering Soc 24*(9), 717–727.

Mulgrew, B., P. Grant, and J. Thompson (1999). *Digital Signal Processing: Concepts and Applications*. London: Macmillan Press.

Nyquist, H. (1924). Certain factors affecting telegraph speed. *Bell System Technical Journal 3*, 324–346.

Nyquist, H. (1928). Certain topics in telegraph transmission theory. *Transactions of the AIEE 47*, 617–644.

Ohm, G. S. (1971). On the definition of a tone with the associated theory of the siren and similar sound producing devices. In R. Lindsay (Ed.), *Acoustics: Historical and Philosophical Development*, pp. 243–248. Stroudsburg, PA: Dowden, Hutchinson and Ross.

Oppenheim, A. V., R. W. Schafer, and J. R. Buck (1999). *Discrete-time Signal Processing (2nd Ed.)*. Upper Saddle River, NJ: Prentice-Hall, Inc.

Pampin, J. (2004). ATS: A system for sound analysis transformation and synthesis based on a sinusoidal plus critical-band noise model and psychoacoustics. In *Proceedings of the 2004 International Computer Music Conference, ICMC 2004, Miami, Florida, USA, November 1–6, 2004*, pp. 1–4. Michigan Publishing.

Papoulis, A. and S. U. Pillai (2002). *Probability, Random Variables, and Stochastic Processes* (4th ed.). Boston: McGraw-Hill.

Penrose, R. (2005). *The Road to Reality*. London: Vintage Books.

Pickles, J. (2013). *An Introduction to the Physiology of Hearing*. Bingley: Emerald Publ.

Pinkston, R. (2000). FM synthesis in Csound. Cambridge, MA: MIT Press.

Pope, S. T. (1993). Machine tongues xv: Three packages for software sound synthesis. *Computer Music Journal 17*(2), 23–54.

Portnoff, M. (1976). Implementation of the digital phase vocoder using the fast Fourier transform. *IEEE Transactions on Acoustics, Speech, and Signal Processing 24*, 243–248.

Puckette, M. (1995a). Formant-based audio synthesis using nonlinear distortion. *Journal of the Audio Engineering Society 43*(1/2), 40–47.

Puckette, M. (1995b). Phase-locked vocoder. In *IEEE ASSP Workshop on Applications of Signal Processing to Audio and Acoustics*, pp. 222–225.

Putnam, M. (1980). A thirty-five year history and evolution of the recording studio. *AES Preprint 1*(1661), 1–10.

Rask-Andersen, H., W. Liu, E. Erixon, A. Kinnefors, K. Pfaller, A. Schrott-Fischer, and R. Glueckert (2012). Human cochlea: Anatomical characteristics and their relevance for cochlear implantation. *The Anatomical Record 295*(11), 1791–1811.

Rayleigh, J. W. S. (1945). *The Theory of Sound*. New York Dover.

Regalia, P. and S. Mitra (1987). Tunable digital frequency response equalization filters. *IEEE Transactions on Acoustics, Speech and Signal Processing 35*(1), 118–120.

Riemann, H. (1893). *Vereinfachte Harmonielehre*. London: Augner and Co.

Risset, J. C. (1969). *An Introductory Catalogue of Computer Synthesized Sounds*. Murray Hill, NJ: Bell Telephone Lab.

Risset, J. C. (2009). Max Mathews' influence on (my) music. *Computer Music Journal 33*(3), 26–34.

Roads, C. and M. Mathews (1980). Interview with Max Mathews. *Computer Music Journal 4*(4), 15–22.

Roebel, A. (2005). Efficient spectral envelope estimation and its application to pitch shifting and envelope preservation. In *Proceedings of the 8th International Conference on Digital Audio Effects*, pp. 30–35.

Russolo, L. (1967). *The Art of Noise*. New York: Something Else Press.

Sauveur, J. (1701). *Système Général des intervalles des sons*. Paris: L'Académie Royale des Sciences.

Schaeffer, P. (2006). Acousmatics. In C. Cox and D. Warner (Eds.), *Audio Culture, Readings in Modern Music*, pp. 76–81. London: Continuum Press.

Schaeffer, P. (2017). *Treatise on Musical Objects, An Essay Across Disciplines.* Oakland: University of California Press.

Schaeffer, P. and M. Chion (1983). *Guide des Objets Sonores*. Paris: Buchet Chastel/INA-GRM.

Scharf, B. (1970). Critical bands. In J. Tobias (Ed.), *Foundations of Modern Auditory Theory, Vol. 1*, pp. 159–202. London: Academic Press.

Schoenberg, A. (1983). *Theory of Harmony*. Berkeley: University of California Press.

Schottstaedt, B. (2019). An introduction to FM. https://ccrma.stanford.edu/software/snd/snd/fm.html.

Schouten, J. E. (1940). The residue and the mechanism of hearing. In *Proc. Kon. Ned. Ahad. Wetenschap 43*, pp. 991–999.

Serra, X. (1989). *A System for Sound Analysis/Transformation/Synthesis based on a Deterministic plus Stochastic Decomposition*. Ph.D. thesis, Stanford University, Stanford, CA.

Sethares, W. (2005). *Tuning, Timbre, Spectrum, Scale*. London: Springer.

Shannon, C. E. (1949). Communication in the presence of noise. *Proceedings Institute of Radio Engineers 37*(1), 10–21.

Smalley, D. (1986). Spectro-morphology and structuring processes. In S. Emmerson (Ed.), *The Language of Electroacoustic Music*, pp. 61–96. London: Macmillan Press.

Smith, J. O. and J. B. Angell (1982). A constant-gain digital resonator tuned by a single coefficient. *Computer Music Journal 6*(4), 36–40.

Soo, J. and K. K. Pang (1990). Multidelay block frequency domain adaptive filter. *IEEE Transactions on Acoustics, Speech, and Signal Processing 38*(2), 373–376.

Steiglitz, K. (1996). *A Signal Processing Primer*. Menlo Park, CA: Addison-Wesley.

Stockham, Jr., T. G. (1966). High-speed convolution and correlation. In *Proceedings of the April 26–28, 1966, Spring Joint Computer Conference*, AFIPS '66, New York, pp. 229–233. ACM.

Sweitzer, K., N. Bishop, and V. Genberg (2004). Efficient computation of spectral moments for determination of random response statistics. In *Proceedings of the International Conference, International Conference on Noise and Vibration Engineering 2004*, Leuven, Belgium, pp. 2667–2692.

Tenney, J. (1963). Sound generation by means of a digital computer. *Journal of Music Theory 7*(1), 24–70.

Teruggi, D. (2007). Technology and musique concrète: The technical developments of the Groupe de Recherches Musicales and their implication in musical composition. *Organised Sound 12*(3), 213–231.

Timoney, J. and V. Lazzarini (2009). Exponential FM bandwidth criterion for virtual analogue applications. In *Proceedings of the 14th Conference on Digital Audio Effects*, Paris, pp. 115–118.

Timoney, J., V. Lazzarini, J. Pekonen, and V. Valimaki (2011). Adaptive phase distortion synthesis. In *Proceedings of the 12th Conference on Digital Audio Effects*, Como, Italy, pp. 1–8.

Tomlyn, B. and J. Aikin (1985). How to program the DX7. *Keyboard Magazine 1985*(6), 66–71.

Torger, A. and A. Farina (2001). Real-time partitioned convolution for ambiophonics surround sound. In *2001 IEEE Workshop on Applications of Signal Processing to Audio and Acoustics*, New Paltz, New York.

Truax, B. (1988). Real-time granular synthesis with a digital signal processor. *Computer Music Journal 12*(2), 14–26.

Truesdell, C. (1960). *The Rational Mechanics of Elastic and Fluid Bodies (1638–1788)*. Basel: Orell Füssli.

Vercoe, B. (1983). Computer system and languages for audio research. In *The New World of Digital Audio (Audio Engineering Society Special Edition)*, Rye, NY, pp. 245–250.

von Békesy, G. (1961). *Experiments in Hearing*. New York: McGraw-Hill.

Wallis (1677). Dr. Wallis's letter to the publisher, concerning a new musical discovery; written from Oxford, March 14. *Phil. Trans. R. Soc. 12*(1676/7), 1–10.

Waters, S. (2000). Beyond the acousmatic: hybrid tendencies in electroacoustic music. In S. Emmerson (Ed.), *Music, Electronic Media, and Culture*, pp. 58–83. London: Ashgate.

Watkins, N. (2017). Mandelbrot's 1/f fractional renewal models of 1963–67: The non-ergodic missing link between change points and long range dependence. In I. Rojas, H. Pomares, and O. Valenzuela (Eds.), *Advances in Time Series Analysis and Forecasting*, pp. 197–208. Cham: Springer International Publishing.

Watson, G. (1944). *A Treatise of the Theory of Bessel Functions*. Cambridge: Cambridge Univ. Press.

Williams, P. (1980). *A New History of the Organ*. London: Faber and Faber.

Wiltshire, T. (2016). Adventures in top octave generation. https://electricdruid.net/adventures-in-top-octave-generation/.

Wishart, T. (1988). The composition of Vox-5. *Computer Music Journal 12*(4), 21–27.

Wishart, T. (1994). *Audible Design*. York: Orpheus the Pantomime.

Wishart, T. (1996). *On Sonic Art*. London: Harwood Academic Publishers.

Wishart, T. (2000). Sonic composition in *Tongues of Fire*. *Computer Music Journal 24*(2), 22–30.

Wyss, W. (1991). Fractional noise. *Foundations of Physics Letters 4*(3), 235–246.

Xenakis, I. (1992). *Formalized Music: Thought and Mathematics in Composition*. Hillsdale, NY: Pendragon Press.

Zattra, L. (2007). The assembling of Stria by John Chowning: A philological investigation. *Computer Music Journal 31*(3), 38–64.

Zucker, R. (1972). Elementary transcendental functions. In M. Abramowitz and I. Stegun (Eds.), *Handbook of Mathematical Functions*, pp. 65–226. Washington, DC: US of America Department of Commerce.

INDEX

The manufacturer's authorised representative in the EU for product safety is Oxford University Press España S.A. of El Parque Empresarial San Fernando de Henares, Avenida de Castilla, 2 - 28830 Madrid (www.oup.es/en or product.safety@oup.com). OUP España S.A. also acts as importer into Spain of products made by the manufacturer.
Printed and bound by CPI Group (UK) Ltd, Croydon, CR0 4YY
14/09/2026
02216298-0004